The Concise Oxford Dictionary of

Mathematics

SIXTH EDITION

RICHARD EARL AND
JAMES NICHOLSON

OXFORD
UNIVERSITY PRESS

OXFORD
UNIVERSITY PRESS

Great Clarendon Street, Oxford, OX2 6DP,
United Kingdom

Oxford University Press is a department of the University of Oxford.
It furthers the University's objective of excellence in research, scholarship,
and education by publishing worldwide. Oxford is a registered trade mark of
Oxford University Press in the UK and in certain other countries

First edition 1990
Second edition 1996
Third edition 2005
Fourth edition 2009
Fifth edition 2014
Sixth edition 2021

Published in the United States of America by Oxford University Press
198 Madison Avenue, New York, NY 10016, United States of America

British Library Cataloguing in Publication Data
Data available

Library of Congress Control Number: 2021936612

ISBN 978-0-19-884535-5

Printed and bound by
Clays Ltd, Elcograf S.p.A.

Contents

Contents

Contributors to the First and Second Editions

C. R. J. Clapham (Editor of the First and Second Editions)
J. R. Pulham
University of Aberdeen

C. Chatfield
R. Cheal
J. B. Gavin
University of Bath

D. P. Thomas
University of Dundee

Preface to Second Edition*

This dictionary is intended to be a reference book that gives reliable definitions or clear and precise explanations of mathematical terms. The level is such that it will suit, among others, sixth-form pupils, college students, and first-year university students who are taking mathematics as one of their courses. Such students will be able to look up any term they may meet and be led on to other entries by following up cross-references or by browsing more generally.

The concepts and terminology of all those topics that feature in pure and applied mathematics and statistics courses at this level today are covered. There are also entries on mathematicians of the past and important mathematics of more general interest. Computing is not included. The reader's attention is drawn to the appendices which give useful tables for ready reference.

Some entries give a straight definition in an opening phrase. Others give the definition in the form of a complete sentence, sometimes following an explanation of the context. An asterisk is used to indicate words with their own entry, to which cross-reference can be made if required.

This edition is more than half as large again as the first edition. A significant change has been the inclusion of entries covering applied mathematics and statistics. In these areas, I am very much indebted to the contributors, whose names are given on page iii. I am most grateful to these colleagues for their specialist advice and drafting work. They are not, however, to be held responsible for the final form of the entries on their subjects. There has also been a considerable increase in the number of short biographies, so that all the major names are included. Other additional entries have greatly increased the comprehensiveness of the dictionary.

The text has benefited from the comments of colleagues who have read different parts of it. Even though the names of all of them will not be given, I should like to acknowledge here their help and express my thanks.

Christopher Clapham

---

* The formatting for cross-referencing has been changed since the second edition, and the above text has been modified to be consistent with the new format.

Preface to Third Edition

Since the second edition was published the content and emphasis of applied mathematics and statistics at sixth-form, college, and first-year university levels has changed considerably. This edition includes many more applied statistics entries as well as dealing comprehensively with the new decision and discrete mathematics courses and a large number of new biographies on 20th-century mathematicians. I am grateful to the Headmaster and Governors of Belfast Royal Academy for their support and encouragement to take on this task, and to Louise, Joanne, and Laura for transcribing my notes.

James Nicholson

Preface to Fourth Edition

Since the third edition was published there has been a dramatic increase in both access to the Internet and the amount of information available. The major change to this edition is the introduction of a substantial number of web links, many of which contain dynamic or interactive illustrations related to the definition.

James Nicholson

Preface to Fifth Edition

The fifth edition introduces much fuller coverage of mathematical logic and of topology, and has extended the appendix on trigonometry and increased the number of appendices to include summaries of geometry, algebra, and probability distributions, along with a description of the Millennium Prize problems.

James Nicholson

Preface to Sixth Edition

Any editor of a mathematics dictionary needs to consider its role when there are large amounts of information about mathematics available on the Internet. Certainly, as a dictionary, it has to remain a source of reference, but I think the strengths of a dictionary in this format are that it has a target audience and the benefits it might bring that audience as they 'thumb' through it (literally or electronically) cross-referencing from one definition or example to another. I consider this a healthy way for students to find out more about mathematics, and it can be a highly individual one.

The target audience for this dictionary ranges from the A-level mathematician or equivalent through to first and second years in higher education. This means some entries need careful designing so that each reader can learn something from an entry. It is unlikely A-level students will be needing to look up formally what an 'irreducible representation' is and so such a definition can be written more for a higher-education audience, but they might, more generally, have heard of 'representation theory' and want to get a sense of the nature of the subject.

So an entry such as 'sample space' might begin informally with straightforward examples but finish discussing 'probability spaces' or saying that the 'events' make a 'sigma algebra'. I hope the reader looking for formal definitions will understand the need for the earlier informality and those looking for an informal sense of a concept not be put off by subsequent rigour or technical language or even use the cross-references to delve further into the topic.

There are over 800 new entries in the main dictionary and 6 further appendices, including a historical timeline of mathematical discoveries. The new entries mainly cover topics from early higher education, from fluid dynamics and quantum theory to multivariable calculus and differential geometry, from information theory and coding to mathematics education and the philosophy of mathematics. I hope students will gain a sense of the breadth, impact, and currency of mathematics from these new entries.

Richard Earl

A The number 10 in *hexadecimal notation.

a Abbreviation for *atto-.

a- Prefix meaning 'not'. For example, an asymmetric figure is one which possesses no symmetry, which is not symmetrical.

AAS *See* SOLUTION OF TRIANGLES.

A_n The *alternating group of the set $\{1,2,\ldots,n\}$.

abacus A counting device consisting of rods on which beads can be moved so as to represent numbers.

***abc* conjecture** A significant and open *conjecture in *number theory first made in the 1980s. Given $\varepsilon > 0$, the conjecture says that there are finitely many triples (a,b,c) of *coprime positive integers such that $a + b = c$ and $c > \text{rad}\,(abc)^{\varepsilon + 1}$. Here $\text{rad}(n)$ denotes the product of the distinct *prime factors of n. If true, it would imply other theorems and conjectures including the *Mordell conjecture, *Catalan's conjecture, and *Fermat's Last Theorem.

Abel, Niels Henrik (1802–29) Norwegian mathematician who, at the age of 19, proved that the general equation of degree greater than 4 is not *solvable by radicals. In other words, there can be no formula for the roots of such an equation similar to the familiar formula for a *quadratic equation. He was also responsible for fundamental developments in the theory of algebraic functions. He died in some poverty at the age of 26, just a few days before he would have received a letter announcing his appointment to a professorship in Berlin.

abelian group Suppose that G is a *group with the operation $\circ$. Then G is abelian if the operation $\circ$ is commutative; that is, if, for all elements a and b in G, $a \circ b = b \circ a$. In an abelian group, the operation is usually denoted $+$.

abelian group (classification theorem) A finite *abelian group is *isomorphic to the *product group $C_{d_1} \times C_{d_2} \times \ldots \times C_{d_k}$ for unique integers $d_1, d_2, \ldots \; d_k \geqslant 2$ such that $d_1|d_2|\ldots|d_k$. Thus, there are five abelian groups of order $240 = 2^4 \cdot 3 \cdot 5$, namely C_{240}, $C_2 \times C_{120}$, $C_4 \times C_{60}$, $C_2 \times C_2 \times C_{60}$, $C_2 \times C_2 \times C_2 \times C_{30}$.

Abelianization *See* COMMUTATOR.

Abel Prize A prestigious annual prize awarded to mathematicians by the king of Norway, named in honour of Niels Henrik *Abel and modelled on the Nobel Prizes. It was first awarded in 2003 to Jean-Pierre *Serre.

Abel's Limit Theorem For a *convergent series $\{a_k\}$, the limit assigned by the Abel summation method exists and is equal to the sum of the series.

Abel's partial summation formula For two arbitrary sequences $\{a_k\}$ and $\{b_k\}$ with $A_k = \sum_{r=1}^{k} a_r$, the result that $\sum_{r=j}^{k} a_r b_r = \sum_{r=j}^{k} A_j(b_j - b_{j+1}) + A_k b_{k+1} - A_{j-1} b_j$. This has some similarities with the formula for *integration by parts.

Abel summation A method of computing the sum of a possibly *divergent series of *complex numbers $\{a_k\}$ as the limit, as z approaches 1 from below, of the *power series $\sum(a_k z^k)$, assuming the series $\{a_k\}$ has *radius of convergence $= 1$. Abel summation is consistent with, and more general than, *Cesàro summation.

Abel's test A test for the *convergence of an *infinite series which states that if $\sum a_n$ is a convergent series, and $\{b_n\}$ is bounded and monotically decreasing, i.e. $b_{n+1} \leq b_n$ for all n, then $\sum a_n b_n$ is also convergent.

above Greater than. The limit of a function at a from above is the limit of $f(x)$ as $x \to a$ for values of $x > a$. It can be written as $f(a^+)$ or $\lim_{x \searrow a} f(x)$. A function $f(x)$ is continuous at a if and only if the limit from above and the limit from *below exist and are equal.

abscissa The x-coordinate in a Cartesian coordinate system in the plane. *Compare* ORDINATE.

absolute error *See* ERROR.

absolute frequency The number of occurrences of an event. For example, if a die is rolled 20 times and 4 sixes are observed, the absolute frequency of sixes is 4 and the *relative frequency is 4/20.

absolute geometry A synonym for NEUTRAL GEOMETRY.

absolutely continuous A stronger condition than *continuous or *uniformly continuous. A function f on an interval I (of the real line) is absolutely continuous if, for every positive number ε, there exists a positive number δ so that for every finite sequence of *pairwise disjoint subintervals of I with $\sum_k |y_k - x_k| < \delta$ then $\sum_k |f(y_k) - f(x_k)| < \varepsilon$. Functions satisfying the *Lipschitz condition are absolutely continuous. Absolutely continuous

functions have a *derivative almost everywhere (*see* ALMOST ALL) which satisfies the *Fundamental Theorem of Calculus. *See also* CANTOR DISTRIBUTION.

absolutely convergent series A series $\{a_n\}$ is said to be absolutely convergent if $\sum_{n=1}^{\infty} |a_n|$ is *convergent. Absolutely convergent series are convergent. If $a_n = (-1)^{n-1} \times \frac{1}{n}$ then the series is convergent but not absolutely convergent, whereas $a_n = (-1)^{n-1} \times \frac{1}{n^2}$ is absolutely convergent. A series which is convergent but not absolutely convergent is called *conditionally convergent.

absolutely summable A synonym for ABSOLUTELY CONVERGENT.

absolute value For any real number a, the absolute value (also called the *modulus) of a, denoted by $|a|$, is a itself if $a \geq 0$, and $-a$ if $a \leq 0$. Thus $|a|$ is positive except when $a = 0$. The following properties hold:

(i) $|ab| = |a||b|$.
(ii) $|a + b| \leq |a| + |b|$.
(iii) $|a - b| \geq ||a| - |b||$.
(iv) For $a > 0$, $|x| \leq a$ if and only if $-a \leq x \leq a$.

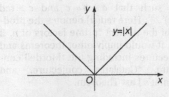

Graph of absolute value

(ii) is the *triangle inequality and (iii) is sometimes known as the *reverse triangle inequality.

absorbing set A subset S of a *vector space X is an absorbing set if for any point x in X then tx lies in S if $t \in \mathbb{R}$ is small enough and positive. An alternative name is a 'radial set', which reflects the notion that the set S will generate all elements of X in the sense that any $x \in X$ can be written as $x = \alpha s$ for some $s \in S$ and some real α.

absorbing state See RANDOM WALK.

absorption laws For all sets A and B, $A \cap (A \cup B) = A$ and $A \cup (A \cap B) = A$. These are the absorption laws.

abstract algebra The area of mathematics concerned with *algebraic structures, such as *groups, *rings, and *fields, involving sets of elements with particular operations satisfying certain axioms. The theory of certain algebraic structures is highly developed; in particular, the theory of *vector spaces is so extensive that its study, known as *linear algebra, is essentially a separate subject from abstract algebra.

abstraction The process of studying the underlying rules and structures that often connect seemingly different results, and distilling that structure by means of formal *axioms. The purpose is to derive, from a set of axioms, general results that are then applicable to any example of that structure. This was a common theme of much of 20th-century pure mathematics, applying equally in *algebra (e.g. *groups), *analysis (e.g. *Banach spaces), *geometry (e.g. *manifolds), *topology (*metric spaces).

abundant number An integer that is smaller than the sum of its positive divisors, not including itself, For example, 12 is divisible by 1, 2, 3, 4, and 6, and $1 + 2 + 3 + 4 + 6 = 16 > 12$. See DEFICIENT NUMBER, PERFECT NUMBER.

acceleration Suppose that a particle is moving in a straight line, with a point O on the line taken as origin and one direction taken as positive. Let x be the

*displacement of the particle at time t. The acceleration of the particle is equal to $\ddot{x}$ or d^2x/dt^2, the *rate of change of the *velocity with respect to t. If the velocity is positive (that is, if the particle is moving in the positive direction), the acceleration is positive when the particle is speeding up and negative when it is slowing down. However, if the velocity is negative, a positive acceleration means that the particle is slowing down and a negative acceleration means that it is speeding up.

In the preceding paragraph, a common convention has been followed, in which the unit vector $\mathbf{i}$ in the positive direction along the line has been suppressed. Acceleration is in fact a vector quantity, and in the 1-dimensional case above it is equal to $\ddot{x}\mathbf{i}$.

When the motion is in two or three dimensions, vectors are used explicitly. The acceleration $\mathbf{a}$ of a particle is a vector equal to the rate of change of the velocity $\mathbf{v}$ with respect to t. Thus $\mathbf{a} = d\mathbf{v}/dt$. If the particle has *position vector $\mathbf{r}$, then $\mathbf{a} = d^2\mathbf{r}/dt^2 = \ddot{\mathbf{r}}$. When Cartesian coordinates are used, $\mathbf{r} = x\mathbf{i} + y\mathbf{j} + z\mathbf{k}$, and then $\ddot{\mathbf{r}} = \ddot{x}\mathbf{i} + \ddot{y}\mathbf{j} + \ddot{z}\mathbf{k}$.

If a particle is travelling in a circle with constant speed, it still has an acceleration because of the changing direction of the velocity. This acceleration is towards the centre of the circle and has magnitude $\frac{v^2}{r}$ where v is the speed of the particle and r is the radius of the circle.

Acceleration has the *dimensions LT^{-2}, and the *SI unit of measurement is the metre per second per second, abbreviated to ms^{-2}.

acceleration–time graph A graph that shows *acceleration plotted against time for a particle moving in a straight line. Let $v(t)$ and $a(t)$ be the *velocity and acceleration, respectively, of the particle at time t. The acceleration-time graph is the graph $y = a(t)$, where the t-axis is horizontal and the y-axis is vertical with the positive direction upwards. With the convention that any area below the horizontal axis is negative, the area under the graph between $t = t_1$ and $t = t_2$ is equal to $v(t_2) - v(t_1)$.

acceptance region See HYPOTHESIS TESTING.

acceptance sampling A method of quality control where a *sample is taken from a batch and a decision whether to accept the batch is made on the basis of the quality of the sample. The most simple method is to have a straight accept/reject criterion, but a more sophisticated approach is to take another sample if the evidence from the existing sample, or a set of samples, is not clearly indicating whether the batch should be accepted or rejected. One of the main advantages of this approach is reducing the cost of taking samples to satisfy quality control criteria.

accessibility In a *connected *graph, the accessibility of a *vertex equals the sum of total distances from that vertex to all other vertices.

accumulation point For a set S in a *topological space X, a point x (which is in X but not necessarily in S) is an accumulation point of S if every *neighbourhood of x in X also contains an element of S distinct from x. Also called cluster point or limit point. The *closure of a set is the *union of the set and the set of accumulation points. See DERIVED SET, ADHERENT POINT.

accuracy A measure of the precision of a numerical quantity, usually given to n *significant figures (where the proportional accuracy is the important aspect) or n *decimal places (where the absolute accuracy is more important).

accurate (correct) to n decimal places Rounding a number with the accuracy specified by the number of *decimal places given in the rounded value. So $e = 2.71828\ldots = 2.718$ to three decimal places and $= 2.72$ to two decimal places. $\sqrt{86.56} = 9.30076\ldots$ is 9.30 correct to two decimal places. Where a number of quantities are being measured and added or subtracted, using values correct to the same number of decimal places ensures that they have the same degree of accuracy. However, if the units are changed, for example between centimetres and metres, then the accuracy of the measurements will be different if the same number of decimal places is used in the measurements.

accurate (correct) to n significant figures Rounding a number with the accuracy specified by the number of *significant figures given in the rounded value. So $e = 2.71828\ldots = 2.718$ to four significant figures and $= 2.72$ to three significant figures. $e^{-3} = 0.049787\ldots = 0.0498$ correct to three significant figures. Rounding to the same number of significant figures ensures all the measurements have about the same proportionate accuracy. If the units are changed, for example between centimetres and metres, then the accuracy of the measurements will not be changed if the same number of significant figures is used in the measurements.

Acheson, David (1946–) Author of several books popularizing mathematics, most notably *1089 and All That* and *The Calculus Story*.

Achilles paradox The paradox of *Zeno which arises from considering how overtaking takes place. Achilles gives a tortoise a head start in a race. To overtake, he must reach the tortoise's initial position, then where the tortoise had moved to, and so on *ad infinitum*. However, the conclusion that he cannot overtake because he has to cover an infinite sum of well-defined non-zero distances is false, because the infinite series of distances travelled and times taken both have finite sums.

acnode A synonym for ISOLATED POINT (OF A CURVE).

action (in dynamics) See LEAST ACTION, PRINCIPLE OF.

action, acts See GROUP ACTION.

action limits The outer limits set on a *control chart in a production process. If the observed value falls outside these limits, then action will be taken, often resetting the machine. For the means of samples of sizes n in a process with *standard deviation σ with target *mean μ, the action limits will be set at $\mu \pm 3.09 \frac{\sigma}{\sqrt{n}}$.

active constraint An inequality such as $y + 2x \geq 13$ is said to be active at a point on the boundary, i.e. where equality holds, for example (6, 1) and (0, 13).

activity networks (edges as activities) The *networks used in *critical path analysis where the edges (arcs) represent activities to be performed. Paths in the network represent the precedence relations between the activities, and *dummy activities are required to link paths where common activities appear, but the paths are at least partly independent. While this is a complication, each activity appears on only one edge, and the sequence of activities needed is easier to follow than when the vertices are used to represent the activities. Once the activity network has been constructed from the precedence table, the *critical path algorithm (edges as activities) can be applied.

activity networks (vertices as activities) The *networks used in *critical path analysis where the vertices (nodes) represent activities to be performed. The edges (arcs) coming out from any vertex X join X to any vertex Y whose activity cannot start until X has been completed, and the edge is labelled with the time

taken for activity X. Note that there will often be more than one such activity, in which case each edge will carry the same time. The construction of an activity network requires a listing of the activities, their duration, and the precedence relations which identify which activities are dependent on the prior completion of other activities. With this structure, *dummy activities are not needed, but the same activity will be represented by more than one edge when more than one other activity depends on its prior completion, and the sequence of activities is less easy to follow than the alternative structure *activity networks (edges as activities).

actual infinity See POTENTIAL INFINITY.

acute angle An angle that is less than a *right angle. An acute triangle is one where all its angles are acute.

acyclic Not cyclic. In particular a *graph is acyclic if it contains no *cycles.

addend One of the numbers combined when forming a *sum.

adders (in combinatorial logic) The half-adder and full adders are sections of circuits which use a system of *logic gates to add binary digits, by using a combination for which the truth table output is identical to the output required by the binary addition.

（🌐）SEE WEB LINKS

• An article demonstrating how a simple adder works.

addition A *binary operation present in many *algebraic structures, which is invariably *commutative and *associative. The inputs are addends, and the output is their sum. Addition of *natural numbers might concretely be considered as the accumulation of numbers of objects, but formal definitions are needed for more complicated mathematical objects (*see* ADDITION (of complex numbers), ADDITION (of matrices), ADDITION (of vectors)) and especially for infinite *series. *See also* COMPONENTWISE, POINTWISE ADDITION, ALGEBRA OF LIMITS, ADDITIVE FUNCTION.

addition (of complex numbers) Let the *complex numbers z_1 and z_2, where $z_1 = a + bi$ and $z_2 = c + di$, be represented by the points P_1 and P_2 in the *complex plane. Then $z_1 + z_2 = (a + c) + (b + d)i$, and $z_1 + z_2$ is represented in the complex plane by the point Q such that OP_1QP_2 is a parallelogram; that is, such that $\overrightarrow{OQ} = \overrightarrow{OP_1} + \overrightarrow{OP_2}$. In this way the addition of complex numbers corresponds exactly to the addition of *vectors.

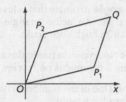

The parallelogram OP_1QP_2

addition (of matrices) Let $\mathbf{A}$ and $\mathbf{B}$ be $m \times n$ *matrices, with $\mathbf{A} = [a_{ij}]$ and $\mathbf{B} = [b_{ij}]$. The operation of addition is defined by taking the sum $\mathbf{A} + \mathbf{B}$ to be the $m \times n$ matrix $\mathbf{C}$, where $\mathbf{C} = [c_{ij}]$ and $c_{ij} = a_{ij} + b_{ij}$. The sum $\mathbf{A} + \mathbf{B}$ is not defined if $\mathbf{A}$ and $\mathbf{B}$ are not of the same order. This operation $+$ of addition on the set of all $m \times n$ matrices is *associative and *commutative.

addition (of vectors) Given *vectors $\mathbf{a}$ and $\mathbf{b}$, let OA and OB be *directed line segments that represent $\mathbf{a}$ and $\mathbf{b}$, with the same initial point O. The sum of OA and OB is the directed line segment OC, where $OACB$ is a *parallelogram, and the sum $\mathbf{a} + \mathbf{b}$ is defined to

be the vector $\mathbf{c}$ represented by $\overrightarrow{OC}$. Alternatively, the sum of vectors $\mathbf{a}$ and $\mathbf{b}$ can be defined by representing $\mathbf{a}$ by a directed line segment OP and $\mathbf{b}$ by PQ where the final point of the first directed line segment is the initial point of the second. Then $\mathbf{a} + \mathbf{b}$ is the vector represented by OQ. Addition of vectors has the following properties, which hold for all $\mathbf{a}$, $\mathbf{b}$, and $\mathbf{c}$:

(i) $\mathbf{a} + \mathbf{b} = \mathbf{b} + \mathbf{a}$, the commutative law.
(ii) $\mathbf{a} + (\mathbf{b} + \mathbf{c}) = (\mathbf{a} + \mathbf{b}) + \mathbf{c}$, the associative law.
(iii) $\mathbf{a} + \mathbf{0} = \mathbf{0} + \mathbf{a} = \mathbf{a}$, where $\mathbf{0}$ is the zero vector.
(iv) $\mathbf{a} + (-\mathbf{a}) = (-\mathbf{a}) + \mathbf{a} = \mathbf{0}$, where $-\mathbf{a}$ is the negative of $\mathbf{a}$.

Addition of *coordinate vectors in *n-dimensional space is done *componentwise. So if $\mathbf{a} = (a_1, a_2, \ldots, a_n)$ and $\mathbf{b} = (b_1, b_2, \ldots, b_n)$, then $\mathbf{a} + \mathbf{b} = (a_1 + b_1, a_2 + b_2, \ldots, a_n + b_n)$.

Parallelogram law for addition

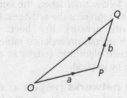

Triangle law for addition

addition formulae (in trigonometry) *See* COMPOUND ANGLE FORMULAE.

addition law (probability) If A and B are two events then the addition law states that the $\Pr(A \text{ or } B) = \Pr(A) + \Pr(B) - \Pr(A \text{ and } B)$, or using set notation $P(A \cup B) = P(A) + P(B) - P(A \cap B)$. In the special case where A and B are *mutually exclusive events this reduces to $P(A \cup B) = P(A) + P(B)$.

additive function A function for which $f(x + y) = f(x) + f(y)$. For example, $f(x) = 3x$ is an additive function since $f(x + y) = 3(x + y) = f(x) + f(y)$ but $g(x) = \sqrt{x}$ is not additive since $\sqrt{x + y}$ is not in general equal to $\sqrt{x} + \sqrt{y}$. *Compare* SUBADDITIVE, SUPERADDITIVE.

additive group A *group with the operation +, called *addition, may be called an additive group. The operation in a group is normally denoted by addition only if it is *commutative, so an additive group is *abelian.

additive identity The *identity element under an operation of *addition, usually denoted by 0, so $a + 0 = 0 + a = a$.

additive inverse *See* INVERSE ELEMENT.

adherent point A point of the *closure of a set. *Compare* ACCUMULATION POINT.

ad infinitum Continuing infinitely many times.

adj Abbreviation for *adjoint.

adjacency matrix For a *graph G, with n vertices $v_1, v_2, \ldots, v_n$, the adjacency matrix $\mathbf{A}$ is the $n \times n$ matrix $[a_{ij}]$ with a_{ij} equalling the number of edges connecting v_i to v_j. The matrix $\mathbf{A}$ is *symmetric if G is not *directed and the sum of any row's (or column's) entries is equal to the *degree of the corresponding vertex. An example of a graph and its adjacency matrix $\mathbf{A}$ is shown in the figure.

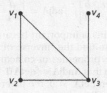

A graph G

$$\mathbf{A} = \begin{bmatrix} 0 & 1 & 1 & 0 \\ 1 & 0 & 1 & 0 \\ 1 & 1 & 0 & 1 \\ 0 & 0 & 1 & 0 \end{bmatrix}$$

The adjacency matrix of G

adjacent angles A pair of angles that share a common vertex and edge.

adjacent edges A pair of edges in a *graph joined by a common vertex.

adjacent side The side of a right-angled triangle between the right angle and the given angle. *Compare* OPPOSITE SIDE.

adjacent vertices A pair of vertices in a *graph joined by a common edge. *See* NEIGHBOURHOOD.

adjoint (adjugate) (of a matrix) The adjoint of a square matrix $\mathbf{A}$, denoted by adj$\mathbf{A}$, is the transpose of the matrix of *cofactors of $\mathbf{A}$. For $\mathbf{A} = [a_{ij}]$, let A_{ij} denote the *cofactor of the entry a_{ij}. Then the matrix of cofactors is the matrix $[A_{ij}]$ and adj $\mathbf{A} = [A_{ij}]^T$. For example, a 3×3 matrix $\mathbf{A}$ and its adjoint can be written

$$\mathbf{A} = \begin{bmatrix} a_{11} & a_{12} & a_{13} \\ a_{21} & a_{22} & a_{23} \\ a_{31} & a_{32} & a_{33} \end{bmatrix}, \text{adj}\mathbf{A} = \begin{bmatrix} A_{11} & A_{21} & A_{31} \\ A_{12} & A_{22} & A_{32} \\ A_{13} & A_{23} & A_{33} \end{bmatrix}.$$

In the 2×2 case, a matrix $\mathbf{A}$ and its adjoint has the form

$$A = \begin{bmatrix} a & b \\ c & d \end{bmatrix}, \quad \text{adj}A = \begin{bmatrix} d & -b \\ -c & a \end{bmatrix}.$$

The adjoint is important because it can be used to find the *inverse of a matrix. From the properties of cofactors, it can be shown that $A\text{adj}A = (\det A)I$. It follows that, when $\det A \neq 0$, the inverse of A is $(1/\det A)\,\text{adj}A$.

adjoint (of a linear map) For $T:V{\to}V$, a *linear map of an *inner product space, the adjoint $T^*:V{\to}V$ satisfies $\langle Tv,w\rangle = \langle v,T^*w\rangle$ for all $v,w \in V$. If M is the *matrix of T with respect to an *orthonormal basis then the *transpose M^T (real) or *conjugate transpose M^* (complex) is the matrix for T^*.

adjoint equation For a system of *linear differential equations $y' = Ay$, where A is a matrix with *adjoint A^*, the adjoint equation is $z' = -A^*z$.

adjugate A synonym for ADJOINT (of a matrix).

a.e. Abbreviation for ALMOST EVERYWHERE.

aerodynamic drag A *body moving through the air, such as an aeroplane flying in the Earth's atmosphere, experiences a *force due to the flow of air over the surface of the body. The force is the sum of the aerodynamic drag, which is tangential to the flight path, and the lift, which is normal to the flight path.

aerodynamics See FLUID MECHANICS.

affine geometry The study of those geometric properties which are invariant under *affine transformations. So *distance is not preserved, nor *angle, but *parallel lines and *ellipses are. *Points at infinity remain at infinity, and in fact affine geometry can be considered as *projective geometry that preserves the line (or more generally the hyperplane) at infinity.

affine hull The affine hull of a set S is the smallest *affine set containing S. This is equivalent to the intersection of all

*affine sets containing S. *Compare* CONVEX HULL.

affine set Contains all the points on the entire line through any two distinct points of a set S. *Compare* CONVEX SET.

affine space The affine subspaces of 3-dimensional *Euclidean space are the *points, *lines, *planes, and whole space. The translations of these subspaces are all *vector spaces. Formally an affine space is a map $S \times X \to S$ where S is a set and X is a vector space X, and where we write $s + x$ for the image of $(s, \mathbf{x})$, satisfying

(i) $s + \mathbf{0} = s$,
(ii) $(s + \mathbf{x}) + \mathbf{y} = s + (\mathbf{x} + \mathbf{y})$.
(iii) given $s,\ t \in S$, there is a unique $\mathbf{x} \in X$ such that $s + \mathbf{x} = t$.

The solutions of an *inhomogeneous *linear equation form an affine space. For example, the solutions to the ODE $y''-y = 1$ has a solution set S and the vector space X is the solution space of $y''-y = 0$. Any vector space V is an affine space with $S = X = V$.

affine subspace A subset of a *vector space which contains all lines between points of the subset.

affine transformation A *transformation which preserves *collinearity and therefore the straightness and *parallel nature of lines and the ratios of distances. The affine transformations of $\mathbb{R}^n$ are of the form $\mathbf{v} \mapsto A\mathbf{v} + \mathbf{b}$ where A is *invertible and $\mathbf{b} \in \mathbb{R}^n$.

affine variety See VARIETY.

a fortiori From the Latin, meaning 'with even stronger reason', this is often used to refer to a general statement from which another follows immediately; for example, since 11 is a *prime number, it is *a fortiori* not *divisible by 5.

aggregate Census returns are used to construct aggregate *statistics concerning a wide range of characteristics such as economic, social, health, education, often grouped by geographical region, gender, age, etc. The name derives from the way they are compiled as counts from a large number of individual census returns. *See also* INDEX (IN STATISTICS).

agree If $f(x)$ and $g(x)$ are functions defined on a set S, and $f(x) = g(x)$ for all $x \in S$, then f and g agree on the set S.

air resistance The resistance to motion experienced by an object moving through the air caused by the flow of air over the surface of the object. It is a *force that affects, for example, the speed of a drop of rain or of a parachutist falling towards the Earth's surface. As well as depending on the nature of the object, air resistance depends on the speed of the object. Possible *mathematical models assume that the magnitude of the air resistance is proportional to the speed (laminar flow) or to the square of the speed (turbulent flow). *See* TERMINAL SPEED.

Airy function The differential equation $\phi''(t) - t\,\phi(t) = 0$ has the Airy function, a *special function, $\phi(t) = \frac{1}{\pi}\int_0^\infty \cos\left(tx + \frac{x^3}{3}\right)dx$ as its solution.

Aitken's method (in numerical analysis) If an iterative formula $x_{r+1} = f(x_r)$ is to be used to solve an equation, Aitken's method of accelerating convergence uses the initial value and the first two values obtained by the formula to calculate a better approximation than the iterative formula would produce. This can then be used as a new starting point from which to repeat the process until the required accuracy has been reached. While this is computationally intensive, it is the sort of process which spreadsheets handle very easily.

If x_0, x_1, x_2 are the initial value and the first two iterations and $\Delta x_r = x_{r+1} - x_r$, $\Delta^2 x_r = \Delta x_{r+1} - \Delta x_r$ are the forward differences then $x_4 = x_3 - \dfrac{(\Delta x_2)^2}{\Delta^2 x_1}$. More generally this will be expressed as $x_{r+1} = x_r - \dfrac{(\Delta x_{r-1})^2}{\Delta^2 x_{r-2}}$.

Alan Turing Institute The Alan Turing Institute (*see* TURING, ALAN MATHISON) is the UK's national institute for *data science and artificial intelligence, founded in 2015. The institute's director is currently Sir Adrian *Smith.

aleph The first Hebrew letter ℵ, commonly used to denote infinite *cardinal numbers. *See also* TRANSFINITE NUMBER.

aleph-null The smallest infinite *cardinal number. The *cardinality of any set which can be put in *one-to-one correspondence with the set of *natural numbers. Such sets are said to be *countably infinite or *denumerable. Paradoxically the set of rational numbers between 0 and 1, the set of all *rational numbers, and the set of *natural numbers all have this same cardinality. The symbol ℵ$_0$ is used.

Alexander, James (Waddell) (1888–1971) American mathematician and pioneering topologist, who did much to develop and make rigorous *Poincaré's visions for *algebraic topology and *combinatorial topology, and who also first defined cohomology. In 1928 he introduced a polynomial *knot invariant, now known as the Alexander polynomial.

algebra Elementary algebra relates to the general properties of *arithmetic and solutions of *polynomial equations involving unknown *variables. More advanced algebra focuses on *algebraic structures that are usually generalizations or abstractions of more concrete problems. The use of algebra, to some extent, permeates all of mathematics, and algebraic techniques have become important in other areas such as *algebraic topology. The word algebra derives from a work *al-jabr* of *Khwārizmī. *Viète made significant advances in the use of algebraic notation in the 16th century. *See also* ABSTRACT ALGEBRA and LINEAR ALGEBRA.

algebra A *vector space equipped with a *bilinear product. An example is the space of $n \times n$ real matrices, with product $(A,B) \mapsto AB$. As here, the product need not be commutative. *See also* BANACH ALGEBRA, DIVISION ALGEBRA, GROUP ALGEBRA.

algebraically closed (algebraic closure) A *field is algebraically closed if every root of every polynomial with coefficients in the field is in the field. The *Fundamental Theorem of Algebra shows that $\mathbb{C}$ is algebraically closed. The algebraic closure of a field is the smallest *extension of the field which is algebraically closed. So, the algebraic closure of $\mathbb{R}$ is $\mathbb{C}$ and the algebraic closure of $\mathbb{Q}$ is the field of *algebraic numbers.

algebraic basis *See* HAMEL BASIS.

algebraic curve A planar algebraic curve is one that can be represented by a *polynomial equation in two variables. The degree of the curve is the *degree of the polynomial. Such a curve defined over the *rational numbers might consist of a finite number of points, whilst a complex algebraic curve (*see* COMPLEX NUMBER) is a *Riemann surface.

algebraic geometry The area of mathematics related to the study of *geometry by algebraic methods and to the study of geometric objects that can be defined algebraically, e.g. by *polynomial equations.

algebraic integer An *algebraic number with a monic *minimal polynomial that has *integer *coefficients. The algebraic integers form a *subring of the *complex numbers.

algebraic multiplicity The *multiplicity of an *eigenvalue as a *root of the *characteristic polynomial. The algebraic multiplicity of an eigenvalue is always at least its *geometric multiplicity.

algebraic number A real or complex number that is the root of a non-zero *polynomial equation with integer coefficients. All *rational numbers are algebraic, since a/b is the root of the equation $bx-a = 0$. Some *irrational numbers are algebraic; for example, $\sqrt{2}$ is the root of the equation $x^2-2 = 0$. An irrational number that is not algebraic (such as π) is called a *transcendental number. The algebraic numbers are a *denumerable *subfield of $\mathbb{C}$. *See* MINIMAL POLYNOMIAL.

algebraic structure The term used to describe an abstract concept defined as consisting of certain elements with *binary operations satisfying given axioms. Thus *groups, *rings, *fields, *vector spaces and *modules are algebraic structures. The aim is to recognize similarities that appear in different contexts within mathematics and to encapsulate these by means of a set of *axioms.

algebraic topology The area of mathematics which uses *abstract algebra to study topology, and *algebraic structures as topological invariants. *See* FUNDAMENTAL GROUP.

algebra of limits A collection of results relating to algebraic rules of convergence. If real (or complex) *sequences a_n and b_n converge to *limits A and B, then $a_n + b_n$ converges to $A + B$, $a_n b_n$ converges to AB and a_n/b_n converges to A/B provided $B \neq 0 \neq b_n$. Similar results apply to the sums of *series and limits of *functions.

algebra of sets The *power set $\wp(E)$ of all subsets of a *universal set E is closed under the *binary operations $\cup$ (*union) and $\cap$ (*intersection), $+$ (*symmetric difference) and the *unary operation $'$ (*complementation). The following are some of the properties, or laws, that hold for subsets A, B, and C of E:

(i) $A \cup (B \cup C) = (A \cup B) \cup C$, $A + (B + C) = (A + B) + C$ and $A \cap (B \cap C) = (A \cap B) \cap C$, the *associative properties.

(ii) $A \cup B = B \cup A$, $A + B = B + A$, and $A \cap B = B \cap A$, the *commutative properties.

(iii) $A \cup \emptyset = A$, $A + \emptyset = A$ and $A \cap \emptyset = \emptyset$, where $\emptyset$ is the *empty set.

(iv) $A \cup E = E$ and $A \cap E = A$.

(v) $A \cup A = A$, $A + A = \emptyset$ and $A \cap A = A$.

(vi) $A \cap (B \cup C) = (A \cap B) \cup (A \cap \cap C)$, $A + (B \cap C) = (A + B) \cap (A + C)$, and $A \cup (B \cap C) = (A \cup B) \cap (A \cup C)$, the *distributive properties.

(vii) $A \cup A' = E$, $A + A' = E$,   and $A \cap A' = \emptyset$.

(viii) $E' = \emptyset$ and $\emptyset' = E$.

(ix) $(A')' = A$.

(x) $(A \cup B)' = A' \cap B'$ and $(A \cap B)' = A' \cup B'$, *De Morgan's laws.

The application of these laws to subsets of E is known as the algebra of sets. Despite some similarities with the algebra of numbers, there are important and striking differences. If $|E| = n$ then $(\wp(E), +, \cap)$ is *isomorphic as a *ring to $(\mathbb{Z}_2^n, +, \times)$. *See* BOOLEAN ALGEBRA.

algorithm A precisely described routine procedure that can be applied and systematically followed through to a conclusion. The word derives from the mathematician *Khwārizmī.

algorithmic complexity The number of steps required in an algorithm varies with the size of the input variables. For example, the *Euclidean Algorithm can find the *greatest common divisor of m, n, where $m < n$, in $O(\ln n)$ steps (in the *big O notation). Preferentially, an algorithm will terminate in polynomial time (such as *quick sort), which means that the number of steps is *bounded above by a *polynomial, and consequently the algorithm does not quickly become unmanageable as the input's size increases. But some important problems have no known solution in polynomial time such as *factorization of integers and the *travelling salesman problem. *See* MILLENNIUM PRIZE PROBLEMS.

aliquant part A number or expression which is not a *divisor of a given number or expression. For example, 2 is an aliquant part of any odd number, and $x + 1$ is an aliquant part of $x^2 + 1$.

aliquot part A number or expression which is a *divisor of a given number or expression, and is usually required to be a proper divisor. For example, 2 is an aliquot part of any even number larger than 2, and $x + 1$ is an aliquot part of $x^2 - 1$.

al-Khwārizmī *See* KHWĀRIZMĪ.

almost all (almost everywhere) Holding for all values except on a *null set. Thus, if two (Lebesgue) *integrable functions agree almost everywhere then their *integrals are equal.

almost disjoint Two sets are almost disjoint if their *intersection is finite (or small in some other defined sense). A collection of subsets is said to be almost disjoint if they are *pairwise almost disjoint.

almost surely almost everywhere (*see* ALMOST ALL) in a *probability measure.

almost uniform convergence A sequence of functions which has *uniform convergence except on sets of arbitrarily small *measure has almost uniform convergence. In general this is *weaker than the uniform convergence almost everywhere but does imply convergence almost everywhere.

alphabet In *coding, the collection of symbols used to create *words. In a *binary code the alphabet is $\{0, 1\}$.

alternate angles *See* TRANSVERSAL.

alternating function A synonym for SKEW-SYMMETRIC FUNCTION.

alternating group The *subgroup of the *symmetric group, S_n, containing all the *even permutations of n objects. It is denoted A_n and has order $n!/2$ for $n > 1$. For $n > 4$, it is the only proper, non-trivial, *normal subgroup of S_n and is a *simple group.

alternating series A series in which the sign alternates between positive and negative, i.e. any series in the form $a_n = (-1)^n p_n$ or $(-1)^{n-1} p_n$ where all $p_n > 0$ for all n.

alternating series test The result that an *alternating series *converges if the terms decrease *monotonically to zero in *absolute value. It is attributed to Gottfried *Leibniz.

alternative hypothesis *See* HYPOTHESIS TESTING.

altitude (geometry) A line through one vertex of a triangle and perpendicular to the opposite side. The three altitudes of a triangle are concurrent at the *orthocentre.

altitude (astronomy) The angle above the horizon that a star is located at. *See also* AZIMUTH.

AM-GM An abbreviation for ARITHMETIC-GEOMETRIC MEAN INEQUALITY.

ambiguous case The case where two sides of a triangle are known, and an acute angle which is not the angle between the known sides. There can often be two possible triangles which satisfy all the given information, hence the name.

If you know the length of AB, and of BC, and the acute angle at A, you can construct the triangle as follows. Draw AB. From A draw a line making the required angle to AB. Place the point of a compass at B and draw an arc of a circle with radius equal to the required length of BC. Where the line from A and the arc intersect is the position of C. Unless the angle at C is exactly $90°$ or the given information is inconsistent, there will be two different points of intersection and hence two possible triangles.

amicable numbers A pair of numbers with the property that each is equal to the sum of the *proper positive divisors of the other. For example, 220 and 284 are amicable numbers because the proper divisors of 220 are 1, 2, 4, 5, 10, 11, 20, 22, 44, 55, and 110, whose sum is 284, and the proper divisors of 284 are 1, 2, 4, 71, and 142, whose sum is 220.

These numbers, known to the Pythagoreans, were used as symbols of friendship. The amicable numbers 17 296 and 18 416 were found by *Fermat, and a list of 64 pairs was produced by *Euler. In 1867, a 16-year-old Italian boy found the second smallest pair, 1184 and 1210, overlooked by Euler. More than a billion pairs are now known. It is unknown

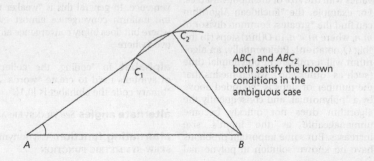

ABC_1 and ABC_2 both satisfy the known conditions in the ambiguous case

whether or not there are infinitely many pairs of amicable numbers.

amplitude Suppose that $x = A\sin(\omega t + \alpha)$, where A (> 0), ω and α are constants. This may, for example, give the displacement x of a particle, moving in a straight line, at time t. The particle is thus oscillating about the origin O. The constant A is the amplitude and gives the maximum distance in each direction from O that the particle attains. (ω is the *angular frequency and α is the *phase.)

The term may also be used in the case of *damped oscillations to mean the corresponding coefficient, even though it is not constant. For example, if $x = 5e^{-2t}$ $\sin 3t$ the oscillations are said to have amplitude $5e^{-2t}$, which tends to zero as t tends to infinity.

AMS Acronym for the American Mathematical Society, a society of professional mathematicians founded in New York in 1888.

analog device A measuring instrument of a continuous quantity which shows the measurement on a continuous scale. For example, the hands of a clock (provided that the second hand moves smoothly). *Compare* DIGITAL.

analysis The area of mathematics generally taken to include those topics that involve the use of limit processes. Such limit processes apply to the definitions of *continuity, *derivatives, *integrals, *infinite series, *Fourier series, and *infinite products. Such ideas also apply to the *complex numbers and complex-valued functions in *complex analysis. Abstractions of these ideas lead to ideas of *topological spaces, *metric spaces, and *measure spaces. *Linear algebra focuses mainly on finite-dimensional *vector spaces, and techniques of analysis are usually applied to the study of infinite-dimensional vector spaces (*see* BANACH SPACES, FUNCTION SPACES, HILBERT SPACES). Related areas such as

*differential equations and *numerical analysis would also be considered part of applied analysis.

analysis of variance A general procedure for partitioning the overall variability in a set of *data into components due to specified causes and random variation. It involves calculating such quantities as the 'between-groups sum of squares' and the 'residual sum of squares' and dividing by the *degrees of freedom to give so-called 'mean squares'. The results are usually presented in an ANOVA table, the name being derived from the opening letters of the words 'analysis of variance'. Such a table provides a concise summary from which the influence of the *explanatory variables can be estimated and hypotheses can be tested, usually by means of *F-tests.

(((⊕))) SEE WEB LINKS

- A description of the analysis and its interpretation in a medical context (subscription).

analysis situs A now obsolete name for *topology. Also the name of a seminal paper of 1895 by *Poincaré in which he defined the *fundamental group.

analytic A *function is analytic if it can be defined by a *convergent *Taylor series in a *neighbourhood of any point of the *domain. An analytic function necessarily has derivatives of all orders, but the converse is not true. The real function, defined by $f(x) = \exp(-1/x^2)$ for $x \neq 0$ and $f(0) = 0$, is infinitely differentiable at $x = 0$, with all derivatives being 0; thus, the Taylor series at $x = 0$ only converges to the function at 0. The space of analytic functions is denoted C^ω.

analytic (of a complex function) As a *holomorphic function is analytic, then the terms 'complex analytic' and 'holomorphic' are used synonymously.

analytic continuation A *holomorphic function $F:V \to \mathbb{C}$ is the analytic continuation of $f:U \to \mathbb{C}$ where $\varnothing \neq U \subseteq V$ if f and F *agree on U. If V is *connected, then the analytic continuation to V is unique, but analytic continuations to two sets V_1 and V_2 may not agree on $V_1 \cap V_2$.

analytic geometry The area of mathematics relating to the study of *coordinate geometry. *Compare* SYNTHETIC GEOMETRY.

analytic number theory An area of *number theory that largely employs methods from *analysis. Its inception is usually credited to *Dirichlet and his theorem about *primes among *arithmetic progressions. Other results employing analytical means include the *prime number theorem, *Chebyshev's proof of *Bertrand's postulate, and the circle method of *Hardy and *Littlewood.

analytics The branch of *data science applying mathematics to large data sets with the aims of devising strategies for competitive advantage and value creation, particularly in relation to commerce.

and *See* CONJUNCTION.

angle (angular) A configuration of two half-lines, or *arms, emanating from a point; the angle is said to be *subtended by its arms. The angle is measured as a fraction of a *full angle equating to 360 *degrees or 2π *radians; centring a unit circle at the point, the angle's measure in radians is equal to the length of the arc between the two arms. Angles may also have a positive (anticlockwise) *sense or a negative (clockwise) sense. *See also* ACUTE ANGLE, ADJACENT ANGLES, COMPLEMENTARY ANGLES, CONJUGATE ANGLES, EXTERIOR ANGLE, INTERIOR ANGLE, OBTUSE ANGLE, REFLEX ANGLE, RIGHT ANGLE, STRAIGHT ANGLE, SUPPLEMENTARY ANGLES. *Compare* SOLID ANGLE.

angle (between lines in space) Given two lines in space, let $\mathbf{u}_1$ and $\mathbf{u}_2$ be vectors with directions along the lines. Then the angle between the lines, even if they do not meet, is equal to the angle between the vectors $\mathbf{u}_1$ and $\mathbf{u}_2$ (*see* ANGLE (between vectors)), with the directions of $\mathbf{u}_1$ and $\mathbf{u}_2$ chosen so that the angle θ satisfies $0 \leq \theta \leq \pi/2$ (θ in radians), or $0 \leq \theta \leq 90$ (θ in degrees). If l_1, m_1, n_1 and l_2, m_2, n_2 are *direction ratios for directions along the lines, the angle θ between the lines is given by

$$\cos\theta = \frac{|l_1 l_2 + m_1 m_2 + n_1 n_2|}{\sqrt{l_1^2 + m_1^2 + n_1^2}\sqrt{l_2^2 + m_2^2 + n_2^2}}.$$

angle (between lines in the plane) In coordinate geometry of the plane, the angle α between two lines with gradients m_1 and m_2 is given by

$$\tan\theta = \frac{m_1 - m_2}{1 + m_1 m_2}$$

This is obtained from the $\tan(A-B)$ formula. In the special cases when $m_1 m_2 = -1$ or when m_1 or m_2 is infinite, it has to be interpreted as follows: if $m_1 m_2 = -1$, $\theta = 90°$; if m_i is infinite, $\theta = 90° - |\tan^{-1} m_j|$ where $j \neq i$; and if m_1 and m_2 are both infinite, $\theta = 0$.

angle (between planes) Given two planes, let $\mathbf{n}_1$ and $\mathbf{n}_2$ be vectors *normal to the two planes. Then a method of obtaining the angle between the planes is to take the angle between $\mathbf{n}_1$ and $\mathbf{n}_2$ (*see* ANGLE (between vectors)), with the directions of $\mathbf{n}_1$ and $\mathbf{n}_2$ chosen so that the angle θ satisfies $0 \leq \theta \leq \pi/2$ (θ in radians), or $0 \leq \theta \leq 90$ (θ in degrees).

angle (between vectors) Given non-zero vectors $\mathbf{a}$ and $\mathbf{b}$, let OA and OB be *directed line segments representing $\mathbf{a}$ and $\mathbf{b}$. Then the angle θ between the vectors $\mathbf{a}$ and $\mathbf{b}$ is the angle $\angle AOB$, where θ is taken to satisfy $0 \leq \theta \leq \pi$ (θ in radians), or $0 \leq \theta \leq 180$ (θ in degrees). It can be expressed in terms of the *scalar product by

$$\cos \theta = \frac{\mathbf{a} \cdot \mathbf{b}}{|\mathbf{a}||\mathbf{b}|}.$$

angle of depression/elevation The angle a line makes below/above a plane, usually the horizontal, especially for a line of sight, or the direction of projection of a projectile.

angle of friction The angle λ such that $\tan \lambda = \mu_s$, where μ_s is the coefficient of static *friction. Consider a block resting on a horizontal plane, as shown in the figure. In the limiting case when the block is about to move to the right on account of an applied *force of magnitude P, $N = mg$, $P = F$, and $F = \mu_s N$. Then the *contact force, whose components are N and F, makes an angle λ with the vertical.

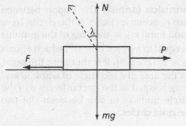

λ is the angle of friction

angle of inclination See INCLINED PLANE.

angle of projection The angle that the direction in which a particle is projected makes with the horizontal. Thus it is the angle that the initial velocity makes with the horizontal.

angular acceleration Suppose that the particle P is moving in the plane, in a circle with centre at the origin O and radius r_0. Let (r_0, θ) be the *polar coordinates of P. At an elementary level, the angular acceleration may be defined

to be $\ddot{\theta}$, where a dot denotes *differentiation with respect to time.

At a more advanced level, the angular acceleration α of the particle P is the vector defined by $\alpha = \dot{\omega}$, where ω is the *angular velocity. Let $\mathbf{i}$ and $\mathbf{j}$ be unit vectors in the directions of the positive x- and y-axes and let $\mathbf{k} = \mathbf{i} \times \mathbf{j}$. Then, in the case above of a particle moving along a circular path, $\omega = \dot{\theta} \mathbf{k}$ and $\alpha = \ddot{\theta} \mathbf{k}$. If $\mathbf{r}$, $\mathbf{v}$ and $\mathbf{a}$ are the position vector, velocity and acceleration of P, then

$$\mathbf{r} = r_0 \mathbf{e}_r, \quad \mathbf{v} = \dot{\mathbf{r}} = r_0 \dot{\theta} \mathbf{e}_\theta,$$
$$\mathbf{a} = \ddot{\mathbf{r}} = -r_0 \dot{\theta}^2 \mathbf{e}_r + r_0 \ddot{\theta} \mathbf{e}_\theta$$

where $\mathbf{e}_r = \mathbf{i} \cos \theta + + \mathbf{j} \sin \theta$ and $\mathbf{e}_\theta = -\mathbf{i} \sin \theta + \mathbf{j} \cos \theta$ (see CIRCULAR MOTION). Using the fact that $\mathbf{v} = \omega \times \mathbf{r}$, it follows that the acceleration $\mathbf{a} = \alpha \times \mathbf{r} + \omega \times (\omega \times \mathbf{r})$.

angular frequency The constant ω in the equation $\ddot{x} = -\omega^2 x$ for *simple harmonic motion. The angular frequency ω is usually measured in radians per second. The *frequency of the oscillations is equal to $\omega/2\pi$ and the *period is $2\pi/\omega$.

angular measure There are two principal ways of measuring angles: by using *degrees, in more elementary work, and by using *radians, essential in more advanced work, in particular, when *calculus is involved.

angular momentum Suppose that the particle P of mass m has position vector $\mathbf{r}$ and is moving with velocity $\mathbf{v}$. Then the angular momentum $\mathbf{L}$ of P about the point A with position vector $\mathbf{r}_A$ is the vector defined by $\mathbf{L} = (\mathbf{r} - \mathbf{r}_A) \times m\mathbf{v}$. It is the *moment of the *linear momentum about the point A. See also CONSERVATION OF ANGULAR MOMENTUM.

Consider a *rigid body rotating with *angular velocity ω about a fixed axis, and let $\mathbf{L}$ be the angular momentum of the rigid body about a point on the fixed axis. Then $\mathbf{L} = I\omega$, where I is the *moment of inertia of the rigid body about the fixed axis.

To consider the general case, let $\boldsymbol{\omega}$ and $\mathbf{L}$ be column vectors representing the angular velocity and the angular momentum of a rigid body about a fixed point (or the centre of mass). Then $\mathbf{L} = \mathbf{I}\boldsymbol{\omega}$, where $\mathbf{I}$ is a 3×3 matrix, called the *inertia matrix, whose elements involve the moments of inertia and the *products of inertia of the rigid body relative to axes through the fixed point (or centre of mass).

The rotational motion of a rigid body depends on the angular momentum of the rigid body. In particular, the rate of change of the angular momentum about a fixed point (or centre of mass) equals the sum of the *moments of the forces acting on the rigid body about the fixed point (or centre of mass).

angular speed The magnitude of the *angular velocity.

angular velocity Suppose that the particle P is moving in the plane, in a circle with centre at the origin O and radius r_0. Let (r_0, θ) be the *polar coordinates of P. At an elementary level, the angular velocity may be defined to be $\dot{\theta}$, where a dot denotes *differentiation with respect to time.

At a more advanced level, the angular velocity $\boldsymbol{\omega}$ of the particle P is the vector defined by $\boldsymbol{\omega} = \dot{\theta}\mathbf{k}$, where $\mathbf{i}$ and $\mathbf{j}$ are unit vectors in the directions of the positive x- and y-axes, and $\mathbf{k} = \mathbf{i} \times \mathbf{j}$. If $\mathbf{r}$ and $\mathbf{v}$ are the position vector and velocity of P, then

$$\mathbf{r} = r_0 \mathbf{e}_r, \quad \mathbf{v} = \dot{\mathbf{r}} = r_0 \dot{\theta} \mathbf{e}_\theta,$$

where $\mathbf{e}_r = \mathbf{i} \cos\theta + \mathbf{j} \sin\theta$ and $\mathbf{e}_\theta = -\mathbf{i} \sin\theta + \mathbf{j} \cos\theta$ (*see* CIRCULAR MOTION). By using the fact that $\mathbf{k} = \mathbf{e}_r \times \mathbf{e}_\theta$, it follows that the velocity $\mathbf{v}$ is given by $\mathbf{v} = \boldsymbol{\omega} \times \mathbf{r}$.

Consider a rigid body rotating about a fixed axis, and take coordinate axes so that the z-axis is along the fixed axis. Let (r_0, θ) be the polar coordinates of some point of the rigid body, not on the axis, lying in the plane $z = 0$. Then the angular velocity $\boldsymbol{\omega}$ of the rigid body is defined by $\boldsymbol{\omega} = \dot{\theta}\mathbf{k}$.

In general, for a rigid body that is rotating, such as a top spinning about a fixed point, the rigid body possesses an angular velocity $\boldsymbol{\omega}$ whose magnitude and direction depend on time, though the formula $\mathbf{v} = \boldsymbol{\omega} \times \mathbf{r}$ still applies.

annihilator (modules) Given a subset S of a *module M over a commutative *ring R, then the annihilator of S is $\mathrm{Ann}_R(S) = \{r \in R : rs = 0 \text{ for all } s \in S\}$. It is an *ideal of R.

annihilator (vector spaces) Given a subset S of a *vector space V, then the annihilator S° is the set of linear *functionals f such that $f(S) = \{0\}$. Then S° is a *subspace of the *dual space V^*; if S is itself a subspace and V is finite-dimensional, then $\dim S + \dim S^\circ = \dim V$. The correspondence $S \leftrightarrow S^\circ$ is crucial to *duality in *projective geometry.

annulus (annuli) The region between two concentric circles. If the circles have radii r and $r + w$, the area of the annulus is equal to $\pi(r + w)^2 - \pi r^2$, which equals $w \times 2\pi(r + \frac{1}{2}w)$. It is therefore the same as the area of a rectangle of width w and length equal to the circumference of the circle midway in size between the two original circles.

An annulus

anonymity of survey data People will often answer personal questions differently depending on how anonymous they believe their answers will be. Since there is a high premium on gaining truthful responses, methods of ensuring anonymity or at least strict confidentiality in a survey are important.

ANOVA *See* ANALYSIS OF VARIANCE.

antecedent The clause in a conditional statement which expresses the condition. So in 'if n is divisible by 2, n is even', the antecedent is 'n is divisible by 2'. *Compare* CONSEQUENT.

anti- A prefix variously denoting the *inverse, as in *antilogarithm, or an opposite or contrast, such as *antipodal or *antisymmetric.

anticlastic A *surface is anticlastic at a point if the *Gaussian curvature is negative there, such as with a *saddle-point of a graph $z = f(x,y)$.

anticlockwise Movement in the opposite direction to that which the hands of a clock normally take. In compass terms, $N \rightarrow W \rightarrow S \rightarrow E$ is anticlockwise. In *polar coordinates, or when orienting curves, this is taken to be the positive direction.

antiderivative Given a *real function f, any function ϕ such that $\phi'(x) = f(x)$, for all x (in the domain of f), is an antiderivative of f. If ϕ_1 and ϕ_2 are both antiderivatives of a *continuous function f, defined on an interval, then $\phi_1(x)$ and $\phi_2(x)$ differ by a constant. In that case, the notation

$$\int f(x)\,dx$$

may be used for an antiderivative of f, with the understanding that an *arbitrary constant can be added to any antiderivative. Thus,

$$\int f(x)\,dx + c$$

where c is an arbitrary constant, is an expression that gives all the antiderivatives of f.

antidifferentiate The reverse process of *differentiation used as a method of *integration, as opposed to defining integrals as the limits of sums of incremental elements.

antilogarithm The antilogarithm of x is the number whose *logarithm is equal to x. If y is the number whose logarithm (to base a) is x, then $\log_a y = x$. This is equivalent to $y = a^x$. So, if base a is being used, $\text{antilog}_a x$ is identical with a^x; for common logarithms, $\text{antilog}_{10} x$ is just 10^x, and this notation is preferable.

antiparallel A pair of *directed lines or *vectors whose directions are *parallel but having the opposite *sense.

antipodal points Two points on a *sphere that are at opposite ends of a *diameter.

antiprism A *convex *polyhedron with two 'end' *faces that are *congruent polygons lying in parallel planes, with each *vertex of one polygon joined by an edge to two vertices of the other polygon, so that the remaining faces are triangles. If the end faces are regular, the antiprism is said to be right-regular. If the end faces are regular and the triangular faces are equilateral, the antiprism is a *semi-regular polyhedron.

anti-realism *See* REALISM.

antisymmetric matrix = SKEW-SYMMETRIC MATRIX.

antisymmetric relation A *binary relation $\sim$ on a set S is antisymmetric if, for all a and b in S, whenever $a \sim b$ and $b \sim a$, then $a = b$. For example, the relation $\leq$ on the set of integers is antisymmetric, as is any *partial order. *Compare* ASYMMETRIC RELATION.

aperiodic In a *Markov chain, the ith state has period d where

$$d = gcd\{n : [\mathbf{M}^n]_{ii} > 0\},$$

where **M** denotes the transition matrix. So in the two state Markov chain with

$$\mathbf{M} = \begin{pmatrix} 0 & 1 \\ 1 & 0 \end{pmatrix}$$

each state has period 2. A state is aperiodic if $d = 1$.

Apéry's constant The number $\zeta(3) = \sum_1^\infty n^{-3}$, where ζ denotes the *zeta function. Exact values of $\zeta(n)$, for even positive n, were determined by *Euler and are rational multiples of π^n. But very little is known for odd n. In 1978 Roger Apéry showed $\zeta(3)$ to be *irrational, but it remains unknown whether the constant is *transcendental.

apex (apices) See BASE (of a triangle), PYRAMID.

aphelion See APSE.

apogee See APSE.

Apollonius of Perga (about 262–190 BC) Greek mathematician whose most famous work *The Conics* was, until modern times, the definitive work on the *conic sections: the *ellipse, *parabola, and *hyperbola. He proposed the idea of epicyclic motion for the planets. His methods were close to introducing *Cartesian coordinates, but at the time the Greeks had no notion of *zero or *negative numbers. *Euclid, *Archimedes, and Apollonius were pre-eminent in the period of the third century BC known as the Golden Age of Greek mathematics.

Apollonius' circle Given two points A and B in the plane and a positive constant k, the *locus of all points P such that $AP/PB = k$ is a circle. Such a circle is an Apollonius' circle. Taking $k = 1$ gives a straight line, so either this value must be excluded or, in this context, a straight line must be considered to be a special case of a circle (*see* CIRCLINE). In the figure, $k = 2$.

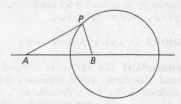

A circle of Apollonius with $k = 2$

a posteriori Not able to be known without experience, i.e. the truth or otherwise cannot be argued by logic from agreed definitions. For example, *Newton's laws of motion describe the way bodies are observed to behave, rather than being deduced from principles in the way that the probabilities of the score when tossing a fair die are.

apothem The line, or the length of the line, joining the centre of a regular polygon to the midpoint of one of its sides. This is also the radius of the *inscribed circle.

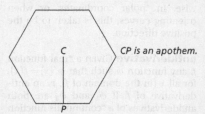

CP is an apothem.

applied mathematics The area of mathematics related to the study of natural phenomena, which includes *mechanics of all types, *probability, *statistics, *mathematical biology, *decision theory, and *game theory, as well as applications of pure mathematics such as in *crytography.

approximate To find a value or expression for a quantity within a specified degree of *accuracy.

approximation When two quantities X and x are approximately equal, written

$X \approx x$, one of them may be used in suitable circumstances in place of, or as an approximation for, the other. For example, $\pi \approx \frac{22}{7}$ and $\sqrt{2} \approx 1.414$.

approximation theory The area of *numerical analysis related to finding a simpler function $f(x)$ which has approximately the same values as $F(x)$ over a specified interval. Analysis can then be carried out on the more accessible $f(x)$ knowing that the difference is small.

a priori Able to be known without experience, i.e. the truth or otherwise can be argued by logic from agreed definitions. For example, the probabilities of the score when tossing a fair die are deduced from the principle of equally likely outcomes.

a priori (in statistics) Another term for *prior probability.

apse (apsis) A point in an *orbit at which the body is moving in a direction perpendicular to its *position vector. In an elliptical orbit in which the centre of attraction is at one focus there are two apses, the points at which the body is at its nearest and its farthest from the centre of attraction. When the centre of attraction is the Sun, the points at which the body is nearest and farthest are the perihelion and the aphelion. When the centre of attraction is the Earth, they are the perigee and the apogee.

Arabic numeral See NUMERAL.

arbitrarily large/small To say that there are arbitrarily large *prime numbers means that the set of prime numbers is not *bounded above, and this phrase is generally used for sets that are not bounded above. Similarly, to say that there are arbitrarily small positive *rationals means that given any $\varepsilon > 0$ there exists a rational number q in the range $0 < q < \varepsilon$. *Compare* SUFFICIENTLY LARGE/SMALL.

arbitrary Otherwise unconstrained; for example, the phrase 'let x be an arbitrary real number' introduces x as a real number, selected without constraint. The arbitrariness of the choice may only be implicit—let $x \in \mathbb{R}$—encapsulates the same. The term is commonly used to describe the *constant of integration. The term might also be combined with constraints such as 'let x be an arbitrary positive number', so that, aside from being positive, x should be considered as any positive number.

arc (of a curve) The part of a curve between two given points on the curve. If A and B are two points on a circle, there are two arcs AB. When A and B are not at opposite ends of a diameter, it is possible to distinguish between the longer and shorter arcs by referring to the major arc AB and the minor arc AB.

arc (of a digraph) See DIGRAPH.

arc-connected A *topological space in which there is a simple path (*see* SIMPLE CURVE) between any two points. An arc-connected space is then *path-connected and a *Hausdorff path-connected space is arc-connected.

arccos, arccosec, arccot, arcsin, arcsec, arctan See INVERSE TRIGONOMETRIC FUNCTION.

arccosh, arccosech, arccoth, arcsinh, arcsech, arctanh See INVERSE HYPERBOLIC FUNCTION.

Archimedean property The Archimedean property of the *natural numbers says that $\mathbb{N}$ is not *bounded above; that is, for any *real number x there exists a natural number n such that $x < n$.

Archimedean solid A *convex *polyhedron is called semi-regular if the faces are regular polygons, though not all congruent, and if the vertices are all alike, in the sense that the different kinds of face are arranged in the same order around each vertex. Right-regular *prisms with square side faces and right-regular *antiprisms whose side faces are equilateral triangles are semi-regular. Apart from these, there are thirteen semi-regular polyhedra, known as the Archimedean solids. These include the *truncated tetrahedron, the *truncated cube, the *cuboctahedron, and the *icosidodecahedron.

SEE WEB LINKS

• Links to the thirteen Archimedean solids, which can be manipulated to see the shapes fully.

Archimedean spiral A curve whose equation in polar coordinates is $r = a\theta$, where $a\ (>0)$ is a constant. In the figure, $|OA| = a\pi$, $|OB| = 2a\pi$, and so $|OB| = 2|OA|$. *Compare* EQUIANGULAR SPIRAL.

SEE WEB LINKS

• An animation showing the Archimedean spiral.

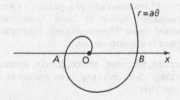

Archimedean spiral

Archimedes (287–212 BC) Greek mathematician who is rated one of the greatest of all time. He made considerable contributions to geometry, for example finding the surface area and the volume of a sphere and the area of a segment of a *parabola. His work on hydrostatics and equilibria was also fundamental. One of his important works, *The Method of Mechanical Theorems*, was rediscovered as recently as 1906. He may or may not have shouted

'Eureka' and run naked through the streets, but he was certainly murdered by a Roman soldier in Syracuse, an event that marks the end of a golden era of Greek mathematics.

Archimedes' constant A synonym for PI.

Archimedes' method Archimedes' *method of exhaustion* involves finding the area of a region by *inscribing a *sequence of polygons inside the region and *circumscribing a sequence of polygons outside the region. The region's area then lies between the inscribed areas and the circumscribed areas. If the polygons are chosen appropriately, as the two sequences progress, all other possible values of the region's area, except the correct value, are ultimately 'exhausted'. Archimedes used this method to show that $3\frac{1}{7} < \pi < 3\frac{10}{71}$.

arc length Let $y = f(x)$ be the graph of a function f such that f' is *continuous on $[a, b]$. The length of the arc, or arc length, of the curve $y = f(x)$ between $x = a$ and $x = b$ equals

$$\int_a^b \sqrt{1 + (f'(x))^2}\,dx.$$

Parametric form For the curve $x = x(t)$, $y = y(t)$ ($t \in [\alpha, \beta]$), the arc length equals

$$\int_\alpha^\beta \sqrt{\left(\frac{dx}{dt}\right)^2 + \left(\frac{dy}{dt}\right)^2}\,dt.$$

Polar form For the curve $r = r(\theta)$ ($\alpha \le \theta \le \beta$), the arc length equals

$$\int_\alpha^\beta \sqrt{r^2 + \left(\frac{dr}{d\theta}\right)^2}\,d\theta.$$

are A unit of area equalling 100 square metres. *See* HECTARE.

area A measure of the extent of a surface, or the part of surface enclosed by some specified boundary. For simple geometrical shapes such as rectangles, triangles, cylinders, etc., the areas can be calculated by simple formulae based on their dimensions (*see* Appendix 1). Other more complex plane figures or 3-dimensional surfaces require the use of *integration, or numerical approximations to find their areas. (*See* SURFACE AREA.)

area of a surface of revolution Let $y = f(x)$ be the graph of a function f such that f' is *continuous on $[a,b]$ and $f(x) \geq 0$ for all x in $[a,b]$. The area of the surface obtained by rotating, through one revolution about the x-axis, the arc of the curve $y = f(x)$ between $x = a$ and $x = b$, equals

$$2\pi \int_a^b f(x) \sqrt{1 + (f'(x))^2}\, dx.$$

Parametric form For the curve $x = x(t)$, $y = y(t)$ ($t \in [\alpha, \beta]$), the surface area equals

$$2\pi \int_\alpha^\beta y \sqrt{\left(\frac{dx}{dt}\right)^2 + \left(\frac{dy}{dt}\right)^2}\, dt.$$

Polar form For the curve $r = r(\theta)$ ($\alpha \leq \theta \leq \beta$), the surface area equals

$$2\pi \int_\varepsilon^\beta r \sin\theta \sqrt{r^2 + \left(\frac{dr}{d\theta}\right)^2}\, d\theta.$$

area under a curve Suppose that the curve $y = f(x)$ lies above the x-axis, so that $f(x) \geq 0$ for all x in $[a, b]$. The area under the curve, that is, the area of the region bounded by the curve, the x-axis, and the lines $x = a$ and $x = b$, equals

$$\int_a^b f(x)\, dx.$$

The definition of *integral is made precisely in order to achieve this result.

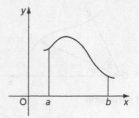

When the function is positive

If $f(x) \leq 0$ for all x in $[a, b]$, the integral above is negative. However, it is still the case that its absolute value is equal to the area of the region bounded by the curve, the x-axis and the lines $x = a$ and $x = b$. If $y = f(x)$ crosses the x-axis, appropriate results hold. For example, if the regions A and B are as shown in the figure below, then

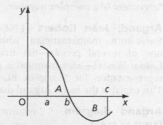

When the function changes sign

area of region $A = \int_a^b f(x)\, dx$ and area of region $B = -\int_b^c f(x)\, dx$.

It follows that

$$\int_a^c f(x)\, dx = \int_a^b f(x)\, dx + \int_b^c f(x)\, dx$$

= area of region A − area of region B.

The combined areas of regions A and B is given by the integral $\int_a^b |f(x)|\, dx$. *See* SIGNED AREA.

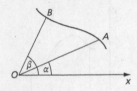

Area of a polar sector

Polar areas If a curve has an equation $r = r(\theta)$ in polar coordinates, there is an integral that gives the area of the region bounded by an arc AB of the curve and the two radial lines OA and OB. Suppose that $\angle xOA = \alpha$ and $\angle xOB = \beta$. The area of the region described equals

$$\int_{\alpha}^{\beta} \frac{1}{2} r^2 d\theta.$$

arg Abbreviation and symbol for the *argument of a complex number.

Argand, Jean Robert (1768–1822) Swiss-born mathematician who was one of several people—the first was Caspar Wessel—who invented a planar representation for *complex numbers. This explains the name Argand diagram.

Argand diagram A synonym for COMPLEX PLANE.

argument (of a complex number) Suppose that the *complex number z is represented by the point P in the *complex plane. The argument of z, denoted by arg z, is the angle θ (in radians) that OP makes with the positive real axis Ox, with the angle given a positive sense anticlockwise from Ox. *Principal values of θ may be taken so that $0 \le \arg z < 2\pi$, or $-\pi < \arg z \le \pi$, or alternatively arg z may be used to denote any of the values $\theta + 2n\pi$, where n is an integer.

argument (of a function) See FUNCTION.

Aristarchus of Samos (about 270 BC) Greek astronomer, noted for being the first to affirm that the Earth rotates and travels around the Sun. He treated astronomy mathematically and used geometrical methods to calculate the relative sizes of the Sun and Moon and their relative distances from the Earth.

Aristotelian logic The formal deductive method of *logic dealing with the relations between *categorical propositions in their form as distinct from their content, especially *syllogisms.

Aristotle (384–322 BC) Greek philosopher who made important contributions to mathematics through his work on deductive *logic.

arithmetic The area of mathematics relating to numerical calculations involving only the basic operations of addition, subtraction, multiplication, division, and simple powers. The term 'higher arithmetic' refers to elementary *number theory.

arithmetic function (number-theoretic function) Arithmetic functions are common and important in *number theory. At their most general, arithmetic functions may be defined as real- or complex-valued functions defined for positive integers. Commonly such functions will also be assumed to be multiplicative in the sense that $f(mn) = f(m)f(n)$ when m and n are coprime. See DIVISOR FUNCTION, MÖBIUS FUNCTION, SIGMA FUNCTION, TOTIENT FUNCTION.

arithmetic-geometric mean The limit of the series obtained by the *arithmetic-geometric mean iteration. This mean can be expressed in terms of *elliptic integrals.

arithmetic-geometric mean inequality The arithmetic mean of a set of non-negative numbers is never less than their geometric mean. So for any $a, b \ge 0$, $\frac{a+b}{2} \ge \sqrt{ab}$. Generally, for a set of numbers $\{a_i\}$, $\frac{1}{n} \sum a_i \ge \sqrt[n]{\prod a_i}$ with equality if and only if all a_i take the same value. This inequality is often abbreviated to AM-GM.

arithmetic-geometric mean iteration If $a < b$ are two positive real numbers, let $a_0 = a$, $b_0 = b$ and $a_{n+1} = \frac{a_n + b_n}{2}$, $b_{n+1} = \sqrt{a_n b_n}$ for $n \geq 0$. Then (a_n) is *increasing and (b_n) is *decreasing and $\lim_{n \to \infty} (a_n) = \lim_{n \to \infty} (b_n)$ is the *arithmetic-geometric mean.

arithmetic mean *See* MEAN.

arithmetic progression A synonym for ARITHMETIC SEQUENCE.

arithmetic sequence A finite or infinite sequence of terms $a_1, a_2, a_3, \ldots$ with a common difference d, so that $a_2 - a_1 = d$, $a_3 - a_2 = d$, and so on. The first term is usually denoted by a. For example, 2, 5, 8, 11, ... is the arithmetic sequence with $a = 2$, $d = 3$. In such an arithmetic sequence, the n-th term a_n is given by $a_n = a + (n-1)d$.

arithmetic series A series $a_1 + a_2 + a_3 + \ldots$ (which may be finite or infinite) in which the terms form an *arithmetic sequence. Let s_n be the nth *partial sum (that is, the sum of the first n terms). Then s_n is given by the formula $s_n = \frac{1}{2}n(a + l)$ where a is the first term and l is the last (nth) term. In particular, when $a = 1$ and the common difference $d = 1$:

$$\sum_{r=1}^{n} r = 1 + 2 + \ldots + n = \frac{1}{2}n(n+1).$$

arm One of the two lines forming an angle.

arrangement *See* PERMUTATION.

array An ordered collection of elements, usually numbers. A *vector is an example of a 1-dimensional array, and a *matrix is an example of a 2-dimensional array. Three or more dimensional arrays are more difficult to represent on paper, though the subscript notation $a_{ij} =$ the element in the i-th row and j-th column of a matrix easily generalizes to higher dimensions. Arrays are equal if all corresponding elements are equal. *Compare* TENSOR.

arrow *See* DIGRAPH.

arrow paradox One of *Zeno's paradoxes which relate to the issue of whether time and space are made up of minute indivisible parts. At one of these moments in time, the arrow occupies a well-defined space. It is moving neither to where it is nor to where it is not (because no time elapses in which it can move), so there is no motion during this instant. If time is composed entirely of instants at which no motion can occur, then all motion is impossible.

Arrow's Impossibility Theorem The surprising result that when voters have three or more distinct options to choose from, there is no rank-order voting system which can aggregate the individual preferences of two or more individuals so that four apparently reasonable conditions are met.

( ) SEE WEB LINKS
• An illustration of the theorem.

Āryabhata (about 476–550) Indian mathematician, author of one of the oldest Indian mathematical texts. Written in verse, the *Āryabhaṭīya* is a summary of miscellaneous rules for calculation and mensuration. It deals with, for example, the areas of certain plane figures, values for π, and the summation of *arithmetic series. Also included is the equivalent of a table of sines, based on the half-chord rather than the whole *chord of the Greeks.

ASA *See* SOLUTION OF TRIANGLES.

ASCII (American Standard Code for Information Interchange) A binary *code representing characters used by monitors, printers, etc.

assignment problem A problem in which things of one type are to be matched with the same number of things of another type in a way that is, in a specified sense, the best possible. For example, when n workers are to be assigned to n jobs, it may be possible to specify the value v_{ij} to the company, measured in suitable units, if the i-th worker is assigned to the j-th job. The values v_{ij} may be displayed as the entries of an $n \times n$ matrix. By introducing suitable variables, the problem of assigning workers to jobs in such a way as to maximize the total value to the company can be formulated as a *linear programming problem.

association A synonym for CORRELATION.

associative The *binary operation ○ on a set S is associative if, for all a, b, and c in S, $(a \circ b) \circ c = a \circ (b \circ c)$.

assumption A statement that is presumed to be true in the particular circumstances. Any conclusions reached are dependent on the validity of the assumption.

astroid A *hypocycloid in which the radius of the rolling circle is a quarter of the radius of the fixed circle. It has

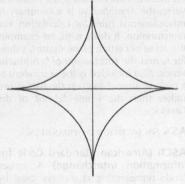

An astroid

*parametric equations $x = a \cos^3 t$, $y = a \sin^3 t$, where a is the radius of the fixed circle.

asymmetric A plane figure is asymmetric if it is neither *symmetrical about a line nor symmetrical about a point.

asymmetric relation A binary relation $\sim$ on a set S is *asymmetric if, for all a and b in S, whenever $a \sim b$, then $b \sim a$ does not hold. For example, the relation $<$ on the set of integers is asymmetric. *Compare* ANTISYMMETRIC RELATION.

asymptote A line l is an asymptote to a curve if the distance from a point P to the line l tends to zero as P tends to infinity along some unbounded part of the curve. Consider the following examples:

(i) $y = \dfrac{x+3}{(x+2)(x-1)}$, (ii) $y = \dfrac{3x^2}{x^2+x+1}$,

(iii) $y = \dfrac{x^3}{x^2+x+1}$.

Example (i) has $x = -2$ and $x = 1$ as vertical asymptotes and $y = 0$ as a horizontal asymptote. Example (ii) has no vertical asymptotes and $y = 3$ as a horizontal asymptote. To investigate example (iii), it can be rewritten as

$$y = x - 1 + \frac{1}{x^2+x+1}.$$

Then it can be seen that $y = x-1$ is a slant (or oblique) asymptote, that is, an asymptote that is neither vertical nor horizontal.

asymptotically equal Two functions $f(x)$ and $g(x)$ are asymptotically equal if $f(x)/g(x)$ has limit 1 as x tends to ∞, written $f(x) \sim g(x)$. For example $(x+1)^3(x+2)^4 \sim x^7$. Being asymptotically equal implies something about the relative size of two functions and does not imply they are approximately equal; in the example, the difference between $(x+1)^3(x+2)^4$ and x^7 becomes arbitrarily large as x increases. *See* BIG O NOTATION, LITTLE O NOTATION.

asymptotically stable *See* LYAPUNOV STABILITY.

asymptotic distribution The limiting *distribution of an infinite *sequence of *random variables. *See* CENTRAL LIMIT THEOREM, CONVERGENCE IN DISTRIBUTION.

asymptotic functions A pair of functions which are *asymptotically equal.

asymptotic series A series $s_n(x) = a_0 + \frac{a_1}{x} + \frac{a_2}{x^2} + \ldots + \frac{a_n}{x^n}$ is called an asymptotic series for a function $f(x)$ if $\lim_{|x| \to \infty} \{x^n[f(x) - s_n(x)]\} = 0$ for every n or equivalently if $f(x) = s_n(x) + o(x^{-n})$ in the *little o notation.

Atiyah, Sir Michael Francis (1929–2019) British mathematician who made important contributions in *algebraic topology, *algebraic geometry, index theory, and quantum field theory. His index theorem (with Isadore Singer) shows the deep connections between the *topology of *compact *manifolds and *elliptic differential equations on such spaces. Awarded the *Fields Medal in 1966 and the Abel Prize in 2004.

atlas *See* CHART.

atmospheric pressure The *pressure at a point in the atmosphere due to the *gravitational force acting on the air. This pressure depends on position and time and is measured by a barometer. The standard atmospheric pressure at sea level is taken to be 101 325 *pascals, which is the definition of 1 atmosphere. In general, the atmospheric pressure decreases with height.

atom In *measure theory, a set which has strictly positive measure, with the property that any subset either has equal measure or has zero measure.

attainable set The set of vertices which can be reached from a given vertex in a *digraph in one or more steps.

atto- Prefix used with *SI units to denote multiplication by 10^{-18}. Abbreviated as a.

attracting fixed point *See* FIXED-POINT ITERATION.

attractor A set towards which a *dynamical system tends to evolve or remain close. So a stable *equilibrium is an attractor, but more generally an attractor can be a point, or a finite set of points, or a *curve, *surface, or higher-dimensional *mainfolds. *See* LORENZ ATTRACTOR.

augmented matrix For a given set of m linear equations in n unknowns $x_1, x_2, \ldots x_n$,

$$a_{11}x_1 + a_{12}x_2 + \ldots + a_{1n}x_n = b_1,$$
$$a_{21}x_1 + a_{22}x_2 + \ldots + a_{2n}x_n = b_2,$$
$$\vdots$$
$$a_{m1}x_1 + a_{m2}x_2 + \ldots + a_{mn}x_n = b_m,$$

the augmented matrix is the matrix

$$\begin{bmatrix} a_{11} & a_{12} & \ldots & a_{1n} & b_1 \\ a_{21} & a_{22} & \ldots & a_{2n} & b_2 \\ \vdots & \vdots & \ddots & \vdots & \vdots \\ a_{m1} & a_{m2} & \ldots & a_{mn} & b_m \end{bmatrix}$$

obtained by adjoining to the matrix of coefficients an extra column of entries taken from the right-hand sides of the equations. The solutions of a set of linear equations may be investigated by transforming the augmented matrix to *echelon form or *reduced echelon form by *elementary row operations. *See* GAUSSIAN ELIMINATION, GAUSS-JORDAN ELIMINATION.

Aut *See* AUTOMORPHISM.

autocorrelation (serial correlation) If $\{x_i\}$ is an ordered sequence of observations then the product moment *correlation coefficient between pairs (x_i, x_{i+1}) is the autocorrelation of lag 1, and between pairs (x_i, x_{i+k}) is the autocorrelation of lag k. The values of k for which the autocorrelation is not negligible can provide important information as to the underlying structure of a *time series.

autocovariance Similar to *autocorrelation but measuring the *covariance between the sequences.

automatic repeat request An error control method for transmitting *data that uses acknowledgements by the receiver within a prescribed period of time. If the sender has not received an acknowledgement in the time agreed, then the sender automatically repeats the transmission.

automorphic function Given a *group G acting on a set X, an automorphic function f is invariant under the group action, that is $f(g.x) = f(x)$ for all $x \in X$ and $g \in G$.

automorphism (automorphism group) An *isomorphism of a mathematical object with itself—so a *one-to-one correspondence from a set onto itself, preserving any further structure (usually algebraic) that is present. For example, *complex conjugation is an automorphism of $\mathbb{C}$ as a *group or *field or real *vector space but not as a complex vector space, as it does not respect *scalar multiplication. The set of automorphisms of an object X, under *composition, then form the automorphism group denoted $\mathrm{Aut}(X)$.

autonomous differential equation A *differential equation where the *independent variable does not explicitly occur. For example $\frac{dv}{dt} = -kv$ is autonomous, but $\frac{dv}{dt} = 3t^2 + 4t - 3$ is not.

auxiliary equation See LINEAR DIFFERENTIAL EQUATION WITH CONSTANT COEFFICIENTS, DIFFERENCE EQUATION.

average deviation A synonym for MEAN DEVIATION.

average of a rate of a quantity The constant value for which the total change in the interval would have been the same as that observed. For example, a car may vary its *instantaneous speed in the duration of a journey, but the average *speed = total *distance travelled/time of journey. In the case of *velocity, the average is a vector, namely total *displacement divided by total time taken.

average of a set of data Usually intended to refer to the *arithmetic mean.

axial plane One of the planes containing two of the coordinate axes in a 3-dimensional Cartesian coordinate system. For example, one of the axial planes is the yz-plane, containing the y-axis and the z-axis, and it has equation $x = 0$.

axiom A statement whose truth is either to be taken as self-evident or to be assumed. Many areas of *pure mathematics, such as *group theory, involve assuming a set of axioms and discovering what results can be logically deduced from them. See ABSTRACTION.

axiomatic set theory The presentation of *set theory as comprising axioms together with rules of inference. See ZERMELO-FRAENKEL AXIOMS.

axiom of choice States that for any set of sets there is at least one set that contains exactly one element in common with each of the non-empty sets. This is equivalent to *Zorn's lemma or to the statement that every set can be *well ordered.

axis (axes) See COORDINATES.

axis (of a cone) See CONE.

axis (of a cylinder) See CYLINDER.

axis (of a parabola) See PARABOLA.

axis of rotation A *line about which a *curve or planar region is *rotated to form a surface or a solid. *Compare* AXIS OF SYMMETRY; *see* SURFACE OF REVOLUTION.

axis of symmetry An axis of *symmetry might refer to reflective symmetry of a planar figure or to rotational symmetry. A line in the plane is an axis of symmetry of a figure if the figure is an *invariant set under *reflection in that line. Or an axis of symmetry might refer to a line such that an object is an invariant set for some non-trivial *rotations about that line. There may be finitely many such rotations, as with a *cube around its various axes of symmetry, or infinitely many, as with a *cylinder about its axis.

azimuth A term used in astronomy to define the bearing of a star relative, usually, to the north. Sometimes south is used as the reference direction. *See also* ALTITUDE (in astronomy).

B The number 11 in *hexadecimal notation.

Babbage, Charles (1792–1871) British mathematician and inventor of mechanical calculators. His 'analytical engine' was designed to perform mathematical operations mechanically using a number of features essential in the design of today's computers, but, partly through lack of funds, the project was not completed.

backward difference If $\{(x_i, f_i)\}$, $i = 0$, 1, 2, . . . is a given set of function values with $x_{i+1} = x_i + h$, $f_i = f(x_i)$ then the backward difference at f_i is defined by $f_i - f_{i-1} = f(x_i) - f(x_{i-1})$. *Compare* FORWARD DIFFERENCE.

backward induction The *contrapositive form of *mathematical induction which shows that if the proposition $P(n)$ fails then the proposition must fail for some $P(k)$, for some $k \le n-1$. As with the standard form of induction, if $P(1)$ is true, this means that $P(2)$ must also be true and so on.

backward scan In an *activity network (edges as activities), the backward scan identifies the latest time for each vertex (node). Starting with the sink, work backwards through the network, for each vertex calculate the sum of edge plus total time on next vertex for all paths leaving that vertex. Put in the largest of these times onto that vertex.

backward substitution *See* LU DECOMPOSITION.

Baire Category Theorem The theorem involves two statements, neither of which implies the other. (i) Every *complete metric space is a *Baire space. (ii) Every locally compact *Hausdorff space is a Baire space. The theorem has many important applications in *functional analysis, including implying an algebraic (or *Hamel) basis of an infinite-dimensional *Banach space must be *uncountable and so motivating the need for *Schauder bases.

Baire space A *topological space in which every *intersection of any *countable collection of *open *dense sets is dense.

balanced (of a set) A subset A of a *vector space where for each element x in A, tx belongs to A if $|t| < 1$. For example, a sphere centred at the origin in Cartesian space is a balanced set.

balanced block design *See* BLOCK DESIGN.

ball An *open ball in a *metric space containing all the points which are strictly less than a given distance from a fixed point. A *closed ball contains all the points which are not more than a given distance from a fixed point. The open balls form a *basis for the *topology in a metric space.

Banach, Stefan (1892–1945) Polish mathematician who was a major contributor to the subject known as *functional analysis. Much subsequent work was inspired by his monograph of the theory in 1932. He also proved the *contraction mapping theorem.

Banach algebra An *algebra on a *Banach space that satisfies the inequality $\|xy\| \le \|x\| \times \|y\|$ for all elements x, y of the space, where $\| \ \|$ is the *norm of the space.

Banach measure *See* MEASURE.

Banach space A *complete *normed vector space over the real or complex numbers. *Compare* HILBERT SPACE.

Banach-Tarski paradox A solid three-dimensional sphere can be partitioned into five sets which can be moved using *isometries and reassembled to create two identical solid spheres. The result is paradoxical, as it seems volume ought to be an invariant of the process, but the five parts are not *measurable and cannot be assigned volumes themselves. The paradox cannot be proved using the *Zermelo-Fraenkel axioms but is true assuming the *axiom of choice.

banded matrix Where the elements are zero except in a band of diagonals around the leading diagonal.

bar The superscript symbol, as in $\bar{z}$ or $\bar{x}$, is used to denote the *conjugate of a complex number or the *mean of a statistical variable.

bar chart A diagram representing the *frequency distribution for nominal or discrete *data. It consists of a sequence of bars, or rectangles, corresponding to possible values, and the length of each is proportional to the frequency. The figure shows the kinds of vehicles recorded in a small traffic survey. *Compare* HISTOGRAM.

barrel A subset of a *normed space, or a *topological vector space, which is *closed, *convex, *absorbing, and *balanced. A topological vector space is said to be barrelled if every barrel contains a *neighbourhood of the *origin.

Barrow, Isaac (1630–77) English mathematician whose method of finding *tangents, published in 1670, is essentially that now used in differential *calculus. He may have been the first to appreciate that the problems of finding tangents and areas under curves are inversely related. When he resigned his chair at Cambridge, he was succeeded by *Newton, on Barrow's recommendation.

barycentric coordinates If a triangle's vertices have *position vectors **a,b,c**, the position vector of any point of the plane containing the triangle can be written uniquely as $\alpha\mathbf{a} + \beta\mathbf{b} + \gamma\mathbf{c}$ where $\alpha + \beta + \gamma = 1$. The point's barycentric coordinates are α,β,γ. Points inside the triangle have positive barycentric coordinates. At the triangle's *centroid $\alpha = \beta = \gamma = 1/3$. Barycentric coordinates can similarly be defined for a *simplex.

base (of an exponential function) *See* EXPONENTIAL FUNCTION TO BASE A.

base (of an isosceles triangle) *See* ISOSCELES TRIANGLE.

base (of logarithms) *See* LOGARITHMS.

base (of natural logarithms) *See* E.

base (of a pyramid) *See* PYRAMID.

base (for representation of numbers) The integer represented as 4703 in standard decimal notation is written in this way because

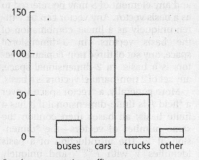

Bar chart representing traffic survey

$$4703 = (4 \times 10^3) + (7 \times 10^2)$$
$$+ (0 \times 10) + 3.$$

The same integer can be written in terms of powers of 8 as follows:

$$4703 = (1 \times 8^4) + (1 \times 8^3) + (1 \times 8^2)$$
$$+ (3 \times 8) + 7.$$

The expression on the right-hand side is abbreviated to 11137_8 and is the *representation of this number to base 8 (or *octal). In general, if g is an integer greater than 1, any positive integer a can be written uniquely as

$$a = c_n g^n + c_{n-1} g^{n-1} + \ldots + c_1 g + c_0,$$

where each c_i is a non-negative integer less than g. This is the representation of a to base g and is abbreviated to $c_n c_{n-1} \ldots c_1 c_{0g}$. Real numbers, not just integers, can also be written to any base, by using figures after a 'decimal' point, just as familiar *decimal representations of real numbers are written to base 10. *See also* BINARY REPRESENTATION, DECIMAL, HEXADECIMAL.

base (of a triangle) It may be convenient to consider one side of a *triangle to be the base of the triangle. The *vertex opposite the base may then be called the apex of the triangle, and the distance from the apex to the line of the base is the height of the triangle.

base angles *See* ISOSCELES TRIANGLE.

base field The *field containing the *scalars over which a *vector space is defined. Also the field that is being extended in a *field extension.

baseline risk The risk at the beginning of a time period or under specific conditions.

Basel problem First posed in 1650, the Basel problem was to evaluate the sum $1^{-2} + 2^{-2} + 3^{-2} + 4^{-2} + \ldots$. The problem was resolved by *Euler, whose hometown was Basel, in 1734; he showed the sum equals $\pi^2/6$. He further evaluated the sum of the reciprocals of all even powers.

base period The time used as a reference point for comparison of variables such as with indices (*see* INDEX (in statistics)). An index of 100 is usually taken at the base period and the value at other periods is expressed proportionally, so if prices have risen 17% the index will record 117.

base unit *See* SI UNITS.

basic solution When *slack variables have been introduced into a *linear programming problem there are more equations than there are variables. If there are n more variables than equations the basic solutions are found by setting various combinations of n variables to zero, thus reducing the number of variables to the number of equations. If one of the slack variables is negative then that solution is not *feasible. For each solution, the n variables set to zero are called non-basic variables and the others are called basic variables.

basic variables *See* BASIC SOLUTION.

basis (bases) A set S of *vectors is a spanning set if any vector can be written as a *linear combination of elements in S. If, in addition, the vectors in S are *linearly independent, then S is a basis, and any element of S may be referred to as a basis vector. Any vector can be written uniquely as a linear combination of the basis vectors. In 3-dimensional space, any set of three non-coplanar vectors is a basis. In 2-dimensional space, any set of 2 non-parallel vectors is a basis.

More generally, a *vector space V over a *field F is finite-dimensional if it has a finite basis; all bases then contain the same number of vectors—the *dimension of V. The introduction of a basis identifies V with $F^{\dim V}$ and uniquely assigns *coordinates to each vector. Any linearly independent set can be extended to a basis, and any spanning set contains a basis. Assuming the *axiom of choice, any vector can be shown to have a basis. *Compare* COMPLETE ORTHONORMAL BASIS, DUAL BASIS, HAMEL BASIS, ORTHONORMAL BASIS, SCHAUDER BASIS.

basis (of a topology) Given a *topology on a set X, a basis for the topology is a collection of subsets of X such that every *open set can be written as a union of sets in the basis. In a *metric space the *open balls form a basis. *See also* PRODUCT SPACE.

basis vector *See* BASIS.

Bayes, Thomas (1702–61) English mathematician remembered for his work in *probability, which led to a method of statistical *inference based on the theorem that bears his name. His paper on the subject was not published until 1763, after his death.

Bayesian inference In Bayesian *statistics, parameters have probability distributions, while in frequentist statistics parameters have fixed values. In Bayesian inference, a *prior distribution is proposed for a *parameter and after further data is collected *Bayes' Theorem is used to calculate a *posterior distribution in light of the new data. Frequentist inference, in contrast, would use a *hypothesis test to assign a *p-value to a null hypothesis that the data came from a population with a suggested parameter value.

Bayes' Theorem Let $A_1, A_2, \ldots, A_k$ be *mutually exclusive *events whose union is the whole sample space of an experiment and let B be an event with $\Pr(B) \neq 0$. Then $\Pr(A_i|B)$ equals

$$\frac{\Pr(B|A_i)\Pr(A_i)}{\Pr(B|A_1)\Pr(A_1) + \ldots + \Pr(B|A_k)\Pr(A_k)}.$$

For example, let A_1 be the event of tossing a double-headed coin and A_2 the event of tossing a normal coin. Suppose that one of the coins is chosen at random so that $\Pr(A_1) = \frac{1}{2}$ and $\Pr(A_2) = \frac{1}{2}$. Let B be the event of obtaining 'heads'. Then $\Pr(B|A_1) = 1$ and $\Pr(B|A_2) = \frac{1}{2}$. So

$$\Pr(A_1|B)$$
$$= \frac{\Pr(B|A_1)\Pr(A_1)}{\Pr(B|A_1)\Pr(A_1) + \Pr(B|A_2)\Pr(A_2)}$$
$$= \frac{1 \times \frac{1}{2}}{1 \times \frac{1}{2} + \frac{1}{2} \times \frac{1}{2}} = \frac{2}{3}.$$

This says that, given that 'heads' was obtained, the probability that it was the double-headed coin that was tossed is 2/3.

Here, $\Pr(A_i)$ is a *prior probability and $\Pr(A_i | B)$ is a *posterior probability.

bearing The direction of the course upon which a ship is set or the direction in which an object is sighted may be specified by giving its bearing, the angle that the direction makes with north. The angle is measured in degrees in a clockwise direction from north. For example, north-east has a bearing of 45°, and west has a bearing of 270°.

Bellman's principle of optimality Any part of an optimal path is itself optimal. This is one of the fundamental principles of *dynamic programming by which the length of the known optimal path is extended step by step until the complete path is known.

bell-shaped curve *See* NORMAL DISTRIBUTION.

belongs to If x is an element of a set S, then x belongs to S and this is written $x \in S$. Naturally, $x \notin S$ means that x does not belong to S.

below (of functions) Less than. The limit of a function at a from below is the limit of $f(x)$ as $x \to a$ for values of $x < a$. It can be written as $f(a^-)$ or $\lim_{x \to a} f(x)$. *Compare* ABOVE.

bending moment The effect at a point which is the sum of all the *moments of forces acting on the structure.

Benford's law An observation relating to the leading decimal digits (*see* DECIMAL REPRESENTATION) of many *data sets, in particular that the leading digit is more likely to be small. According to the law, 1 is the leading digit for around 30% of data and 9 only 5%. The law applies best to data that are distributed over several *orders of magnitude, e.g. city populations, but would not apply to human heights. The law has been applied to demonstrate likely fraudulence in some data sets.

Berkeley, George (1685–1753) An Irish philosopher, commonly referred to as Bishop Berkeley. In mathematics, he is best known for his critique of *calculus in *The Analyst*, published in 1734. He rightly criticized the lack of any rigorous foundations of calculus, a situation which would not be fully resolved for at least another century.

Bernoulli distribution The discrete probability *distribution whose *probability mass function is given by $\Pr(X=0) = 1-p$ and $\Pr(X=1) = p$. It is the *binomial distribution B$(1, p)$.

Bernoulli equation A *differential equation which can be written in the form $y' + f(x)y = g(x)y^n$. The transformation $z = y^{1-n}$ gives $z' = \frac{dz}{dx} = (1-n)y^{-n}y'$. Dividing the original equation by y^n gives $y^{-n}y' + f(x)y^{1-n} = g(x)$ which reduces to $\frac{1}{1-n}z' + f(x)z = g(x)$ which is in linear form and can be solved by use of an *integrating factor.

Bernoulli family A family from Basel that produced a stream of significant mathematicians. The best known are the brothers Jacques (or James or Jakob) and Jean (or John or Johann), and Jean's son Daniel. **Jacques**

Bernoulli (1654–1705) did much work on the newly developed calculus but is chiefly remembered for his contributions to probability theory: his *Ars conjectandi* was published posthumously in 1713. The work of **Jean Bernoulli** (1667–1748) was more definitely within calculus: he discovered *l'Hôpital's rule and proposed the *brachistochrone problem. He was one of the founders of the *calculus of variations. In the next generation, the work of **Daniel Bernoulli** (1700–82) was mainly in hydrodynamics.

Bernoulli number A sequence B_n of *rational numbers defined by the generating function

$$\frac{x}{e^x - 1} = \sum_{n=0}^{\infty} \frac{B_n x^n}{n!}.$$

As $x/(e^x-1) + x/2$ is an *even function, then $B_1 = -\frac{1}{2}$ and B_n is zero for all other odd n. For $n = 0, 2, 4, 6, \ldots$ they are $1, \frac{1}{6}, -\frac{1}{30}, \frac{1}{42} \ldots$.

Bernoulli polynomial The nth Bernoulli polynomial is

$$B_n(x) = \sum_{k=0}^{n} \binom{n}{k} B_k x^{n-k},$$

where B_k denotes the kth *Bernoulli number. The formulae for sums of *powers of *natural numbers then read as

$$\sum_{k=1}^{n} k^m = \frac{B_{m+1}(n+1) - B_{m+1}}{m+1}.$$

Bernoulli's inequality The inequality

$$(1+x)^\alpha \geq 1 + \alpha x$$

valid for $x \geq -1$ and $\alpha \geq 0$.

Bernoulli's Theorem (probability) The *relative frequency of an *event happening is m/n where m is the number of occurrences in n trials. If p is the *probability of the event, then the relative frequency tends towards the probability as the *sample size increases to infinity. Precisely, the theorem states that for any $\varepsilon > 0$, $\lim\limits_{n \to \infty} \mathrm{Pr}\left\{ \left| \dfrac{m}{n} - p \right| < \varepsilon \right\} = 1$.

This is a special case of the *weak law of large numbers.

Bernoulli's Theorem (fluid dynamics) In an *inviscid, *incompressible, *steady fluid flow,

$$H = \frac{p}{\rho} + \frac{1}{2} |\mathbf{u}|^2 + gz$$

is constant along *streamlines. Here p denotes *pressure, ρ denotes *density, $\mathbf{u}$ is the flow velocity, and g denotes gravity. If, further, the flow is *irrotational, then H is constant throughout the flow.

Bernoulli trial One of a sequence of independent experiments, each of which has an outcome considered to be success or failure, all with the same probability p of success (*see* BERNOULLI DISTRIBUTION). The number of successes in a sequence of Bernoulli trials has a *binomial distribution. The number of experiments required to achieve the first success has a *geometric distribution.

Bertrand's paradox For a given circle, the *mean length of a *chord, chosen at *random, is not *well defined. This paradox results from the fact that there are several reasonable ways of randomly choosing a chord that lead to different answers. Bertrand highlighted this issue to show that the specific method of choosing randomly was needed for a *probability to be well defined.

Bertrand's postulate For any positive integer n, there is always at least one prime number p satisfying $n < p \leq 2n$. For example, for $n = 7$ the prime 11 lies between 7 and 14. This was conjectured by Bertrand in 1845 and proved by *Chebyshev in 1852.

Bessel, Friedrich Wilhelm (1784–1846) German astronomer and mathematician who made contributions in applied mathematics and introduced what are now called *Bessel functions. These functions satisfy a certain *differential equation, called *Bessel's equation, and variously arise in applied mathematics and physics.

Bessel function Functions that provide solutions to *Bessel's equation. Bessel's functions of the first kind are of the form $J_n(z) = \sum\limits_{r=0}^{\infty} \dfrac{(-1)^r}{r! \Gamma(n+r+1)} \left(\dfrac{z}{2}\right)^{n+2r}$ where Γ denotes the *gamma function. In particular, $J_0(z) = \sum\limits_{r=0}^{\infty} \dfrac{(-1)^r}{(r!)^2} \left(\dfrac{z}{2}\right)^{2r}$.

Bessel's equation Second-order differential equation in the form $x^2 y'' + xy' + (x^2 - n^2)y = 0$. This occurs in physics and engineering applications, including in the *harmonics of a vibrating circular drum.

Bessel's inequality For a vector v and an *orthonormal set $v_1, \ldots, v_k$, in an *inner product space V, the *inequality $\|v\|^2 \geq \sum_1^k |\langle v, v_i \rangle|^2$. Equality holds for all v, if and only if $v_1, \ldots, v_k$, is an orthonormal *basis. *Compare* PARSEVAL'S IDENTITY.

best approximation A point of a subset in a *metric space that is closest to a given point outside the subset. In general such points are not guaranteed to exist but will if the subset is *compact.

best unbiased estimator *See* ESTIMATOR.

beta distribution If X is a random variable with *probability density function given by $f(x, \alpha, \beta) = \dfrac{1}{B(\alpha, \beta)} x^{\alpha-1} (1-x)^{\beta-1} 0 \leq x \leq 1$ where $B(\alpha, \beta)$ is a *beta function and λ, v, x all > 0, then we say that X has a beta distribution with parameters α, β. The beta distribution has mean $\dfrac{\alpha}{\alpha+\beta}$ and variance $\dfrac{\alpha\beta}{(\alpha+\beta)^2(\alpha+\beta+1)}$.

beta function The function $B(p,q) = \int_0^1 x^{p-1}(1-x)^{q-1}\,dx = \frac{\Gamma(p)\Gamma(q)}{\Gamma(p+q)}$ where $\Gamma(a)$ is the *gamma function. For positive integers m, n, $B(m,n) = \frac{(n-1)!(m-1)!}{(m+n-1)!}$.

between-groups design (between-subjects design) An experiment designed to measure the value of a (dependent) variable for distinct groups, under each of the experimental conditions.

Bézout's lemma Given two positive *integers a and b with *greatest common divisor h, there exist integers u and v such that $ua + vb = h$. The integers u and v can be constructed using the *Euclidean algorithm. Bézout's lemma has significance for the *Chinese remainder theorem and for linear *Diophantine equations.

Bézout's theorem Two *algebraic curves in the complex *projective plane, defined by two *coprime *polynomials of *degrees m and n, have mn intersections counting *multiplicities of intersections. For example, two *parallel lines still meet in one (1×1) *point at infinity, and a *parabola and one of its *tangent lines meet in a multiplicity two (2×1) intersection.

Bhāskara (1114–85) Eminent Indian mathematician who continued in the tradition of *Brahmagupta, making corrections and filling in many gaps in the earlier work. He solved examples of *Pell's equation and grappled with the problem of division by zero.

bi- Prefix denoting two, for example biannual happens twice a year.

bias A prejudice or a lack of objectivity or randomness resulting in an imbalance that makes it likely that the outcome will tend to be distorted. In *statistics this is when a process contains some systematic imbalance so that, on average, the outcome of the process is not equal to the true value. Randomization techniques are employed to try and remove bias that may result from

other sample *estimator selection methods requiring choices to be made. *See also* NON-RESPONSE BIAS, RESPONSE BIAS, SELECTION BIAS, SELF-SELECTED SAMPLES.

biased estimator The distribution of an *estimator is biased if its *expected value does not equal the population *mean. Individual values returned by estimators will usually not be exactly equal to the true value, and the determination of bias is a property of the sampling distribution.

biased sample A sample whose composition is determined not only by the *population from which it has been taken but also by some property of the sampling method which has a tendency to cause an over-representation of some parts of the population. It is a property of the sampling method rather than the individual sample.

bi-conditional A logical statement in the form 'p if and only if q', meaning that p and q are *equivalent.

BIDMAS (BODMAS) A mnemonic for remembering the priorities with which a calculation should be determined. BIDMAS stands for *brackets, *indices, *division-*multiplication, *addition-*subtraction. BODMAS is alternatively used with 'orders' replacing 'indices'. So $2 + 3\times4 = 14$ as the multiplication is done before addition. Note that division and multiplication have equal priority, likewise addition and subtraction, despite the order the letters appear in the acronym. So $2-3+3 = 2$ rather than -4 (which would be the answer if addition preceded subtraction).

Bieberbach conjecture Ludwig Bieberbach, in 1916, made the conjecture that if a *holomorphic function

$$f(z) = z + \sum_{n=2}^{\infty} a_n z^n$$

is *univalent on the *disc $|z| < 1$, then $|a_n| \le n$ for all $n \ge 2$. This was proved by Louis de Branges in 1985.

bifurcation The values of a parameter at which the nature of an *equilibrium or *attractor changes in a *dynamical system. Commonly, a stable equilibrium might become unstable but bifurcate (i.e. split) into two stable equilibria. *See* LOGISTIC MAP.

big data *See* DATA SCIENCE.

big O notation Notation for comparing the relative order of functions. We write $f(x) = O(g(x))$ if $f(x)/g(x)$ remains *bounded as $x \to \infty$, so, for example, $(x+1)(x+2) = O(x^2)$. The notation is also used in the *neighbourhood of finite values; for example $\sin x = x + O(x^3)$ denotes the fact that $(\sin x - x)/x^3$ remains bounded (in fact *converges) as $x \to 0$. *See also* LITTLE O NOTATION.

biholomorphic (biholomorphism) A *bijection between two open subsets (*see* OPEN SET) of $\mathbb{C}$ is biholomorphic (or a biholomorphism) if both it and its *inverse are *holomorphic. This is synonymous with the map being a *conformal equivalence. The definition can be further extended to *Riemann surfaces. *See* RIEMANN MAPPING THEOREM.

bijection (bijective mapping) A *one-to-one *onto mapping, that is, a mapping that is both *injective and *surjective. Also known as a *one-to-one correspondence.

bilateral shift The *linear operator T which moves every element of a sequence by one position, so if $v = T(u)$ then $v_k = u_{k-1}$ for every k.

bilateral symmetry *See* SYMMETRICAL ABOUT A LINE.

bilinear A *function of two variables is bilinear if it is *linear with respect to each variable independently, e.g.

$f(x, y) = 3xy + x$ is bilinear but $g(x, y) = 3xy + x^2$ is not.

billion A thousand million (10^9). This is now standard usage, though in Britain billion used to mean a million million (10^{12}).

bimodal A frequency distribution is said to be bimodal if it shows two clear peaks. The two (local) *modes need not be exactly equal, in which case the modes are referred to as major and minor modes.

binary Base 2. *See* BINARY REPRESENTATION.

binary code A *code using the *alphabet {0,1}.

binary digit The values 0 and 1 are the only digits used in binary and form the basis of computer instructions. The term 'bit' comes from **B**inary dig**IT**.

binary operation A binary operation $\circ$ on a set S is a rule that associates with any elements a and b of S an element denoted by $a \circ b$. This is equivalent to saying that S is closed under $\circ$. Common examples include *addition, *multiplication, and *composition (on certain sets). *See* ASSOCIATIVE, COMMUTATIVE, OPERATION.

binary relation It is most natural to think of a binary relation, such as $\le$, as taking an ordered pair of elements, such as $(3,4)$ or $(6,2)$, and returning true for the former and false for the latter, as $3 \le 4$ is true and $6 \le 2$ is false. More formally, a binary relation on a set S is a subset R of the *Cartesian product $S \times S$. We then write aRb if $(a,b) \in R$. *See* EQUIVALENCE RELATION, PARTIAL ORDER.

binary representation The representation of a number to *base 2. It uses just the two *binary digits 0 and 1, and this is the reason why it is important in computing. For example, $37 = 100101_2$, since

$$37 = (1 \times 2^5) + (0 \times 2^4) + (0 \times 2)^3$$
$$+ (1 \times 2^2) + (0 \times 2) + 1.$$

Real numbers, not just integers, can also be written in binary notation, by using binary digits after a 'binary' point, just as familiar *decimal representations of real numbers are written to base 10. For example, the number $\frac{1}{10}$, which equals 0.1 in decimal notation, has, in binary notation, a recurring representation: $0.00011001100110011\ldots_2$.

binary search algorithm Used to search a list to see if it contains a particular item and to locate it if it does. The list needs to be in alphabetical (or some other) order, and the binary search algorithm identifies the middle term in the list and compares it with the search item. If it is the same the search is complete; if not it identifies the half of the list in which the item would be, and the other half is discarded. The process is repeated until the item is located or the list consists of only one item, in which case the item was not contained in the original list.

binary tree See TREE.

binary word A binary *word of length n is a string of n binary digits, or bits. For example, there are 8 binary words of length 3, namely, 000, 100, 010, 001, 110, 101, 011 and 111.

binding constraint A synonym for ACTIVE CONSTRAINT.

Binet's formula See FIBONACCI NUMBERS.

binomial An algebraic expression containing two distinct terms connected by *addition or *subtraction.

binomial coefficient The number, denoted by $\binom{n}{r}$, where n is a positive integer and r is an integer such that $0 \le r \le n$, defined by the formula

$$\binom{n}{r} = \frac{n(n-1)\ldots(n-r+1)}{1 \times 2 \times \ldots \times r} = \frac{n!}{r!(n-r)!}.$$

Since, by convention, $0! = 1$, we have $\binom{n}{0} = \binom{n}{n} = 1$. These numbers are called binomial coefficients because they occur as coefficients in the *Binomial Theorem. They are sometimes denoted by ^nC_r; this notation arose as the binomial coefficient equals the number of ways of selecting r objects out of n (see SELECTION). The numbers have the following properties:

(i) $\binom{n}{r}$ is an integer (this is not obvious from the definition).

(ii) $\binom{n}{n-r} = \binom{n}{r}$.

(iii) $\binom{n+1}{r} = \binom{n}{r-1} + \binom{n}{r}$.

(iv) $\binom{n}{0} + \binom{n}{1} + \binom{n}{2} + \ldots + \binom{n}{n} = 2^n$.

It is instructive to see the binomial coefficients laid out in the form of *Pascal's triangle.

binomial distribution The discrete probability *distribution for the number of successes when n independent *Bernoulli trials are carried out, each with the same probability p of success. The *probability mass function is given by $\Pr(X = r) = {}^nC_r p^r (1-p)^{n-r}$, for $0 \le r \le n$. This distribution is denoted $B(n, p)$ and has *mean np and *variance $np(1-p)$.

binomial series (expansion) The series

$$1 + \frac{\alpha}{1!}x + \frac{\alpha(\alpha-1)}{2!}x^2 + \ldots$$
$$+ \frac{\alpha(\alpha-1)\ldots(\alpha-n+1)}{n!}x^n + \ldots,$$

being the *Taylor series for the function $(1+x)^\alpha$. In general, it is valid for $-1 < x < 1$. In fact, the series *converges for *complex x with *modulus less than 1, but then care is needed interpreting the *sum as $(1 + x)^{\alpha}$ (*see* BRANCH). If α is a non-negative integer, the expansion is a finite series and so is a *polynomial, and then it is equal to $(1 + x)^{\alpha}$ for all x.

Binomial Theorem A generalization of formulae such as $(x + y)^2 = x^2 + 2xy + y^2$ and $(x + y)^3 = x^3 + 3x^2y + 3xy^2 + y^3$ used in elementary algebra. The Binomial Theorem states: for all positive integers n,

$$(x + y)^n = \sum_{r=0}^{n} \binom{n}{r} x^{n-r} y^r$$

$$= x^n + \binom{n}{1} x^{n-1} y + \binom{n}{2} x^{n-2} y^2 + \ldots$$

$$+ \binom{n}{r} x^{n-r} y^r + \ldots + y^n,$$

where $\binom{n}{r} = \frac{n!}{r!(n-r)!}$ (*see* BINOMIAL COEFFICIENT).

When $y = 1$, the theorem agrees with the terminating *binomial series. There is a *multinomial theorem for more general expansions such as $(x_1 + x_2 + \cdots x_k)^n$.

binormal *See* SERRET-FRENET FORMULAE.

bin packing problem The scheduling problem of how to pack a number of boxes of different volumes into a finite number of identical bins using as few bins as possible. The problem has been shown to be NP-hard, which means that, if P ≠ NP, it cannot be solved in *polynomial time.

biology *See* MATHEMATICAL BIOLOGY.

biometry (biostatistics) The development and application of statistical methods to biological problems, including the design of experiments and the analysis and interpretation of experimental *data. **Biometrics** may also refer to the collection of data relating to living

organisms, such as DNA, or for the purposes of identification, as with fingerprints or retinal scans.

bipartite graph A *graph in which the *vertices can be *partitioned into two subsets V_1 and V_2, so that no two vertices in V_1 are joined and no two vertices in V_2 are joined. The *complete bipartite graph $K_{m,n}$ is the bipartite graph with m vertices in V_1 and n vertices in V_2, with every vertex in V_1 joined to every vertex in V_2 by a single *edge.

$K_{2,3}$

Birkhoff, George David (1884–1944) American mathematician who made important contributions to the study of *dynamical systems and in *ergodic theory.

birth-death process A continuous-time *Markov process modelling population changes due to births and deaths and possibly more generally due to immigration and emigration.

bisect To divide into two equal parts.

bisection method A numerical method for finding a *root of an equation $f(x) = 0$. If values a and b are found such that $f(a)$ and $f(b)$ have opposite signs and f is *continuous on the interval $[a, b]$, then (by the *intermediate value theorem) the equation has a root in (a, b). The method is to bisect the interval and replace it by either one half or the other, thereby closing in on the root.

Let $c = \frac{1}{2}(a + b)$. Calculate $f(c)$. If $f(c)$ has the same sign as $f(a)$, then take c as a new value for a; if not (so that $f(c)$ has the same sign as $f(b)$), take c as a new value for b. (If it should happen that $f(c) = 0$, then c is already a root.) Repeat this whole process until the length of the interval $[a, b]$ is less than 2ε, where

ε is the desired *accuracy, specified in advance. The midpoint of the interval can then be taken as an *approximation to the root, and the error will be less than ε. *Compare* BINARY SEARCH ALGORITHM.

(⊕) SEE WEB LINKS

• An interactive page which demonstrates the method for finding a root of a cubic equation.

bisector The line that divides an *angle into two equal angles. *See also* EXTERNAL BISECTOR, INTERNAL BISECTOR, PERPENDICULAR BISECTOR, TRISECTION OF AN ANGLE, TRISECTOR.

bit Abbreviation for *binary digit, i.e. either 0 or 1. A bit is the unit of *information and (Shannon) *entropy.

bivariate Relating to two variables, especially to two *random variables. *See* JOINT CUMULATIVE DISTRIBUTION FUNCTION, JOINT DISTRIBUTION, JOINT PROBABILITY DENSITY FUNCTION, JOINT PROBABILITY MASS FUNCTION.

Black-Scholes equation A *partial differential equation *modelling the dynamics of a financial market and which provided estimates for (European-style) stock options. It states

$$\frac{\partial V}{\partial t} + \frac{1}{2}\sigma^2 S^2 \frac{\partial^2 V}{\partial S^2} + rS\frac{\partial V}{\partial S} - rV = 0,$$

where V is the option's price, S is the stock's price, r is the interest rate, and σ represents volatility of the stock price.

blancmange function A *continuous real function $b(x)$, defined for $0 \leq x \leq 1$, which is *differentiable at no point. It is defined as

$$b(x) = \sum_{n=1}^{\infty} \frac{a(2^n x)}{2^n}$$

where $a(x)$ is the minimum distance of the *real number x from an integer. Its graph is:

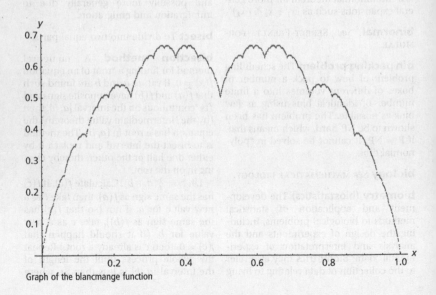

Graph of the blancmange function

blinding (in design of experiments) In *experimental design, especially in evaluating the effectiveness of drugs or other medical interventions, there is the possibility that beneficial effects accrue because the patient psychologically expects to feel better, which will interfere with identifying any actual benefits of the treatment. In a single-blind experiment the patient does not know which treatment they have been assigned, and in a double-blind experiment neither the patient nor the doctors or others dealing with the patient know which treatment they have been assigned.

Where the nature of the treatments means they cannot be interchanged without the subject being aware, a double-dummy may be used where patients are apparently given both treatments, but one or other will be a *placebo, and the *control group will be given both as placebos.

⊕ SEE WEB LINKS

• A full discussion of the ethics and principles of blinding in clinical trials (subscription).

block code A *code in which all the codewords have the same *length.

block design An experimental design where experimental units with similar characteristics are grouped together as a block and treated as though they are indistinguishable. *Repeated measures designs are an example where the same individual is subject to measurement under different experimental treatments. A balanced block design requires the blocks to be the same size and each experimental treatment to be applied the same number of times. A completely balanced block design further requires each treatment to be applied the same number of times within each block.

block diagonal matrix A square matrix in which the only non-zero elements form square *submatrices arranged along the main diagonal. A very simple case is shown below where a 2×2 matrix

and a 1×1 matrix make up a block diagonal 3×3 matrix.

$$\begin{bmatrix} 3 & -2 & 0 \\ 4 & 2 & 0 \\ 0 & 0 & -1 \end{bmatrix}$$

See also JORDAN NORMAL FORM, RATIONAL CANONICAL FORM.

block multiplication *Multiplication of matrices where the *entries are *submatrices rather than single elements; the submatrices still multiply as if they were entries. For example, given two 3×3 matrices of the form

$$A_1 = \begin{pmatrix} M_1 & v_1 \\ w_1 & c_1 \end{pmatrix} \ and \ A_2 = \begin{pmatrix} M_2 & v_2 \\ w_2 & c_2 \end{pmatrix}$$

where the M_i, v_i, w_i, c_i are respectively 2×2, 2×1, 1×2, 1×1 matrices, then their *product equals

$$A_1 A_2 = \begin{pmatrix} M_1 M_2 + v_1 w_2 & M_1 v_2 + v_1 c_2 \\ w_1 M_2 + c_1 w_2 & c_1 c_2 \end{pmatrix}.$$

BODMAS *See* BIDMAS.

body An object in the real world idealized in a *mathematical model as a *particle, a *rigid body or an *elastic body, for example.

Bohr, Niels Henrik David (1885–1962) Danish mathematician and theoretical physicist whose work on the structure of atoms and on radiation won him the Nobel Prize for physics in 1922. He made further substantial contributions in the new field of *quantum theory, together with *Heisenberg formulating the *Copenhagen interpretation. During the Second World War he escaped from occupied Denmark and worked in the UK and the USA on the nuclear bomb.

Boltzmann, Ludwig (1844–1906) Austrian physicist who pioneered *statistical mechanics and gave the modern definition of entropy.

Bolyai, János (1802–60) Hungarian mathematician who, in a work published in 1832 but probably dating from 1823, announced his discovery of *non-Euclidean geometry. His work was independent of *Lobachevsky. He had persisted with the problem, while serving as an army officer, despite the warnings of his father, an eminent mathematician and friend of *Gauss, who had himself spent many years on it without success. Later, János was disheartened by lack of recognition.

Bolzano's Theorem *See* INTER-MEDIATE VALUE THEOREM.

Bolzano-Weierstrass Theorem A *bounded *sequence of real (or complex) numbers has a *convergent *subsequence. This is equivalent to saying that a *closed, *bounded *interval is *sequentially compact.

Bombelli, Rafael (1526–72) Italian mathematician who, in his book *L'Algebra*, was among the first to treat complex numbers seriously, rather than as just a means to solving cubic equations.

Bondi, Sir Hermann (1919–2005) Austrian mathematician, physicist and astronomer who worked in the UK as an academic and government scientist. He helped develop the steady state theory of the universe as an alternative to the Big Bang theory.

Boole, George (1815–64) British mathematician who was one of the founding fathers of mathematical *logic. His major work, published in 1854, is his *Investigation of the Laws of Thought*. The kind of symbolic argument that he developed led to the study of so-called *Boolean algebras, of interest in computing and *algebra. His work, together with that of *De Morgan and others, helped to pave the way for the development of modern algebra.

Boolean algebra A set of elements defined with two *binary operations ($\vee$ and $\wedge$), a *unary operation ($\neg$) and two elements 0 and 1 which possess the following properties:

(i) Both operations $\vee$ and $\wedge$ are *commutative and *associative.
(ii) 0 is an identity for addition and 1 is an identity for multiplication.
(iii) Each operation $\vee$ and $\wedge$ is *distributive over the other.
(iv) $a \vee (\neg a) = 1$ and $a \wedge (\neg a) = 0$ for all a.

For example, a *power set $\wp(X)$, where $\vee$ is *union, $\wedge$ is *intersection, $\neg$ is complementation in X, and 0 is the empty set and 1 is X, is a Boolean algebra. *Compare* BOOLEAN RING; *see also* STONE SPACE.

Boolean ring A ring in which $x^2 = x$ for all elements. A *Boolean algebra $(R, \vee, \wedge, \neg)$ can be associated with a Boolean ring $(R, +, \times)$ with a 1 by defining:

$$x \vee y = x + y + x \times y, \quad x \wedge y = x \times y,$$
$$\neg x = 1 + x.$$

The *power set $\wp(X)$ is such a Boolean ring with *symmetric difference, *intersection, *complement, $1 = X$, and the Boolean algebra associated with it has operations *union, intersection, and complement.

Boolean-valued function A *function which takes the values false and true, or possibly 0 and 1, with 0 representing false and 1 representing true.

bootstrapping A process, due to Bradley Efron, of generating *confidence intervals for *statistics which has the advantage of not making assumptions about the underlying distribution of the population. It involves taking repeated random samples from the actual *sample of observations which has been taken and producing confidence intervals for the statistic based on the distribution of the statistic in that process. Because it involves large numbers of repetitions of the resampling process it is dependent on computer power.

bordering A bordered matrix is a matrix to which extra rows (on the bottom) and columns (on the right) have been added. This can be done algorithmically so that properties of the original matrix can be determined, in computationally efficient ways, from study of the bordered matrix.

Borel, Félix Édouard Justin Émile (1871–1956) French mathematician who proved important results in *set theory, *probability, and *measure theory.

Borel–Cantelli Lemma A general result in *measure theory that is a particularly useful application when applied to a probability measure. If $\{A_n\}$ is an infinite sequence of *measurable sets, where the sum of the measures $\sum_{n=1}^{\infty} \mu(A_n)$ is finite, then the set of points which lie in an infinite number of the sets must have measure zero. If the A_n are events in a *probability space, then the probability that infinitely many of the events occur is zero. A related result shows that if the events are independent and the sum of their probabilities is infinite, then the probability that infinitely many of the events occur is one.

Borel measure For a *topological space, a Borel measure is any measure defined on the *Borel sets. It may further be assumed that the measure of any *compact subset is finite.

Borel set In a *topological space, a Borel set is a set which can be obtained, from the set of *open sets, by *countable *unions and *relative complement. Consequently, by *De Morgan's Laws, the set of Borel sets is closed under countable *intersections. The Borel sets form a *sigma algebra, in fact the sigma algebra generated by the *topology.

Born, Max FRS (1882–1970) German mathematician and theoretical physicist who collaborated with many of the best minds of the day including *Heisenberg, *Pauli, Fermi and *Dirac. As a result of those collaborations he published important work on the foundations of *quantum theory, and then his own studies providing a statistical interpretation of the *wave function, for which he was awarded the Nobel Prize for Physics in 1954.

bottleneck problems A class of *constrained optimization problems involving restrictions on *network flows.

bound Let S be a non-empty subset of $\mathbb{R}$. The real number b is said to be an upper bound for S if $s \leq b$ for every $s \in S$. If S has an upper bound, then S is bounded above. Moreover, b is the supremum (or least upper bound) of S if b is an upper bound for S and no upper bound for S is less than b; this is written $b = \sup S$. For example, if $S = \{0.9, 0.99, 0.999, \dots\}$ then $\sup S = 1$. Note here that 1 is not itself in the set S. Similarly, the real number c is a lower bound for S if $c \leq s$ for every $s \in S$. If S has a lower bound, then S is bounded below. Moreover, c is the infimum (or greatest lower bound) of S if c is a lower bound for S and no lower bound for S is greater than c; this is written $c = \inf S$. A set is bounded if it is bounded above and below. More generally, in a *metric space M, a subset S is bounded if S is contained in some *ball.

Any non-empty set that is bounded above has a supremum—this is the completeness *axiom of the real numbers—and likewise any non-empty set that is bounded below has an infimum.

boundary (of a set) Given a *metric space M and a subset S, then the boundary of S is the complement of the *interior of S in the *closure of S. So a *half-plane has a line as its boundary. A *closed curve or closed surface has no boundary.

boundary condition A condition applying to a *function on the *boundary of its *domain; it might specify the function's value on the boundary or provide information about its derivative(s) there. *See* BOUNDARY VALUE PROBLEM; *compare* INITIAL CONDITIONS.

boundary layer In *fluid dynamics, a boundary layer is a thin layer of fluid close to a surface where the effects of viscosity are important. The *Navier-Stokes equations can then be separately analysed in the boundary layer, where the thinness of the layer is to some extent negligible, and outside the boundary layer, where the effects of viscosity are negligible.

boundary value problem A *differential equation to be satisfied over a region together with a set of *boundary conditions. *See* DIRICHLET PROBLEM, NEUMANN CONDITION, ROBIN BOUNDARY CONDITION.

bounded above *See* BOUND.

bounded below *See* BOUND.

bounded function A real *function f, defined on a domain S, is bounded if there is a number M such that $|f(x)| < M$ for all x in S. More generally a function whose *codomain is a *metric space is bounded if its image is a *bounded set. *See* WEIERSTRASS' THEOREM.

bounded sequence The (real or complex) sequence $a_1, a_2, a_3, \ldots$ is bounded if there is a number M such that $|a_n| < M$ for all n.

bounded set *See* BOUND.

bounded space A *metric space is bounded if some *ball equals the entire metric space.

bounded variation A real function f on an *interval $[a,b]$ is said to have bounded variation if the set of values

$$\sum_{i=1}^{n} |f(x_i) - f(x_{i-1})|$$

is bounded above where $a = x_0 < x_1 < \ldots < x_n = b$ is any *partition of the interval. *Differentiable functions with *bounded derivatives have bounded variation. $x\cos(1/x)$ is an example of a function which does not have bounded variation.

Bourbaki, Nicolas The pseudonym used by a group of mathematicians, of changing membership, mostly French, who since 1939 have been publishing volumes intended to build into an encyclopaedic survey of pure mathematics, the *Éléments de mathématique*. Its influence is variously described as profound or baleful, but it is undoubtedly extensive. Bourbaki has been the standard-bearer for what might be called the Structuralist School of modern mathematics.

Box-Jenkins model A mathematical model first proposed by Box and Jenkins in 1967 for forecasting and prediction in *time series analysis based on the variable's past behaviour. It produces very accurate short-term forecasts but requires a large amount of past *data. The method involves determining what type of model is appropriate by analysing the *autocorrelations and partial autocorrelations of the stationary data and comparing the patterns with the standard behaviour of the various types. The parameters of the model can then be estimated to provide the best fit to the data.

box plot (box-and-whisker diagram) A diagram constructed from a set of numerical data showing a box that indicates the middle 50% of the ranked observations together with lines, sometimes called 'whiskers', showing the maximum and *minimum observations in the sample. The *median is marked on the box by a line. Box plots are particularly useful for comparing several samples.

The figure shows box plots for three samples, each of size 20, drawn uniformly from the set of integers from 1 to 100.

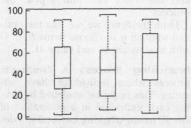

Box plot with whiskers

bra *See* KET.

braces *See* BRACKETS.

brachistochrone Suppose that A and B are points in a vertical plane, where B is lower than A but not vertically below A. Imagine a particle starting from rest at A and travelling along a curve from A to B under the force of gravity. The curve with the property that the particle reaches B as soon as possible is called the brachistochrone (from the Greek for 'shortest time'). The straight line from A to B does not give the shortest time. The required curve is a *cycloid, vertical at A and horizontal at B. The problem

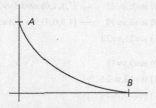

Graph of the brachistochrone

was posed in 1696 by Jean *Bernoulli, and his solution, together with others by *Newton, *Leibniz and Jacques Bernoulli, was published the following year.

brackets The symbols used to enclose a group of symbols or numbers that are to be taken together. Brackets can be used to change the order in which operations are to be done. For example, $7×2 + 3$ will be $14 + 3 = 17$. If the sum intended is $7×5 = 35$ then brackets can be put round $2 + 3$ so that the sum is taken before the product, i.e. $7×(2 + 3)$. Brackets may appear as round brackets or parentheses (. . .), as square brackets [. . .], or as braces { . . . }, commonly used in set-theoretic notation. Brackets can be nested so that the whole contents of the inner bracket are treated as a single term in the larger bracket. The number of ways n pairs of brackets can be nested equals the nth *Catalan number. *See also* BIDMAS.

Brahmagupta (about 598–665) Indian astronomer and mathematician whose text on astronomy includes some notable mathematics for its own sake, such as the areas of *quadrilaterals and the solution of certain *Diophantine equations.. Here, the systematic use of negative numbers and zero occurs for probably the first time.

Brahmagupta's formula A *cyclic *quadrilateral, with sides of length a,b,c,d, has *area

$$A = \sqrt{(s - a)(s - b)(s - c)(s - d)},$$

where $s = (a + b + c + d)/2$. *Compare* HERO'S FORMULA.

branch (holomorphic branch) Given a *multifunction f on an *open subset U of $\mathbb{C}$, a holomorphic branch of f is a *holomorphic function on U which selects from the possible choices of $f(z)$ for each $z \in U$. Essentially, the branch is a holomorphic choice of *principal values for f. For example, every $z \in \mathbb{C}$ which is not a positive real can be written uniquely as $z = re^{i\theta}$ on a *cut plane where $r > 0$ and $0 < \theta < 2\pi$. A holomorphic branch for the *complex logarithm on this cut plane is then $\log r + i\theta$. Note that 0 is a *branch point and that there is a *discontinuity across the positive *real axis of $2\pi i$.

branch (of a hyperbola) The two separate parts of a *hyperbola are called the two branches.

branch and bound method (of solving the knapsack problem) This procedure constructs a branching method that terminates each branch once the constraint limit has been reached and by allowing items to be added in an order which is determined at the start. This reduces considerably the number of combinations that have to be tried. The following simple example will be used to illustrate the detailed method.

A knapsack has maximum weight of 15, and the following items are available: A has weight (w) 5 and value (v) 10, written A (5, 10) with B (7, 11), C (4, 8) and D (4, 12).

Method. Place the items in decreasing order of value per unit weight, i.e. D, A, C, B (A and C could be reversed). Then construct a vector (x_1, x_2, x_3, x_4) where $x_i = 0$ if the item is not being taken and $x_i = 1$ if it is. Start with $(0, 0, 0, 0)$. At each stage, if a branch has not been terminated, and there are n zeros at the end of the vector, construct n branches which change exactly one of those zeros to a one and for which the total weight does not exceed the limit so the first stage will have four branches $(1, 0, 0, 0)$, $(0, 1, 0, 0)$, $(0, 0, 1, 0)$ and $(0, 0, 0, 1)$. Calculate the total weight (w) for each new branch created, and the value (v), and repeat the process.

From the figure we see that the optimal solution is to choose items B, C, D with total weight 15 and value 31.

branching process A *stochastic process where individuals have offspring. This might be modelled by having each individual in a generation of size Z_n having offspring identically, independently distributed with *pgf $G(s)$. The pgf of Z_n is then $G^n(s)$, where G^n denotes n applications of G. The probability that the population becomes extinct equals the smallest non-negative solution of the equation $s = G(s)$.

branch point (complex analysis) A point a is *not* a branch point of a *multifunction f if there is a holomorphic *branch of f on some punctured disc $0 < |z-a| < \varepsilon$ around a and otherwise a is a branch point of f. Thus, 0 is a branch point of $\sqrt{z}$ and also of $\log z$.

break-even point The point at which revenue begins to exceed cost. If one graph is drawn to show total revenue plotted against the number of items

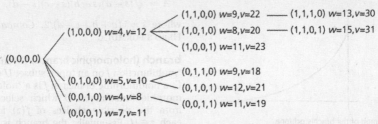

Example illustrating the branch and bound method

made and sold and another graph is drawn with the same axes to show total costs, the two graphs normally intersect at the break-even point. To the left of the break-even point, costs exceed revenue and the company runs at a loss; to the right, revenue exceeds costs and the company runs at a profit.

Brianchon's theorem See PASCAL'S THEOREM.

bridges of Königsberg In the early 18th century, there were seven bridges in the town of Königsberg (now Kaliningrad). They crossed the different branches of a river as shown in the figure. The question was asked whether it was possible, on a single walk, to cross each bridge exactly once. This prompted *Euler to consider the problem more generally and publish arguably the first research paper in *graph theory. The original question asked, essentially, whether the graph shown is *traversable. It can be shown that a *connected graph is traversable if and only if there are 0 or 2 vertices with odd *degree. *Compare* EULERIAN TRAIL.

(((⊕))) SEE WEB LINKS

• An interactive page in which you can construct your own problem like the bridges of Königsberg.

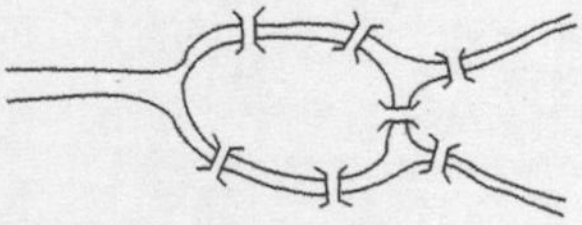

A map of the bridges

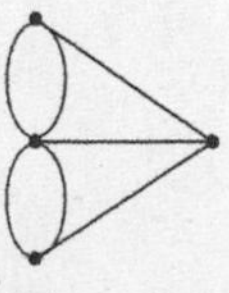

The bridges as a graph

Briggs, Henry (1561–1630) English mathematician responsible for introducing common *logarithms (base 10). Following the publication of tables of logarithms by *Napier, Briggs consulted him and proposed an alternative definition using base 10. Briggs published, in 1624, tables including 30 000 logarithms to 14 decimal places.

Brouwer, Luitzen Egbertus Jan (1881–1966) Dutch mathematician, who proved significant results in the areas of *set theory, *topology, and *measure theory He was also the founder of the doctrine known as *intuitionism, which insists on *constructive proofs and so rejects proofs which make use of the *principle of the excluded middle. This resulted in much controversy, particularly with *Hilbert and his *formalist philosophy.

Brouwer's Fixed Point Theorem Any *continuous function $f:D{\rightarrow}D$ from the closed disc $D = \{(x,y) : x^2 + y^2 \leq 1\}$ to itself has a *fixed point.

Brownian motion Brownian motion, named after the Scottish botanist Robert Brown (1773–1858), describes the random motions of a particle in a medium. Within mathematics, the term relates to a *stochastic process introduced by Norbert *Wiener. The Wiener process W_t for $t \geq 0$, models the movement of such a particle beginning at the origin at $t = 0$ and is characterized by the following properties:

(i) $W_0 = 0$
(ii) Future displacements $W_{t+u} - W_t$ (where $u \geq 0$) are *independent of W_s where $s \leq t$.
(iii) $W_{t+u} - W_t$ (where $u \geq 0$) has *normal distribution with mean 0 and variance u.
(iv) W_t is continuous in t.

bubble sort algorithm Makes repeated passes through a list of numbers, and on each pass adjacent numbers are compared and reversed if they are not in the required order. Each pass will put at least one more element into the correct position, like bubbles rising, and the process is complete when a complete set of comparisons is made which required no switching.

(⊕) SEE WEB LINKS

• A demonstration of the bubble sort in action.

Buffon's needle Suppose that a needle of length l is dropped at random onto a set of parallel lines a distance d apart, where $l < d$. The probability that the needle lands crossing one of the lines can be shown to equal $2l/(\pi d)$. The experiment in which this is repeated many times to estimate the value of π is called Buffon's needle. It was proposed by Georges Louis Leclerc, Comte de Buffon (1707–88).

(⊕) SEE WEB LINKS

• A simulation of Buffon's needle experiment.

Buridan's ass The paradox in *logic where an ass is exactly halfway between two buckets of water and dies of thirst because it has no logical basis on which to decide to move to either one. Although the paradox bears the name of a medieval philosopher, this is because it satirizes his philosophical arguments, and the paradox has been around at least since Aristotle.

Burnside's lemma See ORBIT-COUNTING FORMULA.

butterfly effect See CHAOS.

byte A block of *bits, usually 8, used in computing as the code for a single character (as in *ASCII).

c Abbreviation for *centi-.

c The *Banach space of *convergent (real or complex) *sequences with *norm $\|(x_n)\| = \sup|x_n|$. The Banach space of sequences which converge to 0 is denoted c_0. Both are *closed *subspaces of l^∞.

c The symbol for the *cardinality of the set of *real numbers.

ℂ The set of *complex numbers. The symbol also denotes the *field of complex numbers, and $\mathbb{C}^*$ denotes the multiplicative *group of non-zero complex numbers.

ℂ∞ The *extended complex plane.

C The Roman *numeral for 100, and the number 12 in *hexadecimal representation.

C^0 $C^0(X)$ denotes the *vector space of real-valued *continuous functions on a space X.

C^* $C^*(X)$ denotes the *vector space of real-valued *continuous, bounded functions on a space X.

C^1 $C^1(X)$ denotes the *vector space of real-valued continuously *differentiable functions on a space X.

C^∞ $C^\infty(X)$ denotes the *vector space of real-valued functions on a space X which have *continuous *derivatives of all orders.

C^ω $C^\omega(X)$ denotes the *vector space of real-valued *analytic functions on a space X.

Caesar cipher *See* CIPHER.

calculate To work out the value of a mathematical or arithmetical procedure, or the output of an *algorithm.

calculator (calculating machine) A device for performing arithmetical calculations or algebraic manipulations. The earliest example is the *abacus. Hand-held electronic calculators now have more computational power than the early mainframe *computers.

calculus A branch of mathematics dealing with rates of change (*differential calculus) and aggregation of such changes (*integral calculus). Its development is usually dated to the work of *Newton and *Leibniz, though the work of their predecessors such as *Gregory and *Fermat played an essential role. Especially via the role of *differential equations in *mathematical modelling, calculus has, since its inception, been the principal quantitative language of science. *See also* FUNDAMENTAL THEOREM OF CALCULUS.

calculus of variations A development of calculus concerned with problems in which a *function is to be determined such that some related definite *integral achieves a maximum or minimum value. Applications include *brachistochrone problem and finding *geodesics. *See* EULER'S EQUATION, FUNCTIONAL.

cancel To eliminate terms from an expression, usually through one of the four basic arithmetical operations, to produce a simplified form. For example, $x^2 + 3x - 2 = x^2 - 2x + 8$ can be reduced $3x - 2 = -2x + 8$ by cancelling x^2 from both sides, which is effectively subtracting from both sides.

cancellation laws Let ∘ be a *binary operation on a set S. The cancellation laws are said to hold if, for all a, b, and c in S,

(i) if $a \circ b = a \circ c$, then $b = c$,
(ii) if $b \circ a = c \circ a$, then $b = c$.

It can be shown, for example, that in a *group the cancellation laws hold.

canonical basis The set of *orthonormal vectors associated with *coordinates in n-dimensional *Cartesian space. In 3-dimensional space, the vectors **i**, **j**, **k** in the directions Ox, Oy, Oz.

canonical form (normal form) Standard format of expression. For example, $y = mx + c$ and $ax + by + c = 0$ are canonical forms for the equation of a straight line in the plane. See also NORMAL FORM OF CONICS, JORDAN NORMAL FORM, QUADRIC, RATIONAL CANONICAL FORM.

Cantor, Georg (Ferdinand Ludwig Philipp) (1845–1918) Mathematician

responsible for the establishment of *set theory and for profound developments in the notion of the infinite. He was born in St Petersburg but spent most of his life at the University of Halle in Germany. In 1873, he showed that the set of *rational numbers is *denumerable. He also showed that the set of *real numbers is *uncountable. Later he fully developed his theory of infinite sets (see CARDINAL ARITHMETIC, CARDINAL NUMBER). The latter part of his life was clouded by repeated mental illness.

Cantor's Diagonal Theorem (Cantor's Theorem) The set of all *subsets of any set cannot be put into *one-to-one correspondence with the *elements of the set. Consequently, there are infinitely many infinite *cardinal numbers.

Cantor distribution (Cantor function) The Cantor function $C:[0,1] \to [0,1]$ is an example of a *continuous *cumulative distribution function which has no *probability density function. To define C, we write $x \in [0,1]$ in *ternary. If x's

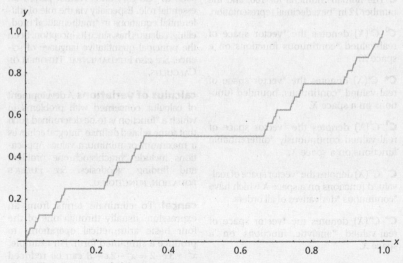

The Devil's Staircase, the cdf of the Cantor distribution

expansion includes the digit 1, then all digits after the first 1 are made 0; any 2s are then made into 1s, and $C(x)$ is this expansion, evaluated in binary. For example, any x in the range $7/9 < x < 8/9$ has expansion $x = 0.21\ldots_3$ and so $C(x) = 0.11_2 = 3/4$. See the sketch. The graph of C is sometimes called the Devil's Staircase.

C is continuous everywhere and *differentiable *almost everywhere with derivative 0. In fact, C is not differentiable only on the *Cantor set, which is a *null set. So any pdf would be 0 almost everywhere, and hence no pdf exists. Note that $\int_0^1 C'(x)\mathrm{d}x = 0 \neq 1 = C(1) - C(0)$. This does not contradict the *Fundamental Theorem of Calculus, as C' is not continuous. C is *uniformly continuous but is not *absolutely continuous.

Cantor-Schröder-Bernstein Theorem For *sets X and Y, if $|X| \leq |Y|$ and $|Y| \leq |X|$, then $|X| = |Y|$. We write $|X| \leq |Y|$ if there is a *one-to-one map from X to Y. So the theorem states that if there is an *injection from X to Y and an injection from Y to X, then there is a *bijection between X and Y. The theorem proves that the relation $\leq$ for *cardinal numbers is *anti-symmetric.

Cantor set Take the closed *interval $[0, 1]$. Remove the open interval that forms the middle third, that is, the open interval $(\frac{1}{3}, \frac{2}{3})$. From each of the remaining intervals again remove the open interval that forms the middle third. The Cantor set is the set that remains when this process is continued indefinitely. It consists of those numbers with *ternary representation $(0.d_1d_2d_3\ldots)_3$ where each d_i equals 0 or 2. The set is *uncountable, is *null, and has *fractal dimension $\log2/\log3 = 0.63$ to 2 d.p.

Cantor's Intersection Theorem In a *complete metric space, a nested sequence (*see* NESTED SETS) of subsets $\{A_n\}$ for which

the *diameters of A_n tend to zero and contain a unique point of *intersection.

Cantor's paradox Suppose there exists an infinite set A containing the largest possible number of elements. *Cantor's Diagonal Theorem shows that its *power set has more elements than A had. This proves there is no largest *cardinal number.

cap The operation $\cap$ (*see* INTERSECTION) is read as 'cap'. *Compare* CUP.

capacity A synonym for VOLUME.

capacity (of a cut) The sum of all the maximum flows allowable across all *edges intersecting the *cut (*see* NETWORK).

capacity (of an edge) The maximum flow possible along an *edge (*see* NETWORK).

Carathéodory measurable *See* OUTER MEASURE.

Cardano, Girolamo (1501–76) Italian physician and mathematician, whose *Ars magna* contained the first published solutions of the general cubic equation (*see* CUBIC POLYNOMIAL) and the general *quartic equation, even though these were due to *Tartaglia and Cardano's assistant Ludovico Ferrari respectively. Cardano was an outstanding mathematician of the time in the fields of *algebra and *trigonometry.

cardinal arithmetic Given two (possibly infinite) *cardinal numbers $|X|$, $|Y|$, where X, Y are sets, it is possible to define sums, products, and powers. The sum $|X| + |Y|$ is defined as $|X \sqcup Y|$, the cardinality of the *disjoint union. The product $|X||Y|$ is defined as $|X \times Y|$, the cardinality of the *Cartesian product. The power $|X|^{|Y|}$ is defined as the cardinality of the set of *functions from Y to X. In particular, $2^{|Y|}$ is the cardinality of the *power set of Y. All these definitions are theorems for finite sets, so it makes sense to extend these definitions to infinite sets. *Cantor showed that $2^{|\mathbb{N}|} = |\mathbb{R}|$. *See also*

CANTOR-SCHRÖDER-BERNSTEIN THEOREM, CONTINUUM HYPOTHESIS.

cardinality For a finite set A, the cardinality of A, denoted by $|A|$, is the number of elements in A. The notation $\#A$ or $n(A)$ is also used. For sets A, B,

$$|A \cup B| = |A| + |B| - |A \cap B|.$$

See also INCLUSION-EXCLUSION PRINCIPLE.

cardinal number The number of elements in a set. If two sets can be put in *one-to-one correspondence with one another they have the same cardinal number or *cardinality. For finite sets the cardinal numbers are 0, 1, 2, 3, ..., but infinite sets require new symbols to describe their cardinality. The cardinal number of the *natural numbers, *integers, and *rational numbers is *aleph-null. The cardinal number of the real numbers is denoted c (for continuum).

cardioid The curve traced out by a point on the circumference of a circle rolling round another circle of the same radius. Its equation, in which a is the radius of each circle, may be taken in polar coordinates as $r = 2a(1 + \cos\theta)(-\pi < \theta \leq \pi)$. In the figure, $OA = 4a$ and $OB = 2a$.

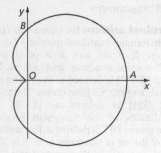

The cardioid $r = 2a(1 + \cos\theta)$

Carmichael number A synonym for PSEUDOPRIME.

Cartesian Relating to the work of René *Descartes, especially in representing *geometry by an algebraic framework.

Cartesian coordinates *See* COORDINATES (in Euclidean space).

Cartesian distance The distance between two points expressed in Cartesian coordinates, so the distance between (x_1, y_1, z_1) and (x_2, y_2, z_2) is given by $\sqrt{(x_1 - x_2)^2 + (y_1 - y_2)^2 + (z_1 - z_2)^2}$.

Cartesian plane The 2-dimensional space defined by Cartesian coordinates (x,y).

Cartesian product The Cartesian product $A \times B$, of sets A and B, is the set of all *ordered pairs (a, b), where $a \in A$ and $b \in B$. More generally, for sets A_1, $A_2, \ldots, A_n$, the Cartesian product $A_1 \times A_2 \times \ldots \times A_n$ can be defined as the set of ordered n-tuples $(a_1, a_2, \ldots, a_n)$ where $a_i \in A_i$ for each i.

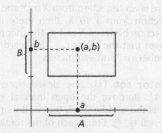

The Cartesian product $A \times B$

Cartesian space The 3-dimensional space defined by Cartesian coordinates (x,y,z). *Compare* EUCLIDEAN SPACE.

Cartesian tensor *See* TENSOR.

cascade charts (Gantt charts) A graphical way to represent the solution following a *critical path analysis. The critical activities are drawn across the top,

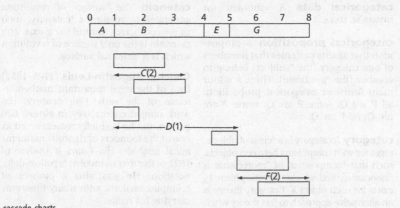

cascade charts

and non-critical activities are displayed in the figure, showing pairs of bars where the activity is carried out as early and as late as possible without causing a delay.

case control studies A study in which 'cases' that have a particular condition are compared with 'controls' that do not, to see how they differ on an *explanatory variable of interest. In the same way that *matched pairs are useful in designing an experiment, matching patients by other possible *explanatory variables can improve the effectiveness of the study. There are still inherent problems due to the possibility of *confounding variables.

casting out nines A way of checking the plausibility of *arithmetic calculations. The sum of the digits of a product equals the product of the sum of the digits of the numbers multiplied. This is because 10^n is 1 *modulo 9 for all n. In all cases the summing of digits is repeated until a single digit (i.e. between 0 and 9) is obtained, called the digital root. For example, 347 has a sum of digits of 14, then 5. Also, 514 has a sum of digits 10, then 1. Then $347 \times 514 = 178\,358$ with a sum of digits of 32, reducing to 5 which is

5×1. The name arises as any 9s can be omitted, or cast out, in the process of finding the digital root.

Catalan numbers Numbers of the form $C_n = \frac{1}{n+1} \binom{2n}{n} = \frac{(2n)!}{(n+1)!n!}$ for $n \geq 0$. The first few Catalan numbers are 1, 2, 5, 14, 42 . . . . These numbers arise in various combinatorial ways.

Catalan's conjecture The statement that the only integers $x, y, m, n \geq 2$ which satisfy $x^m - y^n = 1$ are $x = 3$, $m = 2$ and $y = 2$, $n = 3$. The conjecture was proved true in 2002 by Preda Mihăilescu.

Catalan's constant $\frac{1}{1^2} - \frac{1}{3^2} + \frac{1}{5^2} - \frac{1}{7^2} + \ldots$. It appears in problems in *combinatorial analysis. It is not known whether the number is *irrational.

catastrophe theory A topological theory of *dynamical systems, developed by René *Thom, modelling the sudden, rapid transitions that some systems are capable of. It has been variously proposed as a model for an attacking animal turning to flight or when a stock market crashes.

categorical data A synonym for NOMINAL DATA.

categorical proposition A proposition that asserts or denies that members of one category (the *subject) belong to another (the *predicate). There are four main forms of categorical proposition: all P are Q; some P are Q; some P are not Q; no P are Q.

category A category is a *class of objects, together with morphisms between objects, such that *composition of *morphisms is *associative, and an identity morphism 1_X exists for each object X. Category theory is an alternative approach to *set theory when studying the *foundations of mathematics. As an example, the class of sets as the objects, with morphisms being *functions, makes a category; note there is no 'set of all sets'.

category error (category mistake) An erroneous answer which is not only incorrect but not of the correct type or category. The answer '2' to the question 'How many moons does Earth have?' is simply false; the answer 'Mars' is a category error as clearly a *natural number is needed as the answer.

catenary The *curve in which a flexible heavy uniform chain between two points. With suitable axes, the equation of the curve is $y = c\cosh(x/c)$ (*see* HYPERBOLIC FUNCTION).

(⊕) SEE WEB LINKS
• An interactive exploration of the shape of a catenary.

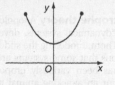

The graph of a catenary

catenoid The *surface of revolution generated by rotating a *catenary, such as $y = c\cosh(x/c)$, around the x-axis. The catenoid is the only surface of revolution which is a *minimal surface.

Cauchy, Augustin-Louis (1789–1857) One of the most important mathematicians of the early 19th century. His work ranged enormously in almost 800 papers, but he is chiefly remembered as one of the founders of rigorous mathematical analysis. His *Cours d'Analyse* of 1821 is the first to involve *epsilon-delta notation. He was also a pioneer of *complex analysis, with many theorems carrying his name.

Cauchy convergence criterion That a real (or complex) sequence is *convergent if and only if it is a *Cauchy sequence.

Cauchy distribution The Cauchy distribution has *probability density function $f(x) = \frac{1}{\pi(1+x^2)}$ for $x \in \mathbb{R}$. It has undefined *expectation, *variance, and *moment generating function. If *random variables X and Y have the standard *normal distributions, then X/Y has the Cauchy distribution.

Cauchy-Frobenius lemma *See* ORBIT-COUNTING FORMULA.

Cauchy integral test *See* INTEGRAL TEST.

Cauchy–Riemann equations For a *holomorphic function $f(z) = u + iv$ of the *complex variable $z = x + iy$ the Cauchy-Riemann equations state
$$\frac{\partial u}{\partial x} = \frac{\partial v}{\partial y}, \frac{\partial u}{\partial y} = -\frac{\partial v}{\partial x}.$$
If these *partial derivatives are *continuous and satisfy the Cauchy-Riemann equations, then conversely $f(z)$ is holomorphic.

Cauchy–Schwarz inequality In an *inner product space V then $|\langle v,w \rangle| \leq \|v\| \times \|w\|$ for any $v,w \in V$, with equality if and only if v, w are *linearly dependent. Thus, the *angle between non-zero vectors v, w can be defined as arccos $(\langle v, w \rangle / (\|v\| \times \|w\|))$.

Cauchy–Schwarz inequality for integrals If $f(x)$, $g(x)$ are real functions, then $\{\int f(x)g(x)dx\}^2 \leq \{\int f(x)^2 dx\} \{\int g(x)^2 dx\}$ if all these integrals exist. This the *Cauchy-Schwarz inequality applied to the space of *square-integrable functions.

Cauchy–Schwarz inequality for sums If a_i and b_i are real numbers, $i = 1, 2, \ldots, n$, then $\sum_{i=1}^{n} a_i b_i \leq \sqrt{(\sum_{i=1}^{n} a_i^2)(\sum_{i=1}^{n} b_i^2)}$ This is the *Cauchy-Schwarz inequality for $V = \mathbb{R}^n$ with the *scalar product.

Cauchy sequence A real (or complex) *sequence $\{a_n\}$ is Cauchy if $|a_n - a_m|$ tends to 0 as m, n tend to infinity. A real (or complex) sequence is Cauchy if and only if it is *convergent. The definition has the merit of implying convergence without explicit knowledge of the *limit. Generally, in a *metric space, Cauchy sequences can be defined using the distance $d(a_n, a_m)$ instead. *See* COMPLETE METRIC SPACE.

Cauchy's formula for derivatives Let f be a *holomorphic function defined on an *open set $U \subseteq \mathbb{C}$ and let $a \in U$. Then f has *derivatives of all *orders and the nth derivative at a is given by

$$f^{(n)}(a) = \frac{n!}{2\pi i} \int_C \frac{f(z)}{(z-a)^{n+1}} \, dz,$$

where C is a *simple, *closed, *positively oriented, continuous, piecewise-smooth curve in U and a is inside C. When $n = 0$ this result is known as Cauchy's integral formula. *See also* TAYLOR'S THEOREM.

Cauchy's integral formula *See* CAUCHY'S FORMULA FOR DERIVATIVES.

Cauchy's Integral Theorem For a *closed curve C and a *holomorphic function $f(z)$, $\int_c f(z) \, dz = 0$. *Compare* GREEN'S THEOREM.

Cauchy's mean value theorem Let f and g be *continuous functions on the interval $[a, b]$ which is *differentiable on (a, b). Then there exists c in (a, b) such that

$$f'(c)\Big(g(b) - g(a)\Big) = g'(c)\Big(f(b) - f(a)\Big)$$

The *mean value theorem follows from this by setting $g(x) = x$.

Cauchy's Residue Theorem For a function f, which is *holomorphic *inside and on a *simple, *closed, *positively oriented, continuous, piecewise-smooth curve C in $\mathbb{C}$, except at finitely many *isolated *singularities $a_1, a_2, \ldots, a_n$ inside C

$$\int_C f(z) \, dz = 2\pi i \sum_{k=1}^{n} \text{Res}(f; a_k),$$

where $\text{Res}(f; a_k)$ denotes the *residue of f at a_k. *See* CONTOUR INTEGRATION.

Cauchy's test *See* ROOT TEST.

Cauchy's Theorem (complex analysis) A synonym for CAUCHY'S INTEGRAL THEOREM.

Cauchy's Theorem (group theory) If G is a *finite *group and p is a *prime number which divides the *order of G, then G contains an element of order p, and hence a *subgroup of order p. This is a partial converse to *Lagrange's Theorem. *See also* SYLOW'S THEOREMS.

causation Causing or producing an effect. It is often assumed that the existence of *correlation between two variables indicates causation, but this is often not the case. They may both be related to a third *confounding variable.

Cavalieri, Bonaventura (1598–1647) Italian mathematician known for his method of 'indivisibles' for calculating areas and volumes. In his method, an area is thought of as composed of lines and a volume as composed of areas. Here can be seen the beginnings of the ideas of *integral calculus.

Cayley, Arthur (1821–95) British mathematician who contributed greatly to the resurgence of *pure mathematics in Britain in the 19th century. He published over 900 papers on many aspects of *geometry and *algebra. He conceived and developed the theory of *matrices and was the first to list the *axioms of an abstract *group in 1849.

Cayley–Hamilton Theorem If the *characteristic polynomial $p(\lambda) = \det(\lambda\mathbf{I} - \mathbf{A})$ of an $n \times n$ matrix $\mathbf{A}$ is written

$$p(\lambda) = \lambda^n + b_{n-1}\lambda^{n-1} + \ldots + b_1\lambda + b_0$$

then

$$\mathbf{A}^n + b_{n-1}\mathbf{A}^{n-1} + \ldots + b_1\mathbf{A} + b_0\mathbf{I} = \mathbf{O}.$$

The theorem implies that the *minimal polynomial of $\mathbf{A}$ divides the characteristic polynomial.

Cayley Representation Theorem Every *group G is *isomorphic to a group of *permutations. Every element can be identified with the effect of pre-multiplication by that element on G. This gives an *injective *homomorphism $G \to \text{Sym}(G)$, where $\text{Sym}(G)$ denotes the *symmetry group of G.

Cayley's Theorem (for trees) There are n^{n-2} *trees on n labelled *vertices. This is equal to the number of *spanning trees on the *complete graph with n vertices. See KIRCHOFF'S THEOREM.

Cayley table The multiplication table for a *group. Given a finite group $G = \{g_1, \ldots g_n\}$, then the product g_ig_j appears in the ith row and jth column of the table. As pre-multiplication by g_i and post-multiplication by g_j are both *invertible, then a Cayley table is a *Latin square.

cdf An abbreviation for CUMULATIVE DISTRIBUTION FUNCTION.

ceiling (least integer function) The smallest *integer $\lceil x \rceil$ not less than a given *real number x. So $\lceil 3.2 \rceil = 4$ and $\lceil 5 \rceil = 5$. Compare FLOOR.

Celsius (centigrade) Symbol C. The *temperature scale, and the unit of measurement of temperature, which takes 0C as the freezing point of water, and 100C as the boiling point of water.

centesimal Hundredth or relating to hundredth parts. The *grade is an *angular measure which is one-hundredth of a *right angle.

centi- Prefix used with *SI units to denote multiplication by 10^{-2}. Abbreviated as c.

centile A synonym for PERCENTILE. See QUANTILE.

central angle An *angle whose vertex is the centre of a given *circle.

central conic A non-degenerate *conic with a centre of symmetry, and thus an *ellipse or a *hyperbola.

central difference If $\{(x_i, f_i)\}$, $i = 0, 1, 2, \ldots$ is a given set of function values with $x_{i+1} = x_i + h$, $f_i = f(x_i)$ then the central difference at f_i is defined by

$$\frac{f_{i+1} - f_{i-1}}{2} = \frac{f(x_{i+1}) - f(x_{i-1})}{2}.$$

central difference approximation The central difference approximation for the *derivative of a *function $f(x)$ uses the *gradient of the *chord joining the two points whose x values are a small amount h from the value x_0, giving

$$f'(x_0) \approx \frac{f(x_0 + h) - f(x_0 - h)}{2h}.$$

central force A *force acting on a particle, directed towards or away from a fixed point. O, the centre. A *conservative central force $\mathbf{F}$ on a particle P is given by $\mathbf{F} = f(r)\mathbf{r}$, where $\mathbf{r}$ is the *position vector of P and $r = |\mathbf{r}|$.

Examples are the *gravitational force $\mathbf{F} = -(GMm/r^3)\mathbf{r}$, and the force $\mathbf{F} = -(k(r-l)/r)\mathbf{r}$ due to an spring of unit natural length (*see* HOOKE'S LAW).

centralizer The centralizer $C_G(a)$ of an element a in a *group G is

$$C_G(a) = \{g \in G \ : ga = ag\}.$$

It is a *subgroup of G and

$$|C_G(a)| \times |C(a)| = |G|,$$

where $C(a)$ denotes the *conjugacy class of a. More generally, for a subset $S \subseteq G$, then

$$C_G(S) = \{g \in G : gs = sg \text{ for all } s \in S\}.$$

Compare NORMALIZER.

Central Limit Theorem A fundamental theorem of *statistics connects the distribution of a growing *sample from a given distribution with the *normal distribution. Formally, let $X_1, X_2, X_3, \ldots$ be a sequence of *independent, identically distributed *random variables with *mean μ and finite *variance σ^2, and let

$$\overline{X}_n = \frac{X_1 + X_2 + \ldots + X_n}{n}, \quad Z_n = \frac{(\overline{X}_n - \mu)}{\sigma/\sqrt{n}}.$$

Then, as n increases, the distribution of Z_n *converges in distribution to the standard normal distribution with mean 0 and variance 1.

It implies that for a large *sample the sample mean has approximately the *normal distribution with *mean μ and *variance σ^2/n.

(((⊕))) SEE WEB LINKS

• An applet that lets you define the population and then take multiple samples of different sizes to explore the behaviour of the Central Limit Theorem.

central quadric A non-*degenerate *quadric with a centre of symmetry;

and thus an *ellipsoid or a *hyperboloid of one or two sheets.

central vertex A vertex whose *eccentricity is the *radius of the *graph. So it is a vertex which is as close to all the other vertices in the graph as is possible.

centre (group theory) The centre $Z(G)$ of a *group G equals

$$\{g \in G \ : gh = hg \text{ for all } h \in G\}.$$

It is a *normal subgroup of G.

centre *See* CIRCLE, ELLIPSE, HYPERBOLA.

centre of curvature *See* CURVATURE.

centre of gravity When a system of *particles or a *rigid body with a total mass m experiences a *uniform gravitational force, the total effect on the system or body as a whole is equivalent to a single force acting at the centre of gravity. This point coincides with the *centre of mass. The centre of gravity moves in the same way as a single particle of mass m would move under the uniform gravitational force. When the gravitational force is given by the *inverse square law of gravitation, the same is not generally true.

centre of mass Suppose that particles $P_1, \ldots, P_n$, with corresponding masses $m_1, \ldots, m_n$, have position vectors $\mathbf{r}_1, \ldots, \mathbf{r}_n$, respectively. The centre of mass (or centroid) is the point with position vector $\mathbf{r}_C$, given by

$$m\mathbf{r}_C = \sum_{i=1}^{n} m_i \mathbf{r}_i, \text{ where } m = \sum_{i=1}^{n} m_i,$$

m being the total mass of the particles.

For a *rigid body, the corresponding definitions involve *integrals. In vector form, the position vector $\mathbf{r}_C$ of the centre of mass is given by

$$m\mathbf{r}_C = \int_R \rho(\mathbf{r})\mathbf{r}\,dV \text{ where } m = \int_R \rho(\mathbf{r})\,dV.$$

where $\rho(\mathbf{r})$ is the density at the point with position vector $\mathbf{r}$, R is the region occupied by the body and m is the total mass of the body.

centre of symmetry *See* SYMMETRICAL ABOUT A POINT.

centrifugal force *See* FICTITIOUS FORCE.

centripetal force Suppose that a particle P of mass m is moving with constant *speed v in a circular path, with centre at the origin O and radius r_0. Let P have polar coordinates (r_0, θ). (*See* CIRCULAR MOTION.) Then $v = r_0 \dot{\theta}$ and the acceleration of the particle is in the direction towards O and has magnitude $r_0 \dot{\theta}^2$. It follows that, if P is acted on by a force **F**, this force is in the direction towards O and has *magnitude mv^2/r_0. It is called the centripetal force.

centroid A synonym for CENTRE OF MASS.

centroid (of a triangle) The geometrical definition of the centroid G of a *triangle ABC is as the point at which the *medians of the triangle are *concurrent. It is, in fact, 'two-thirds of the way down each median', so that, for example, if A' is the midpoint of BC, then $AG = 2GA'$. This is the point at which a triangular *lamina of uniform density has its *centre of mass. It is also the centre of mass of three particles of equal mass situated at the vertices of the triangle.

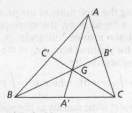

The centroid G of ABC

If A, B, and C have *position vectors **a**, **b**, and **c**, then G has position vector $\frac{1}{3}(\mathbf{a} + \mathbf{b} + \mathbf{c})$.

Cesàro summation A means of summing *series which generalizes the usual definition of convergence. Given a series $\sum_1^\infty a_k$ then define $s_n = \sum_1^n a_k$. The Cesàro sum of the series is S if $S = \lim_{n\to\infty} \frac{1}{n}\sum_1^n s_k$. If a series is convergent to S, then its Cesàro sum is also S, but some *divergent series have Cesàro sums; for example, $1-1+1-1+\ldots$ has a Cesàro sum of ½. The *Abel sum generalizes the Cesàro sum still further.

Ceva's Theorem Due to Giovanni Ceva (1648–1734), the theorem states: let L, M, and N, be points on the sides BC, CA, and AB of a triangle (possibly extended). Then AL, BM, and CN are concurrent if and only if

$$\frac{BL}{LC} \cdot \frac{CM}{MA} \cdot \frac{AN}{NB} = 1.$$

Compare MENELAUS' THEOREM.

chain rule The following rule that gives the *derivative of the *composition of two functions: If $h(x) = (f \circ g)(x) = f(g(x))$ for all x, then $h'(x) = f'(g(x))g'(x)$. For example, if $h(x) = (x^2 + 1)^3$, then $h = f \circ g$, where $f(x) = x^3$ and $g(x) = x^2 + 1$. Then $f'(x) = 3x^2$ and $g'(x) = 2x$. So $h'(x) = 3(x^2 + 1)^2 \, 2x = 6x(x^2 + 1)^2$. Another notation can be used: if $y = f(g(x))$, write $y = f(u)$, where $u = g(x)$. Then the chain rule says that $dy/dx = (dy/du)(du/dx)$. As an example of the use of this notation, suppose that $y = (\sin x)^2$. Then $y = u^2$, where $u = \sin x$. So $dy/du = 2u$ and $du/dx = \cos x$, and hence $dy/dx = 2 \sin x \cos x$.

chain rule (multivariable) More generally, if $f(x_1, \ldots, x_n)$ is a function of n variables x_i which each are functions of m variables $t_1, \ldots, t_m$, then the multivariable version of the chain rule reads

$$\frac{\partial f}{\partial t_i} = \frac{\partial f}{\partial x_1}\frac{\partial x_1}{\partial t_i} + \frac{\partial f}{\partial x_2}\frac{\partial x_2}{\partial t_i} + \ldots + \frac{\partial f}{\partial x_n}\frac{\partial x_n}{\partial t_i}.$$

Chalkdust A popular mathematics magazine begun by students at University College London in 2015.

( ) **SEE WEB LINKS**
• The Chalkdust website.

chance nodes (chance vertices) *See* EMV ALGORITHM.

chance variable A synonym for RANDOM VARIABLE.

change of base (of logarithms) *See* LOGARITHM.

change of basis (change of basis matrix) Given a *linear map $T:V \to W$ between *vector spaces, and a choice of *bases V for V and W for W, then the *matrix of the linear map T sends the *coordinate vector of $v \in V$ to the coordinate vector of Tv in W. Denote this matrix as $_W T_V$. If V' and W' are two other bases, then $_{W'} T_{V'} = (_{W'} I_W)(_W T_V)(_V I_{V'})$, where I denotes the identity map (on V or W). The matrices $_{W'} I_W$ and $_V I_{V'}$ are referred to as change of basis matrices. Respectively, they change the coordinates of a vector with respect to W (or V') to the coordinates of the same vector with respect to W' (or V).

change of coordinates Changing from one coordinate system to another, more suitable to a particular problem, is a powerful method throughout mathematics. In *linear algebra, this might relate to *diagonalization or the normal forms of *matrices (such as *Jordan normal form). In *geometry, a change of coordinates might transform a *conic's equation into normal form. In *calculus a *substitution may simplify an integral. In mechanics, *spherical polar coordinates might be better suited for a certain *rigid body. In all cases, it is either important that the change of coordinates does not affect a calculation—for example, *lengths are *invariant under an *orthogonal change of variable—or the extent of that change is understood—for example, introducing the *Jacobian in multiple integrals.

change of observer (change of reference) Let A and B be observers, possibly moving relative to one another. The same event will be observed differently by A and B. Knowing one observation and the relative position of A and B allows the calculation of what the other observer would see. In classical mechanics *time and *distance are independent of the choice of observer. *See* FRAME OF REFERENCE, GALILEAN RELATIVITY, SPECIAL RELATIVITY.

change of variable (in integration) *See* INTEGRATION, JACOBIAN.

chaos A situation in which a fully deterministic *dynamical system can appear to be random and unpredictable due to the sensitive dependence of the process on its starting values and the wide range of qualitatively different behaviours available to the process. This sensitive dependence is often called the butterfly effect. *See* LOGISTIC MAP, LORENZ ATTRACTOR.

character (character theory) Character theory is an important tool in the study of finite *groups. The character $\chi_\varphi : G \to F$ of a *representation $\varphi : G \to GL(V)$ is the map $\chi_\varphi(g) = \text{trace}\varphi(g)$. Here G is a group and V is a finite-dimensional *vector space over a *field F. The degree of the character is the *dimension of V. A character is called irreducible if the representation is *irreducible. *See* CHARACTER TABLE.

characteristic The *integer part of a logarithm in base 10. The fractional part is called the mantissa.

characteristic equation *See* CHARACTERISTIC POLYNOMIAL.

characteristic function A synonym for INDICATOR FUNCTION.

characteristic function (probability) Given a *random variable X, then its characteristic function is $\varphi_X(t) = E(e^{itX})$, where E denotes the *expectation. If the *probability density function is $f_X : \mathbb{R} \to \mathbb{R}$, then note that $\varphi_X(t) = \hat{f}_X(-t)$, where $\hat{f}_X$ denotes the *Fourier transform. Characteristic functions have the merit of converging, which is not always the case for *moment generating functions (*see* CAUCHY DISTRIBUTION).

characteristic of a field The smallest positive whole number n such that the sum of the multiplicative identity added to itself n times equals the additive identity. If no such n exists, the field is said to have characteristic zero. The characteristic is necessarily *prime or zero.

characteristic polynomial Let $\mathbf{A}$ be a real *square matrix. Then $\det(\lambda\mathbf{I} - \mathbf{A})$ is a polynomial in λ and is called the characteristic polynomial of $\mathbf{A}$. Its degree is equal to order of $\mathbf{A}$. The equation $\det(\lambda\mathbf{I} - \mathbf{A}) = 0$ is the characteristic equation of $\mathbf{A}$, and its real roots are the *eigenvalues of $\mathbf{A}$. These definitions apply equally to square matrices over other *fields and to *linear maps of finite-dimensional *vector spaces. *See also* CAYLEY-HAMILTON THEOREM.

characteristic value A synonym for EIGENVALUE

characteristic vector A synonym for EIGENVECTOR.

character table The irreducible *characters of a finite *group G tend to be listed in a character table. This concisely represents a lot of information about G and its *normal subgroups. As an example the character table of S_4 is:

	e	(123)	$(12)(34)$	(12)	(1234)
χ_1	1	1	1	1	1
χ_2	1	1	1	-1	-1
χ_3	3	-0	-1	-1	-1
χ_4	3	0	-1	-1	1
χ_5	2	-1	2	0	0

Characters are constant on *conjugacy classes, so one representative per class is shown. There are as many *irreducible representations (rows) as there are conjugacy classes (columns). The degrees of the characters appear in the first column, and the *order of G is the sum of their squares; here $24 = 1 + 1 + 9 + 9 + 4$. The first listed character is the trivial character, associated with the representation $\varphi(g) = I_1$ for all g.

chart A diagrammatic or graphical representation of data, as in *pie chart.

chart (coordinate map) Around each point of a *manifold, a *homeomorphism from a *neighbourhood of the point onto its image in Cartesian space $\mathbb{R}^n$. The collection of charts makes an atlas, and *transition maps relate coordinates in overlapping neighbourhoods.

Chebyshev, Pafnuty Lvovich (1821–94) Russian mathematician and founder of a notable school of mathematicians in St Petersburg. His name is remembered in results in *algebra, *analysis, and *probability theory. In *number theory, he proved *Bertrand's postulate.

Chebyshev polynomials Polynomials $T_n(x)$ and $U_n(x)$ which uniquely satisfy $T_n(\cos\theta) = \cos n\theta$ and $U_n(\cos\theta)\sin\theta = \sin(n+1)\theta$ for all θ. They are *orthogonal polynomials on the *interval $(-1,1)$ with weight functions $w(x) = (1 - x^2)^{1/2}$ and $w(x) = (1 - x^2)^{-1/2}$ respectively.

Chebyshev's inequality For a *random variable X with standard deviation σ and mean μ, then $\Pr\{|X - \mu| > k\sigma\} \leq 1/k^2$. While the inequality is a rather weak statement, in that most distributions do not come close to the limit specified, it can be very useful to be able to identify an upper limit that it is impossible for a probability to exceed.

Chebyshev's Theorem *See* BERTRAND'S POSTULATE.

check digit (checksum) A *binary *block code where all words are codewords is in no way *error-detecting. If an additional bit is added to each codeword so that the number of 1s in each new codeword is even, the new code is error-detecting. For example, the codewords 00, 01, 10, and 11 would become the new codewords 000, 011, 101, and 110. The additional bit is called a check digit, and in this case the construction is a *parity check. It would detect any single error in a transmitted codeword. More complicated checksums can be used in order to detect or correct errors; for example, a Hamming code needs only 8 extra bits to be able to correct a single error for a block code of length 247.

Chen's theorem *See* GOLDBACH'S CONJECTURE.

Chinese postman problem (in graph theory) *See* ROUTE INSPECTION PROBLEM.

Chinese remainder theorem For *coprime natural numbers m and n, $\mathbb{Z}_{mn}$ is *isomorphic to $\mathbb{Z}_m \times \mathbb{Z}_n$ as a *ring. More concretely, there is unique x in the range $0 \leq x < mn$ that solves the simultaneous *congruences $x \equiv a$ (mod m) and $x \equiv b$ (mod n). The isomorphism from $\mathbb{Z}_{mn}$ to $\mathbb{Z}_m \times \mathbb{Z}_n$ sends x mod mn to $(x$ mod m, x mod $n)$. If u,v are integers such that $um + vn = 1$ (*see* BÉZOUT'S LEMMA), then the inverse sends $(a$ mod m, b mod $n)$ to $vna + umb$ mod mn.

More generally, if I and J are *coprime ideals of a *ring R then

$$\frac{R}{I \cap J} \cong \frac{R}{I} \times \frac{R}{J}.$$

This can be proved by applying the first *isomorphism theorem to the *homomorphism $r \mapsto (r + I, r + J)$.

chirality An object is chiral, or exhibits chirality, if it is distinguishable from its mirror image. For example the *trefoil knot has left-handed and right-handed forms:

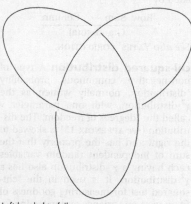

Left-handed trefoil

Right-handed trefoil

See also RIGHT-HANDED SYSTEM.

chi-squared contingency table test Very similar to the *chi-squared test but testing whether the characteristics used to categorize members of a *sample are independent. The expected values in each cell in this case are calculated by

$$\frac{\text{Row Sum} \times \text{Column}}{\text{Grand total}}.$$

See also YATES' CORRECTION.

chi-squared distribution A type of non-negative continuous probability *distribution, normally written as the χ^2-distribution, with one parameter ν called the *degrees of freedom. The distribution (see APPENDIX 15) is skewed to the right and has the property that the sum of independent random variables each having a χ^2-distribution also has a χ^2-distribution. It is used in the *chi-squared test for measuring goodness of fit, in tests on variance and in testing for independence in *contingency tables. It has *mean ν and *variance 2ν.

chi-squared test A test, normally written as the χ^2-test, to determine how well a set of observations fits a particular discrete *distribution or some other given null hypothesis (see HYPOTHESIS TESTING). The observed frequencies in different groups are denoted by O_i, and the expected frequencies from the statistical model are denoted by E_i. For each i, the value $(O_i-E_i)^2/E_i$ is calculated, and these are summed. The result is compared with a *chi-squared distribution with an appropriate number of *degrees of freedom. The number of degrees of freedom depends on the number of groups and the number of parameters being estimated. The test requires that the observations are independent and that the sample size and expected frequencies exceed minimum numbers depending on the number of groups.

Cholesky's decomposition For a *symmetric, *positive definite real matrix

A, there exists an upper *triangular matrix $\mathbf{R}$, with positive diagonal entries, such that $\mathbf{A} = \mathbf{R}^T\mathbf{R}$. More generally, if $\mathbf{A}$ is *Hermitian, then such $\mathbf{R}$ still exists with $\mathbf{A} = \mathbf{R}^*\mathbf{R}$ where *denotes the *Hermitian conjugate.

chord Let A and B be two points on a curve. The straight line through A and B, or the *line segment AB, is called a chord, the word being used when a distinction is to be made between the chord AB and the *arc AB.

Christoffel symbol For a *parameterized surface $\mathbf{r}(u,v)$ the Christoffel symbols Γ_{ij}^k are defined by

$$\mathbf{r}_{uu} = \Gamma_{11}^1\mathbf{r}_u + \Gamma_{11}^2\mathbf{r}_v + L\mathbf{n},$$
$$\mathbf{r}_{uv} = \Gamma_{12}^1\mathbf{r}_u + \Gamma_{12}^2\mathbf{r}_v + M\mathbf{n},$$
$$\mathbf{r}_{vv} = \Gamma_{22}^1\mathbf{r}_u + \Gamma_{22}^2\mathbf{r}_v + N\mathbf{n},$$

where $\mathbf{n}$ denotes a *unit normal and L, M, N are the coefficients of the *second fundamental form. The Christoffel symbols are *intrinsic (and so may be expressed in terms of the *first fundamental form) and a curve $\mathbf{r}(u(s),v(s))$ is a *geodesic parametrized by *arc length s if

$$\ddot{u} + \Gamma_{11}^1\dot{u}^2 + 2\Gamma_{12}^1\dot{u}\dot{v} + \Gamma_{22}^1\dot{v}^2 = 0,$$
$$\ddot{v} + \Gamma_{11}^2\dot{u}^2 + 2\Gamma_{12}^2\dot{u}\dot{v} + \Gamma_{22}^2\dot{v}^2 = 0.$$

chromatic number For a graph G, the maximum number of colours needed so that all regions touching one another (meeting at an edge or a vertex) are in a different colour is the chromatic number, denoted by $\chi(G)$. The *Four Colour Theorem proved that $\chi(G) \leq 4$ for all *planar graphs.

Church, Alonzo (1903–95) A pioneering figure in *logic and computer science who gave a negative answer to *Hilbert's decision problem in 1935, a year before *Turing independently showed the same. He proved this via his 'lambda calculus', a model of computation still important today. Church's thesis (or the

Church-Turing thesis) states that any notion of 'effectively calculable' will lead to the same notion of computable functions that Church and Turing arrived at.

Church's thesis (Church-Turing thesis) The principle that any definition of a *computable function on the *natural numbers—one that can be effectively implemented, for example by an *algorithm—will lead to the same computable functions as in Church's and Turing's work. As such, the thesis is unprovable, though no counterexample has been found amongst subsequent alternative definitions of computable.

Chu Shih-chieh (about AD 1300) One of the greatest Chinese mathematicians, who wrote two influential texts, the more important being *Su-yuan yu-chien* ('Precious Mirror of the Four Elements'). Notable are the methods of solving equations by successive approximations and the summation of series using *finite differences. A diagram of *Pascal's triangle, as it has come to be called, known in China from before Pascal's time, also appears.

cipher A primitive means of *encrypting a message—the 26 letters of the alphabet are permuted and every occurrence of a particular letter in a text is substituted for its replacement to make the ciphertext. A very basic example is Caesar ciphers, where each letter is shifted by some fixed amount to find its replacement, for example ABCD...being replaced by RSTU.... Even though there are 26! > 4×10^{26} ciphers, *frequency analysis can quickly reduce the possibilities, and so ciphers are easy to decrypt.

cipher (cypher) An old term for *zero or the *digit 0.

ciphertext *See* CRYPTOGRAPHY.

circle The circle with centre P and radius r is the *locus of points in the plane whose distance from P is equal to r. If P has Cartesian *coordinates (a,b), this circle has equation $(x-a)^2 + (y-b)^2 = r^2$. An equation of the form $x^2 + y^2 + 2gx + 2fy + c = 0$ represents a circle if $g^2 + f^2 - c > 0$ and is then an equation of the circle with centre $(-g,-f)$ and radius $\sqrt{g^2 + f^2 - c}$.

The *area of a circle of radius r equals πr^2, and the *length of the *circumference equals $2\pi r$.

circle of convergence A *circle in the *complex plane such that the *power series

$$\sum_{k=0}^{\infty} a_k (z - z_0)^k$$

*absolutely converges for all z satisfying $|z-z_0| < R$ and *diverges for all z satisfying $|z-z_0| > R$. Here R is the radius of convergence, and if the series converges for all the *complex plane then R is infinite. The disc with centre z_0 and radius R is called the disc of convergence. For points on the circumference of the circle the series may either converge or diverge. These same terms apply to real power series, though the disc will in that case be an *interval. The radius convergence can commonly be determined using the *ratio test and in general the formula

$$R = \frac{1}{\limsup |a_k|^{1/k}}$$

applies (as a consequence of *Cauchy's test).

circle of curvature *See* CURVATURE.

circle theorems The following is a summary of some of the theorems that are concerned with properties of a circle:

(i) Let A and B be two points on a circle with centre O. If P is any point on the *circumference of the circle and on the same side of the *chord AB as O,

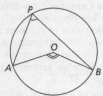

(i) $\angle AOB = 2\angle APB$

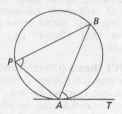

(iv) $\angle APB = \angle BAT$

then $\angle AOB = 2\angle APB$. Hence the 'angle at the circumference' $\angle APB$ is independent of the position of P.

(ii) If Q is a point on the circumference and lies on the other side of AB from P, then $\angle AQB = 180-\angle APB$. Hence opposite angles of a cyclic quadrilateral (*see* CYCLIC POLYGON) add up to 180.

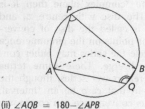

(ii) $\angle AQB = 180-\angle APB$

(iii) When AB is a *diameter, the angle at the circumference is the 'angle in a semicircle' and is a *right angle.

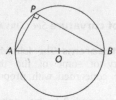

(iii) A diameter subtends a right angle

(iv) If T is any point on the *tangent at A, then $\angle APB = \angle BAT$.

(v) Suppose now that a circle and a point P are given. Let any line through P meet the circle at points A and B. Then $PA.PB$ is constant; that is, the same for all such lines. If P lies outside the circle and a line through P touches the circle at the point T, then $PA.PB = PT^2$.

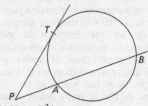

(v) $PA.PB = PT^2$ is constant

circline A *circle or a *line. Under *stereographic projection, circles on the *sphere map to lines and circles in the plane, according to whether the circle does or does not pass through the 'North Pole' N. Likewise, in the *extended complex plane, lines can be considered as circles that pass through *infinity.

circuit *See* COMBINATORIAL LOGIC.

circular argument An argument using the essence of the conclusion to arrive at that conclusion.

circular data *See* DIRECTIONAL DATA.

circular function A term used to describe the *trigonometric functions.

circular measure A synonym for ANGULAR MEASURE.

circular motion Motion of a particle in a circular path. Suppose that the path of the particle P is a circle in the *Cartesian plane, with centre at the origin O and radius r_0. Let $\mathbf{r}$, $\mathbf{v}$, and $\mathbf{a}$ be the *position vector, *velocity, and *acceleration of P. If P has *polar coordinates (r_0, θ), then

$$\mathbf{r} = r_0(\mathbf{i} \cos \theta + \mathbf{j} \sin \theta),$$
$$\mathbf{v} = \dot{\mathbf{r}} = r_0(-\dot{\theta}\mathbf{i} \sin \theta + \dot{\theta}\mathbf{j} \cos \theta),$$
$$\mathbf{a} = \ddot{\mathbf{r}} = r_0(-\ddot{\theta}\mathbf{i} \sin \theta - \dot{\theta}^2\mathbf{i} \cos \theta + \ddot{\theta}\mathbf{j} \cos \theta - \dot{\theta}^2\mathbf{j} \sin \theta).$$

Let $\mathbf{e}_r = \mathbf{i} \cos \theta + \mathbf{j} \sin \theta$ and $\mathbf{e}_\theta = -\mathbf{i} \sin \theta + \mathbf{j} \cos \theta$, so that $\mathbf{e}_r$ and $\mathbf{e}_\theta$ are rotations of $\mathbf{i}$ and $\mathbf{j}$ by θ anticlockwise. Then the previous equations become

$$r = r_0\mathbf{e}_r, \quad \mathbf{v} = \dot{\mathbf{r}} = r_0\dot{\theta}\mathbf{e}_\theta,$$
$$\mathbf{a} = \ddot{\mathbf{r}} = -r_0\dot{\theta}^2\,\mathbf{e}_r + r_0\ddot{\theta}\mathbf{e}_\theta.$$

If the particle, of mass m, is acted on by a force $\mathbf{F}$, where $\mathbf{F} = F_1\mathbf{e}_r + F_2\mathbf{e}_\theta$, then the equation of motion $m\ddot{\mathbf{r}} = \mathbf{F}$ gives $-mr_0\dot{\theta}^2 = F_1$ and $mr_0\ddot{\theta} = F_2$. If the transverse component F_2 of the force is zero, then $\dot{\theta} = $ constant and the particle has constant speed.

See also ANGULAR ACCELERATION, ANGULAR VELOCITY, RADIAL AND TRANSVERSE COMPONENTS.

circulation For a fluid with velocity field $\mathbf{u}$, the circulation around a *closed curve C equals $\Gamma = \int_C \mathbf{u} \cdot d\mathbf{r}$. If C bounds a *surface Σ, then by *Stokes' Theorem $\Gamma = \int_\Sigma \boldsymbol{\omega} \cdot d\mathbf{S}$ where $\boldsymbol{\omega}$ denotes *vorticity. Thus, vorticity can be considered (locally) as circulation per unit area.

circumcentre (circumcircle) The circumcircle of a *triangle is the *circle that passes through the three *vertices. Its centre is the circumcentre, at which the perpendicular *bisectors of the sides of the triangle are concurrent.

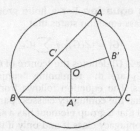

The circumcentre and circumcircle

circumference The circumference of a circle is the boundary of the circle or the length of the boundary, that is, the perimeter. The (length of the) circumference of a circle of radius r is $2\pi r$.

circumscribe To draw a geometric figure outside another so they have points in common. For example, a circle might circumscribe a polygon by passing through all its vertices, or a polygon might circumscribe a circle by having all its edges tangential to the circle.

cis The notation $\text{cis}\,\theta$ is sometimes used for $\cos\theta + i \sin\theta$. *See* DE MOIVRE'S THEOREM, POLAR FORM OF A COMPLEX NUMBER.

class A collection of sets, in particular a collection of sets which is not itself a set, in which case it is referred to as a proper class. There is no 'set of all sets', and so this is a proper class. In *Zermelo-Fraenkel set theory, classes are treated informally, but *Von Neumann formally integrated classes into his set theory. *See also* RUSSELL'S PARADOX.

class boundaries The boundaries for the classes when *data is grouped. For continuous data, measurements will normally be recorded initially to some degree of *accuracy. For example, if intervals are $1.5 \leq x < 2.0$; $2.0 \leq x < 2.4$. .... with observations recorded correct to the nearest 0.1, then the class boundaries are 1.45, 1.95, 2.45, etc.

class equation For a finite group G, the class equation states that

$$|G| = |Z(G)| + \sum |C_k|$$

where $Z(G)$ denotes the *centre of G and the C_k are the remaining *conjugacy classes. The equation follows from the fact that the conjugacy classes *partition G and that a group element has a singleton conjugacy class if and only if it is in the centre.

classical mechanics A term that broadly refers to *mathematical modelling of macroscopic mechanics previous to advances in *quantum theory or *relativity. This includes the mechanics of *Newton, *Lagrange, and *Hamilton.

Classification Theorem for Surfaces Any *connected closed (i.e. *compact) surface is *homeomorphic to precisely one of a list of standard surfaces. The *orientable standard surfaces are the *tori of differing *genus $g \geq 0$; the non-orientable standard surfaces are the sphere with $k > 0$ *Möbius strips sewn in. The standard surfaces are topologically distinct on account of the orientability (or not) and their *Euler characteristic.

(((()))) SEE WEB LINKS

• A well-illustrated description of some types of surfaces.

class interval Numerical data may be *grouped by dividing the set of possible values into class intervals and counting the number of observations in each interval. For example, if possible test marks lie in the range 0 to 99, groups could be defined by the intervals 0–19, 20–39, 40–59, 60–79, and 80–99. It is often best (but not essential) to take the class intervals to be of equal widths.

class mark A value within a *class interval, usually the mid-interval value, which is used to represent the interval.

cleartext A transmitted message or information where no *encryption is employed.

clock arithmetic See MODULO N ARITHMETIC.

clockwise Movement in the same direction of timing as the hands of a clock normally take. In compass terms $N \rightarrow E \rightarrow S \rightarrow W$ is clockwise.

clopen set A subset of a *topological space is clopen if it is both *open and *closed. This is possible because 'closed' is not the opposite of 'open', but rather a subset is closed if its *complement is open. In any topological space X, the *empty set and X are clopen; other subsets may also be.

closed (under an operation) See OPERATION.

closed (in geometry and graph theory) A continuous curve is closed if it has no ends or, in other words, if it begins and ends at the same point. Likewise a *walk, *trail, or *path which finishes at its starting point is closed.

closed disc See DISC.

closed-form A closed-form expression explicitly describes a mathematical object using finitely many operations. For example, the *Fibonacci numbers can be defined recursively but it is Binet's formula that explicitly describes them in closed form.

closed interval The closed interval $[a,b]$ is the set $\{x \mid x \in \mathbb{R} \text{ and } a \leq x \leq b\}$.

closed set (in topology) A subset of a *topological (or *metric) space is closed if its *complement is *open. This is equivalent to the subset containing all its *limit points.

closed surface A surface which is *compact has no boundary or bounding curve. Examples include the sphere, torus, and *Klein bottle but not a *polygon or *Möbius strip. More generally a closed *manifold is compact and without boundary. *See* CLASSIFICATION THEOREM FOR SURFACES.

closure The closure of a subset A of a *topological space is the smallest *closed subset containing A, or equivalently is obtained by including in it all the *limit points of A. For example, if A is the set $\{x \in \mathbb{R}: 1 < x < 2\}$ then the closure of A would include 1 and 2 as the limit points, giving $\{x \in \mathbb{R}: 1 \leq x \leq 2\}$. *See* DENSE SET, KURATOWSKI CLOSURE AXIOMS.

cluster analysis The task of assigning objects to recognizable groups called clusters, according to various measurements. These clusters commonly show *correlation between different attributes. The notion of a cluster cannot be precisely given, and many different algorithms are hence used in cluster analysis.

cluster point *See* ACCUMULATION POINT.

cluster sampling Where a population is geographically scattered it is reasonable to divide it into regions from which a sample is taken, and then a sample of individuals is taken from those regions only. The result is that the individuals in the final sample appear as clusters in the original population, but the costs of taking the sample are much lower than doing a full random sampling process. There are different strategies possible at both stages of the sampling process.

coaxial Having the same axis.

cobweb plot (cobwebbing) *See* FIXED-POINT ITERATION.

code (codeword, coding theory) *Data and *information, which may take many forms, are often encoded for the purpose of storage and/or transmission. Abstractly, this involves turning source words into codewords, which

are usually binary strings of 0s and 1s (*see* BINARY CODE). This may be done optimally to minimize the average length of a codeword (as with *Huffman coding) or may involve redundancy as with *error-correcting and error-detecting codes, used if the chance of transmission errors is not negligible. *See* BLOCK CODES, ENTROPY, HAMMING DISTANCE, INFORMATION THEORY, LINEAR CODES, SHANNON'S THEOREM.

coded data Data which has been translated from the form in which it is collected, or the value it took originally according to some specified rule. Responses to questionnaire data may be in the form of tick boxes, or of open responses, but to make analysis easier these will often be given numerical codes. For measurements, data may be coded to standardize data from different sources in order to facilitate comparisons or transformed, for example by taking logarithms, in order to gain greater insights into the behaviour.

codomain *See* FUNCTION.

coefficient *See* BINOMIAL COEFFICIENT, POLYNOMIAL.

coefficient of correlation *See* CORRELATION.

coefficient of determination The proportion of the *variance of the *dependent variable which is explained by the model used to fit the data. For a set of data $\{x_i, y_i\}$, $1 \leq i \leq n$, if $\hat{y}_i$ is the value of y predicted by the model when $x = x_i$, then the unexplained variance after fitting the model is $\dfrac{\sum (y_i - \hat{y}_i)^2}{n}$ and the total variance is $\dfrac{\sum (y_i - \overline{y})^2}{n}$. The explained variance is total variance-unexplained variance. When a linear model is fitted (by the *least squares line of regression), the coefficient of determination $= \dfrac{\text{explained variance}}{\text{total variance}}$ is the square of the *correlation coefficient.

coefficient of friction See FRICTION.

coefficient of restitution A parameter associated with the behaviour of two bodies during a *collision. Suppose that two billiard balls are travelling in the same straight line and have velocities u_1 and u_2 before the collision and velocities v_1 and v_2 after the collision. If the coefficient of restitution is e, then

$$v_2 - v_1 = -e(u_2 - u_1).$$

This formula is *Newton's law of restitution. The coefficient of restitution always satisfies $0 \leq e \leq 1$. When $e = 0$, the balls remain in contact after the perfectly inelastic collision. When $e = 1$, the collision is perfectly *elastic: there is no loss of *kinetic energy.

coefficient of skewness See SKEWNESS.

coefficient of variation A measure of *dispersion equal to the *standard deviation of a *sample divided by the *mean. The value is a dimensionless quantity, not dependent on the units or scale in which the observations are made, and is often expressed as a *percentage.

cofactor Let **A** be the square matrix $[a_{ij}]$. The cofactor, A_{ij}, of the entry a_{ij} is equal to $(-1)^{i+j}$ times the *determinant of the matrix obtained by deleting the i-th row and j-th column of **A**. If **A** is the 3×3 matrix shown, the factor $(-1)^{i+j}$ has the effect of introducing a + or - sign according to the pattern on the right:

$$\mathbf{A} = \begin{bmatrix} a_{11} & a_{12} & a_{13} \\ a_{21} & a_{22} & a_{23} \\ a_{31} & a_{32} & a_{33} \end{bmatrix} \begin{bmatrix} + & - & + \\ - & + & - \\ + & - & + \end{bmatrix}.$$

So, for example,

$$A_{12} = -\begin{vmatrix} a_{21} & a_{23} \\ a_{31} & a_{33} \end{vmatrix}, A_{31} = +\begin{vmatrix} a_{12} & a_{13} \\ a_{22} & a_{23} \end{vmatrix}.$$

See ADJOINT (OF A MATRIX)

coincident Occupying the same space or time. In particular, where geometrical shapes or functions have all points in common.

Collatz conjecture A *sequence of positive integers can be recursively defined as follows: $x_{n+1} = 3x_n + 1$ if x_n is odd and $x_{n+1} = x_n/2$ if x_n is even. The Collatz conjecture states that that sequence eventually returns to 1 from any initial term x_1, e.g. 17, 52, 26, 13, 40, 20, 10, 5, 16, 8, 4, 2, 1. The conjecture is unproven but has been verified for initial terms up to 10^{20}.

collinear Any number of points are said to be collinear if there is a straight line passing through all of them.

collision A collision occurs when two bodies move towards one another and contact takes place between the bodies. The subsequent motion is often difficult to predict. There is normally a loss of *kinetic energy. A simple example is a collision between two billiard balls, their subsequent behaviour depending upon the *coefficient of restitution.

colourable A *graph is said to be colourable if its *chromatic number is finite. See FOUR COLOUR THEOREM

column equivalence Two *matrices of the same size are column-equivalent if one matrix can be transformed into the other by *elementary column operations. See ROW EQUIVALENCE, EQUIVALENCE (OF MATRICES).

column operation See ELEMENTARY COLUMN OPERATION.

column rank See RANK.

column space The *vector space *spanned by the columns of a matrix.

Its *dimension equals the *column rank. The column space of a matrix **A** is the *image of the map $x \mapsto Ax$.

column vector A *matrix with exactly one column, that is, an $m \times 1$ matrix of the form

$$\begin{bmatrix} a_1 \\ a_2 \\ \vdots \\ a_m \end{bmatrix}.$$

The columns of a general matrix can be considered as individual column vectors.

combination A synonym for SELECTION.

combinatorial analysis (combinatorics) The area of mathematics concerned with counting strategies to determine the ways in which objects can be arranged to satisfy certain conditions.

combinatorial logic Switching arrangements in electronic circuits use *logic gates to control the flow of electrical pulses. The basic logic gates represent the *not, *and, and *or *connectives, with the *nand gate also used commonly to represent the combination *not and.

combinatorial topology The branch of *topology seeking to compute topological invariants using a finite description that captures all relevant topology in a problem, for example using *simplicial complexes. The *tetrahedron is a *triangulation of the *sphere, but a finer-grade triangulation might be needed to capture how a *continuous function maps the sphere to itself.

commensurable Two non-zero *real numbers x and y are commensurable if x/y is *rational.

common denominator An *integer that is exactly divisible by all the *denominators of a group of *fractions. Used in adding or subtracting fractions where the first step is to express each fraction as an *equivalent fraction with a common denominator.

common difference See ARITHMETIC SEQUENCE.

common factor (common divisor) A number or algebraic expression which is a *factor of each of a group of numbers or expressions. *Compare* COMMON MULTIPLE, GREATEST COMMON DIVISOR.

common fraction A synonym for SIMPLE FRACTION.

common logarithm See LOGARITHM.

common multiple A number or algebraic expression which is a *multiple of each of a group of numbers or expressions. *Compare* COMMON DIVISOR, LEAST COMMON MULTIPLE.

common perpendicular The common perpendicular of two *skew lines in space is the line segment between them that is perpendicular to both. Its *length is the distance between the lines.

common ratio See GEOMETRIC SEQUENCE.

common tangent A line which is a tangent to two or more curves.

communicating class In a *Markov chain, two states communicate if there is a positive *probability of going from one state to the other via some sequence of transitions and vice versa. This is an *equivalence relation on the set of states, and the *equivalence classes are the communicating classes. The chain is irreducible if there is one communicating class.

commutative The *binary operation $\circ$ on a set S is commutative if, for all a and b in S, $a \circ b = b \circ a$.

commutative algebra The branch of *algebra concerned with *commutative rings, which provides the rigour and techniques for *algebraic geometry. *Varieties are defined as the zero sets of *polynomials to which commutative rings can be associated. The geometric properties of the varieties (such as *singular points) can be determined from the study of the associated rings. For example, the ring associated with the curve $y^2 = x^2$ in $\mathbb{C}^2$ is

$$\frac{\mathbb{C}[x,y]}{\langle x^2 - y^2 \rangle} \cong \frac{\mathbb{C}[x,y]}{\langle x+y \rangle} \oplus \frac{\mathbb{C}[x,y]}{\langle x-y \rangle}.$$

Note that $\langle x+y \rangle$ and $\langle x-y \rangle$ are *prime ideals of $\mathbb{C}[x,y]$ and correspond to the irreducible components $x+y = 0$ and $x-y = 0$ of the curve. *See also* NOETHERIAN RING.

commutative diagram A commutative diagram is an arrangement of *sets and *functions such that any two paths from one given set to another give equal functions. A simple example is

$$
\begin{array}{ccc}
A & \xrightarrow{\alpha} & B \\
f \downarrow & & \downarrow g \\
X & \xrightarrow{\beta} & Y
\end{array}
$$

where the commutativity of the diagram means $\beta \circ f = g \circ \alpha$, as both these *compositions arise from paths from A to Y

commutative ring *See* RING.

commutator In a *group G, an element $[x, y] = x^{-1} y^{-1} xy$ for x, y in G. Note this is always the *identity if G is *abelian. The commutators generate a *normal subgroup known as the commutator subgroup, or derived subgroup, denoted $[G,G]$. The *quotient group $G/[G,G]$ is abelian and called the abelianization of G. *See also* DERIVED SERIES.

In a *ring, elements x and y have commutator $[x,y] = xy - yx$. Such commutators are important in *quantum theory, with *Heisenberg's uncertainty principle being an *inequality involving a commutator of two *observables.

commute Let $\circ$ be a *binary operation on a set S. The elements a and b of S commute (under the operation $\circ$) if $a \circ b = b \circ a$. For example, multiplication on the set of all real 2×2 matrices is not *commutative, but if A and B are *diagonal matrices, then A and B commute.

compact A *topological space in which any *cover of *open sets contains a finite cover. The continuous image (*see* CONTINUOUS FUNCTION) of a compact space is compact. A *metric space is compact if and only if it is *complete and *totally bounded. *See also* FINITE INTERSECTION PROPERTY, HEINE-BOREL THEOREM, SEQUENTIALLY COMPACT, TYCHONOFF'S THEOREM.

compactification A *compact *topological space Y is a compactification of a topological space X if $X \subseteq Y$ (or X is *embedded in Y) and X is *dense in Y. Compactifications of the plane include the *Riemann sphere and the real projective plane (*see* PROJECTIVE SPACE).

compact operator A *linear map T: $X \to Y$ between two *Banach spaces is a compact operator if the *closure of the *image of the unit *ball in X is a *compact subset of Y.

companion matrix For a *monic polynomial $f(x) = x^n + a_{n-1}x^{n-1} + \ldots + a_1x + a_0$, the companion matrix of f is the $n \times n$ matrix

$$
\begin{bmatrix}
0 & 0 & \cdots & & 0 & -a_0 \\
1 & 0 & 0 & \cdots & 0 & -a_1 \\
0 & 1 & 0 & & 0 & -a_2 \\
\vdots & 0 & 1 & \ddots & \vdots & \vdots \\
\vdots & \vdots & \ddots & \ddots & 0 & -a_{n-2} \\
0 & 0 & \cdots & 0 & 1 & -a_{n-1}
\end{bmatrix}
$$

Its *characteristic polynomial and *minimal polynomial both equal $f(x)$. *See* RATIONAL CANONICAL FORM.

comparison test Assuming $0 \leq a_n \leq b_n$, if the series $\sum_1^\infty b_n$ converges, then $\sum_1^\infty a_n$ converges.

compass A compass, or pair of compasses, is a hinged drawing instrument used for drawing *circles or *arcs of circles. *See* RULER AND COMPASS CONSTRUCTION.

complement (group theory) Let H be a subgroup of a *group G. A complement K of H is a subgroup such that

$$G = HK = \{hk : h \in H, \ k \in K\}$$

and $H \cap K = \{e\}$. If such a complement exists, then every element can be uniquely written as hk. But a complement need not exist, nor be unique if one does.

complement, complementation Let A be a subset of some *universal set E. Then the complement of A is the *set difference $E \backslash A$ and may be denoted A' or A^c or $\bar{A}$ when the universal set is understood or has been specified. Complementation (the operation of taking the complement) is a unary operation on the *power set of E. The following properties hold:

(i) $E' = \emptyset$ and $\emptyset' = E$.
(ii) For all A, $(A')' = A$.
(iii) For all A, $A \cap A' = \emptyset$ and $A \cup A' = E$

complement (probability) In *probability, the complement of an *event A is the event 'not A', that is, the event of A not occurring.

complementary angles Two angles that add up to a right angle. Each angle is the complement of the other. *Compare* CONJUGATE ANGLES, SUPPLEMENTARY ANGLES.

complementary function *See* LINEAR DIFFERENTIAL EQUATION WITH CONSTANT COEFFICIENTS.

complete (in logic) If a set of *axioms describing a system is complete, then any true statement within that system can be deduced from the axioms. *See* GÖDEL'S COMPLETENESS THEOREM, GÖDEL'S INCOMPLETENESS THEOREMS.

complete graph A *simple graph in which every *vertex is joined to every other by a single *edge. The complete graph with n vertices, denoted by K_n, is *regular of degree n-1 and has $\frac{1}{2}n(n-1)$ edges. *See also* BIPARTITE GRAPH.

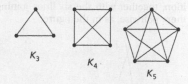

K_3 K_4 K_5

complete lattice *See* LATTICE.

completely balanced block design *See* BLOCK DESIGN.

completely normal space A *normal space X in which every subspace of X is itself a normal space. *See* SEPARATION AXIOMS.

completely regular space A *topological space X in which every *closed subset S of X and a point p of X which is not in S are *functionally separable. *See* SEPARATION AXIOMS.

complete matching A *matching in a *bipartite graph in which all vertices are used. This requires each set to have the same number of vertices, n, and the complete matching will have n edges.

complete metric space A *metric space M in which every *Cauchy sequence is *convergent in M. A *subspace of a complete metric space is complete if and only if it is *closed.

completeness axiom of the real numbers *See* BOUND.

complete orthonormal basis In a *Hilbert space X, an *orthonormal sequence (x_n) which is a *Schauder basis. This is equivalent to *Parseval's identity holding for all $x \in X$, or to the only solution of $\langle x, x_n \rangle = 0$, for all n, being $x = 0$.

complete quadrangle (complete quadrilateral) The configuration in a projective plane (*see* PROJECTIVE SPACE) consisting of four points in *general position, together with the six lines joining them pairwise, as in the figure.

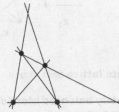

A complete quadrangle

Its *dual is a complete quadrilateral consisting of four lines, no three being *concurrent, together with the six points in which they pairwise intersect.

complete set of residues A set of n integers, one from each of the n *residue classes modulo n. Thus, $\{0, 1, 2, 3, 4\}$ and $\{0,1,3,9,27\}$ are complete sets of residues modulo 5.

complete solution A synonym for GENERAL SOLUTION.

completing the square The *quadratic equation $2x^2 + 5x + 1 = 0$ can be solved by first writing it as

$$x^2 + \frac{5}{2}x = -\frac{1}{2},$$

and then

$$\left(x + \frac{5}{4}\right)^2 = -\frac{1}{2} + \frac{25}{16} = \frac{17}{16}.$$

This step is known as completing the square: the left-hand side is made into an exact square by adding a suitable constant to both sides. The solutions for x can then be arrived at by taking *square roots and rearranging. Applying this method to the general quadratic equation derives the *quadratic formula.

completion A *complete *metric space Y is a completion of a metric space X if $X \subseteq Y$ (or X is *embedded in Y) and X is *dense in Y. The completion of a metric space is unique up to *isometry. The completion of $\mathbb{Q}$ is $\mathbb{R}$.

complex analysis The area of mathematics relating to the study of complex functions, especially *holomorphic and *meromorphic functions. *See* CAUCHY'S INTEGRAL THEOREM, CAUCHY'S RESIDUE THEOREM, CONTOUR INTEGRATION.

complex analytic A synonym for HOLOMORPHIC.

complex conjugate *See* CONJUGATE (of a complex number).

complex exponential The *exponential series converges for all *complex numbers to define a *holomorphic function exp. The derivative of exp is exp, and the identity $\exp(z + w) = \exp z \times \exp w$ is still valid. *Euler's identity $\exp z = \cos z + i \sin z$ is true for all $z \in \mathbb{C}$, though it need not be the case that $\cos z$ and $\sin z$ are the *real and *imaginary parts of $\exp z$. The range of exp is $\mathbb{C} \backslash \{0\}$, and exp is periodic with *period $2\pi i$. *Compare* COMPLEX LOGARITHM.

complex function A function whose *domain and *codomain are subsets of the *complex numbers. *See* HOLOMORPHIC.

complexity (of an algorithm) *See* ALGORITHMIC COMPLEXITY.

complex logarithm The complex *logarithm is a *multifunction. If w is a particular logarithm of $z \neq 0$, meaning $\exp w = z$, then every value of $\log z$ has the form $w + 2n\pi i$ where n is an integer. To define $\log z$ as a function, *principal values can be chosen via a *branch of the logarithm on a *cut plane; for example, every z in the *complement of $[0,\infty)$ can be uniquely written $z = r\exp(i\theta)$, where $r > 0$ and $0 < \theta < 2\pi$ and we set $\log z = \ln r + i\theta$. This defines a *holomorphic function $\log z$ on the cut plane, with derivative $1/z$ but which has a *discontinuity of $2\pi i$ across the cut. Branches of complex *powers might then be defined by $z^\alpha = \exp(\alpha \log z)$, which is again holomorphic and has derivative $\alpha z^{\alpha-1}$.

complex number There is no real number x such that $x^2 + 1 = 0$. The introduction of an 'imaginary' number i such that $i^2 = -1$ gives rise to the complex numbers $a + bi$, where a and b are real. The complex numbers subsume the real numbers (when $b = 0$). The set of all complex numbers is denoted $\mathbb{C}$. (The use of j in place of i is quite common.) Complex numbers can be added, subtracted, and multiplied as expected, remembering $i^2 = -1$; for example:

$$(a + bi) + (c + di) = (a + c) + (b + d)i,$$
$$(a + bi)(c + di) = (ac - bd) + (ad + bc)i.$$

Division by a non-zero complex number is also possible, noting:
$$\frac{a + bi}{c + di} = \frac{(a+bi)(c-di)}{(c+di)(c-di)} = \left(\frac{ac + bd}{c^2 + d^2}\right) + i\left(\frac{bc - ad}{c^2 + d^2}\right)$$ (*see* INVERSE OF A COMPLEX NUMBER). The set $\mathbb{C}$ of complex numbers in fact forms a *field. More formally, the complex number $a + bi$ can be identified with the ordered pair (a,b) of real numbers and arithmetic operations defined formally on such pairs in light of the previous calculations. *See also* ARGUMENT, COMPLEX ANALYSIS, FUNDAMENTAL THEOREM OF ALGEBRA, MODULUS OF A COMPLEX NUMBER, POLAR FORM OF A COMPLEX NUMBER.

complex plane The *Cartesian plane is called the complex plane or *Argand diagram when each point (x, y) represents the *complex number $x + yi$.

complex potential For an inviscid, *irrotational, *incompressible flow $u(z)$ in the *complex plane, with *velocity potential φ and *stream function ψ, the complex potential is $w = \varphi + i\psi$. Then $u(z) = \overline{dw/dz}$ where the bar denotes *conjugate.

complex system A complex system has many constituent parts which interact with one another in various, usually complicated ways. Thus, *modelling complex systems is problematic and usually involves some simplifications.

The Earth's climate is an important example. Biological systems, whether in the case of a single cell or an individual organism or species' interactions, are complex systems. Many aspects of human existence and society—use of the Internet, spread of disease, population movements, economies—are all complex systems.

Complex systems are commonly represented as *networks with vertices representing the parts and edges representing interactions. As the number of individual elements may be too great to treat each as a part, homogeneous subsets may be compartmentalized together; for example, an entire city may be treated as a single component.

component (of a compound statement) *See* COMPOUND STATEMENT.

component (of a graph) Two vertices of a graph are in the same component if and only if there is a *path from one to the other. So a graph is *connected if and only if it has a single component.

component (of a vector) Given a *vector **p**, in *Cartesian space, there are unique real numbers x, y, and z

such that $\mathbf{p} = x\mathbf{i} + y\mathbf{j} + z\mathbf{k}$, where $\mathbf{i}$, $\mathbf{j}$, and $\mathbf{k}$ are the *canonical basis. Then $x\mathbf{i}$, $y\mathbf{j}$, and $z\mathbf{k}$ are the components of $\mathbf{p}$ with respect to $\mathbf{i}$, $\mathbf{j}$, $\mathbf{k}$, and x, y, and z are the *coordinates of $\mathbf{p}$. These can be determined by using the *scalar product: $x = \mathbf{p} \cdot \mathbf{i}$, $y = \mathbf{p} \cdot \mathbf{j}$ and $z = \mathbf{p} \cdot \mathbf{k}$.

More generally, if $\mathbf{u}$, $\mathbf{v}$, and $\mathbf{w}$ are any 3 non-coplanar vectors, then any vector $\mathbf{p}$ in 3-dimensional space can be expressed uniquely as $\mathbf{p} = x\mathbf{u} + y\mathbf{v} + z\mathbf{w}$, and $x\mathbf{u}$, $y\mathbf{v}$, and $z\mathbf{w}$ are called the components of $\mathbf{p}$ with respect to the *basis $\mathbf{u}$, $\mathbf{v}$, $\mathbf{w}$, and x, y, and z are the coordinates of $\mathbf{p}$ with respect to $\mathbf{u}$, $\mathbf{v}$, $\mathbf{w}$. In this case, however, the components cannot be found so simply by using the scalar product. *See also* VECTOR PROJECTION.

componentwise Relating to algebraic operations on *product spaces. For example, *addition of *coordinate vectors is componentwise, which means that the ith coordinate of the sum is the sum of the ith coordinates of the vectors being added. Likewise, the group operation in a *product group $G_1 \times G_2$ involves the two group operations of G_1 and G_2 applied componentwise. (*Compare* POINTWISE.)

composite A positive *integer n is composite if it is neither *prime, nor equal to 1; that is, if $n = hk$, for integers h, $k > 1$.

composition (of functions) Let $f: S \to T$ and $g: T \to U$ be *functions. With each s in S is associated the element $f(s)$ of T, and hence the element $g(f(s))$ of U. This rule gives a function from S to U, which is denoted by $g \circ f$ and is the composition of f and g. Note that f operates first, then g. Thus $g \circ f: S \to U$ is defined by $(g \circ f)(s) = g(f(s))$, and exists if and only if the *domain of g equals the *codomain of f.

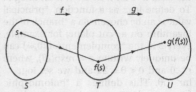

The composition $g \circ f$

For example, suppose that $f: \mathbb{R} \to \mathbb{R}$ and $g: \mathbb{R} \to \mathbb{R}$ are defined by $f(x) = 1 - x$ and $g(x) = x/(x^2 + 1)$. Then $f \circ g: \mathbb{R} \to \mathbb{R}$ and $g \circ f: \mathbb{R} \to \mathbb{R}$ both exist, and

$$(f \circ g)(x) = 1 - \frac{x}{x^2 + 1},$$
$$(g \circ f)(x) = \frac{1 - x}{(1 - x)^2 + 1}.$$

The term 'composition' is also used for the operation $\circ$; from the previous example we see $\circ$ is not generally *commutative. Composition of functions is associative: if $f: S \to T$, $g: T \to U$ and $h: U \to V$ are functions, then $h \circ (g \circ f) = (h \circ g) \circ f$.

composition (of a number) A composition of the positive integer n is obtained by writing

$$n = n_1 + n_2 + \ldots + n_k,$$

where $n_1, n_2, \ldots, n_k$ are positive integers, and the order in which $n_1, n_2, \ldots, n_k$ matters. The number of compositions of n equals 2^{n-1}. For example the compositions of 4 are

$$4 = 3 + 1 = 1 + 3 = 2 + 2 = 2 + 1 + 1 =$$
$$1 + 2 + 1 = 1 + 1 + 2 = 1 + 1 + 1 + 1.$$

See PARTITION.

composition factor *See* COMPOSITION SERIES.

composition series *Subgroups G_0, $G_1, \ldots, G_n$ of a finite *group G such that

(i) $\{e\} = G_0$ and $G = G_n$, (ii) G_k is a *normal subgroup of G_{k+1} for $0 \le k < n$, (iii) the *quotient groups G_{k+1}/G_k, called composition factors, are all *simple. $\{e\} \triangleleft A_3 \triangleleft S_3$ is a composition series for S_3 as $A_3/\{e\} \cong C_3$ and $S_3/A_3 \cong C_2$ are simple. *See* JORDAN-HÖLDER THEOREM.

compound angle formulae (in hyperbolic functions) *See* HYPERBOLIC FUNCTIONS.

compound angle formulae (in trigonometry) The *trigonometric functions of the sum or difference of two angles can be expressed in terms of the functions of the individual angles. *See* APPENDIX 14.

compound fraction A fraction in which the numerator or denominator, or both, contain fractions. For example, $\frac{3}{4}/(1 + \frac{2}{3})$. *Compare* SIMPLE FRACTION.

compound interest Suppose that an amount of money P is invested, attracting interest at i per cent annually. After one year, the amount becomes $P(1 + i/100)$. So, adding on i per cent is equivalent to multiplying by $1 + i/100$. When interest is compounded annually, all the previous year's amount is used to calculate the interest due in the following year. So, after n years, the amount becomes

$$P\left(1 + \frac{i}{100}\right)^n.$$

This is the formula for compound interest. This amount increases *exponentially in contrast to the *linear growth obtained with *simple interest.

(⊕) SEE WEB LINKS

• A compound interest calculator.

compound number A quantity expressed in a mixture of units. For example, 3 metres and 25 centimetres or 3 hours and 10 minutes.

compound pendulum A pendulum consisting of a *rigid body that is free to swing about a horizontal axis. Suppose that the *centre of mass of the rigid body is a distance d from the axis, m is the mass of the body, and I is the *moment of inertia of the body about the axis. The equation of motion can be shown to give $I\ddot{\theta} = -mgd\sin\theta$. As with the *simple pendulum, this gives approximate *simple harmonic motion for small oscillations.

compound statement A statement formed from simple statements by the use of words such as 'and', 'or', 'not', or 'implies' or their corresponding symbols. The simple statements involved are the components of the compound statement. For example, $(p \wedge q) \vee (\neg r)$ is a compound statement built up from the components p, q and r.

compression *See* STRESS.

computability theory (recursion theory) The branch of mathematics and computer science that focuses on the theoretical limits of computers and *algorithms and whether certain problems are *decidable. *See* HALTING PROBLEM, TURING MACHINE.

computable number A number is computable if it can be determined to any required *accuracy by an *algorithm. *Rational numbers are computable, but so is any number expressible as the sum of an infinite *series, such as *pi. As there are *denumerable algorithms (or programs) that are of finite length in a finite alphabet, there are denumerably many computable numbers. As the *real numbers are *uncountable, this means almost all real numbers are not computable.

computation A calculation.

compute A synonym for CALCULATE.

computer Usually a digital electronic device that carries out logical and arithmetical *calculations according to a very precise set of instructions contained within a *program, known as software. It typically comprises a number of different components, though in laptop and handheld computers some of these may be contained within a single unit. These components are known as computer hardware and will include input devices such as keyboard, mouse, microphone for speech recognition software, or tablet and pen; a central processing unit which carries out the actual calculations; memory storage devices such as the working memory, the computer's hard disk drive; output devices such as a monitor and printer; and communication devices which allow computers to connect to one another or to the Internet.

concatenation Concatenation is a *binary operation on *words, which places one word after another. For example, given the binary words $w_1 = 10$ and $w_2 = 1101$, then their concatenation is $w_1 w_2 = 101101$. Concatenation is the group operation for *free groups and groups defined by a *presentation.

concave polygon A *polygon with an *interior angle of more than 180. *Compare* CONVEX SET.

concavity (concave up and down) A real-valued *function f is concave up (or convex) if the *chord between any two points $(a, f(a))$ and $(b, f(b))$ on the graph $y = f(x)$ always lies above the graph. If the second derivative $f''(x)$ exists and is positive, then f is concave up. The graph of a concave up function lies above any *tangent line to the graph. A function is said to be concave up at a value a if it is concave up on some *neighbourhood of a.

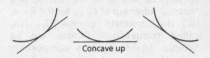
Concave up

f is concave down (or just concave) if the chord between any two points $(a, f(a))$ and $(b, f(b))$ on the graph $y = f(x)$ always lies below the graph. If $f''(x) < 0$ for all x, then f is concave down. The graph of f lies entirely below any of its tangents. A function is said to be concave down at a if it is concave down on some neighbourhood of a.

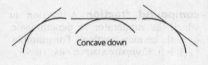
Concave down

concentric Having the same *centre. The term most commonly applies to *circles and *spheres.

conclusion (in logic) A statement derived from starting premises by *proof.

conclusion (in statistics) A decision concerning whether to accept or reject a statistical *hypothesis on the basis of the evidence available.

concrete number A number which counts a specific group of objects, for example four pens. It is the first stage in the appreciation of number as an abstract concept.

concurrent Any number of *lines are concurrent if there is a point through which they all pass.

concyclic A number of points are concyclic if there is a *circle that passes through all of them.

condition, necessary and sufficient The *implication $q \Rightarrow p$ is read as 'if q, then p'. When this is true, q is a sufficient condition for p; that is, the truth of the 'condition' q is sufficient to ensure the truth of p, or equally that 'p is true if q is true'. It is also said that p is a *necessary condition for q; the truth of p is a necessary consequence of the truth of q, or equally 'q is true only if p is true'. When the implication between p and q holds both ways, p is true if and only if q is true, which may be written $p \Leftrightarrow q$. Then q is a *necessary and sufficient condition for p.

conditional A statement that something—the *consequent clause—will be true provided that something-else the *antecedent clause—is true. For example 'if n is divisible by 2, n is even' is a conditional statement, in which the antecedent is 'n is divisible by 2' and the consequent is 'n is even'.

conditional distribution The probability distribution of the *random variable X consisting of the *conditional probabilities $\Pr(X_i \mid B)$ for each outcome X_i. Here B is an *event with $\Pr(B) > 0$.

conditional equation A term used to stress that a particular *equation is not an *identity and as such only holds true for certain values of the *variables appearing in the equation.

conditional expectation The *expected value of the random variable X using the *conditional distribution.

conditionally convergent A *series Σa_n which is *convergent but not *absolutely convergent, such as $\Sigma(-1)^n/n$. *Riemann showed that such a series can be rearranged so as to sum to any *limit.

conditional probability For two events A and B, the probability that A occurs, given that B has occurred, is denoted by $\Pr(A \mid B)$, read as 'the probability of A given B'. This is called a conditional probability. Provided that $\Pr(B) > 0$, $\Pr(A \mid B) = \Pr(A \cap B)/\Pr(B)$.

If A and B are *independent events, $\Pr(A|B) = \Pr(A)$, which rearranges to the product law for independent events: $\Pr(A \cap B) = \Pr(A) \Pr(B)$.

cone A circular cone has a circle (radius r) as base, a vertex lying at height h directly above the centre of the circle, and the curved surface formed by the line segments (length l, the slant height) joining the vertex to the points of the circle. The cone's interior has volume $\frac{1}{3}\pi r^2 h$ and the area of the curved surface equals πrl.

More generally, a cone is the surface consisting of the points of the lines, called generators, drawn through a fixed point V, the vertex, and the points of a fixed curve. The generators may be extended indefinitely in both directions to form a double cone. *See also* QUADRIC CONE.

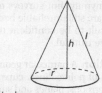

A right circular cone

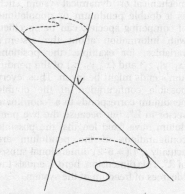

A double cone over a curve

confidence interval (confidence level) An interval, calculated from a *sample, which contains the value of a certain population *parameter with a specified probability. The end-points of the interval are the confidence limits. The specified probability is called the confidence level. An arbitrary but commonly used confidence level is 95%, which means that there is a one-in-twenty chance that the interval does not contain the true value of the parameter. For example, if $\overline{x}$ is the *mean of a sample of n observations taken from a population with a *normal distribution with a known *standard deviation σ, then

$$\left[\overline{x} - \frac{1.96\sigma}{\sqrt{n}}, \ \overline{x} + \frac{1.96\sigma}{\sqrt{n}} \right]$$

is a 95% confidence interval for the population mean μ.

confidentiality (of data) It is easier to try to ensure confidentiality of data collected on sensitive topics than it is to ensure *anonymity, but surveys on sensitive issues are still unreliable because it is very difficult to be confident that the responses are truthful.

configuration A particular geometrical arrangement of points, lines, curves, and, in three dimensions, planes and surfaces.

configuration space The state of a mechanical or *dynamical system, such as a double pendulum or populations of competing species, can be specified with information about the system's variables; for example, the positions (x_1, y_1, z_1) and (x_2, y_2, z_2) of the pendulum's ends might be given. Thus, every possible configuration of the double pendulum corresponds to a *coordinate vector in $\mathbb{R}^6$. But because the two pendulum have fixed lengths, the possible configurations of the pendulum are actually a 4 (= 6-2) dimensional subset of $\mathbb{R}^6$. This dimension, here 4, equals the *degrees of freedom of the system.

confirm (of a statistical experiment) The evidence from a statistical experiment may support belief in a *hypothesis, in which case the term confirm is

sometimes used, though it needs to be used carefully to avoid giving the impression that the hypothesis has been proven beyond doubt.

confocal conics Two *central conics are confocal if they have the same *foci. An *ellipse and a *hyperbola that are confocal intersect at *right angles.

conformable Matrices **A** and **B** are conformable (for multiplication) if the number of columns of **A** equals the number of rows of **B**. Then **A** has order $m \times n$ and **B** has order $n \times p$, for some m, n, and p, and the product **AB**, of order $m \times p$, is defined. *See* MULTIPLICATION (of matrices).

conformal A conformal map is *angle-preserving. A *holomorphic function is conformal at a point if its *derivative is non-zero.

conformally equivalent A synonym for BIHOLOMORPHIC.

confounding variable A variable which is not a factor being considered in an observational study or experiment but which may be at least partially responsible for the observed outcomes. Experimental design methods use randomization to minimize the effect of confounding variables, but that is not possible in observational studies. The possibility of one or more confounding variables is one of the biggest problems in trying to makes inferences based on observational studies.

congruence (modulo n) For each integer $n \geq 2$, the *relation of congruence between integers is defined as a is congruent to b modulo n if $a - b$ is a multiple of n. This is written $a \equiv b \, (\mathrm{mod} \, n)$. Then $a \equiv b \, (\mathrm{mod} \, n)$ if and only if a and b have the same remainder (or *residue) upon division by n. For example, 19 is congruent to 7 modulo 3. The following properties hold, if $a \equiv b \, (\mathrm{mod} \, n)$ and $c \equiv d \, (\mathrm{mod} \, n)$:

(i) $a + c \equiv b + d \, (\mathrm{mod} \, n)$,
(ii) $a - c \equiv b - d \, (\mathrm{mod} \, n)$,
(iii) $ac \equiv bd \, (\mathrm{mod} \, n)$.

That is, addition, subtraction, and multiplication are well defined; division in general is not. The integer n is the modulus of the congruence. Congruence modulo n is an *equivalence relation, with the n *equivalence classes usually represented by $0, 1, 2, \ldots, n-1$. The set of these residue classes is denoted $\mathbb{Z}_n$ and form a commutative *ring for all n and a *field if n is prime. See MODULO N ARITHMETIC.

congruence class A synonym for RESIDUE CLASS.

congruence equation An equation involving *congruences. As *division is not always well defined in *modulo n arithmetic, not all standard algebraic operations apply; for example, the congruence congruence $ax \equiv b$ (mod n) has a unique solution in $\mathbb{Z}_n$ if and only if a and n are *coprime. Also $x^2 - 1 \equiv 0$ (mod 8) is a *quadratic equation with four solutions $x \equiv 1, 3, 5,$ or 7 (mod 8), factorizable in two different ways: $x^2 - 1 \equiv (x-1)(x+1) \equiv (x-3)(x-5)$. When the modulus n is *prime, $\mathbb{Z}_n$ is a *field and these issues do not arise. See also CHINESE REMAINDER THEOREM.

congruent figures Two geometrical figures are congruent if they are identical in shape and size. Equivalently there is an *isometry that maps one figure onto the other. This includes the case when one of them is a mirror image of the other, and so the three triangles shown here are all congruent to each other.

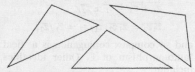
Three congruent triangles

conic (conic section) A curve that can be obtained as the plane section of a double *cone. The figure shows how an *ellipse, *parabola, and *hyperbola can be obtained.

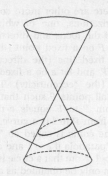
An ellipse

A parabola

A hyperbola

But there are other more convenient characterizations, one of which is by means of the focus and directrix property. Let F be a fixed point (the focus) and l a fixed line (the directrix), not through F, and let e be a fixed positive number (the *eccentricity). Then the locus of all points P such that the distance from P to F equals e times the distance from P to l is a curve, called a conic. The conic is called an *ellipse if $e < 1$, a *parabola if $e = 1$ and a *hyperbola if $e > 1$. Note that a circle is a conic, but it can only be obtained as a limiting form of an ellipse as $e \to 0$ and the directrix moves to infinity.

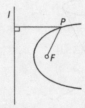

A parabola

In a Cartesian coordinate system, a conic has an equation of the form $ax^2 + 2hxy + by^2 + 2gx + 2fy + c = 0$. For an ellipse, parabola, or hyperbola it is the case that $h^2 - ab$ is negative, zero, or positive respectively. This general degree two equation can also represent *degenerate conics.

The *polar equation of a conic is normally obtained by taking the origin at a focus of the conic and the direction given by $\theta = 0$ perpendicular to the directrix. Then the equation can be written $l/r = 1 + e \cos \theta$ (all θ such that $\cos \theta \neq -1/e$), where e is the eccentricity and l is another constant.

(⊕) **SEE WEB LINKS**

• An interactive demonstration of a conic section.

conical pendulum A pendulum in which a particle, attached by a string of constant length to a fixed point, moves in a circular path in a horizontal plane. Suppose that the string has length l and that the string makes a constant angle α with the vertical. The particle moves in a circular path in a horizontal plane a distance d below the fixed point, where $d = l \cos \alpha$. The period of the conical pendulum is equal to $2\pi \sqrt{d/g}$.

conjecture A mathematical statement which has yet to be proved or disproved but which is suspected to be true on the basis of special cases of the conjecture and/or on the basis of computations. Many famous *open problems involve the term conjecture, such as *Goldbach's conjecture, and some still involve the term even though they are now proved, such as the *Weil conjectures and the *Poincaré conjecture.

conjugacy class The set of all elements of a *group that are *conjugate to an element a. The conjugacy classes partition G and the *cardinality of a conjugacy class divides the *order of G.

conjugate (of an algebraic number) Two *algebraic numbers are conjugate if they have the same *minimal polynomial.

conjugate (of a complex number) For any complex number z, where $z = x + yi$, its conjugate $\overline{z}$ (also denoted z^*) is equal to $x - yi$. As a point in the *complex plane $\overline{z}$ represents the mirror image of z in the real axis. The complex conjugate satisfies,

$$\overline{z_1 \pm z_2} = \overline{z_1} \pm \overline{z_2},$$
$$\overline{z_1 z_2} = \overline{z_1}\,\overline{z_2}, \ \overline{z_1/z_2} = \overline{z_1}/\overline{z_2},$$

and so complex conjugation is a *field *automorphism of $\mathbb{C}$. Other identities include

$$z\overline{z} = |z|^2, \ z + \overline{z} = 2\mathrm{Re}z, \ z - \overline{z} = 2i\mathrm{Im}z.$$

It is an important fact that if the complex number α is a root of a polynomial equation

$$z^n + a_1 z^{n-1} + \dots + a_{n-1}z + a_n = 0,$$

where $a_1, \ldots, a_n$ are real, then $\bar{\alpha}$ is also a root of this equation.

conjugate (conjugate elements)
Elements x and y in a group G are said to be conjugate if there is an element a in G for which $y = a^{-1}xa$. Conjugacy is an *equivalence relation (see CONJUGACY CLASS).

conjugate angles Two angles that add up to four right angles. *See* COMPLEMENTARY ANGLES, SUPPLEMENTARY ANGLES.

conjugate axis *See* HYPERBOLA.

conjugate diameters If l is a *diameter of a *central conic, the midpoints of the chords parallel to l lie on a straight line l', which is also a diameter. The midpoints of the chords parallel to l' lie on l. Then l and l' are conjugate diameters. In the case of a circle, conjugate diameters are perpendicular.

conjugate subgroups Subgroups X and Y of a group G are said to be *conjugate if there is an element a in G for which $Y = a^{-1}Xa$.

conjugate surds For a number of the form $a + \sqrt{b}$ where a, b are *rational and $\sqrt{b}$ is not rational, its conjugate is $a - \sqrt{b}$. Both these numbers have *minimal polynomial $(x - a)^2 - b$. More generally, the conjugates of an *algebraic number are the algebraic numbers with the same minimal polynomial. For example, the *primitive fifth roots of unity are all conjugate, having the minimal polynomial $x^4 + x^3 + x^2 + x + 1$.

conjugation For a group G, a map which sends g to $a^{-1}ga$, where a is a fixed element of G. Such maps are always *automorphisms of G and together form a subgroup under composition of Aut(G).

conjunction If p and q are statements, then the statement 'p and q', denoted by $p \wedge q$, is the conjunction of p and q. For example, if p is 'It is raining' and q is 'It is Monday', then $p \wedge q$ is 'It is raining and it is Monday'. The conjunction of p and q is true only when p and q are both true, and so the *truth table is as follows:

p	q	$p \wedge q$
T	T	T
T	F	F
F	T	F
F	F	F

conjunctive normal form *See* DISJUNCTIVE NORMAL FORM.

connected (of a relation) A *binary relation $\sim$ is connected if for all pairs of distinct elements x, y, either $x \sim y$ or $y \sim x$. So, for example, in the set of real numbers the relation 'is greater than' is connected.

connected (of a space) A *topological space (or *metric space) is connected if it cannot be *partitioned into two nonempty open subsets (*see* OPEN SET). The continuous image (*see* CONTINUOUS FUNCTION) of a connected space is connected. The *product of two connected spaces is connected. *Path-connected spaces are connected, but the converse is not true.

connected component In a *topological space, the connected components are the largest connected subsets. Connected components are necessarily *closed but need not be open; for example, the connected components of $\mathbb{Q}$ are the *singleton sets.

connected graph A *graph in which there is a *path from any vertex to any other.

connected sum In *topology, a means of combining *surfaces or, more generally, *manifolds of the same dimension. Given two closed surfaces S_1 and S_2, each with a *triangulation, their connected sum $S_1 \# S_2$ is formed by removing a triangle from each and gluing the surfaces along the missing triangles' boundaries (as an *identification space). The *Euler characteristic of $S_1 \# S_2$ satisfies

$$\chi(S_1 \# S_2) = \chi(S_1) + \chi(S_2) - 2.$$

For example, the connected sum of two *tori is a torus with two holes, as in the figure.

Connected sum of two tori

If the two manifolds are oriented, then their connected sum is unique up to *homeomorphism. In a similar manner, the knot sum of two *knots can be made. If both knots are oriented, then their knot sum is well defined up to *isotopy. *See* CLASSIFICATION THEOREM FOR SURFACES, PRIME KNOT.

connective In *logic, a *binary operator on the set of sentences, such as *or, *and, *xor. So, a connective combines two sentences into a single sentence.

consequent The part of a conditional statement which expresses the necessary outcome if the *antecedent is true. So in 'if n is divisible by 2, n is even', the consequent is 'n is even'. *Compare with* ANTECEDENT.

conservation condition (in network flows) Nothing is allowed to build up at

any vertex between a *source and a *sink, i.e. at each intermediate vertex the total inflow=total outflow.

conservation of angular momentum When a particle is acted on by a *central force, the *angular momentum of the particle about the centre of the field of force is constant for all time. This fact is called the principle of conservation of angular momentum.

Similarly, when the sum of the *moments of the forces acting on a *rigid body about a fixed point (or the centre of mass) is zero, the angular momentum of the rigid body about the fixed point (or the centre of mass) is conserved. *See also* NOETHER'S THEOREM.

conservation of energy When all the *forces acting on a system are *conservative forces, $T + V =$ constant, where T is the *kinetic energy and V is the *potential energy. This fact is known as the energy equation or the principle of conservation of energy.

From *Newton's second law, $m\mathbf{a} = \mathbf{F}$ for a particle with mass m moving with acceleration $\mathbf{a}$, it follows that $m\,\mathbf{a} \cdot \mathbf{v} = \mathbf{F} \cdot \mathbf{v}$ and hence $(d/dt)(\frac{1}{2}m\mathbf{v} \cdot \mathbf{v}) = \mathbf{F} \cdot \mathbf{v}$. When $\mathbf{F}$ is a conservative force, the energy equation follows by integration with respect to t. *See* NOETHER'S THEOREM, WORK.

conservation of linear momentum When the total *force acting on a system is zero, the *linear momentum of the system remains constant for all time. This fact is called the principle of conservation of linear momentum. *See* NOETHER'S THEOREM.

conservative force (conservative field) A *force is conservative if the *work done by the force as the point of application moves around any *closed path is zero. Consequently, the work done as the point of application moves from one point to another does not depend on the path taken, only on the end points. *Gravity and the *tension in a spring satisfying *Hooke's law are conservative forces. Only for a conservative force can *potential energy be defined. More generally, a *vector field is conservative if $\mathbf{F} = \nabla\varphi$ for a *scalar field φ, called a *potential, and hence $\mathrm{curl}\mathbf{F} = \mathbf{0}$. The work done $\int_C \mathbf{F}\cdot d\mathbf{r}$, along an oriented curve C from $\mathbf{a}$ to $\mathbf{b}$, then equals $\varphi(\mathbf{b})-\varphi(\mathbf{a})$.

conservative strategy In the *matrix game given by the matrix $[a_{ij}]$, suppose that the players R and C use *pure strategies. Let m_i be equal to the minimum entry in the ith row. A maximin strategy for R is to choose the rth row, where $m_r = \max\{m_i\}$. In doing so, R ensures that the smallest *payoff possible is as large as can be. Similarly, let M_j be equal to the maximum entry in the jth column. A minimax strategy for C is to choose the sth column, where $M_s = \min\{M_j\}$. These are called conservative strategies for the two players.

Now let $E(\mathbf{x},\mathbf{y})$ be the *expectation when R and C use *mixed strategies $\mathbf{x}$ and $\mathbf{y}$. Then, for any $\mathbf{x}$, $\min_\mathbf{y} E(\mathbf{x},\mathbf{y})$ is the smallest expectation possible, for all mixed strategies $\mathbf{y}$ that C may use. A maximin strategy for R is a strategy $\mathbf{x}$ that maximizes $\min_\mathbf{y} E(\mathbf{x},\mathbf{y})$. Similarly, a minimax strategy for C is a strategy $\mathbf{y}$ that minimizes $\max_\mathbf{x} E(\mathbf{x},\mathbf{y})$. By the *Fundamental Theorem of Game Theory, when R and C use such strategies, the expectation takes a certain value, the value of the game.

consistent A set of equations is consistent if there is a solution.

consistent (in logic) A consistent theory in logic is one which does not contain a *contradiction.

consistent estimator See ESTIMATOR.

constant (in physical laws) Relating to universal physical constants, such as the *speed of light, *Planck's constant, and the *gravitational constant. See APPENDIX 4.

constant acceleration See EQUATIONS OF MOTION WITH CONSTANT ACCELERATION.

constant function A constant function is a *function which takes the same value on its entire domain.

constant of integration If φ is a particular *antiderivative of a *continuous function f, then any antiderivative of f differs from φ by a constant. It is common, therefore, to write $\int f(x)\ dx = \varphi(x) + c$ where c, an arbitrary constant, is the constant of integration.

constant of proportionality See PROPORTION.

constant speed A particle is moving with constant speed if the *magnitude of the *velocity is independent of time. Thus $\mathbf{v}\cdot\mathbf{v} = $ constant and this gives $2\mathbf{v}\cdot(d\mathbf{v}/dt)=0$. Hence there are three possibilities: $\mathbf{v} = \mathbf{0}$, or $d\mathbf{v}/dt = \mathbf{0}$, or $\mathbf{v}$ is perpendicular to $d\mathbf{v}/dt$. The third possibility shows that the velocity is not necessarily constant. One such example is when a particle is moving in a circle with constant speed, and then the *acceleration is directed to the circle's centre.

constants (mathematical constants) Certain numbers, notably $\pi = 3.1415926\ldots$, $e = 2.718281\ldots$ and the *golden ratio $= \dfrac{1+\sqrt{5}}{2} \approx 1.618\ldots$ which occur repeatedly in mathematical theorems. See APPENDIX 4.

constant term See POLYNOMIAL.

constrained optimization Optimization in circumstances where constraints exist. See LAGRANGE MULTIPLIER, LINEAR PROGRAMMING.

constraint A condition that causes a restriction. For example, if $f(t)$ describes the motion of a pendulum which is released at $t = 0$, then $f(t)$ is only valid when $t \geq 0$. In a probability distribution if $p_i = \Pr\{X = x_i\}$ then the values p_i are subject to two axiomatic constraints: $p_i \geq 0$ for each i and $\sum p_i = 1$.

constructible number (constructible point) A point in the plane is constructible from two given points if it can be drawn using a *construction with ruler and compass. Assigning the two points *coordinates $(0,0)$ and $(1,0)$, a *real number α is constructible if the point $(\alpha,0)$ is constructible. The constructible numbers form a *field, which contains the *rational numbers and is contained in the *algebraic numbers. However, not every algebraic number is constructible. Because a construction involves, at each step, intersecting lines and circles, algebraically nothing more complicated than a *quadratic equation is being solved. Consequently (by the *tower law), it follows that if α is constructible, then the degree of the *field extension $\mathbb{Q}(\alpha) : \mathbb{Q}$ is a power of 2. (The *converse is false.) Thus, $\sqrt[3]{2}$, though algebraic, is not constructible as $\mathbb{Q}(\sqrt[3]{2}) : \mathbb{Q}$ has degree 3; moreover, this fact means it is impossible to duplicate the cube. Trisecting the angle, 60 degrees would be equivalent, algebraically to solving the cubic $t^3 - 3t - 1 = 0$. As this cubic is irreducible, $\mathbb{Q}(t) : \mathbb{Q}$ has degree 3 for any root, and so it is impossible to trisect the angle 60 degrees. Pierre Wantzel demonstrated these facts in 1837 and also showed that if a regular n-gon is constructible, then n is a product of powers of 2 and distinct *Fermat primes, the converse of a famous result of *Gauss. The impossibility of squaring the circle would become clear in 1882, when it was shown that *pi is *transcendental by *Lindemann. *See* CONSTRUCTION WITH RULER AND COMPASS.

construction with ruler and compass Ancient Greek geometers, in particular, were interested in what geometrical constructions were possible using a ruler (strictly a straight edge, so no increments on the edge) and a pair of compasses. Many constructions were known classically, such as *bisecting an angle, reflecting a point in a line, drawing the circle through three non-collinear points, but famously they were unable to resolve the (im)possibility of the following: (i) given a square base of a cube, drawing a second square, a cube on which would have twice the volume; (ii) trisecting a general angle; (iii) given a circle, drawing a square with the same area as the circle. All these are now known to be possible, though the solutions lay more in *algebra and *analysis than in *geometry. *See* CONSTRUCTIBLE NUMBER.

constructive Of a *proof or method that uses an *algorithm or other effective procedure to prove the result or yield answers. By contrast, a non-constructive proof may demonstrate the existence of a mathematical object without providing any method to determine that object. As an example, the formula for solving *quadratic equations explicitly describes the roots but the *Fundamental Theorem of Algebra only shows existence of roots and nothing more.

constructivism The *philosophy of mathematics which asserts that a mathematical object needs to be determined or constructed to demonstrate its existence. *Intuitionism is one such philosophy.

contact Two real *functions $f(x)$ and $g(x)$ (or the *graphs they define) have contact of order k at the value $x = a$ if

$$f^{(i)}(a) = g^{(i)}(a) \quad \text{for} \quad 0 \leq i \leq k$$
$$\text{and} \quad f^{(k+1)}(a) \neq g^{(k+1)}(a).$$

The graphs intersect if the order of contact is at least 0, touch if the order of contact is at least 1, and the order of contact is at least two if the graphs touch and have equal *curvature.

contact force When two bodies are in contact, each exerts a contact force on the other, the two contact forces being equal and opposite. The contact force is the sum of the frictional force, which is *tangential to the surfaces at the point of contact, and the normal reaction, which is normal to the surfaces at the point of contact. If the frictional force is zero, the contact is said to be *smooth, and otherwise the contact is said to be *rough. If the two bodies are moving, or tending to move, relative to each other, the frictional forces will act to oppose the motion. *See also* FRICTION.

contained in (contains) If A is a *subset of the *set B, that is, every *element of A is an element of B, then we say that 'A is contained in B' and that 'B contains A'. These are written $A \subseteq B$ and $B \supseteq A$. It is common, but technically incorrect, to say that B contains b when b is an element of B.

content The content of a *polynomial with *integer coefficients is the *greatest common divisor of its coefficients. So a *primitive polynomial has content 1. By *Gauss' lemma, the content of a product equals the product of the contents.

contingency table A method of presenting the *frequencies of the outcomes from an experiment in which the observations in the *sample are categorized according to two criteria. Each cell of the table gives the number of occurrences for a particular combination of categories. A contingency table in which individuals are categorized by gender and hair colour is shown below. A final column giving the row sums and a final row giving the column sums may be added. Then the sum of the final column and the sum of the final row both equal the number in the sample.

	hair colour			
gender	black	blonde	brown	red
male	30	6	22	6
female	24	9	18	5

If the sample is categorized according to three or more criteria, the information can be presented similarly in a number of such tables.

contingent (in logic) A statement or proposition is contingent if it is neither always true nor always false. For example 'x is divisible by 2' is true when $x = 2$, 4, 6, ... but is not true when $x = 3, 4.3, \sqrt{7}$.

continued fraction An expression of the form $q_1 + 1/b_2$, where $b_2 = q_2 + 1/b_3$, $b_3 = q_3 + 1/b_4$, and so on, where q_1, q_2, ... are positive integers, with the possible exception of q_1. This can be written

$$q_1 + \cfrac{1}{q_2 + \cfrac{1}{q_3 + \cfrac{1}{q_4 + \cdots}}}$$

or, in a form that is easier to print,

$$q_1 + \frac{1}{q_2 +} \frac{1}{q_3 +} \frac{1}{q_4 + \cdots}.$$

If the continued fraction terminates, it gives a rational number. The expression of any given positive rational number as a continued fraction can be found by using the *Euclidean algorithm. For example, 1274/871 is found, by using the steps which appear in the entry on the Euclidean algorithm, to equal

$$1 + \frac{1}{2 +} \frac{1}{6 +} \frac{1}{5}.$$

When the continued fraction continues indefinitely, it represents a real number that is the limit of the sequence

$$q_1,\; q_1 + \frac{1}{q_2},\; q_1 + \frac{1}{q_2 + \frac{1}{q_3}},\; q_1 + \frac{1}{q_2 + \frac{1}{q_3 + \frac{1}{q_4}}}, \ldots.$$

and every real number can be uniquely represented by a continued fraction. For example, representation of $\sqrt{2}$ as a continued fraction is

$$1 + \cfrac{1}{2+}\cfrac{1}{2+}\cfrac{1}{2+\dots}.$$

continuity correction The addition or subtraction of 0.5 when a discrete *distribution, taking only integer values, is used to approximate a continuous distribution, so as to improve the agreement between the two distributions.

continuity equation In *fluid dynamics, the equation

$$\frac{\partial \rho}{\partial t} + \text{div}(\rho \mathbf{u}) = 0,$$

which is a consequence of conservation of *mass. Here ρ denotes *density and $\mathbf{u}$ is the fluid's *velocity. Consequently, if the fluid is incompressible and ρ is constant, then $\text{div}\mathbf{u} = 0$. The term is more generally used for similar equations that arise in *applied mathematics as a consequence of a quantity being conserved.

continuous data *See* DATA.

continuous function The *function $f:S \to \mathbb{R}$ is continuous at a in S if $f(x) \to f(a)$ as $x \to a$ (*see* LIMIT). Rigorously, this means that, given any $\epsilon > 0$, there exists $\delta > 0$ such that $|f(x) - f(a)| < \epsilon$ whenever $|x - a| < \delta$ and $x \in S$. This is equivalent to saying that whenever a *sequence x_n converges to a, $f(x_n)$ converges to $f(a)$. The function f is continuous on S if it is continuous at each point of S. The following *algebra of limits holds:

(i) The sum of two continuous functions is continuous.
(ii) The product of two continuous functions is continuous.
(iii) The quotient of two continuous functions is continuous provided the denominator is not zero.

(iv) Suppose that f is continuous at a, that $f(a) = b$ and that g is continuous at b. Then the *composition $g \circ f$ is continuous at a.
(v) Any *differentiable function is continuous. The converse is not true as the *modulus function is continuous but not differentiable.

See BLANCMANGE FUNCTION, INTERMEDIATE VALUE THEOREM, WEIERSTRASS' THEOREM.

This notion of continuity generalizes readily to *metric spaces: a function $f: M \to N$ between metric spaces is continuous at $a \in M$ if, given any $\epsilon > 0$, there exists $\delta > 0$ such that $d_N(f(m), f(a)) < \epsilon$ whenever $d_M(m,a)| < \delta$ and $m \in M$. This further generalizes to *topological spaces: a function $f:M \to N$ between topological spaces is continuous at $a \in M$ if, given any open subset (*see* OPEN SET) U containing $f(a)$, the *pre-image $f^{-1}(U)$ is open in M.

continuously differentiable A function is continuously differentiable if its *derivative is continuous. The function, defined by $f(x) = x^2 \sin(1/x)$ for $x \neq 0$ and $f(0) = 0$, is differentiable but not continuously differentiable.

continuous random variable *See* RANDOM VARIABLE.

continuum A somewhat archaic word for the set of *real numbers. Also, in *general topology, a non-empty *compact, *connected, *Hausdorff space.

continuum hypothesis The conjecture, made by *Cantor, that there is no set with *cardinality strictly between *aleph-null, the cardinality of the set of *natural numbers, and the cardinality of the set of *real numbers. It is now known to be *independent of the *Zermelo-Fraenkel axioms.

contour integration A contour integral is a *path integral $\int_C f(z)\,dz$ of a function f in the complex plane along an oriented, usually *closed curve C. Such integrals can be evaluated using *Cauchy's Residue Theorem; using appropriately chosen contours and integrands, and then taking real or imaginary parts, a vast range of *integrals and *series can be determined. For example, the substitution $z = e^{i\theta}$ changes the integral $\int_0^{2\pi} \dfrac{d\theta}{(2 + \cos\theta)^2}$ into the path integral $\int_C \dfrac{-4iz\,dz}{(z^2 + 4z + 1)^2}$ where C is the *positively oriented unit circle centred at 0. The new integrand has a double *pole inside C, the *residue of which can be determined and so the integral evaluated.

contour line Physical geography maps show height contours—curves of equal height—and weather charts show pressure isobars—curves of equal pressure. More generally, for a function $f(x,y)$, the contour lines are the *level sets with equation $f(x,y) = c$, where c is a constant value (in the image of f). A sketch of such contour lines is called a **contour plot**.

contractible space A *topological space X that is *homotopy equivalent to a one-point space. Any *convex is necessarily contractible.

contraction mapping A *function, f, on a *metric space, X, which decreases the distance, d, between any two points in X by at least some given factor, i.e. there exists $a < 1$, $d(f(x), f(y)) \leq a \times d(x, y)$ for all x, y in X.

contraction mapping theorem If f is a *contraction mapping on a nonempty, *complete metric space X, then there is a unique fixed point x in X under the mapping, i.e. $f(x) = x$ for exactly one point x. This fixed point is the *limit of the *Cauchy sequence $x_{n+1} = f(x_n)$ beginning from any x_1.

contradiction The simultaneous assertion of both the truth of a proposition and its *negation. Since both cannot be true there must necessarily be a flaw in either the reasoning leading to the simultaneous assertion or in the assumptions on which the deductive reasoning is based. It is this latter situation which provides the basis for *proof by contradiction.

contrapositive The contrapositive of an *implication $p \Rightarrow q$ is the implication $\neg q \Rightarrow \neg p$. An implication and its contrapositive are *logically equivalent, so that one is true if and only if the other is. For example, the theorem that if n^2 is odd then n is odd could be proved by showing instead that if n is even then n^2 is even.

contravariant functor See FUNCTOR.

contravariant tensor See TENSOR.

control To rule out the effects of variables other than the factors the experiment wishes to explore. This may be done by ensuring certain variables are the same, either by directly controlling, for example, temperature or by matching pairs of subjects, for example by weight. Randomization is then normally used so that any *confounding variables which had not been controlled for by either of these techniques should not introduce a systematic source of *bias.

control chart A chart on which *statistics from samples, taken at regular intervals, are plotted so that a production-line operator can monitor the output. The chart will normally have *warning limits and *action limits displayed on it.

(⊕) SEE WEB LINKS

• Construct a control chart for your own choice of parameters.

control group In experimental design, the control group provides a baseline assessment of any change which might happen without the treatment under investigation. For example, in looking at the effects of using a particular diet on a child's growth between 10 and 14 we need to be able to make a comparison with the growth of other children not following that regime. The control group and experimental groups should be either allocated randomly or balanced if that is possible, to provide a fair test of the treatment.

convective derivative For a fluid flow, the *partial derivative $\partial/\partial t$ measures the rate of change at a fixed point as different material flows through that point. Also known as the Lagrangian derivative or the material derivative, the convective derivative measures the change of a quantity q while following the fluid. It is given by $\frac{Dq}{Dt} = \frac{\partial q}{\partial t} + \mathbf{u} \cdot \nabla q$ where $\mathbf{u}$ is the fluid's velocity.

converge (sequence) *See* LIMIT OF A SEQUENCE.

converge (series) The infinite *series $a_1 + a_2 + a_3 + \ldots$ is said to converge if its *partial sums converge as a *sequence; that is, there exists a limit L such that for all $n > N, |\sum_{i=1}^{n} a_i - L| < \epsilon$. Note that for a series to converge the sequence made up of its terms must converge to 0, though the converse is not true as demonstrated by the *harmonic series.

convergence almost surely A sequence of *random variables X_n converges almost surely to a random variable X if

$$\Pr(\lim_{n \to \infty} X_n = X) = 1.$$

This is written $X_n \xrightarrow{a.s} X$ and is also known as strong convergence. Almost sure convergence implies *convergence in probability. *See* STRONG LAW OF LARGE NUMBERS.

convergence in distribution A sequence of *random variables X_n

converges in distribution to a random variable X if $F_n(x) \to F(x)$ for all real x, where F_n and F denote the respective *cumulative distribution functions. This is also known as weak convergence and is written $X_n \xrightarrow{d} X$.

convergence in mean square A sequence of *random variables X_n converges in mean square to a random variable X if $E(|X_n - X|^2) \to 0$ as $n \to \infty$, where E denotes *expectation. This is written $Xn \xrightarrow{L2} X$. Convergence in mean square implies *convergence in probability.

convergence in probability A sequence of *random variables X_n converges in probability to a random variable X if for all $\epsilon > 0$, $\Pr(|X_n - X| > \epsilon)$ tends to 0 as n tends to infinity. This is written $X_n \xrightarrow{P} X$. Convergence in probability implies *convergence in distribution. *See* WEAK LAW OF LARGE NUMBERS.

convergence of functions There are two main types of convergence of *functions: *pointwise convergence and *uniform convergence, which is a stronger condition.

convergence test (for series) *See* APPENDIX 12.

converse The converse of an *implication $p \Rightarrow q$ is the implication $q \Rightarrow p$. If an implication is true, then its converse may or may not be true.

converse relation Given a *binary relation R on a set S, the converse relation R^T is the relation defined by xR^Ty if and only if yRx. Thus, the converse of $<$ is $>$ and vice versa.

convert To change the units of a quantity or the form of expressing something. For example, an *angle of 180 is the same as π *radians and a *vector of magnitude d in a given direction can be converted into *Cartesian coordinates, usually for ease of calculation.

convex function A synonym for CONCAVE UP, CONVEX DOWN.

convex hull The convex hull of a subset S of n-dimensional space is the smallest *convex set containing S. The convex hull of 3 not *collinear points is a *triangle, of 4 not *coplanar points is a *tetrahedron. (*See* SIMPLEX).

convex set A subset of n-dimensional space, such as a *polygon or *polyhedron, is convex if the *line segment joining any two points inside it lies wholly inside it. *Compare* AFFINE SET.

convex up and down Some authors say that a curve is convex up when it is concave down, and convex down when it is concave up (*see* CONCAVITY).

convolution Convolution-like products occur variously in mathematics. For the *Fourier transform and *Laplace transform they are respectively defined by

$$(f * g)(x) = \int_{-\infty}^{\infty} f(t)g(x - t) \ dt,$$

$$(f * g)(x) = \int_{0}^{\infty} f(t)g(x - t) \ dt,$$

and each is defined so that the transform of the convolution is the product of the transforms. The convolution product is *commutative and the *sifting property of the *Dirac delta function shows it to be the *identity. In *probability, if f and g are the *pdfs of two continuous distributions, X and Y on $\mathbb{R}$, then the pdf of their sum $Z = X + Y$ is the first integral above.

Conway, John Horton (1937–2020) English mathematician, perhaps best remembered for his *Game of Life*, a cellular automated zero-player game, but who made important contributions diversely in *number theory, *game theory, *group theory, and *knot theory. He also published a good deal on recreational mathematics and frequently collaborated with Martin *Gardner.

Conway circle For any triangle ABC, six new points can be constructed by extending each of BA and CA by |BC|, each of AB and CB by |AC|, each of AC and BC by |AB|. These six points then lie on the Conway circle. Its centre is the *incentre.

coordinate geometry (analytic geometry) The area of mathematics where geometrical relationships are described algebraically by reference to *coordinates. *Compare* SYNTHETIC GEOMETRY.

coordinates (in Euclidean space) Properly appreciated, a *line, *plane, and 3-dimensional space are each geometric objects with no assigned coordinates, *origin, or *axes. A line, though, may be identified with the set of *real numbers by choosing a point O as the origin (which has coordinate *zero), a *direction for the line, and a choice of unit length. Using the *axioms of *synthetic geometry (or Euclidean geometry) every point can then be uniquely assigned a coordinate. There is a unique point P such that the *line segment OP has length 1 and $\overrightarrow{OP}$ is in the direction of the line; another point Q is assigned coordinate x when $\overrightarrow{OQ} = x \overrightarrow{OP}$.

Likewise, in a plane, an origin O is chosen (which has coordinates $x = y = 0$), and a first line in the plane is chosen as the x-axis, and a point on the line is assigned an x-coordinate as previously, and a zero y-coordinate. A second line, perpendicular to the first, can be constructed and each point of that line assigned a y-coordinate as previously (using the same unit length) and an x-coordinate of zero. For any point Q in the plane, let M and N be points on the x-axis and y-axis such that QM is parallel to the y-axis and QN is parallel to the x-axis. If M has coordinates $(x,0)$ and N has coordinates $(0,y)$ then Q is assigned coordinates (x,y).

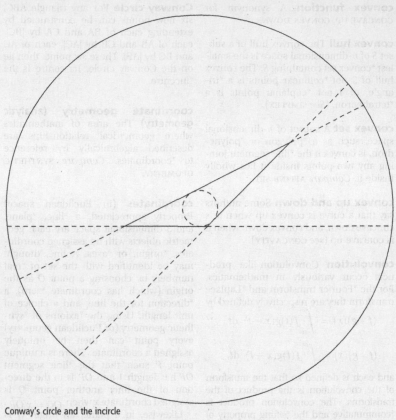

Conway's circle and the incircle

This process can be extended to any finite-dimensional *Euclidean space. Here the coordinates have been defined in such a way that the *canonical basis is an *orthonormal basis, but a *basis can more generally be used to assign coordinates; the axes would not generally be perpendicular and standard formulae (for example, for the *scalar product) would not be valid.

coordinates (in projective space) Given three distinct points in a projective line, a unique system of *homogeneous coordinates exists such that the points have coordinates [1:0], [0:1], and [1:1]. Given a *complete quadrangle of points in a projective plane, these can be uniquely assigned coordinates [1:0:0], [0:1:0], [0:0:1], and [1:1:1]. More generally, homogeneous coordinates can be assigned in this manner for $n + 2$ points in *general position in a n-dimensional *projective space.

coordinate system A system for identifying points on a plane or in space by their coordinates, for example *Cartesian or *polar coordinates.

coordinate vector A *vector represents a movement of space and can be defined in the absence of any *coordinate system. When coordinates are chosen (*see* BASIS), a vector can be represented as a list of coordinates, usually presented as a *row vector or *column vector. If two coordinate systems are being used, the same vector may be represented by different coordinate vectors in the two systems.

Copenhagen interpretation In *quantum theory, the view that a particle's state is represented by a complex *wave function (*see* SCHRÖDINGER'S EQUATION). This wave function provides a probabilistic description of the particle's state, which has no particular state until measured, with the act of *measurement collapsing the wave function.

coplanar A number of points and lines are coplanar if there is a plane in which they all lie. Three points are always coplanar: indeed, any three points that are not collinear determine a unique plane that passes through them.

coplanar vectors A vector **p** is coplanar with non-zero, non-parallel vectors **a** and **b** if and only if there exist *scalars λ and μ such that $\mathbf{p} = \lambda\mathbf{a} + \mu\mathbf{b}$. Note that this is equivalent to the three points with *position vectors **a**, **b**, **p**, together with the origin, being coplanar. *See* LINEARLY DEPENDENT.

coprime *Integers a and b are coprime if their *greatest common divisor is 1. Two *ideals I and J, in a commutative *ring R, are coprime if $I + J = R$. By *Bézout's lemma $a\mathbb{Z}$ and $b\mathbb{Z}$ are coprime ideals in $\mathbb{Z}$ if and only if a and b are coprime integers.

Coriolis force *See* FICTITIOUS FORCE.

corollary A result that follows from a theorem almost immediately, often without further proof.

correction An alteration made to the result of an *observation or calculation in order to improve its accuracy. For example, government estimates of the number of homeless people will be higher than the recorded totals in the census, because it is known that the census will not be able to accurately record all homeless people.

correlation Between two *random variables, the correlation is a measure of the extent to which a change in one tends to correspond to a change in the other. The correlation is high or low depending on whether the relationship between the two is close or not. If the change in one corresponds to a change in the other in the same direction, there is positive correlation, and there is a negative correlation if the changes are in opposite directions. *Independent random variables have zero correlation. One measure of correlation between the random variables X and Y is the correlation coefficient ρ defined by

$$\rho = \frac{\mathrm{Cov}(X, Y)}{\sqrt{\mathrm{Var}(X)\ \mathrm{Var}(Y)}}$$

(*see* COVARIANCE, VARIANCE). This satisfies $-1 \leq \rho \leq 1$. If X and Y are linearly related, then $\rho = -1$ or $+1$.

For a *sample of n paired observations $(x_1, y_1), (x_2, y_2), \ldots (x_n, y_n)$, the (sample) correlation coefficient is equal to

$$\frac{\sum(x_i - \overline{x})(y_i - \overline{y})}{\sqrt{\left(\sum(x_i - \overline{x})^2\right)\left(\sum(y_i - \overline{y})^2\right)}}$$

Note that the existence of some correlation between two variables need not imply *causation.

correlation matrix The $n \times n$ matrix in which $a_{ij} = \mathrm{corr}(X_i, X_j)$ so that the correlations between all pairs of variables are contained in it. All elements a_{ii} in the leading diagonal will be 1 and the matrix is a *symmetric matrix.

correspondence *See* ONE-TO-ONE CORRESPONDENCE.

corresponding angles *See* TRANSVERSAL.

cosecant *See* TRIGONOMETRIC FUNCTION.

cosech, cosh *See* HYPERBOLIC FUNCTION.

coset (group) If H is a *subgroup of a *group G, then for any element, x of G there is a left coset xH consisting of all the elements xh, where h is an element of H. Similarly, there is a right coset, Hx, with elements hx. *See also* LAGRANGE'S THEOREM, NORMAL SUBGROUP, QUOTIENT GROUP.

coset (ring) If I is an *ideal of a *ring R, then the coset of an element x of R is

$$x + I = \{x + i : i \in I\}.$$

See QUOTIENT RING.

cosine *See* TRIGONOMETRIC FUNCTION.

cosine rule *See* HYPERBOLIC PLANE, SPHERICAL TRIANGLE, TRIANGLE.

cotangent *See* TRIGONOMETRIC FUNCTION.

coth *See* HYPERBOLIC FUNCTION.

count To enumerate. To assign, in order, the positive *integers 1,2,3,... to finitely (or *denumerably) many elements of a set.

countable A set X is countable if there is a *one-to-one correspondence between X and a subset of the set of natural numbers. Thus a countable set is either finite or *denumerable. Some authors use 'countable' to solely mean denumerable.

countably infinite A synonym for DENUMERABLE.

counterexample Let $P(x)$ be a mathematical sentence involving a symbol x,

so that, when x is a particular element of some set, $P(x)$ is a statement that is either true or false. What may be of concern is proving or disproving that $P(x)$ is true for all x. This can be shown to be false by producing just one particular x for which $P(x)$ is false. Such a particular x is a counterexample. *See* DE MORGAN'S LAWS (in logic).

counting numbers A synonym for NATURAL NUMBERS.

counts A particular *statistic which simply records the number of instances of various events, such as the number of cars passing a specific position on a road, often collected by some automated process. These simple measures often then form the basis of more sophisticated statistical processes, such as analysing *contingency tables.

couple A system of *forces whose sum is zero. The simplest example is a pair of equal and opposite forces acting at two different points, B and C. It can be shown that the *moment of a couple about any point A is independent of the position of A.

coupled equations A pair of *simultaneous equations or *differential equations, such as the *predator–prey equations.

covariance The covariance of two *random variables X and Y, denoted by Cov(X, Y), is equal to $E((X-\mu_X)(Y-\mu_Y))$, where μ_X and μ_Y are the population *means of X and Y respectively (*see* EXPECTED VALUE). If X and Y are *independent random variables, then Cov$(X, Y) = 0$. For computational purposes, note $E((X-\mu_X)(Y-\mu_Y)) = E(XY) - \mu_X \mu_Y$. For a sample of n paired observations $(x_1, y_1), (x_2, y_2), \ldots, (x_n, y_n)$, the sample covariance is equal to

$$\frac{\sum (x_i - \bar{x})(y_i - \bar{y})}{n}.$$

covariance matrix *See* HYPOTHESIS TESTING.

covariant derivative Let **v** be a smooth *vector field on a *surface X in $\mathbb{R}^3$ and let $\gamma(t)$ be a parameterized curve in X. Then the covariant derivative $D\mathbf{v}/dt$ is the *orthogonal projection of $d(\mathbf{v}(\gamma(t))/dt$ into the *tangent space at $\gamma(t)$. The covariant derivative is in fact *intrinsic to the surface and can be expressed in terms of *Christoffel symbols. *See* PARALLEL TRANSPORT.

covariant functor *See* FUNCTOR.

covariant tensor *See* TENSOR.

cover A cover of a set X is a collection of sets whose *union contains X as a subset. *See* COMPACT.

covering space A covering space for a *topological space X is a topological space C together with a continuous map (*see* CONTINUOUS FUNCTION) $\pi:C \to X$, called the covering map, such that about each point $p \in X$ is an open set U such that $\pi^{-1}(U)$ is a disjoint union of open sets in C which are homeomorphically mapped onto U by π. For example, the real line $\mathbb{R}$ is a covering space for the unit circle S^1 with $\pi:\mathbb{R} \to S^1$ given by $\pi(x) = (\cos x, \sin x)$. The covering map winds $\mathbb{R}$ repeatedly onto S^1. The pre-image of the point $p = (1,0)$ is the set $2\pi\mathbb{Z}$ and so a small arc U around $(1,0)$ exists which has pre-image made up of small disjoint intervals $(2n\pi - \epsilon, 2n\pi + \epsilon)$.

A useful property of covering spaces is that if γ is a path in X there is a unique path Γ in C such that $\gamma = \pi \circ \Gamma$. The path Γ is called the lift of γ.

A universal covering space is a covering space which is *simply connected. So in the example above, $\mathbb{R}$ is a universal covering space for S^1. A universal covering space of X is a covering space for any *connected covering space of X. If a universal covering space exists, it is unique up to *homeomorphism.

Cox, Sir David (1924–) British statistician distinguished for his work in *statistics and applied *probability, especially in relation to industrial and *operational research. His varied contributions include the design and analysis of statistical experiments, analysis of binary data and of point stochastic processes, and the development of new methods in quality control and operational research, particularly in relation to queuing, congestion, and renewal theory.

Coxeter, (Harold Scott MacDonald) 'Donald' (1907–2003) British-born Canadian geometer who made important contributions to the study of *polytopes, *non-Euclidean geometry, and geometric *group theory as well as writing many popular and technical books on geometry.

Cramer, Gabriel (1704–52) Swiss mathematician whose introduction to algebraic curves, published in 1750, contained *Cramer's rule. The rule was known earlier to *Maclaurin.

Cramer's rule Consider a set of n linear equations in n unknowns $x_1, x_2, \dots, x_n$, written in matrix form as $\mathbf{Ax} = \mathbf{b}$. When $\mathbf{A}$ is invertible, the set of equations has a unique solution $\mathbf{x} = \mathbf{A}^{-1}\mathbf{b}$. Since $\mathbf{A}^{-1} = (1/\det\mathbf{A})\,\text{adj}\mathbf{A}$, where $\text{adj}\mathbf{A}$ is the *adjoint of $\mathbf{A}$, this gives the solution

$$\mathbf{x} = \frac{(\text{adj}\,\mathbf{A})\mathbf{b}}{\det\mathbf{A}},$$

which may be written

$$x_j = \frac{b_1 A_{1j} + b_2 A_{2j} + \dots + b_n A_{nj}}{\det\mathbf{A}}$$

$$(j = 1, \dots, n),$$

using the entries of $\mathbf{b}$ and the *cofactors of $\mathbf{A}$. This is Cramer's rule. Computationally Cramer's rule is highly inefficient compared with *Gaussian elimination.

Crelle's Journal One of the earliest mathematical journals, particularly as one not being solely the proceedings of some academy or learned society. It was begun by August Crelle in Berlin in 1826 and published many noteworthy papers, including *Abel's proof that the general *quintic equation is not *solvable by radicals.

critical damping *See* DAMPED OSCILLATIONS.

critical events (critical activities) Events or activities on the *critical path.

critical path A path on an *activity network where any delay will delay the overall completion of the project.

critical path algorithm A *forward scan determines the earliest time for each vertex in an activity network, and a *backward scan determines the latest time for each vertex. The *critical path is any path on which the latest time=the earliest time at every vertex, i.e. on which any delays would mean the delay of the completion of the project.

critical path analysis Suppose that the vertices of a *network represent steps in a process, and the weights on the arcs represent the times that must elapse between steps. Critical path analysis is a method of determining the longest path in the network and hence of finding the least time in which the whole process can be completed.

(⊕) SEE WEB LINKS
- An article including an example of critical path analysis.

critical point A synonym for STATIONARY POINT.

critical region *See* HYPOTHESIS TESTING.

critical value A synonym for STATIONARY VALUE.

cross-correlation The correlation between a pair of *time series variables where the values are paired by occurring at the same time.

crossing number A *knot in three dimensions can be represented, as if laid flat, with various over- and under-crossings. There is a minimum number of crossings for how a given knot can be so represented, for example 3 for the *trefoil. This minimum crossing number is invariant under *isotopy. *See* KNOT THEORY.

cross-multiply Where an equation has *fractions on both sides, it may be simplified by cross-multiplying. In fact both sides are multiplied by the product of the two denominators, so that an equation $a/b = c/d$ rearranges to $ad = bc$.

cross product A synonym for VECTOR PRODUCT.

cross ratio For distinct points z_1, z_2, z_3, z_4 in the *complex plane, their cross ratio equals $\dfrac{(z_1 - z_3)(z_2 - z_4)}{(z_1 - z_4)(z_2 - z_3)}$. The definition in the complex plane can be extended to the *Riemann sphere by allowing any of the z_i to be infinite. The cross ratio in fact can be defined for any distinct four points in a projective line (*see* PROJECTIVE SPACE), the Riemann sphere here being the complex projective line. *See also* HARMONIC RANGE.

crude data A synonym for RAW DATA.

cryptogram *See* CRYPTOGRAPHY.

cryptography The area of mathematics concerning the secure scrambling of a message or information so that, if intercepted, the original message is, in practice, irretrievable without knowledge of the *key. The original message is called the plaintext, and the scrambled message is called the ciphertext or cryptogram. The process of scrambling is called encryption and is achieved via a private or public key. The security of the encryption often relies on the *prime factorization of very large numbers or a similar *one-way function. *See* FREQUENCY ANALYSIS, RSA.

crystallographic groups A planar crystallographic group or wallpaper group is a *group of *isometries of the plane which has a *subgroup of *translations that is *isomorphic to $\mathbb{Z}^2$. (*Compare* FRIEZE GROUPS.)

The symmetry group of the pattern in the figure is one such wallpaper group. It contains translations and *glide reflections but has no non-trivial rotational symmetry. In all there are 17 wallpaper groups (up to isomorphism).

In three dimensions there are 230 crystallographic groups and in four dimensions there are 4783.

cube One of the *Platonic solids, a *polyhedron with 6 square faces, 8 vertices, and 12 edges.

cube root *See* N-TH ROOT.

cube root of unity A complex number z such that $z^3 = 1$. The three cube roots of unity are 1, ω and ω^2, where

$$\omega = e^{2\pi i/3} = \cos \frac{2\pi}{3} + i\sin \frac{2\pi}{3}$$

$$= -\frac{1}{2} + \frac{\sqrt{3}}{2}i,$$

$$\omega^2 = e^{4\pi i/3} = \cos \frac{4\pi}{3} + i\sin \frac{4\pi}{3}$$

$$= -\frac{1}{2} - \frac{\sqrt{3}}{2}i.$$

Properties: (i) $\omega^2 = \overline{\omega}$ (*see* CONJUGATE), (ii) $1 + \omega + \omega^2 = 0$.

cubic polynomial A polynomial of degree three. A cubic equation is a polynomial equation of degree three. *See* SOLVABLE BY RADICALS, TARTAGLIA, VIÈTE'S SUBSTITUTION.

cuboctahedron One of the *Archimedean solids, with 6 square faces and 8 triangular faces. It can be formed by cutting off the corners of a cube to obtain a polyhedron whose vertices lie at the

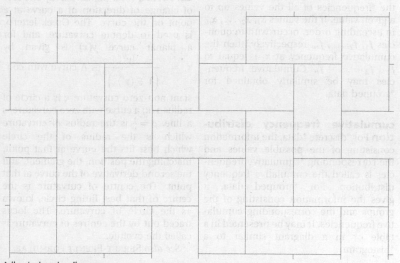

A 'herringbone' wallpaper

midpoints of the edges of the original cube.

cuboid A *parallelepiped all of whose faces are *rectangles.

cumulative distribution function
For a *random variable X, the cumulative distribution function (or cdf) is the function F_X defined by $F_X(x) = \Pr(X \leq x)$. Thus, for a discrete random variable,

$$F_X(x) = \sum_{x_i \leq x} p(x_i),$$

where p_X is the *probability mass function, and, for a continuous random variable, with a *probability density function f_X, then $F_X(x) = \int_{-\infty}^{x} f_X(t) \, dt$.

cumulative frequency The sum of the *frequencies of all the values up to a given value. If the values $x_1, x_2, \ldots, x_n$, in ascending order, occur with frequencies $f_1, f_2, \ldots, f_n$, respectively, then the cumulative frequency at x_i is equal to $f_1 + f_2 + \ldots + f_i$. Cumulative frequencies may be similarly obtained for *grouped data.

cumulative frequency distribution For *discrete *data, the information consisting of the possible values and the corresponding *cumulative frequencies is called the cumulative frequency distribution. For *grouped data, it gives the information consisting of the groups and the corresponding cumulative frequencies. It may be presented in a table or in a diagram similar to a *histogram.

cup The operation $\cup$ (*see* UNION) is read as 'cup'. *Compare* CAP.

curl For a vector function of position $\mathbf{V}(\mathbf{r}) = V_x \, \mathbf{i} + V_y \, \mathbf{j} + V_z \, \mathbf{k}$, the curl of $\mathbf{V}$ is the *vector product of the operator del, $\nabla = \mathbf{i} \frac{\partial}{\partial x} + \mathbf{j} \frac{\partial}{\partial y} + \mathbf{k} \frac{\partial}{\partial z}$ with $\mathbf{V}$ giving curl $\mathbf{V} = \nabla \times \mathbf{V} = \mathbf{i} \times \frac{\partial \mathbf{V}}{\partial x} + \mathbf{j} \times \frac{\partial \mathbf{V}}{\partial y} + \mathbf{k} \times \frac{\partial \mathbf{V}}{\partial z}$ which can be written in determinant form as $\begin{vmatrix} \mathbf{i} & \mathbf{j} & \mathbf{k} \\ \frac{\partial}{\partial x} & \frac{\partial}{\partial y} & \frac{\partial}{\partial z} \\ V_x & V_y & V_z \end{vmatrix}$. If $\mathbf{V} = \nabla \varphi$ then curl$\mathbf{V} = \mathbf{0}$ and if the *domain of $\mathbf{V}$ is *simply connected then curl$\mathbf{V} = \mathbf{0}$ implies $\mathbf{V} = \nabla \varphi$ for a scalar field φ. Curl relates to the local rotation of a field; for example, if $\mathbf{V}$ is the velocity field of a fluid flow, a small ball of fluid will rotate in the direction of curl$\mathbf{V}$, and with an *angular speed half the *magnitude of curl$\mathbf{V}$. *Compare* DIVERGENCE, GRADIENT.

curvature (of a planar curve) The rate of change of direction of a curve at a point on the curve. The Greek letter κ is used to denote curvature, and for a planar curve $y(x)$ is given by $\kappa = \dfrac{|y''|}{\left(1 + (y')^2\right)^{3/2}}$. A curve with constant non-zero curvature κ is a circle of radius $1/\kappa$; a curve with zero curvature is a line. $\rho = \frac{1}{\kappa}$ is the radius of curvature which is the radius of the circle which best fits the curve at that point, matching the position, the gradient, and the second derivative of the curve at that point. The centre of curvature is the centre of that best-fitting circle, known as the circle of curvature. The locus traced out by the centres of curvature is called the evolute.

See also SERRET-FRENET FORMULAE.

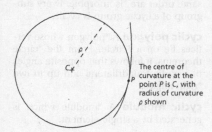

The centre of curvature at the point P is C, with radius of curvature ρ shown

curvature (of a surface) See GAUSSIAN CURVATURE.

curve The *graph of a *continuous function $y = f((x)$ defines a curve, or the variables x, y might be implicitly related as in $x^2 + xy + y^2 = 1$, which defines an *ellipse, or a curve might be defined parametrically: $x = \cos t$, $y = \sin t$ for $0 \le t < 2\pi$ defines a *circle. Classically, curves were defined by their *geometry, but, with the introduction of *Cartesian coordinates, they are easily defined by equations constraining the coordinates. Most generally, the term might refer to a 1-dimensional *manifold, but it also refers to curves with *singularities, such as $y^2 = x^3$, which has a *cusp at the origin, or $y^2 = x^2(x + 1)$, which has a node at the origin. If the variables x and y are *complex numbers, then the *Riemann surfaces defined may be referred to as curves, even though they have 2 real dimensions. See also JORDAN CURVE, PEANO CURVE.

curve sketching When a graph $y = f(x)$ is to be sketched, what is generally required is a sketch showing the general shape of the curve and the behaviour at points of special interest. It is normal to investigate the following: *symmetry, *stationary points, intervals in which the function is always increasing or always decreasing, *asymptotes (vertical,

horizontal, and slant), *concavity, *points of inflexion, and points of intersection with the *axes.

curvilinear coordinates For three commonly used coordinate systems we note

Cartesian coordinates: $\mathbf{r} = x\mathbf{i} + y\mathbf{j} + z\mathbf{k}$, $d\mathbf{r} = dx\mathbf{i} + dy\mathbf{j} + dz\mathbf{k}$.

Cylindrical polar coordinates: $\mathbf{r} = (r\cos\theta, \quad r\sin\theta, z) \quad d\mathbf{r} = dr\mathbf{e}_r + rd\theta\mathbf{e}_\theta + dz\mathbf{e}_z$ where $\mathbf{e}_r = (\cos\theta, \sin\theta, 0)$, $\mathbf{e}_\theta = (-\sin\theta, \cos\theta, 0)$, $\mathbf{e}_z = (0, 0, 1)$.

Spherical polar coordinates: $\mathbf{r} = (r\sin\theta\cos\varphi, r\sin\theta\sin\varphi, r\cos\theta)$, $d\mathbf{r} = dr\mathbf{e}_r + rd\theta\mathbf{e}_\theta + r\sin\theta d\varphi\mathbf{e}_\varphi$ where $\mathbf{e}_r = (\sin\theta\cos\varphi, \sin\theta\sin\varphi, \cos\theta)$, $\mathbf{e}_\theta = (\sin\theta\cos\varphi, \sin\theta\sin\varphi, \cos\theta)$, $\mathbf{e}_\varphi = (-\sin\varphi, \cos\varphi, 0)$.

Denoting the coordinates u_1, u_2, u_3, we see in each case that

$$d\mathbf{r} = h_1 du_1\mathbf{e}_1 + h_2 du_2\mathbf{e}_2 + h_3 du_3\mathbf{e}_3,$$

where each $h_i > 0$ and $\mathbf{e}_1, \mathbf{e}_2, \mathbf{e}_3$ is a *right-handed *orthonormal basis. Such coordinates are called orthogonal curvilinear coordinates, and general expressions for *div, *grad, and *curl exist for such coordinates. (See APPENDIX 16.) In particular the *Jacobian equals $h_1 h_2 h_3$.

cusp A *singular point of a *curve, which has a repeated tangent line. A classic example is as $y^2 = x^3$, which has a cusp at $(0,0)$. Approximating, up to second order terms around $(0,0)$, gives the equation $y^2 = 0$, and so $y = 0$ is a repeated tangent line at the origin.

cut (in a network) A set of edges whose removal would create two disconnected components with the *source in one section and the *sink in the other.

cut plane The *complex plane with a cut, usually a half-line, removed. See BRANCH.

cut point A cut point x of a *connected *metric (or *topological) space X is such that $X \backslash \{x\}$ is disconnected. Thus, the interval $(0,1)$ is not homeomorphic to $(0,1]$ as every point of the former space is a cut point, but 1 is not a cut point of the latter.

cycle (in graph theory) A *closed path with at least one edge. In a *graph, a cycle is a sequence $v_0, e_1, v_1, \ldots, e_k, v_k$ $(k \geq 1)$ of alternately vertices and edges (where e_i is an edge joining v_{i-1} and v_i), with all the edges different and all the vertices different, except that $v_0 = v_k$. *See* HAMILTONIAN GRAPH, TREE.

cycle (permutations) A *permutation f of a set S is a cycle if there are elements $s_1, s_2, \ldots, s_n$ such that $f(s_1) = s_2, f(s_2) = s_3, \ldots, f(s_{n-1}) = s_n, f(s_n) = s_1$ and f fixes all other elements. n is called the length of f and equals the *order of f as an element of the *symmetry group of S. The cycle f is denoted as $(s_1\, s_2 \ldots s_n)$.

cycle decomposition Any *permutation of a finite set may be written as a product of *disjoint *cycles. This expression is unique up to the order of the cycles, and cycling terms within cycles. For example, the permutation of $\{1, \ldots, 8\}$ given by $1 \mapsto 7,\ 2 \mapsto 5,\ 3 \mapsto 2,\ 4 \mapsto 8,\ 5 \mapsto 3,\ 6 \mapsto 1,\ 7 \mapsto 6,\ 8 \mapsto 4$ can be written as $(176)(253)(48)$ or equally as $(532)(84)(761)$. The *order of a permutation is the *least common multiple of the length of its cycles, here $\mathrm{lcm}(3,3,2) = 6$.

cyclic data *See* DIRECTIONAL DATA.

cyclic group Let a be an element of a group G. The elements a^r, where r is an integer, form a *subgroup of G, called the *subgroup generated by a and denoted $\langle a \rangle$. A group G is cyclic if there is an element a in G such that $\langle a \rangle = G$. If G is a finite cyclic group with *identity e, then $G = \{e, a, a^2, \ldots, a^{n-1}\}$, where n is the *order of a. If G is an infinite cyclic group, then $G = \{\ldots, a^{-2}, a^{-1}, e, a, a^2, \ldots\}$. Any two cyclic groups of the same order are *isomorphic. Every subgroup of a cyclic group is cyclic.

cyclic polygon A *polygon whose vertices lie on a *circle. From the *circle theorems, it follows that opposite angles of a cyclic *quadrilateral add up to two *right angles.

cyclic module A *module which is generated by a single element.

cyclic redundancy checks *Error-detecting codes used in networks and storage devices to detect accidental corruption of *data. Blocks of data have a check value inserted at the end of the block. Since these check values add no information to the message, they are *redundant, and the value is based on cyclic coding; hence the name.

cyclic vector Given a *linear map T: $V \rightarrow V$ of an n-dimensional *vector space, a vector v is cyclic if $v, Tv, T^2v, \ldots, T^{n-1}v$ is a basis for V. The *matrix for T with respect to this basis is the *companion matrix of the *minimal polynomial of T.

cycling The behaviour of an *iterative method when a sequence of values repeats itself with some *period.

cycloid The curve traced out by a point P on the circumference of a circle of radius a that rolls without slipping along a straight line. The cycloid has *parametric equations $x = a(t - \sin t)$, $y = a(1 - \cos t)$ $(t \in \mathbb{R})$. In the figure, $OA = 2\pi a$. (*See* BRACHISTOCHRONE.)

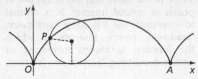

Tracing a cycloid

cyclotomic polynomial The nth cyclotomic polynomial $\Phi_n(z)$ is the *monic polynomial whose roots are the *primitive n-th roots of unity. $\Phi_4(z) = z^2 + 1$ as i and $-i$ are the primitive fourth roots of unity.

When n is prime

$$\Phi_n(z) = z^{n-1} + z^{n-2} + \ldots + z + 1$$

cylinder A cylinder is a *surface with a circular base, a circular top of the same size, and the curved surface formed by the vertical line segments joining them. For a cylinder with base of radius r and height h, the cylinder's interior has volume $\pi r^2 h$, and the area of the curved surface equals $2\pi rh$.

A right circular cylinder

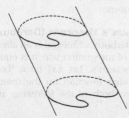

A cylinder over a curve

More generally, a cylinder is a surface, consisting of the points of the lines, called generators, drawn through the points of a fixed curve and parallel to a fixed line. The generators may be extended indefinitely in both directions. *Compare* PRISM.

cylindrical polar coordinates A *coordinate system for 3-dimensional space which is a useful alternative to *Cartesian coordinates when describing a problem with a *line of symmetry. With the z-axis as that line of symmetry, cylindrical polar coordinates are related to Cartesian coordinates by

$$x = \rho \cos\varphi, \quad y = \rho \sin\varphi, \quad z = z.$$

Conversely, the cylindrical polar coordinates can be found from (x, y, z) by:

$$\rho = \sqrt{x^2 + y^2}, \quad \cos\varphi = \frac{x}{\sqrt{x^2 + y^2}},$$

$$\sin\varphi = \frac{y}{\sqrt{x^2 + y^2}}, \quad z = z$$

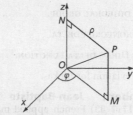

Cylindrical polar coordinates of P

cylindroid A *cylinder with *elliptical cross-section.

cypher *See* CIPHER.

d The symbol used to denote the differential operator. So if y is a real function of x then $y' = \frac{dy}{dx}$ is the *differential of y with respect to x. Since it is $\frac{d}{dx}(y)$, when the second differential is taken, it is written as $\frac{d^2y}{dx^2}$ coming from $\frac{d}{dx}\left(\frac{dy}{dx}\right)$.

See also EXTERIOR DERIVATIVE.

d Abbreviation for *deci-.

d *See* DIVISOR FUNCTION.

D The Roman *numeral for 500. The number 13 in hexadecimal notation.

D$_{2n}$ *See* DIHEDRAL GROUP.

δ$_{ij}$ *See* KRONECKER DELTA.

δ(x) *See* DIRAC DELTA FUNCTION.

da Abbreviation for *deka-.

D'Alembert, Jean-Baptiste le Rond (1717–83) French applied mathematician and music theorist, who derived and solved the *wave equation. The *ratio test is also due to him.

damped oscillations *Oscillations in which the amplitude decreases with time. Consider the equation of motion $m\ddot{x} = -kx - c\dot{x}$, where the first term on the right-hand side arises from an elastic restoring force satisfying *Hooke's law, and the second term arises from a *resistive force. The constants k and c are positive. The form of the general solution of this *linear differential equation depends on the auxiliary equation $ma^2 + ca + k = 0$. When $c^2 < 4mk$, the auxiliary equation has non-real roots, and damped oscillations occur. This is a case of weak damping. When $c^2 = 4mk$, the auxiliary equation has equal roots and critical damping occurs: oscillation just fails to take place. When $c^2 > 4mk$ there is strong damping: the resistive force is so strong that no oscillations take place.

Darboux's theorem Let $f(x)$ be a *real function which is continuous (*see* CONTINUOUS FUNCTION) on the interval $[a,b]$ and *differentiable on the interval (a,b). If $f'(a) < K < f'(b)$ then there exists c in (a,b) such that $f'(c) = K$. This may appear to be a consequence of the *intermediate value theorem, but no assumption is made that that $f'(x)$ is itself continuous.

Darboux's theorem (Darboux integral) Darboux's *integral is an alternative theory of integration which is equivalent to *Riemann's. Let $f(x)$ be a *bounded real-valued function on the interval $[a, b]$. Then both of the following are well defined

$$\int_{\underline{a}}^{b} f(x)\ dx = \sup\left\{\int_{a}^{b} \varphi(x)dx \mid \varphi \text{ is a step function}, \varphi \leq f\right\}.$$

$$\int_{a}^{\overline{b}} f(x)\ dx = \inf\left\{\int_{a}^{b} \varphi(x)dx \mid \varphi \text{ is a step function}, \varphi \geq f\right\}.$$

These are respectively referred to as the lower and upper Riemann or Darboux integrals of f. Then f is said to be Darboux integrable if the lower and upper integrals are equal, and this is equivalent to being Riemann integrable.

The *Dirichlet function on $[a,b]$ has lower integral 0 and upper integral $b-a$ and so is not Darboux or Riemann integrable.

dashpot A device consisting of a cylinder containing a liquid through which a piston moves, used for damping vibrations.

data The *observations gathered from an experiment, survey or observational study. Often the data are a randomly selected *sample from an underlying *population. Numerical data are discrete if the underlying population is finite or *denumerable and are continuous if the underlying population forms an *interval, bounded or unbounded. Data are nominal if the observations are not numerical or quantitative but are descriptive and have no natural order. Data specifying country of origin, type of vehicle, or subject studied, for example, are nominal.

Note that the word 'data' is plural. The singular 'datum' may be used for a single observation.

data logging The process of automatically collecting and storing observations, typically as a sequence in time or space. For example, in hospital an electronic monitor may record a patient's temperature, heart rate, blood pressure, etc. at regular intervals; and in aircraft, the flight data recorders, or 'black box', constantly monitor and record data on all the aircraft's systems.

data science The interdisciplinary study and investigation of big data (extremely large data sets), involving theory and methods from mathematics, *statistics, *computer science, and *machine learning, in particular where classical techniques are not able to deal with the extreme size of the data sets. Large data sets have become increasingly prevalent and important in the information age, for example in the social sciences, behavioural sciences, genomics,

complex scientific experiements, and via the Internet. *See* E-SCIENCE, TOPOLOGICAL DATA ANALYSIS.

deca- A synonym for DEKA-.

decagon A ten-sided polygon.

deceleration A reduction in *speed. If a particle is moving in a line in the positive direction, then it experiences a deceleration when the *acceleration is negative.

deci- Prefix used with *SI units to denote multiplication by 10^{-1}. Abbreviated as d.

decidable In *logic, a proposition which can be shown to be true or false is decidable. In *computability, a *predicate is decidable if there is an *algorithm which can decide, in each possible case, on the truth or falsity of the predicate. If an algorithm exists which decides on the truth of the predicate in finite time but potentially runs forever if the predicate is false, the predicate is called semidecidable.

decile *See* QUANTILE.

decimal expansion A synonym for DECIMAL REPRESENTATION

decimalize To change a measurement system to one based on multiples of 10. In 1971 the British currency was decimalized from £1 = 20 shillings and 1 shilling = 12 pence to £1 = 100 (new) pence (the 'new' has since been dropped).

decimal places In *rounding or *truncation of a number to n decimal places, the original is replaced by a number with just n digits after the decimal point. When the rounding takes place to the left of the decimal point, a phrase such as 'to the nearest 10' or 'to the nearest 1000' is used. To say that $a = 1.9$ to 1 decimal place means that the exact value of a becomes 1.9 after rounding to 1 decimal place, and so $1.85 \leq a <$

1.95. 'Decimal places' is often abbreviated to 'dp'.

decimal point The separator used between the integer and fractional parts of a number expressed in *decimal representation. In the UK a dot is used, while in much of Europe a comma is used as the separator.

decimal representation (decimal expansion) Any real number a between 0 and 1 has a decimal representation, written $0.d_1 d_2 d_3 \ldots$, where each d_i is one of the digits $0, 1, 2, \ldots, 9$; this means that

$$a = d_1 \times 10^{-1} + d_2 \times 10^{-2} + d_3 \times 10^{-3} + \ldots.$$

This notation can be extended to enable any positive real number to be written as

$$c_n c_{n-1} \ldots c_1 c_0 . d_1 d_2 d_3 \ldots$$

using, for the integer part, the normal representation $c_n c_{n-1} \ldots c_1 c_0$ to base 10 (*see* BASE). If, from some stage on, the representation consists of the repetition of a string of one or more digits, it is called a recurring or repeating decimal. For example, the recurring decimal $0.12748748748 \ldots$ can be written as $0.12\overline{748}$ or as $0.12\dot{7}4\dot{8}$, where the dots above indicate the beginning and end of the repeating string. If the repeating string consists of a single zero, this is generally omitted and the representation is called a terminating decimal.

The decimal representation of any real number is unique except for terminating decimals which have two expansions, as with 0.25 and 0.24$\dot{9}$. The numbers that can be expressed as recurring (including terminating) decimals are precisely the *rational numbers.

decision analysis The branch of mathematics considering strategies to be used when decisions have to be made at stages in a process but the outcomes resulting from the implementation of those decisions are dependent on chance.

decision nodes *See* EMV ALGORITHM.

decision problem The decision problem, or *Entscheidungsproblem*, was posed by *Hilbert in 1928, asking whether there is an *algorithm that can determine whether a given statement is true using given axioms. A negative answer was given independently to the problem by *Church and *Turing in 1936.

decision theory The area of *statistics and *game theory concerned with decision making under uncertainty to maximize expected utility.

decision tree The diagram used to represent the process in a *decision analysis problem. Different symbols are used to denote the different types of node or vertex. For example, decisions may be shown as rectangles, chance events as circles, and *payoffs as triangles.

decision variables The quantities to be found in *linear programming or other *constrained optimization problems.

declination *See* DEPRESSION.

decompose To give a *decomposition of a number or other quantity.

decomposition The breakdown of a quantity or expression into simpler components. For example, $24 = 2 \times 2 \times 2 \times 3$ is the decomposition of 24 into its *prime factors or the expression of a polynomial as a product of factors. More generally, decomposition might refer to the breakdown of a space into significant components, for example a *vector space written as a *direct sum of *eigenspaces.

decreasing function A *real function f is decreasing on an interval I if $f(x_1) \geq f(x_2)$ whenever x_1 and x_2 are in I with $x_1 \leq x_2$. Also, f is strictly decreasing if $f(x_1) > f(x_2)$ whenever $x_1 < x_2$.

decreasing sequence A sequence a_1, $a_2, a_3, \ldots$ is said to be decreasing if $a_i \geq a_j$ whenever $i \leq j$, and strictly decreasing if $a_i > a_j$ whenever $i < j$. *See also* NESTED SETS.

Dedekind, (Julius Wilhelm) Richard (1831–1916) German mathematician who developed a formal construction of the *real numbers from the *rational numbers by means of the so-called Dedekind cut. This new approach to *irrational numbers, contained in the very readable paper *Continuity and Irrational Numbers*, was an important step towards the formalization of mathematics. He made significant contributions in *abstract algebra, defining the notions of *ring and *field. He also proposed a definition of infinite sets that was taken up by *Cantor, with whom he developed a lasting friendship.

Dedekind cut A means of constructing the *real numbers from the *rational numbers, introduced by Richard *Dedekind. A Dedekind cut is a subset A of $\mathbb{Q}$ such that:
(i) $\varnothing \neq A \neq \mathbb{Q}$.
(ii) if $x < y$ are rational numbers and $y \in A$, then $x \in A$.
(iii) if $x \in A$, then there exists $y \in A$ such that $x < y$.

A Dedekind cut A then represents the real number that is its *supremum but without having to make reference to real numbers. *Order can be defined by $A \leq B$ if and only if $A \subseteq B$ and *addition by $A + B = \{a + b : a \in A, b \in B\}$.

The other arithmetic operations can also be defined but are a little more involved as care must be taken with signs.

deduction The process of reasoning from *axioms, premises, or assumptions, using accepted logical steps. The term is also used to refer to the outcome of such a reasoning process.

deficient number An integer that is larger than the sum of its positive divisors (not including itself). Any prime number must be deficient by definition but so is any power of any prime, as well as numbers like 10 whose divisors are 1, 2, 5 giving a sum of 8. *See* ABUNDANT NUMBER, PERFECT NUMBER.

definite integral *See* INTEGRAL.

deformation In solid *mechanics, when *external forces act on a body, its configuration may deform from its initial state to a different state. If the deformation is *elastic, the body will return to its initial configuration when the forces are removed. *See* STRAIN, STRESS.

degenerate Where a family of entities can be defined in terms of parameters, and the limiting case produces an entity which is of a different nature. For example, the general quadratic function is in the form $y = \alpha x^2 + \beta x + \gamma$, and the graph is a *parabola. As α decreases the curvature of the parabola steadily decreases, and the limiting case as $\alpha \to 0$ is a straight line, which is a degenerate parabola.

degenerate conic A *conic that consists of a pair of (possibly coincident) straight lines. The equation $ax^2 + 2hxy + by^2 + 2gx + 2fy + c = 0$ represents a degenerate conic if $\Delta = 0$, where

$$\Delta = \begin{vmatrix} a & h & g \\ h & b & f \\ g & f & c \end{vmatrix}.$$

degenerate quadric The *quadric with equation $ax^2 + by^2 + cz^2 + 2fyz + 2gzx + 2hxy + 2ux + 2vy + 2wz + d = 0$ is degenerate if $\Delta = 0$, where

$$\Delta = \begin{vmatrix} a & h & g & u \\ h & b & f & v \\ g & f & c & w \\ u & v & w & d \end{vmatrix}.$$

The non-degenerate quadrics are the *ellipsoid, the *hyperboloid of one sheet, the *hyperboloid of two sheets, the *elliptic paraboloid, and the *hyperbolic paraboloid.

degree (angular measure) The method of measuring angles in degrees dates back to Babylonian mathematics around 2000 BC. A complete revolution is divided into 360 degrees (°); a right angle measures 90°. Each degree is divided into 60 minutes (′) and each minute into 60 seconds (″). In more advanced work, angles are measured in *radians, which are better suited to *calculus and *Taylor series.

degree (of a map) A *continuous function from a *circle to itself wraps the circle around itself a certain *integer number of times. This is the degree of the map. A negative integer relates to the circle being wrapped in the opposite direction. This fact also shows the *fundamental group of a circle is ℤ. *Brouwer extended the notion of degree to continuous maps between *compact *oriented *manifolds of equal dimensions, and again the number characterizes how many times the domain manifold wraps around the codomain manifold.

degree (of a polynomial) *See* POLYNOMIAL.

degree (of a vertex of a graph) The degree of a vertex V of a *graph is the number of edges ending at V. (If loops are allowed, each loop joining v to itself contributes two to the degree of V.)

In the first graph, the vertices U, V, W and X have degrees 2, 2, 3 and 1. The second graph has vertices V_1, V_2, V_3 and V_4 with degrees 5, 4, 6 and 5.

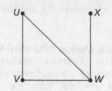

A graph

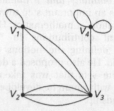

A *multigraph

degree-genus formula A *nonsingular curve in the complex projective plane (*see* PROJECTIVE SPACE) defined by a degree d equation is a *Riemann surface of genus

$$g = \frac{1}{2}(d-1)(d-2).$$

If $d = 1$ or 2, then $g = 0$, and the curve gives the *Riemann sphere; and if $d = 3$, then $g = 1$, an *elliptic curve which is topologically a *torus.

degrees of freedom (in mechanics) The number of degrees of freedom of a body is the minimum number of independent coordinates required to describe the position of the body at any instant, relative to a frame of reference. A particle in straight-line motion or circular motion has one degree of freedom. So too does a *rigid body rotating about a fixed axis. A particle moving in a plane, such as a projectile, or a particle moving on a cylindrical or spherical surface has two degrees of freedom. A rigid body in general motion has six degrees of freedom, say, three to specify the position of the *centre of gravity and three more to orient the body.

degrees of freedom (in statistics) A positive integer normally equal to the number of independent *observations in a *sample minus the number of population parameters to be estimated from the sample. When the *chi-squared test is applied to a *contingency table with h rows and k columns, the number of degrees of freedom equals $(h-1)(k-1)$.

For a number of distributions, the number of degrees of freedom is required to identify which of a family of distributions is to be used. The *chi-squared distribution and *t-distribution each have a single degrees of freedom parameter, and the *F-distribution has two such parameters.

deka- (deca-) Prefix used with *SI units to denote multiplication by 10. Abbreviated as da.

del *See* DIFFERENTIAL OPERATOR.

Delian (altar) problem Another name for the problem of the *duplication of the cube. In 428 BC, an oracle at Delos ordered that the altar of Apollo should be doubled in volume as a means of bringing a certain plague to an end.

delta The Greek letter d, written δ, capital Δ. A small increment in the value of the variable x is usually denoted by δx. When y is a function of x, $\dfrac{\delta y}{\delta x}$ represents the average rate of change of y with respect to x at a point and the derivative of y with respect to x is defined as the limit of that ratio as $\delta x \to 0$. So

$$y' = \frac{dy}{dx} = \lim_{\delta x \to 0} \frac{\delta y}{\delta x}.$$

delta function *See* DIRAC DELTA FUNCTION.

demography In *statistics, the study of human populations, in particular focusing on birth and date rates, age, movement of people, employment, education, ethnicity, religion, etc.

De Moivre, Abraham (1667–1754) Prolific mathematician, born in France, who later settled in England. In *De Moivre's Theorem, he is remembered for his use of *complex numbers in *trigonometry. But he was also the author of two notable early works on *probability. His Doctrine of Chances of 1718, examines numerous problems and develops a number of principles, such as the notion of independent events and the product law. The later work contains the result known as *Stirling's formula and probably the first use of the normal frequency curve.

De Moivre's Theorem From the definition of *multiplication (of a *complex number), it follows that $(\cos\theta_1 + i\sin\theta_1)(\cos\theta_2 + i\sin\theta_2) = \cos(\theta_1+\theta_2) + i\sin(\theta_1+\theta_2)$. This leads to the following result known as De Moivre's Theorem, which is crucial to calculating the powers z^n of a complex number z: **Theorem** For any integer n,

$$(\cos\theta + i\sin\theta)^n = \cos n\theta + i\sin n\theta.$$

De Morgan, Augustus (1806–71) British mathematician and logician who was responsible for developing a more symbolic approach to *algebra and who played a considerable role in the beginnings of *symbolic logic. His name is remembered in *De Morgan's laws, which he formulated. In an article of 1838, he clarified the notion of *mathematical induction.

De Morgan's laws (set theoretic version) For sets A and B, $(A \cup B)' = A' \cap B'$ and $(A \cap B)' = A' \cup B'$, where $'$ denotes *complement. These are De Morgan's laws. These laws extend naturally to more than two sets:

$$\left(\bigcup_{i \in I} A_i\right)' = \bigcap_{i \in I} A_i',$$
$$\left(\bigcap_{i \in I} A_i\right)' = \bigcup_{i \in I} A_i'.$$

De Morgan's laws (logical version) These are the logical equivalent of the set-theoretic version. They capture the fact that for it not to be true that a family of statements all hold only one need be false. Given a family of statements P_i, where $i \in I$, then

$$\neg(\exists i \in I P_i) \Leftrightarrow \forall i \in I(\neg P_i),$$
$$\neg(\forall i \in I P_i) \Leftrightarrow \exists i \in I(\neg P_i).$$

denominator See FRACTION.

dense matrix A matrix which has a high proportion of non-zero entries. *Compare* SPARSE MATRIX.

dense set A set S in a *topological space X is dense if its *closure is X. Informally, this means that every point in X is either in S or arbitrarily close to members of S. The set of *rational numbers is dense in the space of *real numbers. This can be appreciated by considering the truncated *decimal representations of a real number.

density The average density of a body is the ratio of its *mass to its *volume. In general, the density of a body may not be constant throughout the body. The density at a point P, denoted by $\rho(P)$, is equal to the limit as $\Delta V \to 0$ of $\Delta m/\Delta V$, where ΔV is the volume of a small region containing P and Δm is the mass of the part of the body occupying that small region.

Consider a rod of length l, with density $\rho(x)$ at the point a distance x from one end of the rod. Then the mass m of the rod is given by

$$m = \int_0^l \rho(x)\mathrm{d}x$$

denumerable A set X is denumerable if there is a *one-to-one correspondence

between X and the set of *natural numbers. This is equivalent to *countable and infinite. It can be shown that the set of *rational numbers is denumerable but that the set of *real numbers is not.

dependent equations A set of linear equations where at least one of the set may be expressed as a *linear combination of the others.

dependent events See INDEPENDENT EVENTS.

dependent variable (statistics) The variable which is thought might be influenced by certain other *explanatory variables. In *regression, a relationship is sought between the dependent variable and the explanatory variables. The purpose is normally to enable the value of the dependent variable to be predicted from given values of the explanatory variables.

dependent variable See DIFFERENTIAL EQUATION, FUNCTION.

depression The acute angle between the horizontal and a given line (*see* INCLINATION) measured positively in the direction of the downwards vertical. θ is the angle of depression of the boat from the top of the cliff. Sometimes also known as the angle of declination.

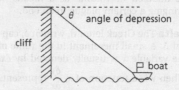

derangement A *rearrangement (or *permutation) in which no object returns to its original place. By the *inclusion-exclusion principle, there are

$$n!\left(1 - \frac{1}{2!} + \frac{1}{3!} - \ldots + \frac{(-1)^{n+1}}{n!}\right)$$

derangements of a set of n elements.

derivative For the *real function f, if $(f(a+h)-f(a))/h$ has a *limit as $h \to 0$, this limit is the derivative of f at a and is denoted by $f'(a)$. (The term 'derivative' may also be used loosely for the *derived function.)

Consider the graph $y = f(x)$. If (x, y) are the coordinates of a general point P on the graph, and $(x + \Delta x, y + \Delta y)$ are those of a nearby point Q on the graph, the quotient $\Delta y / \Delta x$ is the gradient of the chord PQ. Also, $\Delta y = f(x + \Delta x) - f(x)$. So the derivative of f at x is the limit of the quotient $\Delta y / \Delta x$ as $\Delta x \to 0$. This limit can be denoted by dy/dx, which is thus an alternative notation for $f'(x)$. The notation y' is also used.

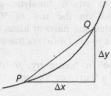

PQ's gradient approximating the curve's

The derivative $f'(a)$ gives the gradient of the curve $y = f(x)$, and hence the gradient of the *tangent to the curve, at $(a, f(a))$. If x is a function of t, where t denotes time, then the derivative dx/dt is referred to as a *rate of change and may be denoted by $\dot{x}$. The *modulus function $f(x) = |x|$ is an example of a function which does not have a *well-defined derivative at $x = 0$ (see also BLANCMANGE FUNCTION). For the derivatives of certain common functions, see APPENDIX 7. See also DIFFERENTIATION, HIGHER DERIVATIVE, LEFT AND RIGHT DERIVATIVE, PARTIAL DERIVATIVE.

derivative test A test to determine the nature of *stationary points of a *function, which uses one or more *derivatives of the function. If $f'(\alpha) = 0$, so there is a stationary point at $x = \alpha$, the first derivative test considers the signs of $f'(x)$ at $x = \alpha^+$, α^-, i.e. whether the gradient is positive, negative, or zero as x approaches the stationary point from above and from below. The various possible combinations are shown below.

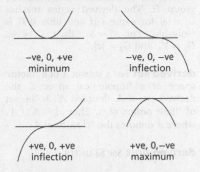

−ve, 0, +ve
minimum

−ve, 0, −ve
inflection

+ve, 0, +ve
inflection

+ve, 0, −ve
maximum

The second derivative test considers the sign of $f''(x)$ at $x = \alpha$. If it is positive, the value of $f'(x)$ is increasing as the function goes through $x = \alpha$, which requires the stationary point to be a *minimum; if $f''(\alpha) < 0$, then it must be a *maximum, while if $f''(\alpha) = 0$, it may be that $x = \alpha$ is any of a minimum $f(x) = (x-\alpha)^4$, a maximum $f(x) = -(x-\alpha)^4$, or a *point of inflection $f(x) = (x-\alpha)^3$. To determine the nature in this case, either the first derivative test may be used or higher derivatives can be considered until the first non-zero higher derivative is found.

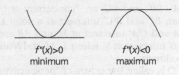

$f''(x) > 0$
minimum

$f''(x) < 0$
maximum

See STATIONARY POINT (in two variables).

derived function The function f', where $f'(x)$ is the *derivative of f at x, is the derived function of f. See also DIFFERENTIATION.

derived series The derived series of a
*group G is defined recursively by $G_0 = G$ and $G_{i+1} = [G_i, G_i]$ where $[H,H]$ denotes the *commutator subgroup of a
group H. The derived series reaches
$G_i = \{e\}$ for some i if and only if G is
*solvable. For $G = S_4$, then $G_1 = A_4$,
$G_2 = V_4$, and $G_3 = \{e\}$.

derived set For a subset A of a *metric
space or a *topological space X, the
derived set of A, denoted A', is the set
of *limit points of A. Then $\bar{A} = A \cup A'$,
where $\bar{A}$ denotes the *closure of A.

derived unit *See* SI UNITS.

Desargues, Girard (1591–1661)
French mathematician and engineer
whose work on *conics and the result
known as *Desargues's Theorem were
to become the basis of the subject
known as *projective geometry. His
1639 book was largely ignored, partly
because of the obscurity of the language.
It was nearly 200 years later that projective geometry developed and the beauty
and importance of his ideas were
recognized.

Desargues' Theorem Consider two
triangles ABC and $A'B'C'$ lying in the
(projective) plane.

The theorem states that if the lines
AA', BB', and CC' are *concurrent at O,
then BC and $B'C'$ intersect at a point L,
CA and $C'A'$ intersect at M, and AB and
$A'B'$ intersect at N, where L, M, and N are
*collinear.

O is called the centre of *perspectivity
of the triangles ABC and $A'B'C'$ and the
line LMN is called the axis of perspectivity. The *dual of Desargues's Theorem is
then its *converse.

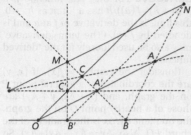
The centre and axis of a perspectivity

Descartes, René (1596–1650) French
philosopher and mathematician who in
mathematics is known mainly for his
methods of applying *algebra to *geometry, from which *analytic geometry
developed via the use of *Cartesian
coordinates, the nascent ideas of which
date back to his *La Géométrie* of 1637.

describe In *geometry, to draw the
shape of a *curve or *figure.

descriptive geometry The area of
mathematics where 3-dimensional shapes
are projected onto a *plane so that spatial
problems can be analysed graphically.

descriptive statistics The part of the
subject of *statistics concerned with
describing the basic statistical features
of a set of *observations. Simple numerical summaries, using notions such as
*mean, *range, and *standard deviation,
together with appropriate diagrams such
as *histograms, are used to present an
overall impression of the data.

design methods *See* EXPERIMENTAL
DESIGN METHODS.

destinations (in transportation problems) *See* TRANSPORTATION PROBLEM.

detachment, law of (in logic) A synonym for MODUS PONENS. *Compare* SYLLOGISM.

determinant For the *square matrix **A**, the determinant of **A**, denoted by det**A** or |**A**|, can be defined as follows.

The determinant of the 1×1 matrix $[a]$ is simply equal to a. If **A** is the 2×2 matrix below, then det**A** $= ad-bc$, and the determinant can also be written as shown:

$$\mathbf{A} = \begin{bmatrix} a & b \\ c & d \end{bmatrix}, \quad \det\mathbf{A} = \begin{vmatrix} a & b \\ c & d \end{vmatrix}.$$

If **A** is a 3×3 matrix $[a_{ij}]$, then det **A**, which may be denoted by

$$\begin{vmatrix} a_{11} & a_{12} & a_{13} \\ a_{21} & a_{22} & a_{23} \\ a_{31} & a_{32} & a_{33} \end{vmatrix},$$

is given by

$$\det \mathbf{A} = a_{11}\begin{vmatrix} a_{22} & a_{23} \\ a_{32} & a_{33} \end{vmatrix} - a_{12}\begin{vmatrix} a_{21} & a_{23} \\ a_{31} & a_{33} \end{vmatrix}$$

$$+ a_{13}\begin{vmatrix} a_{21} & a_{22} \\ a_{31} & a_{32} \end{vmatrix}.$$

Notice how each 2×2 determinant occurring here is obtained by deleting the *row and *column containing the *entry by which the 2×2 determinant is multiplied. This expression for the determinant of a 3×3 matrix can be written $a_{11}A_{11} + a_{12}A_{12} + a_{13}A_{13}$, where A_{ij} is the *cofactor of a_{ij}. This is the evaluation of det**A**, 'by the first row'. In fact, det**A** may be found by using evaluation by any row or column: for example, $a_{31}A_{31} + a_{32}A_{32} + a_{33}A_{33}$ is the evaluation by the third row, and $a_{12}A_{12} + a_{22}A_{22} + a_{32}A_{32}$ is the evaluation by the second column. The determinant of an $n\times n$ matrix **A** may be defined similarly, as $a_{11}A_{11} + a_{12}A_{12} + \ldots + a_{1n}A_{1n}$, and the same value is obtained using a similar evaluation by any row or column. These are

known as the *Laplace expansions of the determinant. However, calculating determinants this way is laborious; using *elementary operations is much more efficient. The following properties hold:

(i) If two rows (or two columns) of a square matrix **A** are identical, then det**A** $= 0$.

(ii) If two rows (or two columns) of a square matrix **A** are interchanged, then only the sign of det**A** is changed.

(iii) The value of det**A** is unchanged if a multiple of one row is added to another row, or if a multiple of one column is added to another column.

(iv) If **A** and **B** are square matrices of the same order, then det(**AB**)= (det**A**) (det**B**).

(v) If **A** is *invertible, then det(**A**$^{-1}$)= (det**A**)$^{-1}$.

(vi) If **A** is an $n\times n$ matrix, then det $k\mathbf{A} = k^n$det**A**.

(vii) det**A**^{T} = det**A**, where **A**^{T} is the *transpose of **A**.

(viii) For an $n\times n$ matrix **A**, the map **x** $\mapsto$ **Ax** scales area/volume by |det**A**| and is *sense-preserving if det**A** > 0 and sense-reversing if det**A** < 0.

determinantal rank The determinantal rank of a *matrix is the order of the largest *square invertible *(see* INVERSE MATRIX*)* *submatrix. It equals the *rank of the matrix.

determine Conditions which are sufficient to specify a mathematical entity uniquely. For example, two points determine a line, as does one point and a gradient in two dimensions.

deterministic model (deterministic system) In contrast with a stochastic model *(see* STOCHASTIC PROCESS*)*, which includes random aspects, a deterministic *mathematical model evolves entirely in a determined fashion, given initial conditions.

developable surface A *surface that can be rolled out flat onto a plane without any distortion. Examples are the *cone and *cylinder. This is equivalent to the surface having zero *Gaussian curvature. On any map of the world, distances are inevitably inaccurate because a sphere is not a developable surface (having positive curvature).

deviation If $\{x_i\}$ is a set of *observations of the *random variable X then $x_i - \bar{x}$ is the deviation of the i-th observation from the *mean.

devil's staircase See CANTOR DISTRIBUTION.

df See DEGREES OF FREEDOM.

diagonal (of a matrix) A synonym for MAIN DIAGONAL.

diagonal (of a polygon) A line joining any two vertices of a polygon that are not connected by an edge and which does not go outside the polygon.

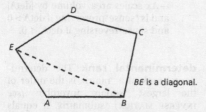

BE is a diagonal.

diagonal (of a polyhedron) A line joining any two vertices of a polyhedron that are not on a common face and which does not go outside the polyhedron.

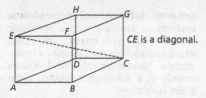

CE is a diagonal.

diagonal argument See CANTOR'S DIAGONAL THEOREM.

diagonal entry For a *square matrix $[a_{ij}]$, the diagonal entries are the entries $a_{11}, a_{22}, \ldots, a_{nn}$, which form the main diagonal.

diagonalizable (diagonalization) A *square matrix is *diagonalizable if it is *similar to a *diagonal matrix; that is, a square matrix $\mathbf{A}$ is diagonalizable if there exists an *invertible matrix $\mathbf{P}$ such that $\mathbf{P}^{-1}\mathbf{AP} = \mathbf{D}$ is diagonal. Equivalently $\mathbf{A}$ has an *eigenbasis (which then forms the columns of $\mathbf{P}$). A real matrix is diagonalizable if the roots of the *characteristic equation are real and distinct, though the converse is not true (e.g. the *identity matrix is diagonal). Generally a real matrix is diagonalizable if the roots of the characteristic equation are real and for each eigenvalue its *algebraic multiplicity equals its *geometric multiplicity. If $\mathbf{P}^{-1}\mathbf{AP} = \mathbf{D}$, then $\mathbf{A}^n = \mathbf{PD}^n\mathbf{P}^{-1}$ and so powers of diagonalizable matrices can be easily calculated.

As examples:

$$\mathbf{A} = \begin{bmatrix} 1 & 1 \\ 0 & 2 \end{bmatrix}, \qquad \mathbf{B} = \begin{bmatrix} 0 & 1 \\ -1 & 0 \end{bmatrix},$$

$$\mathbf{C} = \begin{bmatrix} 1 & 1 \\ 0 & 1 \end{bmatrix}.$$

$\mathbf{A}$ is diagonalizable as it has distinct real eigenvalues (1 and 2); $\mathbf{B}$ is not diagonalizable as a real matrix as its characteristic polynomial $x^2 + 1$ has no real roots, but $\mathbf{B}$ is diagonalizable as a complex matrix as $x^2 + 1$ has distinct complex roots; $\mathbf{C}$ is not diagonalizable as the only eigenvalue 1 has algebraic multiplicity 2 and geometric multiplicity 1.

A *linear map of a finite-dimensional *vector space is diagonalizable if any matrix representing it is diagonalizable. See also MINIMAL POLYNOMIAL, SPECTRAL THEOREM, TRIANGULARIZABLE.

diagonally dominant matrix A *symmetric matrix where the absolute value of any diagonal entry is not less than the sum of the absolute values of the rest of the row that entry appears in, i.e. $|a_{ii}| \geq \sum_{j \neq i} |a_{ij}|$. If the matrix contains complex elements, the same relationship applies with the absolute value function replaced by the modulus function.

diagonal matrix A square matrix in which all the entries not in the *main diagonal are zero.

diagram A representation of relationships or information in graphical or pictorial form. For example, statistical graphs, *force diagrams in mechanics, and *decision trees.

diameter A diameter of a circle or *central conic is a line through the centre. If the line meets the circle or central conic at P and Q, then the line segment PQ may also be called a diameter. The term also applies in both senses to a sphere or *central quadric.

In the case of a circle or sphere, all such line segments have the same length. This length is also called the diameter of the circle or sphere, and is equal to twice the *radius.

diameter (of a graph) The maximum *eccentricity of any vertex in a graph, which is the largest distance between any pair of vertices in the graph.

diameter of a set (metric space) A *bounded *metric space X with metric d has diameter D if $\mu(x, y) \leq D$ and D is the smallest value for which the inequality holds for all pairs x, y in X. So D is the *least upper bound of distances between points in the metric space.

diamond *See* RHOMBUS.

dichotomy A splitting into two non-overlapping parts.

dichotomy paradox One of *Zeno's paradoxes, whose name stems from the basis of the argument, i.e. repeatedly splitting the distance into two parts (the dichotomy) infinitely often. The argument is that motion can never be started because that infinite process will never be completed, because there are an infinite number of milestones to reach before the destination.

die (dice) A small *cube with its faces numbered from 1 to 6. When a *fair die is thrown, the *probability that any particular number from 1 to 6 is obtained on the face landing uppermost is 1/6.

diffeomorphism (**diffeomorphic**) Two open subsets (*see* OPEN SET) U and V of $\mathbb{R}^n$ are diffeomorphic if there is a diffeomorphism $\varphi : U \to V$ between them, that is a *bijection such that both φ and its *inverse φ^{-1} are *differentiable. These ideas then generalize to *manifolds and subspaces of manifolds. *See also* DIFFERENTIAL, TRANSITION MAP; *compare* HOMEOMORPHISM.

difference The difference between two real numbers a, b is $|a - b|$ or the outcome of subtracting the smaller from the larger.

difference equation Let u_0, u_1, $u_2, \ldots, u_n, \ldots$ be a sequence. If the terms satisfy the first-order difference equation $u_{n+1} + au_n = 0$, it is easy to see that $u_n = A(-a)^n$, where $A(= u_0)$ is arbitrary.

Suppose that the terms satisfy the second-order difference equation $u_{n+2} + au_{n+1} + bu_n = 0$. Let α and β be the roots of the quadratic 'auxiliary equation' $x^2 + ax + b = 0$. If $\alpha \neq \beta$, then $u_n = A\alpha^n + B\beta^n$, and if $\alpha = \beta \neq 0$, then $u_n = (A + Bn)\alpha^n$, where A and B are arbitrary constants. For example, the *Fibonacci sequence is given by the difference equation $u_{n+2} = u_{n+1} + u_n$, with $u_0 = 1$ and $u_1 = 1$, and the above method gives Binet's formula

$$u_n = \frac{1}{2}\left(\frac{1+\sqrt{5}}{2}\right)^n - \frac{1}{2}\left(\frac{1-\sqrt{5}}{2}\right)^n.$$

The above theory generalizes to linear constant coefficient difference equations (*see* LINEAR EQUATION) of higher orders, with the solutions forming a *vector space with *dimension equalling the order. The solutions to an inhomogeneous linear difference equation are a particular solution added to the homogeneous solutions. Difference equations, also called recurrence relations, do not necessarily have constant coefficients like those considered above. Such difference equations may be approached using *generating functions.

difference of two sets A synonym for SET DIFFERENCE.

difference of two squares Since $a^2 - b^2 = (a - b)(a + b)$, any expression with the form of the left-hand side, known as the difference of two squares, can be factorized into the form of the right-hand side.

difference quotient A synonym for NEWTON QUOTIENT.

difference sequence If $\{x_i\}$ is a sequence of numbers then $\{x_{i+1} - x_i\}$ is the difference sequence, obtained by subtracting successive terms. If the sequence $\{x_i\}$ is *quadratic, the difference sequence will be linear, i.e. an *arithmetic sequence.

differentiable function The *real function f of one variable is differentiable at a if $(f(a + h) - f(a))/h$ has a limit as $h \to 0$; that is, if the *derivative of f at a exists. Informally a function is differentiable if it is possible to define the *gradient of the *graph $y = f(x)$ and hence define a *tangent at the point. The function f is differentiable in an open interval if it is differentiable at every point in the interval; and f is differentiable on the closed interval $[a, b]$, where $a < b$, if it is

differentiable in (a, b) and if the right derivative of f at a and the *left derivative of f at b exist.

differentiable function (differential) (multivariate case) A function F: $\mathbb{R}^n \to \mathbb{R}^m$ is differentiable at the point $\mathbf{p}$ in $\mathbb{R}^n$ if there exists a *linear map $\mathrm{d}F_{\mathbf{p}}$: $\mathbb{R}^n \to \mathbb{R}^m$ such that

$$\lim_{\mathbf{h} \to 0} \frac{F(\mathbf{p} + \mathbf{h}) - F(\mathbf{p}) - \mathrm{d}F_{\mathbf{p}}(\mathbf{h})}{|\mathbf{h}|} = 0.$$

$\mathrm{d}F_{\mathbf{p}}$ is known as the differential (or derivative) of F at $\mathbf{p}$. When $m = n = 1$, then $\mathrm{d}F_p$ is a 1×1 matrix with entry $F'(p)$. More generally $\mathrm{d}F_{\mathbf{p}}$ is represented by the *Jacobian matrix at p. A function F: $\mathbb{R}^n \to \mathbb{R}^m$ can be expanded as

$$F(x_1, \ldots, x_n) = (f_1(x_1, \ldots, x_n), \ldots, \\ f_m(x_1, \ldots, x_n)).$$

It is *sufficient for F to be differentiable for all the *partial derivatives $\partial f_i / \partial x_j$ to exist and be continuous (*see* CONTINUOUS FUNCTION). More generally, for a differentiable function F: $M \to N$ between smooth *manifolds M and N, the differential at p in M is a linear map $\mathrm{d}F_p : T_p(M) \to T_{F(p)}(N)$ between *tangent spaces. For v in $T_p(M)$ take a curve γ in M such that $\gamma(0) = p$ and $\gamma'(0) = v$. Then $F(\gamma)$ is a curve in N and $\mathrm{d}F_p(v) = (F \circ \gamma)'(0)$.

differential calculus The branch of *calculus dealing with *rates of change and *derivatives generally. The subject has many applications, most notably through *differential equations, which are used to *model a wide range of physical phenomena. *Compare* INTEGRAL CALCULUS.

differential equation Let y be a *real function of x and let $y', y'', \ldots, y^{(n)}$ denote the first, second, ... nth *derivatives of y with respect to x. An ordinary differential equation (ODE for short) is an equation involving $x, y, y', y'', \ldots$. (*compare* PARTIAL

DIFFERENTIAL EQUATION, STOCHASTIC DIFFERENTIAL EQUATION). x is referred to as the independent variable and y as the dependent variable. The order of the differential equation is the order n of the highest derivative $y^{(n)}$ that appears.

Ordinary and partial differential equations are hugely important in *mathematical models describing phenomena in the real world (*see* BESSEL'S EQUATION, BLACK-SCHOLES EQUATION, CAL-CULUS OF VARIATIONS, EULER'S EQUATION, HEAT EQUATION, LAPLACE'S EQUATION, MAXWELL'S EQ-UATIONS, NAVIER-STOKES EQUATIONS, PREDATOR-PREY EQUATIONS, SCHRÖDINGER'S EQUATION, WAVE EQUATION).

The problem of solving a differential equation is to find the general form of functions y whose derivatives satisfy the equation. In certain circumstances, it can be shown that a differential equation of order n has a *general solution for y, involving n arbitrary constants. Such examples include *linear differential equations with constant coefficients. But this is far from the general case (*see* PICARD'S THEOREM). Further information, such as *initial conditions or *boundary conditions, may be given to specify uniquely a solution within the general solution. For example, $y'' - \omega^2 y = 0$, models *simple harmonic motion, and has general solution $y(x) = A\cos\omega x + B\sin\omega x$. If further we know $y(0) = 2$ and $y'(0) = 1$, then this specifies the solution uniquely as $y(x) = 2\cos\omega x + \omega^{-1}\sin\omega x$.

See HOMOGENEOUS FIRST-ORDER DIFFERENTIAL EQUATION, LINEAR DIFFERENTIAL EQUATIONS, LINEAR FIRST-ORDER DIFFERENTIAL EQUATION, LAPLACE TRANSFORM, SEPARABLE FIRST-ORDER DIFFERENTIAL EQUATION.

differential form A differential form of degree 0 on $\mathbb{R}^3$, or a 0-form, is a *differentiable real function $f(x,y,z)$. A 1-form is written

$$f_1 \, dx + f_2 \, dy + f_3 \, dz$$

where f_1, f_2, f_3 are differentiable real functions of x, y, z. The exterior derivative d maps 0-forms to 1-forms via the *chain rule

$$df = \frac{\partial f}{\partial x} dx + \frac{\partial f}{\partial y} dy + \frac{\partial f}{\partial z} dz.$$

Note how the terms here are the same as those that appear in the *gradient of f. More generally, the exterior derivative takes k-forms to $(k+1)$-forms, but the product used is the *exterior product, so that, for example, $dx \, dy = -dy \, dx$ and $dx^2 = 0$. In the case of applying d to a 1-form we get

$$d(F_1 dx + F_2 dy + F_3 dz) =$$

$$\left(\frac{\partial F_1}{\partial x} dx + \frac{\partial F_1}{\partial y} dy + \frac{\partial F_1}{\partial z} dz \right) dx + \dots$$

$$= \left(\frac{\partial F_3}{\partial y} - \frac{\partial F_2}{\partial z} \right) dy \, dz + \left(\frac{\partial F_1}{\partial z} - \frac{\partial F_3}{\partial x} \right)$$

$$dz \, dx + \left(\frac{\partial F_2}{\partial x} - \frac{\partial F_1}{\partial y} \right) dx \, dy,$$

once simplified. Note how the terms here are the same as those that appear in the *curl of (F_1, F_2, F_3). A similar calculation shows that d acts like *divergence when mapping 2-forms to 3-forms.

In general, $d^2 = d \circ d = 0$ generalizing curl(grad) = **0** and div(curl) = 0. Differential forms generalize to $\mathbb{R}^n$ for all n and to *manifolds where they can be expressed in terms of local coordinates. *See* STOKES' THEOREM (GENERALIZED FORM).

differential geometry The area of mathematics which uses *differential calculus in the study of *geometry. Its objects of interest are *curves and *surfaces (and higher dimensional *manifolds) and for example notions of *area, *curvature, *geodesics, and *isometries.

differential operator Generally, any operator involving derivatives or partial derivatives. In particular, the operator del $\nabla = \mathbf{i}\frac{\partial}{\partial x} + \mathbf{j}\frac{\partial}{\partial y} + \mathbf{k}\frac{\partial}{\partial z}$, where $\mathbf{i}$, $\mathbf{j}$, $\mathbf{k}$ are unit vectors in directions Ox, Oy, Oz, and $\frac{\partial}{\partial x}$, $\frac{\partial}{\partial y}$, $\frac{\partial}{\partial z}$ are the *partial derivatives of the function with respect to x, y, z. Note that the operator del does not behave as vectors do (*see* APPENDIX 16). *See also* CURL, DIVERGENCE, GRADIENT, Laplacian.

differential topology In the same way that *general topology focuses on *continuous functions and related spaces, differential topology focuses on spaces and functions that are smooth. Smooth *manifolds are the main spaces of interest, and the notion of equivalence is *diffeomorphic. *Stokes' Theorem and the *Divergence Theorem are theorems of differential topology. The *Classification Theorem for Surfaces which are closed and smooth is essentially the same as the continuous version, but in dimensions 4 and above very different results arise. *See* DONALDSON, SIMON; MILNOR, JOHN.

differentiation The process of obtaining the *derived function f' from the function f, where $f'(x)$ is the *derivative of f at x. (For the derivatives of certain common functions, *see* APPENDIX 7.) From these many other functions can be differentiated using the following rules of differentiation:

(i) If $h = kf$, where k is a constant, then $h' = kf'$.
(ii) If $h = f + g$, then $h = f' + g'$.
(iii) The product rule: If $h = fg$, then $h' = fg' + f'g$.
(iv) The quotient rule: If $h = f/g$ and $g(x) \neq 0$ for all x, then

$$h' = \frac{gf' - fg'}{g^2}.$$

(v) The *chain rule: If $h(x) = (f \circ g)(x) = f(g(x))$ for all x, then

$$h'(x) = f'(g(x))g'(x).$$

differentiation under the integral sign This states that

$$\frac{\mathrm{d}}{\mathrm{d}\alpha}\int_a^b f(t, \alpha)\ \mathrm{d}t = \int_a^b \frac{\partial f}{\partial \alpha}(t, \alpha)\ \mathrm{d}t.$$

This is a special case of *Leibniz's integral rule. As an example, for $\alpha > 0$ define

$$I(\alpha) = \int_0^\infty \frac{\mathrm{e}^{-x} - \mathrm{e}^{-\alpha x}}{x}\ \mathrm{d}x.$$

Then

$$I'(\alpha) = \int_0^\infty \frac{x\mathrm{e}^{-\alpha x}}{x}\ \mathrm{d}x = \int_0^\infty \mathrm{e}^{-\alpha x}\ \mathrm{d}x = \frac{1}{\alpha},$$

and integrating gives $I(\alpha) = c + \ln\alpha$ for some constant c. As $I(1) = 0$ then $I(\alpha) = \ln\alpha$.

diffusion equation *See* HEAT EQUATION.

digit A symbol used in writing numbers in their *decimal representation or to some other *base. In decimal, the digits used are 0-0. In *hexadecimal, the digits are 0-9 and A-F. In *binary the digits are 0 and 1.

digital In numerical form. For example, a digital watch displays the time by numbers rather than the position of the hands in an analogue clock.

digital computer *See* COMPUTER.

digital data *Data that are represented digitally, most commonly in *binary. An analogue source producing continuous data might be encoded digitally, so that they might be stored on a *computer or can more reliably be transmitted. *See* INFORMATION THEORY.

digital root *See* CASTING OUT NINES.

digraph A digraph (or directed graph) consists of a number of *vertices, some of which are joined by arcs, where an arc, or directed edge or arrow, joins one vertex to another and has an arrow on it to indicate its direction. The arc from the vertex u to the vertex v may be denoted by the ordered pair (u, v). The digraph with vertices u, v, w, x and arcs (u, v), (u, w), (v, u), (w, v), (w, x) is shown in the first figure.

As with *graphs there may be multiple arcs and loops, as shown in the second figure.

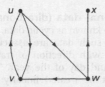

directed graph

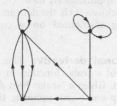

directed *multigraph

dihedral The term, meaning two-sided, also relates to the *angle made by two *half-planes at the bounding line they have in common

dihedral group The *group of *symmetries of a *regular n-sided *polygon; the notation D_{2n} is often used as the group has *order $2n$. If r denotes a *rotation by $360°/n$ and s denotes a *reflection, then r and s are *generators of D_{2n}. The group contains n rotations $e, r, r^2, \ldots, r^{n-1}$ and n reflections $s, rs, r^2s, \ldots, r^{n-1}s$.

Dijkstra's method See SHORTEST PATH ALGORITHM.

dilation (dilatation) A dilation of the plane from O with scale factor $c \neq 0$ is the *transformation of the plane in which the origin O is mapped to itself and a point P is mapped to the point P', where O, P and P' are collinear and $OP' = cOP$. This is given in terms of *Cartesian coordinates by $x' = cx$, $y' = cy$.

dimension (linear algebra) Of a *vector space, the number of elements common to any *basis.

dimension (topology) There are various ways to define the dimension of a *topological space (or *metric space), in such a way that the dimension is invariant under *homeomorphisms. Three approaches are the small inductive dimension, large inductive dimension, and Lebesgue covering dimension. All three only take integer values and agree for *separable metric spaces. *Compare* HAUSDORFF DIMENSION.

dimensions Many physical quantities can be described in terms of the basic dimensions of mass M, length L, and time T, using positive and negative indices. For example, the following have the dimensions given: area, L^2; velocity, LT^{-1}; force, MLT^{-2}; linear momentum, MLT^{-1}; energy, ML^2T^{-2}; and power, ML^2T^{-3}. *Temperature is another dimension, and there are others (e.g. electric current or luminous intensity) within the *SI units system, but these are more relevant to physics than mathematics.

Diophantine equation In *number theory, an algebraic equation in one or more unknowns, with integer coefficients, for which *integer solutions are sought. As examples:

(i) $14x + 9y = 1$ has solutions $x = 2 + 9t$, $y = -3 - 14t$, where t is any integer.
(ii) $x^2 + 1 = 2y^4$ has two solutions $x = 1$, $y = 1$ and $x = 239$, $y = 13$.
(iii) $x^3 + y^3 = z^3$ has only trivial solutions.

See also CATALAN'S CONJECTURE, FERMAT'S LAST THEOREM, HILBERT'S TENTH PROBLEM, PELL'S EQUATION.

Diophantus of Alexandria (about AD 250) Hellenistic Greek mathematician whose work displayed an *algebraic approach to the solution of equations in one or more unknowns, unlike earlier Greek methods that were more *geometrical. In the books of *Arithmetica* that survive, particular numerical examples of more than 100 problems are solved, probably to indicate the general methods of solution. These are mostly of the kind now referred to as *Diophantine equations.

Dirac, Paul Adrien Maurice (1902–84) Born in England to a Swiss father and English mother, he was Lucasian Professor of Mathematics at Cambridge University for 37 years. He is best known for bringing together *relativity theory and *quantum theory and shared the Nobel Prize for physics in 1933 with Erwin *Schrödinger.

Dirac delta function The Dirac delta function $\delta(x)$ has defining properties $\int_{-\infty}^{\infty} \delta(x)\mathrm{d}x = 1$, and $\delta(x) = 0$ for $x \neq 0$. No function, in the traditional sense, has these properties, but $\delta(x)$ can be rigorously understood as the Schwartz *distribution $\delta(\varphi) = \varphi(0)$, where φ is a test function. The delta function is useful for modelling *point masses or charges, or instantaneous *impulses. As a distribution it is the derivative of the *Heaviside function. *See* SIFTING PROPERTY.

directed graph A synonym for DIGRAPH.

directed line A straight line with a specified direction along the line.

directed line segment If A and B are two points on a straight line, the part of the line between and including A and B, together with a specified direction along the line, is a directed line segment. $\overrightarrow{AB}$ denotes the directed line segment from A to B. *See also* VECTOR.

directed number A number with a positive or negative sign when required showing it has a direction from the origin as well as a distance from the origin on the number line.

direction The *orientation of a line. The line will then have two senses which a *directed line will distinguish between.

directional data (directional statistics) Also known as cyclic data, or circular data, such data are associated, for example, with direction of travel, time of day, and day of the year. The data must not be treated linearly; it must instead be recognized that a bearing of 359° is close to a bearing of 1°, or that 23.59 is approximately 00.01 (in the 24-hour clock). As such, the data are treated as being distributed on a circle (or sphere).

directional derivative The rate of change of a scalar function in a given direction, Given a *scalar field $f: \mathbb{R}^3 \rightarrow \mathbb{R}$, a point **p** and a unit vector **u**, then the directional derivative of f at **p** in the direction **u** equals the limit

$$\lim_{t \to 0} \frac{f(\mathbf{p} + t\mathbf{u}) - f(\mathbf{p})}{t}.$$

It also equals $\nabla f \cdot \mathbf{u}$, so that if $\mathbf{u} = \mathbf{i}$, then the directional derivative equals $\partial f / \partial x$.

direction angles The angles that are used in defining *direction cosines

direction cosines In 3-dimensional space, a direction can be specified as follows. Take a point P such that OP has the given direction and $|OP| = 1$. Let α, β and γ be the three angles $\angle xOP$, $\angle yOP$ and $\angle zOP$, measured in *radians with $0 \leq \alpha,\ \beta,\ \gamma\ \leq \pi$. Then $\cos\alpha$, $\cos\beta$, and $\cos\gamma$ are the direction cosines of OP. Point P has coordinates ($\cos\alpha$, $\cos\beta$, $\cos\gamma$), and so $\cos^2\alpha + \cos^2\beta + \cos^2\gamma = 1$.

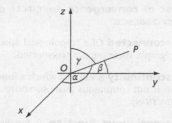

direction angles

direction fields *See* TANGENT FIELDS.

direction ratios Suppose that a direction has *direction cosines $\cos\alpha$, $\cos\beta$, $\cos\gamma$. Any triple of numbers l, m, n, not all zero, such that $l = k\cos\alpha$, $m = k\cos\beta$, $n = k\cos\gamma$, are called direction ratios of the given direction. Since $\cos^2\alpha + \cos^2\beta + \cos^2\gamma = 1$, it follows that

$$\cos\alpha = \frac{\pm l}{\sqrt{l^2+m^2+n^2}}, \cos\beta = \frac{\pm m}{\sqrt{l^2+m^2+n^2}},$$

$$\cos\gamma = \frac{\pm n}{\sqrt{l^2+m^2+n^2}},$$

where either the $+$ sign or the $-$ sign is taken throughout. So any triple of numbers, not all zero, determine two possible sets of direction cosines, corresponding to opposite *directions. The triple l, m, n are said to be direction ratios of a straight line when they are direction ratios of either direction of the line.

direct isometry An *isometry, such as a *rotation or *translation, which is *sense-preserving. Isometries which reverse *orientation, such as *glide reflections, are called indirect.

directly proportional *See* PROPORTION.

direct product *See* PRODUCT GROUP, SEMI-DIRECT PRODUCT.

direct proof For a theorem that has the form $p \Rightarrow q$, a direct proof is one

that supposes p and shows that q follows. *Compare* INDIRECT PROOF.

directrix (directrices) *See* CONIC, ELLIPSE, HYPERBOLA, PARABOLA.

direct sum A *vector space V is a direct sum of *subspaces $X_1, X_2, \ldots, X_k$, written $V = X_1 \oplus X_2 \oplus \cdots \oplus X_k$, if every $v \in V$ can be written uniquely $v = x_1 + x_2 + \ldots + x_k$ where $x_i \in X_i$ for each i. Such a direct sum is referred to as internal. Given vector spaces $V_1, V_2, \ldots, V_k$, the external direct sum $V_1 \oplus V_2 \oplus \cdots \oplus V_k$ is defined on the *Cartesian product $V_1 \times V_2 \times \cdots \times V_k$ with *componentwise *addition and *scalar multiplication. In a similar fashion, the direct sum of *abelian groups, *rings, and *modules can be defined.

direct variation *See* PROPORTION.

Dirichlet, Peter Gustav Lejeune (1805–59) German mathematician who was professor at the University of Berlin before succeeding *Gauss at the University of Göttingen. He proved that in any arithmetic series a, $a + d$, $a + 2d$, ..., where a and d are *coprime, there are infinitely many primes; his proof arguably introduced the field of *analytic number theory. He gave an early version of the modern definition of a *function. He also made contributions to mathematical physics, particularly in potential theory; *see* DIRICHLET PROBLEM.

Dirichlet beta function The *special function

$$\beta(s) = \sum_{n=0}^{\infty}(-1)^n(2n+1)^{-s}$$

which is closely related to the *zeta function.

Dirichlet function Constructed by *Dirichlet as an example of a function which is not Riemann *integrable. It can be defined on any interval $a \le x \le b$ by $f(x) = 1$ if x is *rational and $f(x) = 0$ if x is *irrational. By choosing the points of a *partition to be rational or irrational, a

*Riemann sum can have limit $b-a$ or 0 as the partition's *norm converges to 0. However, f is *Lebesgue integrable as the rationals have *measure 0, and so f has Lebesgue integral 0.

Dirichlet problem Originally the Dirichlet problem involved solving *Laplace's equation inside a region and meeting specified values on the boundary. Now the term might relate to any *PDE. That the solution needs to meet specified values on the boundary is called a Dirichlet boundary condition. *Compare* NEUMANN CONDITION, ROBIN BOUNDARY CONDITION, POISSON'S INTEGRAL FORMULA.

Dirichlet's approximation theorem Let α be a *real number and $Q > 1$ be an integer. Then there are integers p,q with $0 < q < Q$ such that $|q\alpha - p| \le 1/Q$. Consequencetly, there are infinitely many p,q satisfying $|\alpha - p/q| \le 1/q^2$. This is a significant result in Diophantine approximation, the study of approximating real numbers by *rational numbers. *Compare* ROTH'S THEOREM.

Dirichlet series A series in the form $\sum_{n=1}^{\infty} a_n e^{-\lambda_n z}$ where a_n and z are complex and $\{\lambda_n\}$ is a *monotonic increasing sequence of real numbers. When $\lambda_n = \ln n$, the series reduces to $\sum_{n=1}^{\infty} a_n n^{-z}$, known as a Dirichlet L-series.

Dirichlet's test A test for *convergence (*see* CONVERGE (series)) of a series. If $\{a_n\}$ is a series which has *bounded *partial sums, and $\{b_n\}$ is *decreasing and *converges to zero, then $\sum_{1}^{\infty} a_n b_n$ converges.

disc The interior of a *circle is an open disc, and the *union of the interior and the circle itself is referred to as a closed disc.

disc of convergence *See* CIRCLE OF CONVERGENCE.

disconnected Of a *topological space or *graph, meaning not *connected.

discontinuity A value at which a function is not continuous (*see* CONTINUOUS FUNCTION).

discontinuous function A function which is not continuous (*see* CONTINUOUS FUNCTION). A function only requires one *discontinuity to be discontinuous.

discrete Taking only *finitely or *denumerably many values, in contrast to continuous.

discrete data *See* DATA.

discrete Fourier transform The discrete Fourier transform of a vector $\mathbf{x} = (x_0, x_1, \ldots, x_{n-1})$ is the vector $\mathbf{y} = (y_0, y_1, \ldots, y_{n-1})$ obtained by calculating $y_k = \sum_{j=0}^{n-1} \omega^{kj} x_j$ for each $k = 0, 1, \ldots, n-1$ where $\omega = e\left(\frac{-2\pi i}{n}\right) = \cos\left(\frac{2\pi}{n}\right) - i \sin\left(\frac{2\pi}{n}\right)$ This involves a large number of calculations, but the number can be reduced by using an algorithm called the fast Fourier transform.

(((●))) SEE WEB LINKS

• A discrete Fourier transform tool in which choosing different options from the signal menu illustrates a number of signals and the approximating function using a specified number of Fourier coefficients.

discrete logarithm Let p be a prime; then the multiplicative group $\mathbb{Z}_p^*$ of the integers $0 < k < p$ *modulo (*see* MODULO n ARITHMETIC) p is *cyclic, and if r is a *generator (a primitive root), then $k = r^m$ for some $0 \le m < p-1$. This integer m is the discrete logarithm, or index, of k to base r. This is written $m = \text{ind}_r k$. For $p = 11$ and $r = 2$ the powers of r are 1,2,4,8,5,10,9,7,3,6 and so $\text{ind}_2 3 = 8$. In general though, for large primes p, there is no known efficient algorithm for calculating $\text{ind}_r k$; consequently, discrete logarithms have found various uses in *cryptography.

discrete mathematics A broad area of mathematics dealing with *discrete processes and calculations, as opposed to continuous ones. The term commonly refers to areas such as *combinatorics and *graph theory. It finds applications in *computer science due to the discrete nature of computer states.

discrete metric The *metric d defined on a non-empty set X by $d(x, x) = 0$; $d(x, y) = 1$ if $x \neq y$. In a discrete metric space every set is open (*see* OPEN SET).

discrete random variable *See* RANDOM VARIABLE.

discrete space A *topological space is said to be discrete if all the points are *isolated. Such a space is *metrizable using the *discrete metric.

discretization The process of approximating a *continuous function or process by a *discrete alternative.

discriminant For the *quadratic $ax^2 + bx + c = 0$, the quantity $b^2 - 4ac$ is the discriminant. The quadratic has two distinct real roots, repeated roots, or no real roots according to whether the discriminant is positive, zero, or negative.

Note the quadratic's discriminant equals $a^2(\alpha-\beta)^2$ where α and β are the quadratic's root. The discriminant of the cubic $ax^3 + bx^2 + cx + d = 0$ equals $a^4(\alpha-\beta)^2(\beta-\gamma)^2(\gamma-\alpha)^2$ where α, β, γ are the cubic's roots, and higher degree polynomials have similarly defined discriminants.

discriminatory A test is said to be discriminatory if its *power is greater than some previously specified level.

disjoint Sets A and B are disjoint if they have no elements in common; that is, if $A \cap B = \emptyset$.

disjoint union A union where duplicate elements from two sets are separately included (which is not the case with standard *union). For example, given two sets A_1, A_2 their disjoint union, denoted $A_1 \sqcup A_2$, may be identified with $(A_1 \times \{1\}) \cup (A_2 \times \{2\})$. If x is common to both A_1 and A_2, its first version appears as $(x,1)$ in $A_1 \sqcup A_2$ and the second copy as $(x,2)$.

disjunction *See* EXCLUSIVE DISJUNCTION, INCLUSIVE DISJUNCTION. The common usage of the term 'or' where the compound sentence is true if at least one of the parts is true is inclusive disjunction.

disjunctive normal form A formula is in disjunctive normal form if it is expressed as a *disjunction of *conjunctions. This might be visualized using *Venn diagrams as then *union of the regions where the formula holds true, expressed as *intersections of the sets and their *complements.

A formula is in conjunctive normal form if it is expressed as a conjunction of disjunctions. By *De Morgan's laws, the conjunctive normal form of P can be deduced from the *negation of the disjunctive normal form of the negation of P.

dispersion A measure of dispersion is a way of describing how scattered or spread out the *observations in a *sample are. The term is also applied similarly to a *random variable. Common measures of dispersion are the *range, *interquartile range, *mean absolute deviation, *variance, and *standard deviation. The range may be unduly affected by odd high and low values. The mean absolute deviation is difficult to work with algebraically. The standard deviation is in the same units as the data, and it is this that is most often used. The interquartile range may be appropriate when the *median is used as the measure of *location.

displacement The displacement of a point P from an origin O is the vector $\overrightarrow{OP}$. The *rate of change of displacement with time is *velocity. Displacement is different from but not unrelated to *distance. The distance of P from O is the *magnitude of its displacement, but the distance travelled by P might be very different from its final displacement. If P has displacement $\mathbf{r}(t)$ for $0 \le t \le T$, then the final displacement is

$$\mathbf{r}(T) = \int_0^T \frac{d\mathbf{r}}{dt} dt$$

while the distance travelled is.

$$s(T) = \int_0^T \left| \frac{d\mathbf{r}}{dt} \right| dt$$

dissection (of an interval) A synonym for PARTITION (of an interval).

dissipative force A force that causes a loss of *energy. A *resistive force is dissipative because the *work done by it is negative.

distance (in a graph) The distance between two *vertices in a *graph is the length of the shortest *path between them. This defines a *metric on the set of vertices.

distance (in the complex plane) If P_1 and P_2 are points represented by complex numbers z_1 and z_2, the distance $|P_1P_2|$ is equal to $|z_1 - z_2|$, the *modulus of $z_1 - z_2$.

distance between two codewords A synonym for HAMMING DISTANCE.

distance between two lines (in 3-dimensional space) Let l_1 and l_2 be lines in space that do not intersect. There are two cases. If l_1 and l_2 are parallel, the distance between the two lines is the length of any line segment $N_1\,N_2$, with

N_1 on l_1 and N_2 on l_2 perpendicular to both lines. If l_1 and l_2 are not parallel, there are unique points N_1 on l_1 and N_2 on l_2 such that the length of the line segment N_1N_2 is the shortest possible. The length $|N_1N_2|$ is the distance between the two lines. In fact, the line N_1N_2 is the *common perpendicular of l_1 and l_2.

If the two lines are skew and have equations $\mathbf{r} \times \mathbf{a}_1 = \mathbf{b}_1$ and $\mathbf{r} \times \mathbf{a}_2 = \mathbf{b}_2$ where $\times$ denotes the *vector product, then the distance between the lines equals

$$\frac{|\mathbf{a}_1 \cdot \mathbf{b}_2 + \mathbf{a}_2 \cdot \mathbf{b}_1|}{|\mathbf{a}_1 \times \mathbf{a}_2|}$$

distance between two points (in the plane) Let A and B have coordinates (x_1, y_1) and (x_2, y_2). It follows from Pythagoras' Theorem that the distance $|AB|$ is equal to $\sqrt{(x_2 - x_1)^2 + (y_2 - y_1)^2}$.

distance between two points (in Euclidean space) *See* EUCLIDEAN SPACE.

distance from a point to a line (in the plane) The distance from the point P to the line l is the shortest distance between P and a point on l. It is equal to $|PN|$, where N is the point on l such that the line PN is *perpendicular to l. If P has *coordinates (x_1, y_1) and l has equation $ax + by + c = 0$, then the distance from P to l is equal to

$$\frac{|ax_1 + by_1 + c|}{\sqrt{a^2 + b^2}},$$

where $|ax_1 + by_1 + c|$ is the *absolute value of $ax_1 + by_1 + c$.

distance from a point to a plane (in 3-dimensional space) The distance from the point P to the plane p is the shortest distance between P and a point in p, and is equal to $|PN|$, where N is the point in p such that the line PN is normal to p. If P has coordinates (x_1, y_1, z_1) and p has equation $ax + by + cz + d = 0$, the distance from P to p is equal to

$$\frac{|ax_1 + by_1 + cz_1 + d|}{\sqrt{a^2 + b^2 + c^2}}$$

where $|ax_1 + by_1 + cz_1 + d|$ is the *absolute value of $ax_1 + by_1 + cz_1 + d$.

distance-time graph A graph that shows *displacement plotted against time for a particle moving in a straight line. The *gradient at any point is equal to the *velocity of the particle at that time.

distinct Not equal.

distribution The distribution of a *random variable describes the *probability of its taking a certain value, or a value within a certain interval, varies. It may be given by the *cumulative distribution function. More commonly, the distribution of a discrete random variable is given by its *probability mass function and that of a continuous random variable by its *probability density function.

distribution (Schwartz distribution) A (Schwartz) distribution is a generalized function, such as the *Dirac delta function δ. Unlike classical *functions, distributions are not defined at points but by *integrals. A test function φ: $\mathbb{R} \to \mathbb{R}$ is a smooth (i.e. infinitely differentiable) function such that φ is zero outside a bounded interval. If $f: \mathbb{R} \to \mathbb{R}$ is a *continuous function, then the integral

$$\langle f, \varphi \rangle = \int_{-\infty}^{\infty} f(x)\varphi(x)\,\mathrm{d}x$$

exists. Further, knowing all such integrals, it is possible to determine f entirely. A distribution is a linear *functional on the space of test functions (which is continuous in a technical sense); the distributions generalize (locally integrable) functions by means of the previous integrals. As a distribution, $\langle \delta, \varphi \rangle = \varphi(0)$, as expected from the *sifting property. A distribution F is differentiable by the rule $\langle F', \varphi \rangle = -\langle F, \varphi' \rangle$

(compare with *integration by parts). As a distribution, the delta function is the derivative of the *Heaviside function, even though the latter is discontinuous as a function.

distribution-free methods See NON-PARAMETRIC METHODS.

distribution function See CUMULATIVE DISTRIBUTION FUNCTION.

distributive Suppose that ∘ and * are *binary operations on a set S. Then ∘ is distributive over * if, for all a, b, and c in S,

$$a \circ (b * c) = (a \circ b) * (a \circ c),$$
$$(a * b) \circ c = (a \circ c) * (b \circ c)$$

If the two operations are *multiplication and *addition, 'the distributive laws' normally means those that say that multiplication is distributive over addition.

div Abbreviation for *divergence.

diverge (divergent) (of a sequence or series) A *sequence diverges (or is divergent) if it does not *converge to a *limit. A *series diverges (or is divergent) if its partial sums does not converge to a limit.

divergence For a vector function of position $\mathbf{V}(\mathbf{r}) = V_x\mathbf{i} + V_y\mathbf{j} + V_z\mathbf{k}$, the divergence of $\mathbf{V}$ is the *scalar product of the operator *del with $\mathbf{V}$ giving

$$\mathrm{div}\mathbf{V} = \nabla \cdot \mathbf{V} = \frac{\partial V_x}{\partial x} + \frac{\partial V_y}{\partial y} + \frac{\partial V_z}{\partial z}.$$

For a *fluid with flow velocity $\mathbf{u}$, then $\mathrm{div}\mathbf{u}$ measures the rate of expansion of the fluid locally. Compare CURL, GRADIENT.

divergence theorem For a *bounded region R in $\mathbb{R}^3$ with smooth *boundary Σ, and for a continuously differentiable *vector field $\mathbf{F}$ on R,

$$\iiint_R \text{div}\mathbf{F}\,dV = \iint_\Sigma \mathbf{F}\cdot d\mathbf{S}.$$

Here $d\mathbf{S} = \mathbf{n}\,dS$, where S denotes *surface area and $\mathbf{n}$ is the outward-pointing unit *normal. As the *divergence div$\mathbf{F}$ represents the local expansion/contraction of the field $\mathbf{F}$, then the theorem states that the sum of such expansions equals the *flux of $\mathbf{F}$ over the boundary Σ. *See* STOKES' THEOREM (GENERALIZED FORM).

divides Let a and b be *integers. Then a divides b, written $a|b$, if there is an integer c such that $ac = b$. It is said that a is a divisor or factor of b, that b is divisible by a, and that b is a multiple of a.

divisible *See* DIVIDES.

division *See* MULTIPLICATION (OF NUMBERS).

division (of a segment) The construction of a point which divides a line segment in specified proportions. This may be *internal division or *external division.

division algebra An *algebra in which division by any non-zero vector is possible. *Frobenius' theorem states that the only *associative division algebras are the *real numbers, *complex numbers, and *quaternions.

Division Algorithm For *integers a and b, with $b > 0$, there exist unique integers q and r such that $a = bq + r$, where $0 \le r < b$.

In the division of a by b, the number q is the quotient and r is the remainder. *See* EUCLIDEAN DOMAIN.

division ring A *ring which meets all the axioms of being a *field but not necessarily the requirement that multiplication is commutative (i.e. axioms (i)–(x) but not necessarily (vii)). Also called a skew field. Finite division rings are fields.

divisor *See* DIVIDES.

divisor function The number of divisors of n, including 1 and n, denoted by $d(n)$. So $d(6) = 4$ since 1,2,3 and 6 are divisors of 6. For any prime number p, $d(p^k) = k + 1$. The divisor function is an *arithmetic function which is multiplicative.

divisor of zero *See* ZERO-DIVISOR.

dodecagon A twelve-sided *polygon.

dodecahedron (dodecahedra) A *polyhedron with twelve faces, often assumed to be *regular. The regular dodecahedron is one of the *Platonic solids, and its faces are regular *pentagons. It has 20 vertices and 30 edges.

Dodgson, Charles Lutwidge (1832–98) British mathematician and logician, better known under the pseudonym Lewis Carroll as the author of *Alice's Adventures in Wonderland*. He was lecturer in mathematics at Christ Church, Oxford, and wrote a number of minor books on mathematics.

domain *See* FUNCTION.

domain (algebra) A *ring in which the equation $xy = 0$ implies $x = 0$ or $y = 0$ (or both). *See* EUCLIDEAN DOMAIN, INTEGRAL DOMAIN.

domain (analysis) An *open (*see* OPEN SET), *connected subset of a *Euclidean space, the *complex plane, or more generally a *finite-dimensional real *vector space.

dominated convergence theorem Let $f_n:\mathbb{R}\to\mathbb{R}$ be a sequence of (Lebesgue) *integrable functions which converge *pointwise to f. Suppose further than $|f_n| \le g$ for all n where g is integrable. Then f is integrable and

$$\int f\,dx = \lim_{n\to\infty}\int f_n\,dx.$$

The theorem is not true of *Riemann integrals as the *Dirichlet function is a pointwise limit of integrable functions which is not itself Riemann integrable.

Donaldson, Simon (1957–) English mathematician, who was awarded a *Fields Medal in 1986 for his work on differentiable four-dimensional *manifolds. His results showed that certain topological manifolds could have no differentiable structure and that four-dimensional *Euclidean space could be given different 'exotic' structures; this is unique to four dimensions.

dot The symbol '.' used to represent the decimal point and also *multiplication. It is also used to represent the derivative of a function, particularly with respect to time, so that $\dot{y}$ would mean dy/dt.

dot product A synonym for SCALAR PRODUCT.

double-angle formula (in hyperbolic functions) See HYPERBOLIC FUNCTIONS.

double-angle formula (in trigonometry) A formula in trigonometry that expresses a function of a double angle in terms of the single angle. This can be obtained from the corresponding *compound angle formulae by substituting $A = B = x$:

$$
\begin{aligned}
\sin(2x) &= 2\sin x \cos x \\
\cos(2x) &= \cos^2 x - \sin^2 x \\
\tan(2x) &= \frac{2\tan x}{1 - \tan^2 x}.
\end{aligned}
$$

double dummy See BLINDING.

double factorial See FACTORIAL.

double integral The integral of a function with respect to two variables, so if f is a function of two variables x, y, we can have $\iint f(x,y)\, dx\, dy$. See MULTIPLE INTEGRAL.

double negation The proposition that the *negation of the negation of A is equivalent to the proposition A. In English, a double negative is not always precisely a double negation. For example, in *hypothesis testing the conclusion 'there is not sufficient evidence to suggest that the mean is not 340' is not equivalent to saying that there is evidence to suggest that the mean is 340.

double point A point on a curve where the curve intersects itself. If the *tangents are not coincident then the point is a node. If the tangents are coincident, then the point is a tacnode.

double precision See PRECISION.

double root See ROOT.

double sequence A sequence with twin indices. For example, $a_{n,m} = \dfrac{1}{n^2 + m^2}$.

double series A series with two indices. For example, for any *joint distribution with $p_{ij} = \Pr\{X = x_i, Y = y_j\}$,
$$
E(X+Y) = \sum_i \sum_j \Big((x_i + y_j) \times p_{ij} \Big).
$$

double tangent (bitangent) A line which is a tangent to a curve at two distinct points. Also a pair of coincident tangents, as occurs at a *cusp.

dp Short for *decimal places. For example $\pi = 3.14$ to 2 dp.

drag See AERODYNAMIC DRAG.

dual basis Given a basis $e_1, e_2, \ldots e_n$ for a *vector space V, the *functionals $f_1, f_2, \ldots f_n$ defined by $f_i(e_j) = \delta_{ij}$ (the *Kronecker delta) form a basis for the *dual space V^*. For v in V, $f_i(v)$ is the ith coordinate of v with respect to $e_1, e_2, \ldots e_n$.

dual graph The dual graph G^* of a *planar graph G has the faces of G as its vertices, and there is an edge between the vertices of G^* if an edge of G separates the faces of G that those vertices represent. The dual graph of the dual graph is the original graph. An example of a planar graph and its dual are:

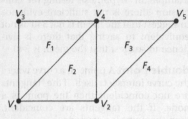

A planar graph G

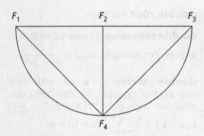

The dual graph of G

duality In many areas of mathematics, for a given problem it is possible to describe a dual problem. Solving the dual problem, rather than the original problem, can sometimes be easier, or recognizing the implications of the dual problem can be of benefit. This is particularly true of *projective geometry. In a projective plane (*see* PROJECTIVE SPACE), *lines are dual to *points, and *concurrency is dual to *collinearity. Thus 'two points lie on a unique line' is dual to 'two lines meet in a unique point'; proving one result immediately implies the other.

dual map For a *linear map $T:V{\rightarrow}W$ of *vector spaces, the dual map $T^*:W^*{\rightarrow}V^*$ between the *dual spaces is defined by $(T^*\varphi)(v) = \varphi(Tv)$ for φ in W^* and v in V. If

the *matrix for T is M, with respect to *bases for V and W, then the matrix for T^* is the *transpose M^T with respect to the *dual bases.

dual space (algebraic) Given a *vector space V over a *field F, the (algebraic) dual space V^* is the vector space of linear *functionals $\varphi:V{\rightarrow}F$ with addition and scalar multiplication defined *pointwise.

$$(\varphi_1+\varphi_2)(v)=\varphi_1(v)+\varphi_2(v),$$
$$(c\varphi)(v)=c(\varphi(v)).$$

If V is finite-dimensional, then V^* has the same dimension (*see* DUAL BASIS).

dual space (analytic) A linear *functional φ on a *normed vector space V is bounded if there exists M such that $|\varphi(v)| \leq M\|v\|$ for all $v{\varepsilon}V$. The analytic dual space V' is the space of all bounded linear functionals on V with *pointwise addition and scalar multiplication. V' is a *Banach space, whether or not V is *complete; the norm on V' is $\|\varphi\| = \sup\{|\varphi(v)| : \|v\| = 1\}$.

dummy activity (in critical path analysis) When two paths in an *activity network (edges as activities) have a common event but are independent, or partly independent of one another, it is necessary to introduce a dummy activity, essentially a logical constraint, linking the two paths at that point, with the duration of a dummy activity being zero. It is usually shown as a dotted line and can be labelled d_i. If R depends on P and Q and S depends on P then the activity network would show:

Representing a dummy activity

Activity networks allow at most one activity to be represented by an edge, so if two activities Q, R have to be carried out after one activity P and before another S a dummy activity is required as shown in the figure below.

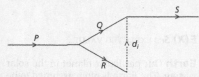

Representing a dummy activity

dummy variable A variable appearing in an expression is a dummy variable if the letter being used could equally well be replaced by another letter. For example, the two expressions

$$\int_0^1 x^2 \mathrm{d}x \text{ and } \int_0^1 t^2 \mathrm{d}t$$

represent the same definite integral, and so x and t are dummy variables. Similarly, the summation

$$\sum_{r=1}^{5} r^2$$

denotes the sum $1^2 + 2^2 + 3^2 + 4^2 + 5^2$ and would still do so if the letter r were replaced by the letter s, say; so r here is a dummy variable.

duplication of the cube See CONSTRUCTION WITH RULER AND COMPASS.

du Sautoy, Marcus (1965–) British mathematician who has done much to improve the communication of mathematical ideas through regular newspaper articles, the *Royal Institution Christmas Lectures in 2006, and books. He was awarded the Berwick Prize of the London Mathematical Society in 2001 for his work on *zeta functions.

(⊕) SEE WEB LINKS
• Marcus du Sautoy's website at Oxford University.

dynamic equilibrium A *body is in dynamic equilibrium if it is moving with constant *velocity and the vector sum of the *forces acting on it is zero.

dynamic programming The area of mathematics relating to the study of *optimization problems where a stepwise decision-making approach is employed. This is often done iteratively.

dynamical system A means of describing fully the state, or evolution of the state, of a system with time. Time may pass *discretely, in which case the system evolves according to a *recurrence relation, or may pass continuously, in which case the system evolves according to a *differential equation. If the system has n *degrees of freedom, the *phase space of possible states is an n-dimensional *manifold. For example, with a double pendulum this space would be a *torus, as the space is entirely determined by the two *angles the pendula make with the downward vertical.

dynamics The area of *mechanics relating to the study of *forces and the motion of *bodies.

e The number that is the base of natural logarithms. There are several ways of defining it. For example, define ln as in the entry the *logarithmic function. Then define exp as the *inverse function of ln (*see* EXPONENTIAL FUNCTION). Then define *e* as equal to exp 1. This amounts to saying that *e* is the number such that

$$\int_1^e \frac{1}{t}\,dt = 1.$$

The number *e* has important properties derived from some of the properties of ln and exp. For example,

$$e = \lim_{h \to 0}(1 + h)^{1/h} = \lim_{n \to \infty}\left(1 + \frac{1}{n}\right)^n,$$

which is how Jacques *Bernoulli first defined the number in 1683. Also, *e* is the sum of the series

$$1 + \frac{1}{1!} + \frac{1}{2!} + \dots + \frac{1}{n!} + \dots.$$

The value of *e* is 2.718 281 83 (to 8 decimal places). The proof that *e* is *irrational is comparatively easy. In 1873, *Hermite proved that *e* is *transcendental.

e (in conics) The symbol for the *eccentricity of a *conic.

e (in group theory) A common notation for the *identity element in a *group.

E The number 14 in *hexadecimal representation. Also abbreviation for *exa-.

E^n Denoting n-dimensional *Euclidean space.

E(X) *See* EXPECTED VALUE.

Earth Our particular planet in the solar system. The Earth is often assumed to be a *sphere with a radius of approximately 6400km. A more accurate accepted value is 6371km. A better approximation is to say that the Earth is in the shape of an oblate *spheroid with an equatorial radius of 6378km and a polar radius of 6357km. The mass of the Earth is 5.976×10^{24}kg.

eccentric Having centres which are distinct, as opposed to *concentric.

eccentricity The ratio of the distances between a point on a *conic and a fixed point (the focus) and between the point and a fixed line (the directrix). For $e = 0$ a *circle is produced, for $0 < e < 1$ an *ellipse is produced, for $e = 1$ a *parabola is produced and for $e > 1$ the conic produced is a *hyperbola.

(SEE WEB LINKS

• Two animations of ellipses with different eccentricities.

eccentricity (graph) For a vertex V in a *graph, the eccentricity is the greatest *distance from V to any other vertex in the graph.

echelon form A row of a matrix is called zero if all its entries are zero. Then a matrix is in echelon form if (i) all the zero rows come below the non-zero rows, and (ii) the first non-zero entry in each non-zero row is 1 and occurs in a column to the right of the

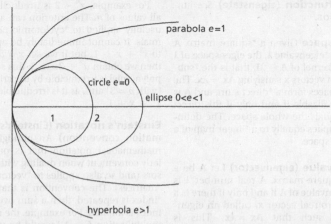

parabola e=1

circle e=0

ellipse 0<e<1

1   2

hyperbola e>1

Conics of different eccentricities

leading 1 in the row above. For example, these two matrices are in echelon form:

$$\begin{bmatrix} 1 & 6 & -1 & 4 & 2 \\ 0 & 0 & 1 & 2 & -3 \\ 0 & 0 & 0 & 1 & 5 \end{bmatrix}, \begin{bmatrix} 1 & 6 & -1 & 4 & 2 \\ 0 & 1 & 2 & -3 & 5 \\ 0 & 0 & 0 & 0 & 0 \end{bmatrix}.$$

Any matrix can be transformed to a matrix in echelon form using *elementary row operations, by a method known as *Gaussian elimination. The solutions of a set of linear equations may be investigated by transforming the *augmented matrix to echelon form. Further elementary row operations may be used to transform a matrix to *reduced echelon form.

ECO An abbreviation for ELEMENTARY COLUMN OPERATION.

EDA An abbreviation for EXPLORATORY DATA ANALYSIS.

edge (edge-set) See GRAPH.

edge-transitive (edge-regular, iso-toxal) A polygon or polyhedron is edge-transitive if there is a *symmetry taking any edge to any other edge. So a

*rhombus is edge-transitive, but not *vertex-transitive, and so not *regular.

education theory See MATHEMATICS EDUCATION.

efficiency See MACHINE.

efficiency (in statistics) A *statistic A is more efficient than another statistic B in estimating a parameter θ if its *variance is smaller. See ESTIMATOR.

effort See MACHINE.

Efron dice See NON-TRANSITIVE DICE.

Egyptian fraction An Egyptian fraction is a sum of reciprocals of positive integers such as $\frac{1}{3} + \frac{1}{5} + \frac{1}{6}$ which sums to $\frac{7}{10}$. All positive *rational numbers may be so expressed, though not uniquely, and a *greedy algorithm (due to *Fibonacci) will determine such expressions.

eigenbasis A *basis consisting of *eigenvectors. A square matrix, or a *linear map of a finite-dimensional *vector space, is *diagonalizable if and only if it has an eigenbasis.

eigenfunction (eigenstate) *See* EIG-
ENVECTOR.

eigenspace Given a *square matrix **A**
with an *eigenvalue λ, the eigenspace of λ
is the *kernel of $\mathbf{A} - \lambda\mathbf{I}$, that is the *subspace of vectors **x** satisfying $\mathbf{Ax} = \lambda\mathbf{x}$. The
eigenspaces form a *direct sum, and **A** is
*diagonalizable if and only if this direct
sum equals the whole space. The definition applies equally to a *linear map of a
*vector space.

eigenvalue (eigenvector) Let **A** be a
real *square matrix. A real number λ is
an eigenvalue of **A** if and only if there is a
non-zero real vector **x**, called an eigenvector, such that $\mathbf{Ax} = \lambda\mathbf{x}$. This is
equivalent to λ satisfying the *characteristic equation $\det(\mathbf{A} - \lambda\mathbf{I}) = 0$. More
generally, these definitions apply to
square matrices and *linear maps over
other *fields; in these cases the eigenvectors may be referred to as eigenfunctions, if the vector space is a *function
space, or as eigenstates, particularly in
*quantum theory.

Einstein, Albert (1879–1955) Outstanding German mathematical physicist, best known for his seminal work
on *relativity. He published in 1905 the
Special Theory of Relativity and in 1915
the General Theory. He made a fundamental contribution to the birth of quantum theory and had an important
influence on thermodynamics. Indeed,
his Nobel Prize of 1921 was for his work
on the photoelectric effect, rather than
for relativity.

Eisenstein's criterion A polynomial
$F(x) = a_nx^n + a_{n-1}x^{n-1} + \ldots + a_1x + a_0$
with integer coefficients is irreducible
over the rationals if a prime number p
exists such that:

(i) p divides each a_i for $i \neq n$;
(ii) p does not divide a_n;
(iii) p^2 does not divide a_0.

For example, $x^n - 2$ is irreducible for
all values of n. The criterion can also be
usefully applied to *cyclotomic polynomials. It cannot immediately be applied
to $x^2 + x + 1$, but if we set $x = u + 1$,
then we obtain $u^2 + 3u + 3$; this second
polynomial is irreducible by the criterion
(with $p = 3$) and, as it is irreducible, so is
$x^2 + x + 1$.

**Einstein's notation (Einstein's sum-
mation convention)** An abridging of
mathematical notation that is particularly convenient when dealing with *tensors (and so also applies to *vectors and
*matrices). The convention is that if an
*index is repeated, then a sum over that
index is implied. For example, the matrix
multiplication $C = AB$ would be written
$c_j^i = a_k^i b_j^k$ with the summation over k
suppressed and with the superscripts
and subscripts denoting that a matrix is
a $(1,1)$-tensor.

elastic A *body is said to be elastic if,
after being deformed by *forces applied
to it, it is able to regain its original shape
as soon as the deforming forces cease to
act. *See* DEFORMATION.

elastic collision A *collision in which
there is no loss of kinetic energy.

elasticity *See* MODULUS OF ELASTICITY.

elastic potential energy *See* POTEN-
TIAL ENERGY.

elastic string A string that can be
extended, but not compressed, and that
resumes its natural length as soon as the
forces applied to extend it are removed.
How the *tension in the string varies with
the *extension may be complicated. If
*Hooke's law holds, the *tension is pro-
portional to the extension, but this will
only apply up to the elastic limit.

electric field A *vector field which
exerts a force on any electrical charge at
a given point. Electric fields which

change with time influence local magnetic fields. *See* MAXWELL'S EQUATIONS.

electromagnetic potentials *Maxwell's equations can be simplified by defining an electric scalar potential ϕ and a magnetic vector potential **A** in terms of which **E**, the *electric field, and **B**, the *magnetic field, can be expressed. **A** and ϕ are given by

$$\mathbf{B} = \nabla \times \mathbf{A} \quad \text{and} \quad \mathbf{E} = -\nabla\phi - \frac{1}{c}\frac{\partial \mathbf{A}}{\partial t}$$

where c is the *speed of light. **A** and ϕ are not uniquely determined by these expressions, and they are normally chosen to satisfy the extra condition $\nabla.\mathbf{A} + \frac{1}{c}\frac{\partial\phi}{\partial t} = 0$.

electromagnetic wave In a vacuum, an electromagnetic wave travels at the *speed of light c. A monochromatic electromagnetic plane wave with *angular frequency ω has *electric field

$$\mathbf{E} = \mathbf{a}\cos\left(\omega\left(t - \frac{\mathbf{r}\cdot\mathbf{e}}{c}\right)\right) + \mathbf{b}\cos\left(\omega\left(t - \frac{\mathbf{r}\cdot\mathbf{e}}{c}\right)\right).$$

Here **e** is a unit vector in the direction of the wave's travel and $\mathbf{r} = (x,y,z)$. Further, **a** and **b** are perpendicular to **e**. The magnetic field is given by $\mathbf{B} = \mathbf{e} \times \mathbf{E}/c$. Then **E** and **B** satisfy *Maxwell's equations in a vacuum and also satisfy the *wave equation.

element An object in a *set is an element of that set.

elementary column operation Commonly abbreviated to ECO. One of the following operations on the columns of a matrix:

(i) interchange two columns,
(ii) multiply a column by a non-zero scalar,
(iii) add a multiple of one column to another column.

An elementary column operation can be produced by post-multiplication by the appropriate *elementary matrix.

elementary function Any of the following real functions: the *rational functions, the *trigonometric functions, the *logarithmic and *exponential functions, the functions f defined by $f(x) = x^{m/n}$ (where m and n are non-zero integers), and all those functions that can be obtained from these by using addition, subtraction, multiplication, division, *composition, and the taking of *inverse functions. *Compare* SPECIAL FUNCTION.

elementary matrix A square matrix obtained from the identity matrix **I** by an *elementary row operation. Thus there are three types of elementary matrix. Examples of each type are:

$$(i) \begin{bmatrix} 1 & 0 & 0 & 0 & 0 & 0 & 0 \\ 0 & 0 & 0 & 0 & 1 & 0 & 0 \\ 0 & 0 & 1 & 0 & 0 & 0 & 0 \\ 0 & 0 & 0 & 1 & 0 & 0 & 0 \\ 0 & 1 & 0 & 0 & 0 & 0 & 0 \\ 0 & 0 & 0 & 0 & 0 & 1 & 0 \\ 0 & 0 & 0 & 0 & 0 & 0 & 1 \end{bmatrix},$$

$$(ii) \begin{bmatrix} 1 & 0 & 0 & 0 & 0 & 0 & 0 \\ 0 & 1 & 0 & 0 & 0 & 0 & 0 \\ 0 & 0 & -3 & 0 & 0 & 0 & 0 \\ 0 & 0 & 0 & 1 & 0 & 0 & 0 \\ 0 & 0 & 0 & 0 & 1 & 0 & 0 \\ 0 & 0 & 0 & 0 & 0 & 1 & 0 \\ 0 & 0 & 0 & 0 & 0 & 0 & 1 \end{bmatrix},$$

$$(iii) \begin{bmatrix} 1 & 0 & 0 & 0 & 0 & 0 & 0 \\ 0 & 1 & 0 & 0 & 4 & 0 & 0 \\ 0 & 0 & 1 & 0 & 0 & 0 & 0 \\ 0 & 0 & 0 & 1 & 0 & 0 & 0 \\ 0 & 0 & 0 & 0 & 1 & 0 & 0 \\ 0 & 0 & 0 & 0 & 0 & 1 & 0 \\ 0 & 0 & 0 & 0 & 0 & 0 & 1 \end{bmatrix}.$$

The matrix (i) is obtained from **I** by interchanging the second and fifth rows, matrix (ii) by multiplying the third row by -3, and matrix (iii) by adding 4 times the fifth row to the second row. Premultiplication of an $m \times n$ matrix **A** by an $m \times m$ elementary matrix produces the result of the corresponding row operation on **A**.

Alternatively, an elementary matrix can be seen as one obtained from the identity matrix by an *elementary column operation; and post-multiplication of an $m \times n$ matrix **A** by an $n \times n$ elementary matrix produces the result of the corresponding column operation on **A**.

elementary row operation One of the following operations on the rows of a matrix:

(i)   interchange two rows,
(ii) multiply a row by a non-zero scalar,
(iii) add a multiple of one row to another row.

An elementary row operation can be produced by a premultiplication by the appropriate *elementary matrix. Elementary row operations are applied to the *augmented matrix of a system of linear equations to transform it into *echelon form or *reduced echelon form. Each elementary row operation corresponds to an operation on the system of linear equations that does not alter the solution set of the equations or the *row space of the matrix. Commonly abbreviated to ERO.

elementary symmetric polynomial
For three variables, α, β, γ, the elementary symmetric polynomials are

$$\sigma_1 = \alpha + \beta + \gamma, \quad \sigma_2 = \alpha\beta + \beta\gamma + \gamma\alpha,$$
$$\sigma_3 = \alpha\beta\gamma.$$

These expressions generalize naturally to n variables, with σ_k being the sum of the $\binom{n}{k}$ products of k of the variables. The elementary symmetric polynomials appear in *Viète's formulae.

The elementary symmetric polynomials are *symmetric functions, and any symmetric polynomial can be expressed in terms of the elementary symmetric polynomials. For example, if $s_k = \alpha^k + \beta^k + \gamma^k$, then

$$s_{k+3} = \sigma_1 s_{k+2} - \sigma_2 s_{k+1} + \sigma_3 s_k$$

gives a general recursive means of expressing s_k in terms of $\sigma_1, \sigma_2, \ldots, \sigma_k$. See NEWTON'S IDENTITIES.

elevation The angle between the horizontal and a straight line, measured positively in the direction of the upwards vertical.

elimination method A method of solving linear simultaneous equations by reducing the number of variables by taking appropriate linear combinations of the equations. For example, when $3x + 2y = 7$ (I) and $5x - 3y = -1$ (II) then $3 \times$ (I) $+ 2 \times$ (II) $\Rightarrow 19x = 19$, so the variable y has been eliminated, allowing the variable x to be identified as 1, and then the substitution back into either of the original equations gives the corresponding value for y.

ellipse A particular 'oval' shape, informally obtained by stretching or squashing a circle. If it has length $2a$ and width $2b$, its area equals πab.

More precisely, we may define an ellipse as a *conic with eccentricity less than 1. Thus, it is the locus of all points P such that the distance from P to a fixed point F_1 (the focus) equals e (< 1) times the distance from P to a fixed line l_1 (the directrix). It turns out that there is another point F_2 and another line l_2 such that the same locus would be obtained with these as focus and directrix. An ellipse is also the conic section that results when a plane cuts a cone in such a way that a bounded section is obtained (see CONIC).

The line through F_1 and F_2 is the major axis, and the points V_1 and V_2 where it cuts the ellipse are the vertices. The length $|V_1 V_2|$ is the length of the major axis and is usually taken to be $2a$. The midpoint of $V_1 V_2$ is the centre of the ellipse. The line through the centre perpendicular to the major axis is the minor axis, and the distance, usually taken to be $2b$, between the points where it cuts the ellipse is the length of the minor axis. The three constants a, b, and e are related by $e^2 = 1 - b^2/a^2$. The eccentricity e determines the shape of the ellipse. The value $e = 0$ is permitted and gives rise to a circle, though this requires the directrices to be infinitely far away, invalidating the focus and directrix description.

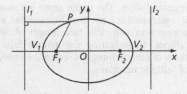

An ellipse in normal form

By taking a *coordinate system with origin at the centre of the ellipse and x-axis along the major axis, the foci have coordinates $(ae, 0)$ and $(-ae, 0)$, the directrices have equations $x = a/e$ and $x = -a/e$, and the ellipse has equation

$$\frac{x^2}{a^2} + \frac{y^2}{b^2} = 1,$$

where $a > b > 0$ (see NORMAL FORM). This ellipse may be parameterized by $x = a\cos\theta$, $y = b\sin\theta$ $(0 \le \theta < 2\pi)$ as *parametric equations.

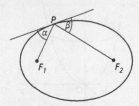

Ellipse highlighting property (ii)

The ellipse has two important properties:

(i) If P is any point of the ellipse with foci F_1 and F_2 and length of major axis $2a$, then $|PF_1| + |PF_2| = 2a$. The fact that an ellipse can be seen as the locus of all such points is the basis of a practical method of drawing an ellipse using a string between two points.

(ii) For any point P on the ellipse, let α be the angle between the tangent at P and the line PF_1, and β the angle between the tangent at P and the line PF_2, as shown in the figure; then $\alpha = \beta$. This property is analogous to that of the parabolic reflector (see PARABOLA).

ellipsis The three dots that appear in finite *sums and *products such as $1 + 2 + \ldots + n$ to represent the omitted terms or at the end of *decimal representations, *infinite series, infinite *sequences, etc. to signify continuation *ad infinitum.

ellipsoid A *quadric whose equation in a suitable coordinate system is

$$\frac{x^2}{a^2} + \frac{y^2}{b^2} + \frac{z^2}{c^2} = 1.$$

The three axial planes are planes of symmetry. All non-empty plane sections of the ellipsoid are *ellipses.

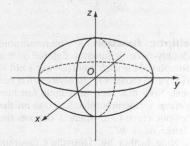

An ellipsoid in normal form

elliptic curve At its simplest an elliptic curve is a *non-singular *curve in $\mathbb{R}^2$ with the equation

$$y^2 = x^3 + px + q.$$

The non-singularity then requires that $4p^3 + 27q^2 \neq 0$. But the study of elliptic curves over the *complex numbers (including *points at infinity, so studying the curves within the complex projective plane (*see* PROJECTIVE SPACE)) has a rich history. Such a curve is then topologically a *torus and as a *Riemann surface can be identified with $\mathbb{C}/\Lambda$, where Λ is some *lattice $a\mathbb{Z} + b\mathbb{Z}$, where a/b is not real. Now $\mathbb{C}/\Lambda$ is an *abelian group, and this group structure on the curve manifests geometrically via a theorem due to *Abel.

Elliptic curves over the *rational numbers are of great interest in *number theory; the Birch-Swinnerton-Dyer conjecture (*see* APPENDIX 23) concerns them. And over *Galois fields, elliptic curves have applications in *cryptography. *See also* DEGREE-GENUS FORMULA.

elliptic cylinder A *cylinder in which the fixed curve is an *ellipse and the fixed line to which the *generators are parallel is *perpendicular to the plane of the ellipse. It is a *quadric and in a suitable coordinate system has the equation

$$\frac{x^2}{a^2} + \frac{y^2}{b^2} = 1.$$

elliptic function A *meromorphic doubly-periodic function defined on the complex plane such that $f(z + ma + nb) = f(z)$ for all integers m, n, where a/b is not real. Note that such an elliptic function defines a *meromorphic function on the *elliptic curve $\mathbb{C}/\Lambda$, where Λ denotes the *lattice $a\mathbb{Z} + b\mathbb{Z}$.

Note further by *Liouville's theorem that *holomorphic elliptic functions are constant. Elliptic functions became of interest as a consequence of *elliptic integrals.

elliptic geometry *See* NON-EUCLIDEAN GEOMETRY, ELLIPTIC PLANE.

elliptic integral A *function $f(x)$ which can be expressed as an *integral of the form

$$f(x) = \int R\left(x, \sqrt{P(x)}\right) \, dx$$

where R is a *rational function and P is a *cubic or *quartic polynomial with no repeated roots. The name arose because integrals in this form were first studied in connection with the *arc length of an *ellipse. Investigating the inverse functions of elliptic integrals led to the study of *elliptic functions.

elliptic paraboloid A *quadric whose equation in a suitable coordinate system is

$$\frac{x^2}{a^2} + \frac{y^2}{b^2} = \frac{2z}{c}.$$

Here the yz-plane and the zx-plane are planes of symmetry. Sections by planes $z = k$, where $k \geq 0$, are ellipses. Sections by planes parallel to the yz-plane and to the zx-plane are parabolas. Planes through the z-axis cut the paraboloid in parabolas with vertex at the origin.

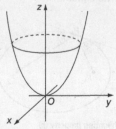

An elliptical paraboloid in normal form

elliptic partial differential equation *See* PARABOLIC PARTIAL DIFFERENTIAL EQUATION.

elliptic plane Spherical geometry (*see* SPHERICAL TRIGONOMETRY), with the *great circles replacing *lines, is not a type of *non-Euclidean geometry; this is because two 'lines' do not meet in a single point but rather in *antipodal points. If, instead, antipodal points of the sphere are identified then we create the elliptic plane, which is a non-Euclidean geometry where each line has no *parallels. The elliptic plane has the *topology of the *real projective plane.

embedding In general, an embedding of X in Y is an injection $f : X \rightarrow Y$ such that the *intrinsic structure of X agrees with the structure inherited *extrinsically from the ambient space Y. This is commonly denoted as $f : X \hookrightarrow Y$.

In *algebra this means that the map is an *isomorphism onto its *image, in *topology a *homeomorphism onto its image, etc. For smooth *manifolds, a smooth embedding is a *diffeomorphism onto its image such that the *differential is injective everywhere. This means that the image will also be a smooth manifold.

empirical Deriving from experience or observation rather than from reasoning.

empirical probability The probability that a fair die will show a four when thrown is 1/6, using an argument based on equally likely outcomes. For a die which is weighted, observations would need to be taken and an empirical probability based on their *relative frequency could be calculated.

empty set The set, denoted by $\emptyset$ with no elements in it. Consequently, its *cardinality is zero.

emv (expected monetary value) Where the *payoff to the player in a game is dependent on chance outcomes, the emv is the average gain the player would achieve per game in the long run. It is calculated as the sum of: each payoff multiplied by the probability of that payoff.

For example:

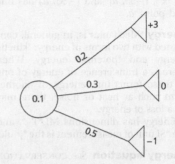

Showing possible payoffs and chances

The emv is $3 \times 0.2 + 0 \times 0.3 + (-1) \times 0.5 = 0.1$ as written in the circle.

emv algorithm For a *decision tree with all *payoffs and the probabilities of all chance outcomes known, determine the *emv for the chance nodes (or vertices) starting from the payoffs. At each decision node the player takes the decision which maximizes the *emv.

The notion of a *utility function develops this type of problem further by recognizing that the same amount of money may be valued differently by people in different financial circumstances.

encrypt *See* CRYPTOGRAPHY.

End *See* ENDOMORPHISM.

endomorphism At its most general, a *morphism from a mathematical object to itself. This might mean a group or ring *homomorphism or a *linear map of a *vector space. If X is an *abelian group, then the endomorphisms of X form a *ring End(X) under *pointwise addition and *composition as multiplication.

End($\mathbb{R}^n$) is isomorphic to the ring of $n \times n$ real *matrices.

end-point A number defining one end of an *interval on the real line. Each of the *bounded intervals $[a, b]$, (a, b), $[a, b)$ and $(a, b]$ has two end-points, a and b. Each of the unbounded intervals $[a, \infty)$, (a, ∞), $(-\infty, a]$, and $(-\infty, a)$ has one end-point, a.

energy *Mechanics is, in general, concerned with two forms of energy: *kinetic energy and *potential energy. When there is a transference of energy of one of these forms into energy of another form such as heat or noise, there may be a loss of energy.

Energy has dimensions ML^2T^{-2}, and the *SI unit of measurement is the *joule.

energy equation *See* CONSERVATION OF ENERGY.

enneagon A synonym for NONAGON.

entering variable *See* SIMPLEX METHOD.

entire function A complex-valued function that is *holomorphic everywhere in the *complex plane. Polynomials, exponentials, and trigonometric and hyperbolic functions are all examples, while the function $f:z \to 1/z$, which is not holomorphic at zero (but is *meromorphic), is not an entire function.

entropy (Shannon entropy) For a *random variable X which takes k values with probabilities $p_1, p_2, \ldots p_k$, the entropy of X is given by $H(X) = -\sum_1^k p_i \log_2 p_i$. This is the *expected value of the *information when sampling X. It is maximal when the probability distribution is uniform. *See* SHANNON'S THEOREM.

entry *See* MATRIX.

Entscheidungsproblem A synonym for DECISION PROBLEM.

enumerable A synonym for COUNTABLE.

enumeration *See* PERMUTATIONS, SELECTIONS.

envelope A *curve or *surface that is *tangential to every curve or surface in a *family. For example, if a family of circles has radius a and centre at a distance $b > a$ from a fixed point C, the envelope will be an annulus with the two circles having radii $b + a$ and $b - a$.

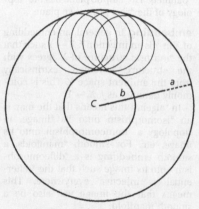

An annulus as an envelope

envelope paradox (exchange paradox) The envelope paradox is based on the following scenario.

You are, at random, given one of two sealed envelopes that contain money. You do not know the sums in the envelopes but do know that one envelope contains twice the amount in the other. You are given the option to swap your envelope for the other. Should you?

It seems clear by symmetry that there is no benefit in swapping—which is the correct answer—but the following argument to the contrary is perhaps initially convincing. Denote by X the amount in your envelope; the other amount is either

$2X$ or $\frac{1}{2}X$. So you have a half chance of gaining X and a half chance of losing $\frac{1}{2}X$ and so an expected gain of $\frac{1}{2}(X - \frac{1}{2}X) = \frac{1}{4}X > 0$ and so you should swap.

Different stances can be taken on what is erroneous about this argument. It may be seen as an abuse of notation by letting X denote different amounts in the two different scenarios or equivalently not recognizing the swapping is not independent of the dealing. *See also* BERTRAND'S PARADOX, MONTY HALL PROBLEM.

epicycloid The *curve traced out by a point on the circumference of a *circle rolling round the outside of a fixed circle. When the two circles have the same radius, the curve is a *cardioid. The epicycloid is a *closed curve if the ratio of the circles' radii is *rational.

epidemiology *See* SIR EPIDEMIC MODEL.

epimorphism A *morphism $f: X \to Y$ with the property that for all morphisms g between Y and Z, $g_1 \circ f = g_2 \circ f \Rightarrow g_1 = g_2$. In *category theory, epimorphisms are the analogue of surjections. *Compare* MONOMORPHISM.

epsilon The Greek letter e, written ε, commonly used in *analysis to represent a small, strictly positive quantity.

epsilon–delta notation The standard notation used to define the concepts of *limits and *continuity.

equality (of complex numbers) *See* EQUATING REAL AND IMAGINARY PARTS.

equality (of sets) Sets A and B are equal if they consist of the same elements. Establishing $A = B$ is equivalent to showing both $A \subseteq B$ and $B \subseteq A$.

equals sign The symbol '$=$' used between two expressions to indicate that two mathematical objects are the same, or to create an *equation which

may hold true for certain values of variables. In a definition, the symbol '$:=$' or '$\overset{\text{def}}{=}$' when an object is being defined as equalling something.

equate To form an *equation by joining two expressions, or an expression and a value, by an equals sign.

equating coefficients Let $f(x)$ and $g(x)$ be *polynomials, and let

$$f(x) = a_n x^n + a_{n-1} x^{n-1} + \ldots + a_1 x + a_0,$$
$$g(x) = b_n x^n + b_{n-1} x^{n-1} + \ldots + b_1 x + b_0,$$

where it is not necessarily assumed that $a_n \neq 0$ and $b_n \neq 0$. If $f(x) = g(x)$ for all values of x, then $a_n = b_n, a_{n-1} = b_{n-1}, \ldots, a_1 = b_1, a_0 = b_0$.

The above is true by definition of the equality of polynomials. Note that, for a prime p, the polynomials x^p and x take the same values for all x mod p (as a consequence of *Fermat's Little Theorem). However x^p and x are distinct polynomials, having different coefficients.

equating real and imaginary parts Complex numbers $a + bi$ and $c + di$ are equal if and only if $a = c$ and $b = d$. Using this fact is called equating real and imaginary parts. For example, if $(a+bi)^2 = 5 + 12i$, then $a^2 - b^2 = 5$ and $2ab = 12$.

equation A statement that asserts that two mathematical expressions are equal in value. If this is true for all values of the variables involved then it is called an *identity, for example $3(x - 2) = 3x - 6$, and where it is only true for some values it is called a *conditional equation; for example $x^2 - 2x - 3 = 0$ is only true when $x = -1$ or 3, which are known as the *roots of the equation.

equation of motion An equation, based on the second of *Newton's laws of motion, that governs the motion of a particle. In vector form, the equation is $m\ddot{r} = \mathbf{F}$, where $\mathbf{F}$ is the total *force acting on the particle of mass m. In *Cartesian coordinates, this is equivalent to the three differential equations $m\ddot{x} = F_1$, $m\ddot{y} = F_2$ and $m\ddot{z} = F_3$, where $\mathbf{F} = F_1\mathbf{i} + F_2\mathbf{j} + F_3\mathbf{k}$. In *cylindrical polar coordinates, taken here to be (r,θ,z), it is equivalent to $m(\ddot{r} - r\dot{\theta}^2) = F_1$, $\frac{m}{2}\frac{d}{dt}(r^2\dot{\theta}) = F_2$, and $m\ddot{z} = F_3$, where now $\mathbf{F} = F_1\mathbf{e}_r + F_2\mathbf{e}_\theta + F_3\mathbf{k}$, and $\mathbf{e}_r = \mathbf{i}\cos\theta + \mathbf{j}\sin\theta$ and $\mathbf{e}_\theta = -\mathbf{i}\sin\theta + \mathbf{j}\cos\theta$ (*see* RADIAL AND TRANSVERSE COMPONENTS).

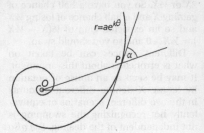

Spiral with constant angle α

equations of motion with constant acceleration Equations relating to an object moving in a straight line with constant *acceleration a. Let u be the initial *velocity, v the final velocity, t the time of travel, and s the displacement from the starting point. Then

$$v = u + at, \quad s = ut + \frac{1}{2}at^2,$$
$$s = \frac{1}{2}(u + v)t, \quad v^2 = u^2 + 2as.$$

Collectively these equations are often referred to as the 'suvat' equations.

equator A *circle which divides a sphere, or other body, into two symmetrical parts.

equi- A prefix denoting *equality.

equiangular (of figures) A pair of figures for which the *angles taken in order in each figure are equal.

equiangular (within a figure) A figure that has all angles equal.

equiangular spiral A curve whose equation in polar coordinates is $r = ae^{k\theta}$, where $a > 0$ and k are constants. Let O be the origin and P be any point on the curve. The curve derives its name from the property that the angle α between OP and the tangent at P is constant. In fact, $k = \cot\alpha$. The curve is also called the logarithmic spiral.

SEE WEB LINKS

- Animations of related curves, and an interactive manipulation of the spiral in Geometer's Sketchpad.

equiareal (equal area) A map is equiareal if it preserves *area. For a *linear map, from $\mathbb{R}^2$ to $\mathbb{R}^2$, this is equivalent to the *determinant equalling ± 1. *Map projections, such as Mollweide's, may be preferred to be equiareal, so that the relative areas of regions are preserved.

equicontinuity A family of functions $f_i : X \rightarrow Y$ between *metric spaces are equicontinuous at x in X if for any $\varepsilon > 0$ there exists $\delta > 0$ such that whenever $d_X(x,y) < \delta$, then $d_Y(f_i(x), f_i(y)) < \varepsilon$ for all i.

The family is uniformly equicontinuous if for any $\varepsilon > 0$ there exists $\delta > 0$ such that whenever $d_X(x,y) < \delta$, then $d_Y(f_i(x), f_i(y)) < \varepsilon$ for all i and all x,y.

equidistant Being the same distance from one or more points or objects. The perpendicular bisector of the *line segment AB is the set of points equidistant from A and B.

The set of points equidistant from a fixed point is a circle in two dimensions and a sphere in three dimensions.

The set of points equidistant from a line segment is formed by a pair of *parallel line segments with two *semicircles

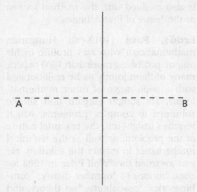

Locus of points equidistant from A and B

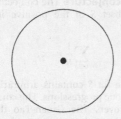

Circle of points equidistant from centre

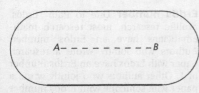

Locus of points equidistant from a line segment

equilateral A *polygon is equilateral if all its sides have the same length. In an equilateral *triangle, the three angles are also all equal, and so the triangle is regular. A *rhombus is equilateral but not *regular.

equilibrium A particle is in equilibrium when it is at rest and the total *force acting on it is zero for all time.

These conditions are equivalent to saying that $\mathbf{r} = \mathbf{r}_0$ (a constant vector), $\dot{\mathbf{r}} = \mathbf{0}$ and $\ddot{\mathbf{r}} = \mathbf{0}$ for all time, where $\mathbf{r}$ is the position vector of the particle. We then say that $\mathbf{r}_0$ is an equilibrium or equilibrium position.

Consider a *rigid body experiencing a system of forces. Let $\mathbf{F}$ be the total force and $\mathbf{M}$ the moment of the forces about the origin. The rigid body is in equilibrium when it is at rest and $\mathbf{F} = \mathbf{0}$ and $\mathbf{M} = \mathbf{0}$ for all time.

A body, in equilibrium, is in *stable* equilibrium if, following a small change in its position, it returns to, or remains close to, the equilibrium position. It is in *unstable* equilibrium if, following a small change in its position, it continues to move further from the equilibrium position. *See* LINEAR THEORY OF EQUILIBRIA, LYAPUNOV STABILITY.

equinumerous (equinumerable) Two sets are equinumerous if there is a *bijection between them. Essentially the two sets contain the same number of elements. *See* CARDINAL NUMBER, CARDINALITY.

equipollent (equipotent) Two statements in *logic are said to be equipollent if either one can be deduced from the other, so in that sense the statements are equivalent.

equiprobable Events with the same *probability. For example, in tossing a fair coin, the two outcomes are equiprobable. In rolling a fair die, the events 'throw an even number' and 'throw a prime number' are equiprobable.

equivalence (of matrices) Two $m \times n$ *matrices M_1, M_2 are equivalent if there exists an invertible $m \times m$ matrix P and an invertible $n \times n$ matrix Q such that $M_1 = PM_2Q$. If the *entries are from a *field, then a matrix M is equivalent to $\begin{pmatrix} I_r & 0 \\ 0 & 0 \end{pmatrix}$ where r is the *rank of M. *See also* SMITH NORMAL FORM.

equivalence class For an *equivalence relation ∼ on a set S, the equivalence class [a], or ā, of a ∈ S is the set of elements of S equivalent to a; that is, [a] = {x | x ∈ S and a ∼ x}. The collection of distinct equivalence classes *partition S.

equivalence relation A *binary relation ∼ on a set S that is *reflexive, *symmetric and *transitive. For an equivalence relation ∼, a is said to be equivalent to b when a ∼ b. It is an important fact that, from an equivalence relation on S, *equivalence classes can be defined to obtain a *partition of S. Conversely, given any partition of S, an equivalence relation can be defined by a ∼ b if and only if a and b are in the same part of the partition.

equivalent *See* EQUIVALENCE (OF MATRICES), EQUIVALENCE RELATION.

equivalent norms Let ‖ ‖ and ‖ ‖′ be two *norms on a real (or complex) *vector space X. They are equivalent if there are positive constants a, b for which a‖x‖ ≤ ‖x‖′ ≤ b‖x‖ for all x in X. Equivalent norms induce the same *topology; all norms on a finite-dimensional vector space are equivalent.

equivariant map If a *group G *acts (*see* ACTION) on sets X and Y, then a map f:X → Y is equivariant if f(g·x) = g·f(x) for all g ∈ G, x ∈ X. For example, if X is the set of *triangles in the plane, Y is the set of points in the plane and G is the group of *isometries, then the map f taking a triangle to its *centroid is equivariant. *Compare* INVARIANT.

Eratosthenes of Cyrene (about 275–195 BC) Greek astronomer and mathematician who was the first to calculate the size of the Earth by making measurements of the *angle of the Sun at two different places a known *distance apart. His other achievements include measuring the tilt of the Earth's axis. He is also credited with the method known as the *sieve of Eratosthenes.

Erdős, Paul (1913–96) Hungarian mathematician who was prolific in his output, publishing more than 1500 papers, many of them jointly as he collaborated with a wide range of other mathematicians. He sought elegant and simple solutions to complex problems, which requires insight into the essential nature of the problem, as well as the technical mathematics to extract the solution. He was awarded the *Wolf Prize in 1983 for contributions to *number theory, *combinatorics, *probability, *set theory, and *analysis.

Erdős conjecture The conjecture that if a subset S of the positive integers satisfies

$$\sum_{n \in S} \frac{1}{n} = \infty,$$

then the set S contains arbitrarily long *arithmetic progressions. The conjecture is unproven. The Green-Tao theorem proves the conjecture true in the special case when S is the set of *prime numbers.

Erdős number Due to Paul *Erdős' prolific research, most research mathematicians have an Erdős number. Authors who jointly wrote a research paper with Erdős have an Erdős number of 1. Other authors who jointly wrote a paper with someone with Erdős number 1 have an Erdős number of 2, and so on.

erf *See* ERROR FUNCTION.

ergodic In a *Markov chain, a state is ergodic if it is *aperiodic, is *recurrent, and has finite mean recurrence time.

ergodic theory Ergodic theory focuses on the long term behaviour of *dynamical systems and the existence of averages of quantities over time when the system is allowed to run for a long time.

ERO An abbreviation for ELEMENTARY ROW OPERATION.

error Let x be an *approximation to a value X. According to some authors, the error is $X - x$; for example, when 1.9 is used as an approximation for 1.875, the error equals -0.025. Others define the error to be $x - X$. Whichever of these definitions is used, the error can be positive or negative. Yet other authors define the error to be $|X - x|$, the difference between the true value and the approximation, in which case the error is always non-negative. When contrasted with *relative error, the error may be called the absolute error.

error analysis In *numerical analysis, error analysis relates to the propagation of errors within calculations and to the bounding of those errors. An *algorithm may involve many iterations, and so error propagation might make a calculated solution meaningless. *See* STABLE NUMERICAL ANALYSIS.

error-correcting (error-detecting) A *code is said to be error-detecting if any one error in a codeword results in a word that is not a codeword, so that the receiver knows that an error has occurred. A code is error-correcting if, when any one error occurs in a codeword, it is possible to decide which codeword was intended. Certain error-correcting codes may not be able to detect errors if more than one error occurs in a codeword; other error-correcting codes can be constructed that can detect and correct more than one error in a codeword. *See* HAMMING DISTANCE.

error function (Gauss error function) The error function $\text{erf}(x)$ is a *special function defined as

$$\text{erf}(x) = \frac{2}{\sqrt{\pi}} \int_0^x e^{-t^2} \, dt.$$

If $\Phi(x)$ denotes the *cumulative distribution function of the *normal distribution $N(0,1)$, then

$$\text{erf}(x) = 2\Phi(x\sqrt{2}) - 1.$$

escape speed (escape velocity) For a celestial body, the minimum *speed with which an object must be projected away from its surface so that the object does not return again because of *gravity. The escape speed for the *Earth is $\sqrt{2gR}$ where R is the radius of the Earth, and is approximately 11.2 kilometres per second. Similarly, the escape speed associated with a spherical celestial body of mass M and radius R is $\sqrt{2GM/R}$ where G is the *gravitational constant.

e-science Computationally intensive science that is achieved via large *computer networks.

escribed circle A synonym for EXCIRCLE.

essential singularity An isolated *singularity a of a complex-valued *function $f(z)$ such that the *Laurent series of $f(z)$ at a includes infinitely many negative powers. For example, the function $\sin(1/z)$ has an essential singularity at $z = 0$. *Compare* POLE, REMOVABLE SINGULARITY.

estimate The value of an *estimator calculated from a particular sample. If the estimate is a single figure, it is a point estimate; if it is an interval, such as a *confidence interval, it is an interval estimate.

estimation The process of determining as nearly as possible the value of a population *parameter by using an *estimator. The value of an estimator found from a particular sample is called an *estimate.

estimator A *statistic used to estimate the value of a population *parameter.

An estimator X of a parameter θ is consistent if the probability of the difference between the two exceeding an arbitrarily small fixed number tends to zero as the *sample size increases indefinitely. An estimator X is an unbiased estimator of the parameter θ if $E(X) = \theta$, and it is a *biased estimator otherwise (see EXPECTED VALUE). The best unbiased estimator is the unbiased estimator with the minimum *variance. The relative *efficiency of two unbiased estimators X and Y is the ratio $\mathrm{Var}(Y)/\mathrm{Var}(X)$ of their variances.

Estimators may be found in different ways, including the method of maximum likelihood (see LIKELIHOOD) and the method of moments (see MOMENT).

Euclid (about 300 BC) Outstanding mathematician of Alexandria, author of what may well be the second most influential book in Western culture: the *Elements*. Little is known about Euclid himself, and it is not clear to what extent the book describes original work and to what extent it is a textbook. The *Elements* develops a large section of elementary geometry by rigorous logic starting from assumed axioms. It includes his proof that there are infinitely many primes, the *Euclidean algorithm, the derivation of the five *Platonic solids, and much more. It served for two millennia as a model of what pure mathematics is about.

(((●))) SEE WEB LINKS

• A fuller biography of Euclid.

Euclidean Algorithm A process, based on the *Division Algorithm, for finding the *greatest common divisor (a, b) of two positive integers a and b. Assuming that $a > b$, write $a = bq_1 + r_1$, where $0 \leq r_1 < b$. If $r_1 = 0$, the gcd (a,b) is equal to b; if $r_1 \neq 0$, then $(a,b) = (b,r_1)$, so the step is repeated with b and r_1 in place of a and b. After further repetitions, the last non-zero remainder obtained is

the required gcd. For example, for $a = 1274$ and $b = 871$, write

$$1274 = 1 \times 871 + 403,$$
$$871 = 2 \times 403 + 65,$$
$$403 = 6 \times 65 + 13,$$
$$65 = 5 \times 13.$$

and then $(1274,871) = (871,403) = (403, 65) = (65,13) = 13$.

The algorithm also enables s and t to be found such that the gcd can be expressed as $sa + tb$ (see BÉZOUT'S LEMMA). We use the previous equations in turn to express each remainder in this form. Thus,

$$403 = 1274 - 1 \times 871 = a - b,$$
$$65 = 871 - 2 \times 403 = b - 2(a - b)$$
$$= 3b - 2a,$$
$$13 = 403 - 6 \times 65 = (a - b) - 6(3b - 2a)$$
$$= 13a - 19b.$$

Euclidean construction A synonym for CONSTRUCTION WITH RULER AND COMPASS.

Euclidean distance (Cartesian distance) The *distance between two points in Euclidean or Cartesian space. In two dimensions this is $\sqrt{(x_1 - x_2)^2 + (y_1 - y_2)^2}$ and in three dimensions it is $\sqrt{(x_1 - x_2)^2 + (y_1 - y_2)^2 + (z_1 - z_2)^2}$ with the obvious notation for the coordinates of the points.

Euclidean domain An *integral domain in which a *Division Algorithm holds. Specifically, an integral domain R with a function $d:R\backslash\{0\} \rightarrow \mathbb{N}$ such that (i) $d(a) \leq d(ab)$ for all $a,b \in R\backslash\{0\}$; (ii) for any $a,b \in R\backslash\{0\}$ there exist $q,r \in R$ such that $a = qb + r$ and $d(r) < d(b)$ or $r = 0$. Examples include $\mathbb{Z}$ with $d(n) = |n|$, the *Gaussian integers with $d(a+ib) = a^2 + b^2$, and *polynomials over a *field with $d(f)$ being the *degree of f. Euclidean domains are *principal ideal domains (see QUADRATIC FIELD).

Euclidean geometry The area of mathematics relating to the study of *geometry based on the definitions and *axioms set out in Euclid's book, the *Elements*. In particular, geometry that relies on the *parallel postulate.

Euclidean group The *group of *isometries of *Euclidean space $\mathbb{R}^n$ under *composition. Every isometry of $\mathbb{R}^n$ has the form $\mathbf{v} \mapsto \mathbf{Av} + \mathbf{b}$, where $\mathbf{A}$ is an $n \times n$ *orthogonal matrix and $\mathbf{b}$ is an $n \times 1$ column vector.

Euclidean norm The *norm defined on $\mathbb{R}^n$ by

$$\|(x_1, x_2, \ldots, x_n)\|$$
$$= \sqrt{(x_1)^2 + (x_2)^2 + \ldots + (x_n)^2}.$$

Equivalent to $p = 2$ in the family of *p-norms.

Euclidean space (Cartesian space) The number line $\mathbb{R}$, plane $\mathbb{R}^2$, and 3-dimensional space $\mathbb{R}^3$ can be generalized to n-dimensional Euclidean or Cartesian space $\mathbb{R}^n$ with *coordinates $(x_1, x_2, \ldots, x_n)$ on which the operations of addition and scalar multiplication are extended in the obvious way. While $\mathbb{R}^n$ is hard to visualize for $n > 3$, it provides a powerful framework in *linear algebra and multivariable *analysis.

If the points P and Q have coordinates $(x_1, x_2, \ldots, x_n)$, and $(y_1, y_2, \ldots, y_n)$ respectively, then the distance PQ can be defined to be equal to

$$\sqrt{(x_1 - y_1)^2 + \ldots + (x_n - y_n)^2}.$$

Sometimes n-dimensional Euclidean space is denoted as E^n, signifying that space has no predetermined coordinates, no natural *origin or axes. Once an origin, perpendicular axes, and unit length have been decided, coordinates may be assigned and E^n can now be identified with $\mathbb{R}^n$.

Euclid numbers Numbers of the form $p_1 p_2 \times \cdots \times p_n + 1$, where $p_1, p_2, \ldots, p_n$, are the first n *prime numbers. Euclid's name is associated with these numbers in reference to his proof that there are infinitely many prime numbers.

Euclid's postulates The postulates (or *axioms) Euclid set out in his famous text, the *Elements*, are:

1. A straight line may be drawn from any point to any other point.
2. A straight line segment can be extended indefinitely at either end.
3. A circle may be described with any centre and any radius.
4. All right angles are equal.
5. If a straight line (the transversal) meets two other straight lines so that the sum of the two interior angles on one side of the transversal is less than two right angles, then the straight lines, extended indefinitely if necessary, will meet on that side of the transversal. Axiom 5 is the famous *parallel postulate.

He also stated definitions of geometrical entities like points and lines, and five 'common notions', which are:

1. Things which are equal to the same thing are also equal to one another.
2. If equals are added to equals, the sums are also equal.
3. If equals are subtracted from equals, the remainders are also equal.
4. Things that coincide with one another are equal to one another.
5. The whole is greater than the part.

Eudoxus of Cnidus (about 380 BC) Greek mathematician and astronomer, one of the greatest of antiquity. All his original works are lost, but it is known from later writers that he was responsible for the work in Book 5 of *Euclid's *Elements*. This work of his was a precise and rigorous development of the *real number system in the language of his day. He also developed methods of determining areas with curved boundaries.

Euler, Leonhard (1707–83) One of the most prolific of famous mathematicians and the greatest mathematician of the 18th century. He was born in Switzerland but is most closely associated with the Berlin of Frederick the Great and the St Petersburg of Catherine the Great. He worked in a highly productive period when the newly developed *calculus was being extended in varied directions, and he made contributions to most areas of mathematics, pure and applied. Euler was also responsible for much notation that is standard today. Among his contributions to the language are the basic symbols, π, e and i, the summation notation $\sum$ and the standard function notation $f(x)$. His *Introductio in analysin infinitorum* was the most important mathematics text of the late 18th century. From the vast bulk of his work, one famous result of which he was justifiably proud is his solution to the *Basel problem:

$$1 + \frac{1}{2^2} + \frac{1}{3^2} + \dots + \frac{1}{n^2} + \dots = \frac{\pi^2}{6}.$$

Euler angles Let **R** be an *orthogonal 3×3 *matrix with *determinant 1; so **R** represents a general *rotation of 3-dimensional space about an *axis through the origin. Then there exist Euler angles α, β, γ in the ranges

$$-180° < \alpha \leq 180°, 0° \leq \beta \leq 180°,$$
$$-180° < \gamma \leq 180°$$

such that $\mathbf{R} = \mathbf{X}(\alpha)\mathbf{Y}(\beta)\mathbf{X}(\gamma)$ where $\mathbf{X}(\theta)$ denotes rotation about the x-axis by θ and $\mathbf{Y}(\theta)$ denotes rotation about the y-axis by θ. Euler angles are useful in describing rotations of *rigid bodies.

Euler characteristic (Euler-Poincaré characteristic) For a *closed *surface S, this is a topological invariant. It is given by the formula $\chi(S) = V - E + F$, where V, E, and F are the numbers of *vertices, *edges, and *faces, respectively, in a

*subdivision of S. For any surface *homeomorphic to a *sphere $\chi(S) = 2$ (see EULER'S THEOREM). *Poincaré generalized this to other closed surfaces: $\chi = 2 - 2g$ for a *torus with g holes and $\chi = 2 - k$ for a sphere with $k > 0$ *Möbius strips sewn in. See CLASSIFICATION THEOREM FOR SURFACES, GAUSS-BONNET THEOREM, POINCARÉ-HOPF THEOREM.

Euler force See FICTITIOUS FORCE.

Eulerian graph One area of graph theory is concerned with the possibility of travelling around a *graph, going along edges in such a way as to use every edge exactly once. A *connected graph is called Eulerian if there is a sequence v_0, e_1, v_1, ..., e_k, v_k alternately of vertices and edges (where e_i is an edge joining v_{i-1} and v_i), with $v_0 = v_k$ and with every edge of the graph occurring exactly once. Simply put, it means that 'you can draw the graph without taking your pencil off the paper or retracing any lines, ending at your starting-point'. The name arises from Euler's consideration of the problem of whether the *bridges of Königsberg could be crossed in this way. It can be shown that a connected graph is Eulerian if and only if every vertex has even *degree. Compare HAMILTONIAN GRAPH, TRAVERSABLE GRAPH.

Eulerian trail A *trail which includes every edge of a graph.

Euler line In a triangle, the *circumcentre O, the *centroid G, and the *orthocentre H lie on a straight line called the Euler line. On this line, $OG:GH = 1:2$. The centre of the *nine-point circle also lies on the Euler line.

Euler-Maclaurin formula Let $a < b$ be integers and $f(x)$ be a *function on $[a,b]$ with *derivatives of all orders. Let $I = \int_a^b f(x) \, dx$ and let S denote the estimate applying the *trapezium rule to this integral using unit intervals. The Euler-Maclaurin formula then states

$$S - I = \sum_{k=1}^{\infty} \frac{B_{2k}}{(2k)!} \left(f^{(2k-1)}(b) - f^{(2k-1)}(a) \right),$$

where B_n denotes the nth *Bernoulli number, and provided that

$$\lim_{n \to \infty} \frac{1}{(2\pi)^n} \int_a^b |f^{(n)}(x)| \ dx = 0.$$

See also BERNOULLI POLYNOMIALS.

Euler multiplier A synonym for MULTIPLYING FACTOR (in differential equations).

Euler number Another name for e, the base of the natural logarithms.

Euler numbers An integer *sequence E_n, related to the *Bernoulli numbers, and defined by the *generating function

$$\frac{1}{\cosh x} = \sum_{n=0}^{\infty} \frac{E_n x^n}{n!}.$$

As $\cosh x$ is an *even function, all odd Euler numbers are zero. The first few even Euler numbers are $E_0 = 1$, $E_2 = -1$, $E_4 = 5$, $E_6 = -61$, $E_8 = 1385$.

Euler's conjecture Euler's conjecture was a generalization of *Fermat's Last Theorem that has since been proved untrue. It stated that $n-1$ nth powers could never add to an nth power. For $n = 3$ this is a special case of Fermat and so true. But a counterexample to $n = 5$ was found in 1966 and the smallest counterexample to $n = 4$ is now known to be $95800^4 + 217519^4 + 414560^4 = 422481^4$.

Euler's constant Let $a_n = 1 + \frac{1}{2} + \frac{1}{3} + \ldots + (1/n) - \ln n$. This sequence has a limit whose value is known as Euler's constant, γ; that is, $a_n \to \gamma$. The value equals 0.57721566 to 8 decimal places. It is not known whether γ is *irrational.

Euler's equation (Euler-Lagrange equation) A central result in the *calculus of variations which addresses problems such as finding curves of minimal length (*geodesics) or solving the *brachistochrone problem. A *functional I associates a real number

$$I(y) = \int_a^b F(x, y, y') \ dx$$

(where a, b are fixed limits) with each of a family of graphs $y(x)$; this functional might represent the *arc length of such graphs, the time taken to travel along the graphs under gravity, etc. Euler's equation then states I is maximal when

$$\frac{d}{dx} \left(\frac{\partial F}{\partial y'} \right) - \frac{\partial F}{\partial y} = 0.$$

As an example, the arc length of a curve $y(x)$ from (a,c) to (b,d) is

$$I(y) = \int_a^b \sqrt{1 + y'(x)^2} \ dx,$$

so that $F(x, y, y') = \sqrt{1 + y'^2}$ and Euler's equation then reads

$$\frac{d}{dx} \left(\frac{y'}{\sqrt{1 + y'^2}} \right) - 0 = 0.$$

Integrating and rearranging we find that y' is constant, and so the graph of y is a straight line.

Euler's equations (fluid dynamics) Euler's equations for the flow of an *inviscid, *incompressible fluid with flow *velocity **u** state

$$\text{div} \mathbf{u} = 0, \quad \frac{D\mathbf{u}}{Dt} = -\frac{1}{\rho} \nabla p + \mathbf{g}.$$

Here ρ denotes the (constant) *density, p denotes *pressure, g denotes acceleration due to *gravity, and D/Dt is the *convective derivative. *Compare* NAVIER-STOKES EQUATIONS.

Euler's formula The name given to the *identity $\cos\theta + i \sin\theta = e^{i\theta}$, a special case of which gives $e^{i\pi} + 1 = 0$.

Euler's phi function A synonym for TOTIENT FUNCTION.

Euler's method A simple numerical method for solving *differential equations. If $\frac{dy}{dx} = f(x, y)$ and an initial condition is known, $y = y_0$ when $x = x_0$, then Euler's method generates a succession of approximations $y_{n+1} = y_n + hf(x_n, y_n)$ where $x_n = x_0 + nh$, $n \geq 1$. This takes the known starting point and moves along a straight line segment with horizontal distance h in the direction of the tangent at (x_0, y_0). The process is repeated from the new point (x_1, y_1) etc. If the step length h is small enough, the *tangents are good approximations to the curve and the method will provide a reasonable accurate estimate. *Compare* RUNGE-KUTTA METHODS.

Euler's Theorem If a finite *planar graph G is drawn in the plane, so that no two edges cross, the plane is divided into a number of regions (including the exterior region) called faces. Euler's Theorem (for planar graphs) is the following:

Theorem Let G be a *connected planar graph drawn in the plane. If there are V vertices, E edges and F faces, then $V - E + F = 2$.

An application of this gives Euler's Theorem (for polyhedra):

Theorem If a *convex polyhedron has V vertices, E edges and F faces, then $V - E + F = 2$.

For example, a cube has $V = 8$, $E = 12$, $F = 6$, and a tetrahedron has $V = 4$, $E = 6$, $F = 4$.

The two theorems are essentially the same as a convex *polyhedron is topologically a *sphere, and a sphere with a point removed is topologically the plane. *See* EULER CHARACTERISTIC.

evaluate To calculate the value of a mathematical object, such as a *function at a particular argument, a definite *integral, or the sum of an *infinite series.

even function The *real function f is an even function if $f(-x) = f(x)$ for all x (in the domain of f). Thus the graph $y = f(x)$ of an even function has the y-axis as a line of symmetry. The following are even functions of x: 5, x^2, $x^6 - 4x^4 + 1$, $1/(x^2 - 3)$, $\cos x$. *Compare* ODD FUNCTION.

even integer *Divisible by two with no remainder.

even permutation A *permutation which can be written as the *composition of an even number of *transpositions. *See* ALTERNATING GROUP, ODD PERMUTATION.

event A (*measurable) subset of the *sample space relating to an experiment. For example, suppose that the sample space for an experiment in which a coin is tossed three times is $\Omega = \{HHH, HHT, HTH, HTT, THH, THT, TTH, TTT\}$, and let $A = \{HHH, HHT, HTH, THH\}$. Then A is the event in which at least two 'heads' are obtained. If, when the experiment is performed, the outcome is one that belongs to A, then A is said to have occurred. The *intersection $A \cap B$ of two events is the event that can be described by saying that 'both A and B occur'. The *union $A \cup B$ of two events is the event that 'either A or B occurs'. The *complement A' of A is the event that 'A does not occur'. The *probability $\Pr(A)$ of an event A is often of interest. The following laws hold:

(i) $\Pr(A \cup B) = \Pr(A) + \Pr(B) - \Pr(A \cap B)$.
(ii) When A and B are *mutually exclusive events, $\Pr(A \cup B) = \Pr(A) + \Pr(B)$.
(iii) When A and B are *independent events, $\Pr(A \cap B) = \Pr(A) \Pr(B)$.
(iv) $\Pr(A') = 1 - \Pr(A)$.

See PROBABILITY SPACE.

eventually A synonym for SUFFICIENTLY LARGE.

evolute *See* CURVATURE.

exa- Prefix used with *SI units to denote multiplication by 10^{18}. Abbreviated as E.

exact The positive solution to the equation $x^2 = 3$ might be reported as 1.73, which it is correct to two *decimal places or to three *significant figures, but the exact answer is $x = \sqrt{3}$.

exact differential If $z = f(x,y)$ is a function of two variables then the *total differential is $dz = \frac{\partial z}{\partial x}dx + \frac{\partial z}{\partial y}dy$. Then $Pdx + Qdy$ is said to be an exact differential if it is the total differential of some function f. Further, $Pdx + Qdy$ is exact if and only if $\partial P/\partial y = \partial Q/\partial x$. Compare TOTAL DIFFERENTIAL.

exact differential equation A *differential equation $Pdx + Qdy = 0$ is exact if $Pdx + Qdy$ is the *total differential of a function $f(x,y)$ so that the *general solution has the form $f = $ constant. Even when $Pdx + Qdy$ is not an *exact differential, an integrating factor $m(x,y)$ may be found so that $mPdx + mQdy$ is exact.

exact divisor A synonym for DIVISOR.

example A particular instance of a generalized statement. A single *counter-example will disprove a generalized claim, but examples do not provide proof. For example, the number of regions created by joining n distinct points on a circle is at most 1, 2, 4, 8, 16 for $1 \leq n \leq 5$ which are in the form 2^{n-1}. However these 5 examples do not verify that result and indeed for $n = 6$, there are at most 31 regions produced, so the general expression cannot be 2^{n-1}. See MOSER'S CIRCLE PROBLEM.

excentre See EXCIRCLE.

exchange paradox See ENVELOPE PARADOX.

excircle An excircle of a triangle is a circle that lies outside the triangle and touches the three sides, two of them

extended. There are three excircles. The centre of an excircle is an excentre of the triangle. Each excentre is the point of intersection of the *bisector of the *interior angle at one vertex and the bisectors of the *exterior angles at the other two vertices.

excluded middle See PRINCIPLE OF THE EXCLUDED MIDDLE.

exclusive See MUTUALLY EXCLUSIVE.

exclusive disjunction If p and q are statements, then the statement 'p or q' where exclusive disjunction is intended, often written as 'p xor q' or $p \veebar q$, is true only if exactly one of p, q is true. The truth table is therefore as follows:

p	q	$p \veebar q$
T	T	F
T	F	T
F	T	T
F	F	F

The set-theoretic equivalent of exclusive disjunction is *symmetric difference. See also INCLUSIVE DISJUNCTION.

exhaustion See ARCHIMEDES' METHOD.

exhaustive A set of events in *probability whose *union is the whole *sample space, or a set of sets whose union is the *universal set under consideration.

existential quantifier See QUANTIFIER.

exp The abbreviation and symbol for the *exponential function.

expand To express in an extended equivalent form. For example, $(a + b)^2 = a^2 + 2ab + b^2$.

expansion A mathematical expression written as a *sum of a number of terms, possibly infinitely many. Powers of *multinomials can be expanded by multiplying out the brackets. Functions might also be be expressed as infinite *power series. *Decimal representations of numbers are often referred to as decimal expansions.

expectation (of a matrix game) Suppose that, in the *matrix game given by the $m \times n$ matrix $[a_{ij}]$, the *mixed strategies $\mathbf{x}$ and $\mathbf{y}$ for the two players are given by $\mathbf{x} = (x_1, x_2, \ldots, x_m)$, and $\mathbf{y} = (y_1, y_2, \ldots, y_n)$. The expectation $E(\mathbf{x}, \mathbf{y})$ is given by

$$E(\mathbf{x}, \mathbf{y}) = \sum_{i=1}^{m} \sum_{j=1}^{n} x_i a_{ij} y_j.$$

If $\mathbf{A} = [a_{ij}]$, then $E(\mathbf{x}, \mathbf{y}) = \mathbf{x} \mathbf{A} \mathbf{y}^T$. The expectation can be said, loosely, to give the average *payoff each time when the game is played many times with the two players using these mixed strategies.

expectation (of a random variable) A synonym for EXPECTED VALUE.

expected monetary value *See* EMV.

expected utility In a context where the outcomes are not determined solely by the decisions an individual takes, the expected utility of a decision is the average value of the *utility function weighted by the probability distribution attached to the possible outcomes of the decision.

expected value The expected value $E(X)$ of a *random variable X is a value that gives the *mean value of the distribution. For a discrete random variable X, $E(X) = \sum p_i x_i$, where $p_i = Pr(X = x_i)$. For a continuous random variable X,

$$E(X) = \int_{-\infty}^{\infty} xf(x) \, dx,$$

where f is the *probability density function of X. The following laws hold:

(i) $E(aX + bY) = aE(X) + bE(Y)$.

(ii) When X and Y are *independent, $E(XY) = E(X)E(Y)$.

(iii) The *law of total expectation*: for a *partition A_1, A_2, $\ldots$, A_n of the *sample space

$$E(X) = \sum_{k=1}^{n} E(X|A_k)P(A_k).$$

The following is a simple example. Let X be the number obtained when a die is thrown. Let $x_i = i$, for $1 \leq i \leq 6$. Then $p_i = \frac{1}{6}$, for all i. So $E(X) = \frac{1}{6} \times 1 + \frac{1}{6} \times 2 + \ldots + \frac{1}{6} \times 6 = 3.5$.

experiment A statistical study in which the researchers make an intervention and measure the effect of this intervention on some outcome of interest, for example then measuring the change in reaction times of people deprived of sleep.

experimental design (in statistics) Observational studies can suggest things for which the explanation lies with some hidden variable. For example, the performance of pupils taught in large classes is better than that of pupils in small classes, not because the large class is a more effective learning environment, but because schools tend to operate larger classes for the more successful. Experimental design methods seek to control the conditions under which observations are made so that any differences in outcome are genuinely attributable to the experimental conditions and not to other *confounding variables. Some of the most common methods are the use of *paired or matched samples, *randomization, and *blind trials.

explanatory variable One of the variables that is thought to influence the value of the *dependent variable in a statistical *model.

explementary angles A synonym for CONJUGATE ANGLES.

explicit function If the *dependent variable y is expressed in the form $y = f(x)$, then y is an explicit function of x. So $y = 5x + 1$ is explicit. But the equation $y^5 + y = x$ is not explicit, even though for every real x the equation uniquely determines a real y. Compare IMPLICIT.

exploratory data analysis (EDA) An approach to data analysis that sets out initially to explore the *data, usually through a variety of mostly graphical techniques, to try to gain insight into the nature of the data and their underlying structure, what the important variables are and to identify outliers. The outcomes can then inform decisions as to what analyses are appropriate. The approach first gained importance with the work of *Tukey in 1977.

exponent A synonym for INDEX. See also FLOATING-POINT NOTATION.

exponential decay If $y = Ae^{kt}$, where $A > 0$ and $k < 0$ are constants, and t denotes time, y is said to be exhibiting exponential decay. Then the length of time it takes for y to be reduced to half its value is the same, whatever the value. This length of time, which equals $-\ln 2/k$, is called the half-life and is a useful measure of the rate of decay. It is applicable, for example, to the decay of radioactive isotopes. Compare EXPONENTIAL GROWTH.

exponential distribution The continuous probability *distribution with *probability density function $f(x) = \lambda \exp(-\lambda x)$, where $\lambda > 0$ and $x \geq 0$. It has *mean $1/\lambda$ and *variance $1/\lambda^2$. The time between events that occur randomly but at a constant rate has an exponential distribution. The distribution is *skewed to the right.

exponential function The function f such that $f(x) = e^x$, or $\exp x$, for all x in $\mathbb{R}$.

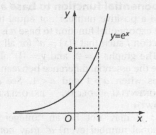

Graph of exponential function

The two notations arise from different approaches described below, but are used interchangeably. Among the important properties that the exponential function has are the following:

(i) $\exp(x + y) = (\exp x)(\exp y)$, $\exp(-x) = 1/\exp x$ and $(\exp x)^r = \exp rx$.

(ii) The exponential function is the *inverse function of the *logarithmic function: $y = \exp x$ if and only if $x = \ln y$.

(iii) $\frac{d}{dx}(\exp x) = \exp x$.

(iv) $\exp x$ is the sum of the series $1 + \frac{x}{1!} + \frac{x^2}{2!} + \dots + \frac{x^n}{n!} + \dots$.

(v) As $n \to \infty$, $\left(1 + \frac{x}{n}\right)^n \to \exp x$.

The exponential function may be defined or characterized in different ways.

1. Define ln as in the *logarithmic function entry, and take exp as its *inverse function. It is then possible to define the value of e as exp 1, establish the equivalence of $\exp x$ and e^x, and prove the other properties above.

2. The *power series in (iv) *converges for all real numbers (and in fact also all complex numbers (see COMPLEX EXPONENTIAL)). This can be seen from the *ratio test. By applying appropriate theorems of *analysis, the other properties can be proved.

3. $\exp x$ may be characterized as the unique function that satisfies the differential equation $dy/dx = y$ with $y(0) = 1$.

exponential function to base *a* Let *a* be a positive number not equal to 1. The exponential function to base *a* is the function *f* such that $f(x) = a^x$ for all *x* in ℝ. The graphs $y = 2^x$ and $y = (\frac{1}{2})^x$ illustrate the essential difference between the cases when $a > 1$ and $a < 1$. *See also* EXPONENTIAL DECAY, EXPONENTIAL GROWTH.

As *x* may not be an *integer (or *rational number even) a^x may not be defined as '*a* multiplied by itself *x* times' and instead is defined by

$$a^x = \exp(x \ln a).$$

This definition still satisfies rules (i)–(vii) of the rules for indices (*see* INDEX). Further, by the *chain rule,

$$\frac{\mathrm{d}}{\mathrm{d}x}(a^x) = a^x \ln a.$$

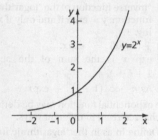

Graph showing exponential growth

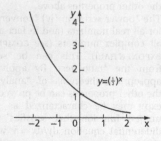

Graph showing exponential decay

exponential growth If $y = Ae^{kt}$, where $A > 0$ and $k > 0$ are constants, and *t* denotes time, *y* is said to be exhibiting exponential growth. This occurs when $\mathrm{d}y/\mathrm{d}t = ky$; that is, when the *rate of change of the quantity *y* at any time is proportional to the value of *y* at that time. Any quantity with exponential growth ultimately outgrows any *polynomial growth. *Compare* EXPONENTIAL DECAY.

exponential of a matrix Given a *square matrix **A**, its exponential is $\exp \mathbf{A} = \mathbf{I} + \mathbf{A} + \mathbf{A}^2/2! + \mathbf{A}^3/3! + \cdots$. This series converges in the space of $n \times n$ matrices. Further, $\exp(\mathbf{A}+\mathbf{B}) = \exp\mathbf{A} \ \exp\mathbf{B}$ for *commuting matrices **A**,**B**, but this identity does not generally hold. Note two simultaneous scalar differential equations $\mathrm{d}x/\mathrm{d}t = 2x + 3y$, $\mathrm{d}y/\mathrm{d}t = x - y$, can be rewritten as $\mathrm{d}\mathbf{r}/\mathrm{d}t = \mathbf{Ar}$ where

$$\mathbf{A} = \begin{pmatrix} 2 & 3 \\ 1 & -1 \end{pmatrix} \quad \text{and} \quad \mathbf{r} = \begin{pmatrix} x \\ y \end{pmatrix}.$$

The solution then equals $\mathbf{r}(t) = \exp(\mathbf{A}t)\mathbf{r}(0)$. This matrix exponential also has an important role in the study of *Lie groups.

exponential series The series $\sum_{n=0}^{\infty} \frac{z^n}{n!} = 1 + z + \frac{z^2}{2} + \frac{z^3}{6} + \dots$ which converges for any real or complex number *z* to the *exponential function $\exp(z)$. *See also* COMPLEX EXPONENTIAL.

exponentiate To raise a number or quantity to a *power.

express To transform into equivalent terms. For example, by *expanding or *factorizing an expression.

extended complex plane The set of complex numbers with a point at *infinity. The set can be denoted by $\mathbb{C}_\infty$ and can be thought of as a *Riemann sphere by means of a *stereographic projection. A complex *sequence z_n tends to ∞ in the real sequence $|z_n|$ tends to ∞.

extended mean value theorem A synonym for CAUCHY'S MEAN VALUE THEOREM.

extended metric In some cases it is useful to allow the *metric in a metric space to attain the value ∞, in which case it is called an extended metric.

extended real numbers The set of *real numbers, with the positive and negative *infinity included.

extension The difference $x-l$ between the actual length x of a string or spring and its natural length l. The extension of a spring is negative when the spring is compressed.

exterior *See* JORDAN CURVE THEOREM.

exterior angle (of a polygon) The angle between one side and the extension of an adjacent side of a polygon.

exterior angle (with respect to a transversal of a pair of lines) *See* TRANSVERSAL.

exterior derivative *See* DIFFERENTIAL FORM.

exterior product For u and v in a *vector space V, their exterior product is denoted $u \wedge v$ and satisfies $u \wedge v = -v \wedge u$. If $v_1, \ldots, v_n$ form a *basis for V, then $v_i \wedge v_j$, where $i < j$, form a basis for the exterior product $\bigwedge^2 V$. Further exterior product spaces $\bigwedge^k V$ can be defined for $1 \le k \le \dim V$.

external bisector The *bisector of the *exterior angle of a triangle (or polygon) is sometimes called the external bisector of the angle of the triangle (or polygon).

external division (of a segment) Let AB be a line segment. Then the point E is

the external division of AB in the ratio $1:k$ if $k\overrightarrow{AE} = \overrightarrow{BE}$ where $\overrightarrow{AB}$ is the vector from A to B.

If A and B have position vectors $\mathbf{a}$ and $\mathbf{b}$, from some origin, then the position vector of E is

$$\mathbf{e} = \frac{k\mathbf{a} - \mathbf{b}}{k - 1},$$

sometimes referred to as the section formula. *Compare* INTERNAL DIVISION.

external force When a system of *particles or a *rigid body is being considered as a whole, an external force is a force acting on the system from outside. *Compare* INTERNAL FORCE.

extrapolate To estimate a value of a quantity beyond the range already known, for example forecasting in *time series. *Compare* INTERPOLATE.

extrapolation Suppose that certain values $f(x_0), f(x_1), \ldots, f(x_n)$ of a function f are known, where $x_0 < x_1 < \ldots < x_n$. A method of finding from these an approximation for $f(x)$, for a given value of x that lies outside the interval $[x_0, x_n]$, is called extrapolation. Such methods are normally far less reliable than *interpolation, in which x lies between x_0 and x_n.

extreme value distribution The distribution of largest and smallest values in a *sample. This area is an important tool in *risk assessment.

extremum A point at which a function achieves a *global maximum or global minimum. The term may also refer to a *local maximum or local minimum.

extrinsic *See* INTRINSIC.

f Commonly used symbol for a *function, as in $f(x) = x^2 + 3$.

f Abbreviation for *femto-.

F The number 15 in hexadecimal notation. Also the symbol for 'false' in *logic and in *truth tables.

$\mathcal{F}$ *See* FOURIER TRANSFORM.

$\mathbb{F}_q$ Notation for the *Galois field with q elements.

face One of the plane surfaces forming a *polyhedron.

face-transitive A *polyhedron is face-transitive (or face-regular or isohedral) if there is a *symmetry taking any face to any other face. For example, two pyramids, with *congruent *regular *hexagons as bases, might be joined at their bases to form a face-transitive polyhedron which is neither *edge-transitive nor *vertex-transitive.

factor *See* DIVIDES.

factor analysis In *statistics, the techniques that aim to reduce the number of *explanatory variables (factors) used to explain observational outcomes. New variables are constructed as combinations of the original variables with the aim of identifying a simpler model structure.

factorial For a positive integer n, the notation $n!$ (read as 'n factorial' or 'n shriek') is used for the product $n(n-1)(n-2)\dots \times 2 \times 1$. Thus $4! = 4 \times 3 \times 2 \times 1 = 24$. Also,

by definition, $0! = 1$. *See also* GAMMA FUNCTION.

For a positive integer n, the 'double factorial' notation $n!!$ denotes the products

$$n \times (n-2) \times (n-4) \times \dots \times 1 = \frac{(2k)!}{2^k k!}$$

when $n = 2k - 1$;

$$n \times (n-2) \times (n-4) \times \dots \times 2 = 2^k k!$$

when $n = 2k$.

factorize To represent a number or polynomial as a product of factors. *See* FUNDAMENTAL THEOREM OF ARITHMETIC, UNIQUE FACTORIZATION DOMAIN.

factor space A synonym for QUOTIENT SPACE.

factor theorem A consequence of the *remainder theorem, the factor theorem states:

Let $f(x)$ be a polynomial. Then $x - h$ is a *factor of $f(x)$ if and only if $f(h) = 0$.

The theorem is helpful investigating factorization over the *integers. For example, to factorize $f(x) = 6x^3 + 19x^2 + 16x + 4$, if $x - h$ is a factor, where h is an integer, then h must divide 4. We may note $f(-2) = -48 + 76 - 32 + 4 = 0$, and so $x + 2$ is a factor. To appreciate that $2x + 1$ is a factor, we need to calculate $f(-1/2)$.

fair Unbiased (*see* BIAS). Thus a fair coin has ½ chance of landing heads or tails and a fair *die has a equal chance of rolling 1-6.

faithful action The *group action of a group G on a set X is faithful if the only $g \in G$ satisfying $g.x = x$ for all $x \in X$ is the group's *identity. Equivalently the *homomorphism $G \rightarrow \mathrm{Sym}(X)$ is *injective, and so every group element is identifiable by its action on X.

fallacy A false statement, an invalid argument, or a specific statement in an argument that invalidates it.

false negative In testing to determine whether a subject has a particular characteristic, especially testing whether a patient has a disease, where the test shows the characteristic is not present when it actually is. Also known as a type II error.

false position (rule of false position, *regula falsi*) An iterative method which usually *converges faster than the *bisection method. Say that $f(a) > 0 > f(b)$ and we wish to approximate a *root of f. The first iterate is $x_1 = (af(b) - bf(a))/(f(b) - f(a))$; this is the x-value where the line connecting $(a, f(a))$ and $(b, f(b))$ crosses the x-axis. Depending on the sign of $f(x_1)$ this process is repeated on the interval $[a, x_1]$ or $[x_1, b]$ to find x_2 and so on.

false positive In testing to determine whether a subject has a particular characteristic, especially testing whether a patient has a disease, where the test shows the characteristic is present when it is not. The *probability of a false positive equals the significance level of the test. Also known as a type I error.

family Typically, a set whose elements are sets or have some further structure. Thus, one might discuss a family of curves. A family may be called an n-parameter family if n parameters need to be specified to identify an element of the family.

fast Fourier transform See DISCRETE FOURIER TRANSFORM.

Fatou's lemma Let $f_n : \mathbb{R} \rightarrow \mathbb{R}$, $n \geq 1$, be a sequence of positive Lebesgue integrable functions (see LEBESGUE INTEGRATION) which converge *almost everywhere to a function f and whose integrals are bounded above. The f is Lebesgue integrable.

***F*-distribution** A non-negative continuous *distribution formed from the ratio of the distributions of two independent random variables with *chi-squared distributions, each divided by its *degrees of freedom. The *mean is $\dfrac{v_2}{v_2 - 2}$ and the *variance is $\dfrac{2v_2^2(v_2 + v_1 - 2)}{v_1(v_2 - 2)^2(v_2 - 4)}$, where v_1 and v_2 are the degrees of freedom of the numerator and denominator respectively. It is used to test the hypothesis that two normally distributed random variables have the same variance and in *regression to test the relationship between an explanatory variable and the dependent variable. The distribution is skewed to the right.

feasible A *constrained optimization problem for which the constraints can be satisfied simultaneously is said to be feasible.

feasible region See LINEAR PROGRAMMING.

Feigenbaum, Mitchell (1944–2019) American mathematical physicist instrumental in developing *chaos theory.

Feit-Thompson theorem Every finite *group of odd *order is *solvable. This was a significant early result in the classification of finite *simple groups. See SPORADIC GROUP.

femto- Prefix used with *SI units to denote multiplication by 10^{-15}. Abbreviated as f.

Fermat, Pierre de (1607–65) A leading mathematician of the 17th century. In number theory he proved *Fermat's Little Theorem and *Fermat's Two Squares

Theorem and famously conjectured *Fermat's Last Theorem. He independently introduced *Cartesian coordinates, and his work on tangents was an acknowledged inspiration to *Newton in the development of *calculus. In a correspondence of 1654 with *Pascal, they laid down the foundations of *probability. Professionally he was a lawyer in Toulouse, and so he was considered the 'Prince of Amateurs'.

Fermat point The point with the minimum total distance to the three vertices of a triangle. If the angle at any of the vertices is more than 120° then that vertex is the Fermat point, otherwise it is found by constructing an *equilateral triangle on each of the three sides of the triangle. For each side of the triangle, join the new vertex of the equilateral triangle to the opposite vertex of the original triangle. These three lines intersect at the Fermat point. In the diagram X is the Fermat point.

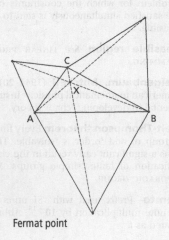

Fermat point

Fermat prime A *prime of the form $2^{2^r} + 1$. *Fermat conjectured all such

numbers to be prime, though *Euler showed that $2^{32} + 1$ is divisible by 641. Currently, the only known Fermat primes are for $r \leq 4$. *See also* CONSTRUCTIBLE POINT.

Fermat's Last Theorem The statement that, for all integers $n > 2$, the equation $x^n + y^n = z^n$ has no solution in positive *integers. Fermat famously noted in a copy of *Diophantus' *Arithmetica* that he had a proof 'too large fit in the margin'. Much research was done over centuries until a proof was completed in 1995 by Andrew *Wiles. *See* PYTHAGOREAN TRIPLE, MODULARITY THEOREM.

Fermat's Little Theorem Let p be a prime and a be an integer not divisible by p. Then $a^{p-1} \equiv 1 \pmod{p}$. Consequently, for a any integer, $a^p \equiv a \pmod{p}$. *See also* TOTIENT FUNCTION.

Fermat's Two Squares Theorem A *prime number, which is 1 more than a *multiple of 4, can be written as the sum of two squares, for example $89 = 5^2 + 8^2$. *See* LAGRANGE'S THEOREM.

Fermat's Theorem The derivative of a *differentiable function at an interior *local minimum or *local maximum equals zero.

Feuerbach's Theorem *See* NINE-POINT CIRCLE.

Feynman, Richard Phillips (1918–88) American mathematician and theoretical physicist who won the Nobel Prize for Physics in 1965 for work on quantum electrodynamics. He worked on the atomic bomb project during the Second World War, when he was already regarded as one of the leading scientists in the field at only 23. He enjoyed considerable success as an author where his intuitive grasp of fundamental physical principles allowed him to communicate to a broad audience.

Fibonacci (about 1170–1250) Pseudonym of an Italian merchant by the name of Leonardo of Pisa, who helped introduce the *Hindu–Arabic number system to Europe. He strongly advocated this system in *Liber abaci*, published in 1202, which also contained problems including one that gives rise to the *Fibonacci numbers.

Fibonacci multiplication Invented by Donald *Knuth, a means of multiplying positive integers, using the *Fibonacci numbers as if they were a *base. By *Zeckendorf's Theorem, every positive integer can be uniquely written as a sum of non-consecutive Fibonacci numbers. Products of two integers so expressed can then be multiplied using the rule $F_m \circ F_n = F_{m+n}$. The operation is *associative and *commutative.

Fibonacci number (**Fibonacci sequence**) One of the numbers in the Fibonacci sequence 1, 1, 2, 3, 5, 8, 13,..., where each number (after the second) is the sum of the preceding two. This sequence has many interesting properties. For instance, the sequence consisting of the ratios of one Fibonacci number to the previous one, $\frac{1}{1}, \frac{2}{1}, \frac{3}{2}$, $\frac{5}{3}, \frac{8}{5}, \frac{13}{8}, \ldots$, has the limit φ the *golden ratio. Binet's theorem states that the nth Fibonacci number equals $\left(\varphi^n - (-\varphi)^{-n}\right)/\sqrt{5}$. *See also* DIFFERENCE EQUATION, GENERATING FUNCTION, LUCAS NUMBERS.

SEE WEB LINKS

• A site with examples of Fibonacci numbers in nature and the relationship to the golden section.

fibre *See* TANGENT BUNDLE.

fictitious force A force that may be perceived to exist by an observer whose *frame of reference is not inertial. Suppose, for example, that a frame of reference with origin O is rotating relative to

an inertial frame of reference with the same origin. A particle P subject to a certain total force satisfies *Newton's second law of motion, relative to the inertial frame of reference. To the observer in the rotating frame, the particle appears to satisfy an equation of motion that is Newton's second law of motion with additional terms. The observer may suppose that these terms are explained by certain fictitious forces.

In the special case where the rotating frame of reference has a constant *angular velocity and the particle is moving in a plane perpendicular to the angular velocity, one fictitious force is in the direction along OP and is called the centrifugal force. This is the force outwards that is believed to exist by a rider on a roundabout. The second fictitious force is perpendicular to the path of P as seen by the observer in the rotating frame of reference and is called the Coriolis force.

In general, if S denotes the rotating frame and ω denotes the angular velocity, then

$$m\mathbf{a} = \mathbf{F} - m\left(\frac{d\omega}{dt}\right)_S \times \mathbf{r} - 2m\omega \times \left(\frac{d\mathbf{r}}{dt}\right)_S$$
$$- m\omega \times (\omega \times \mathbf{r}),$$

where $\mathbf{F}$ is the total force acting, $\mathbf{a}$ denotes the perceived acceleration in S, $\mathbf{r}$ denotes position vector, and the subscript S denotes measurement in the non-inertial frame S. The three terms that appear to distort Newton's second law are, in order, the Euler force (resulting from any *angular acceleration), the Coriolis force, and the centrifugal force.

field A commutative *ring with identity with the following additional property:

(x) For each non-zero a there is an element a^{-1} such that $a^{-1}a = 1$.

(The axiom numbering here follows on from that used for ring and *integral domain.) From the defining properties of a field, Axioms (i)–(viii) and Axiom (x), it can be shown that $ab = 0$ only if

$a = 0$ or $b = 0$. Thus Axiom **(ix)** holds, and so any field is an integral domain. Familiar examples of fields are the *rational numbers $\mathbb{Q}$, the *real numbers $\mathbb{R}$ and the *complex numbers $\mathbb{C}$, each with the usual addition and multiplication. Another example is $\mathbb{Z}_p$, the integers with addition and multiplication modulo p, where p is *prime. *See also* GALOIS FIELD.

field extension A pair of *fields F and K, where the base field F is a *subfield of K. Importantly, K is a *vector space over F, and the degree of the extension is defined as the *dimension of K as such a vector space. For example, $\mathbb{C}$ is a degree 2 extension of $\mathbb{R}$ as $\{1, i\}$ is a basis. *See* SIMPLE EXTENSION, TOWER LAW.

field of force A *vector field defined in a region where a non-contact *force applies. Examples include *gravitational, *electric, and *magnetic fields. A gravitational field represents the strength and direction of gravity, or equivalently gravitational force per unit *mass. A more massive particle is acted on by a greater force, even though the field has not changed.

field of fractions If R is an *integral domain, then a field of fractions F can be defined which naturally contains R. The field F is the set of equivalence classes of (r_1, r_2), where $r_1, r_2 \in R$ and $r_2 \neq 0$ under the relation $(r_1, r_2) \sim (s_1, s_2)$ if $r_1 s_2 = r_2 s_1$. The equivalence class of (r_1, r_2) can be identified with r_1/r_2, so that $(r, 1)$ can be identified with $r \in R$. The field of fractions of $\mathbb{Z}$ is $\mathbb{Q}$ with $(4, 8)$ and $(-2, -4)$ both identified with $\frac{1}{2}$.

Fields Medal A prize awarded for outstanding achievements in mathematics, considered by mathematicians to be equivalent to a Nobel Prize. Medals are awarded to individuals at successive International Congresses of Mathematicians, normally held at four-year intervals. The proposal was made by J. C. Fields to found two gold medals, using funds remaining after the financing of the Congress in Toronto in 1924. The first two medals were presented at the Congress in Oslo in 1936. In some instances, three or four medals have been awarded. It has been the practice to make the awards to mathematicians under the age of 40. For a list of winners, *see* APPENDIX 22.

(🌐) SEE WEB LINKS

• The website of the awarding body of the Fields Medal.

figurate numbers Sequences of numbers associated with geometrical arrays such as the *triangular, *square, *pentagonal, and *tetrahedral numbers.

figure In *geometry, a loosely used term for a combination of points, lines, curves, or surfaces creating a definite form. Also used to refer to a *digit, or more informally to a specified number.

financial mathematics The branch of mathematics that seeks to model financial markets valuing stock derivatives (especially options) which mainly uses methods of stochastic calculus. Following the financial crisis of 2008, the credibility of such models has been damaged, though criticism dates back to *Mandelbrot. *See* BLACK-SCHOLES EQUATION, STOCHASTIC DIFFERENTIAL EQUATION.

finite A set is finite if it is *empty or can be put in *one-to-one correspondence with a set $\{1, 2, \ldots, n\}$ for some positive integer n. *See* CARDINALITY, CARDINAL NUMBER, INFINITE.

finite differences Let $x_0, x_1, \ldots, x_n$ be equally spaced values, so that $x_i = x_0 + ih$, for $1 \leq i \leq n$. Suppose that the values $f_0, f_1, \ldots, f_n$ are known, where $f_i = f(x_i)$, for some function f. The first differences are defined, for $0 \leq i < n$, by $\Delta f_i = f_{i+1} - f_i$. The second differences are defined by

$\Delta^2 f_i = \Delta f_{i+1} - \Delta f_i$ and, in general, the k-th differences are defined by $\Delta^k f_i = \Delta^{k-1} f_{i+1} - \Delta^{k-1} f_i$. For a polynomial of degree n, the $(n+1)$-th differences are zero.

These finite differences may be displayed in a table, as in the following example. Alongside it is a numerical example.

$x_0 \, f_0$			1.0 1.000		
	Δf_0			0.331	
$x_1 \, f_1$	$\Delta^2 f_0$		1.1 1.331		0.066
	Δf_1	$\Delta^3 f_0$		0.397	0.006
$x_2 \, f_2$	$\Delta^2 f_1$		1.2 1.728		0.072
	Δf_2			0.469	
$x_3 \, f_3$			1.3 2.197		

With such tables it should be appreciated that if the values $f_0, f_1, \ldots, f_n$ are *rounded values then increasingly serious errors result in the succeeding columns.

Numerical methods using finite differences have been extensively developed. They may be used for *interpolation, as in the *Gregory–Newton forward difference formula, for finding a polynomial that approximates to a given function, or for estimating *derivatives from a table of values.

finite-dimensional A *vector space is said to be finite-dimensional if it has finite *dimension or equivalently has a finite *basis.

finite element method A numerical method of solving *partial differential equations with *boundary conditions by considering a series of approximations which satisfy the differential equation and boundary conditions within a small region (the finite element of the name).

finite field A synonym for GALOIS FIELD.

finite intersection property A *topological space (or *metric space) has the finite intersection property when every family of *closed subsets, for which all finite subcollections have non-empty intersections, must have a non-empty intersection as an entire family. This property is equivalent to *compactness.

finitely generated A *group G is finitely generated if there is a finite subset S which generates G; that is, the only subgroup of G which contains S is G. A *cyclic group is generated by a single element; $\mathbb{Q}$ is not finitely generated as an additive group. The term similarly applies to other *algebraic structures; a finitely generated *vector space is called *finite-dimensional.

finite population correction The *standard error of the *mean based on a *sample of size n assumes that the *population is infinite or at least so large that the effect of withdrawing items during the sampling process has a negligible effect. If the size of the sample, n, becomes a not insignificant fraction of the population, N, then the finite population correction $\sqrt{\frac{N-n}{N-1}}$ is used to adjust the estimate of the standard error, which becomes $\sqrt{\frac{N-n}{N-1}} \times \frac{\sigma}{\sqrt{n}}$. It is common to use this correction if the sample size is more than 5% of the population, and it reduces the standard error, so resulting in a narrower *confidence interval for the population mean.

finitism A *philosophy of mathematics which only accepts the existence of finite mathematical objects and hence no *actual infinities. *Kronecker was an advocate of finitism. *See also* INTUITIONISM.

first derivative A term used for the *derivative when it is being contrasted with *higher derivatives.

first fundamental form For a surface X in $\mathbb{R}^3$ with $p \in X$, the first fundamental form is the *restriction of the *quadratic form $\mathbf{v} \mapsto |\mathbf{v}|^2$ to the *tangent space at p. If $\mathbf{r}(u,v)$ is a parametrization of X, so that $\mathbf{r}_u = \partial\mathbf{r}/\partial u$ and $\mathbf{r}_v = \partial\mathbf{r}/\partial v$ is a *basis for the tangent space, then the form is given by $\alpha\mathbf{r}_u + \beta\mathbf{r}_v \mapsto E\alpha^2 + 2F\alpha\beta + G\beta^2$, where $E = \mathbf{r}_u \cdot \mathbf{r}_u$, $F = \mathbf{r}_u \cdot \mathbf{r}_v$, $G = \mathbf{r}_v \cdot \mathbf{r}_v$. Any intrinsic (i.e. metric) property of the surface, such as *area, *length of curves, or *Gaussian curvature, can be expressed in terms of E, F, G. See RIEMANNIAN MANIFOLD, SECOND FUNDAMENTAL FORM, THEOREMA EGREGIUM.

first isomorphism theorem See ISOMORPHISM THEOREMS.

first-order differential equation A *differential equation of order 1, such as $3y\frac{dy}{dx} + 5x = 3$.

first-order logic In propositional logic, sentences do not involve *quantifiers; for example, *De Morgan's Laws are laws of propositional logic. Also known as predicate logic, first-order logic is an extension of propositional logic which permits quantifiers over elements; for example, $(\forall x \in \mathbb{R})(\exists y \in \mathbb{R})$ $(x + y = 0)$ states that every *real number has an *additive inverse. But the *completeness axiom for $\mathbb{R}$, that every non-empty, *bounded-above subset of $\mathbb{R}$ has a *supremum, necessarily involves quantification over *subsets of $\mathbb{R}$, and so is a second-order sentence.

first principles Without relying on other theorems for a result. For example, if $f(x) = x^2$, $f'(x) = 2x$ but to show this from first principles requires the following argument:

$$f'(x) = \lim_{\delta x \to 0} \frac{f(x + \delta x) - f(x)}{\delta x}$$
$$= \lim_{\delta x \to 0} \frac{x^2 + 2x\delta x + (\delta x)^2 - x^2}{\delta x}$$
$$= \lim_{\delta x \to 0} (2x + \delta x) = 2x$$

Fisher, Sir Ronald Aylmer (1890–1962) British geneticist and statistician who established methods of designing experiments and analysing results that have been extensively used ever since. His influential book on statistical methods appeared in 1925. He developed the *t-test and the use of *contingency tables and is responsible for the method known as *analysis of variance.

Fisher's exact test A statistical test used to examine the significance of the association between two categorical variables in a 2×2 *contingency table. The test needs the expected values in each cell to be at least 10 but does not depend on the sample characteristics.

fit Mathematical or statistical models are used to describe phenomena in the real world. The fit is the degree of correspondence between the observations and the model's predictions.

fixed point A fixed point of a *function $f : X \to X$ is an element $x \in X$ such that $f(x) = x$.

fixed-point iteration To find a root of an equation $f(x) = 0$ by the method of fixed-point iteration, the equation is first rewritten in the form $x = g(x)$. Starting with an initial approximation x_0 to the root, the values x_1, x_2, x_3, … are calculated using $x_{n+1} = g(x_n)$. The method is said to converge if these values tend to a *limit α. If they do, then $\alpha = g(\alpha)$ and so α is a root of the original equation.

A root of $x = g(x)$ occurs where the graph $y = g(x)$ meets the line $y = x$. It can be shown that, if $|g'(\alpha)| < 1$, then the sequence converges for suitably close initial values x_0, and α is called attracting. If $|g'(\alpha)| > 1$, then the sequence diverges from α for nearby initial values x_0. This is illustrated in the figures; such diagrams are called 'cobweb plots', and the process is called 'cobwebbing'. The equation $x^3 - x - 1 = 0$ has a root α between 1 and 2, so we take $x_0 = 1.5$. The equation can be written in the form $x = g(x)$ in several ways, such as (i) $x = x^3 - 1$ or (ii) $x = (x + 1)^{1/3}$. In case (i), $g'(x) = 3x^2$, $g'(\alpha) > 3 > 1$ and so α is repelling for this iteration; in case (ii), $g'(x) = \frac{1}{3}(x + 1)^{-2/3}$ and $g'(\alpha) < 2^{-2/3}/3 < 1$ and so α is attractive.

More generally a *fixed point $\alpha \in X$ of a function $f:X \rightarrow X$ is attracting if the fixed-point iteration converges to α for initial values in a *neighbourhood of α and is repelling if there is a neighbourhood of α such that the fixed-point iteration eventually moves out of the neighbourhood for all initial values other than α.

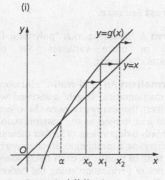

Repelling fixed point $|g'(\alpha)| > 1$

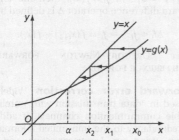

Attracting fixed point $|g'(\alpha)| < 1$

fixed-point notation See FLOATING-POINT NOTATION.

fixed-point theorem Any theorem which gives conditions under which a *function must have a *fixed point (see BROUWER'S FIXED-POINT THEOREM, CONTRACTION MAPPING THEOREM). Such theorems also have importance in economics.

flag A binary variable used to take some action, for example in a conditional if . . . then . . . . else command in a computer program.

flat angle A synonym for STRAIGHT ANGLE.

flex A synonym for POINT OF INFLECTION.

floating-point notation A method of writing real numbers, used in computing, in which a number is written as $a \times 10^n$, where $0.1 \leq a < 1$ and n is an integer. The number a is called the significand, and n is the *order of magnitude. Thus, 634.8 and 0.002 34 are written as 0.6348×10^3 and 0.234×10^{-2}. (There is also a *base 2 version similar to the base 10 version just described.)

This is in contrast to fixed-point notation, in which all numbers are given by means of a fixed number of digits with a fixed number of digits after the decimal point. For example, if numbers are given by means of 8 digits with four of them after the decimal point, the two numbers above would be written (with an approximation) as 0634.8000 and 0000.0023. Compare SCIENTIFIC NOTATION.

float of an activity (in critical path analysis) The amount of time by which the start time of an activity can be varied without delaying the overall completion of the project, i.e. the latest possible time for completion—the earliest possible time for starting—the duration of the activity.

floor (greatest integer function)
The largest *integer $\lfloor x \rfloor$ not greater than
a given *real number x. So $\lfloor 3.2 \rfloor = 3$
and $\lfloor 5 \rfloor = 5$. *Compare* CEILING.

fluent *See* FLUXION.

fluid mechanics The study of fluids—
liquids and gases—whether at rest (fluid
statics) or in motion (fluid dynamics).
The terms hydrostatics and hydrodynamics are also used for liquids at
rest or in motion, and aerodynamics for
gases in motion. *See* EULER'S EQUATIONS,
NAVIER-STOKES EQUATIONS.

flux Given a *vector field **F** and a surface
Σ in $\mathbb{R}^3$, then the flux of **F** across Σ is the
integral $\iint_\Sigma \mathbf{F} \cdot d\mathbf{S}$. If **F** represents the velocity of a fluid, then the flux is the rate at
which fluid is crossing the surface Σ. *See*
FOURIER'S LAW, GAUSS FLUX THEOREM,
SURFACE INTEGRAL.

fluxion In *Newton's work on calculus,
he thought of the variable x as a 'flowing
quantity' or fluent. The rate of change of x
was called the fluxion of x, denoted by $\dot{x}$.

focus The plural is 'foci' and the adjective is 'focal'. *See* CONIC, ELLIPSE,
HYPERBOLA, PARABOLA.

foot of the perpendicular *See*
PROJECTION.

for all The meaning of the universal
*quantifier, symbol $\forall$.

force A dynamic influence that causes a
*particle or *body to *accelerate. (*See*
NEWTON'S LAWS OF MOTION.) Commonly
experienced forces are *gravity, *contact
forces (such as friction and air resistance), forces within a body that deform
or restore its shape, and electrical and
magnetic forces.
　In general, force is a *vector quantity
with a magnitude, a direction, and also a
*point of application. (*See* MOMENT (IN
MECHANICS).) Force has the dimensions

MLT^{-2}, and the *SI unit of measurement is the *newton.

forced oscillations *Oscillations that
occur when a body capable of oscillating
is subject to an applied *force which varies with time. If the applied force is itself
oscillatory, a differential equation such
as $m\ddot{x} + kx = F_0 \sin(\Omega t + \varepsilon)$ may be
obtained. The *natural frequency of this
system equals $\sqrt{k/m}$, and if the forced
frequency Ω agrees with the natural frequency, then *resonance will occur.

(((•))) **SEE WEB LINKS**

• Animations and videos of forced oscillation
  and resonance.

forest *See* TREE.

form A *homogeneous *polynomial in
two or more variables. *See also*
QUADRATIC FORM.

formalism An *anti-realist *philosophy
of mathematics, largely associated with
*Hilbert, that mathematics can be axioma-
tized and treated as the manipulation of
symbols independent of further meaning.
In this view, a mathematical statement has
no truth in reality, its truth being solely a
matter of verification using the *axioms.
See GÖDEL'S INCOMPLETENESS THEOREMS,
INTUITIONISM, LOGICISM, PLATONISM.

forward difference If $\{(x_i, f_i)\}$, $i \geq 0$
is a given set of function values with
$x_{i+1} = x_i + h$, $f_i = f(x_i)$, then the forward difference operator Δ is defined by

$$\Delta f_i = f_{i+1} - f_i = f(x_{i+1}) - f(x_i).$$

See GREGORY–NEWTON FORWARD
DIFFERENCE FORMULA.

forward error correction Widely
used in *data transmission with unreliable communication channels. It adds
*redundancy to the transmitted informa-
tion using an *error-correcting code,
allowing the receiver or retriever to detect

errors. This is important when retransmission is expensive or impossible.

forward scan (in critical path analysis) In an *activity network (edges as activities), the forward scan identifies the earliest time for each vertex. Starting with the *source, work forwards through the network, for each vertex calculate the sum of edge plus total time on previous vertex for all paths arriving at that vertex. Assign the largest of these times to that vertex.

forward substitution See LU DECOMPOSITION.

Foucault pendulum A pendulum consisting of a heavy bob suspended by a long inextensible string from a fixed point, free to swing in any direction, designed to demonstrate the rotation of the Earth. The original experiment, in 1851, was conducted by the French physicist Jean-Bernard-Léon Foucault. If the experiment were set up at the north or south pole, the vertical plane of the swinging bob would appear to an observer fixed on the Earth to precess or rotate once a day. At a location of latitude λ in the northern hemisphere, the vertical plane of the swinging bob would precess in a clockwise direction with an angular speed of $\omega \sin \lambda$, where $\omega = 7.29 \times 10^{-5}$ rad s^{-1}, the angular speed of rotation of the Earth. The period of precession in Oxford, for example, would be about $30\frac{1}{2}$ hours.

foundations of mathematics The study of the logical basis for mathematics, and in particular attempts to establish an axiomatic basis upon which mathematics could be built. Early in the 20th century Bertrand *Russell tried unsuccessfully to produce a unifying set of *axioms for mathematics. See GÖDEL'S INCOMPLETENESS THEOREMS.

Four Colour Theorem It had long been observed that any geographical map (that is, a division of the plane into regions) could be coloured with just four colours in such a way that no two neighbouring regions had the same colour. A proof of this, the Four Colour Theorem, was sought by mathematicians from about the 1850s. In 1890 Heawood proved that five colours would suffice, but it was not until 1976 that Appel and Haken proved the Four Colour Theorem itself. Initially, some mathematicians were sceptical of the proof because it relied on a massive amount of checking of configurations by computer that could not easily be verified independently. However, the proof is now generally accepted.

(((•))) SEE WEB LINKS
• A Shockwave file allowing you to create your own map and colour it in or see an automated solution with no more than four colours.

Fourier, (Jean Baptiste) Joseph, Baron (1768–1830) French engineer and mathematician, best known for his fundamental contributions to the theory of heat conduction and his study of *trigonometric series. These so-called *Fourier series are of immense importance in physics, engineering and other disciplines, as well as being of great mathematical interest.

Fourier analysis The use of *Fourier series and the *Fourier transform in *analysis, in many applications such including signal processing.

Fourier coefficients Given $L>0$, the coefficients a_n and b_n, in the *Fourier series representation of a function of period $2L$, are given by

$$a_n = \frac{1}{L} \int_{-L}^{L} f(x) \cos\left(\frac{n\pi x}{L}\right) \, dx$$

$$b_n = \frac{1}{L} \int_{-L}^{L} f(x) \sin\left(\frac{n\pi x}{L}\right) \, dx.$$

Fourier cosine series Given a function $f(x)$ on the interval $0 < x < L$ the Fourier cosine series is defined as

$$\frac{a_0}{2} + \sum_{n=1}^{\infty} a_n \cos\left(\frac{n\pi x}{L}\right),$$

where $a_n = \frac{2}{L} \int_0^L f(x) \cos\left(\frac{nx}{L}\right) \, dx.$

This defines an *even function of *period 2L.

Fourier series The infinite series $\frac{1}{2} a_0 +$ $\sum_{n=1}^{\infty} \left(a_n \cos(n\pi x / L) + b_n \sin(n\pi x / L)\right),$ where a_n and b_n are the *Fourier coefficients. The Fourier series defines a function of period 2L; the *convergence of the series and the magnitude of the Fourier coefficients are highly related to the order of differentiability of the function f. Fourier series are used to decompose a waveform into component waves of different *frequencies and *amplitudes, allowing identification of different sources from background or random *noise in a signal. Fourier series are also widely used in the solution of *partial differential equations.

Fourier sine series Given a function $f(x)$ on the interval $0 < x < L$, the Fourier sine series is defined as

$$\sum_{n=1}^{\infty} b_n \sin\left(\frac{n\pi x}{L}\right),$$

where $b_n = \frac{2}{L} \int_0^L f(x) \sin\left(\frac{n\pi x}{L}\right) \, dx.$

This defines an *odd function of *period 2L.

Fourier's law In the *modelling of *heat flow, Fourier's law states that $\mathbf{q} = -k \nabla T$, where T denotes *temperature, $\mathbf{q}$ denotes heat *flux, and k is the material's thermal conductivity. See HEAT EQUATION, NEWTON'S LAW OF COOLING.

Fourier transform The *integral transform $\hat{f}(s) = \int_{-\infty}^{\infty} f(x) e^{-isx} dx$ for an *integrable function f on $\mathbb{R}$. The Fourier transform is also denoted by $\mathcal{F}$. The

function $\hat{f}$ is the Fourier transform of f. The Fourier *inverse is given by $f(x) = \frac{1}{2\pi} \int_{-\infty}^{\infty} \hat{f}(s) e^{isx} ds.$ See DISCRETE FOURIER TRANSFORM, CONVOLUTION, LAPLACE TRANSFORM.

four squares theorem See LAGRANGE'S THEOREM.

fourth root of unity A complex number z such that $z^4 = 1$. There are 4 fourth roots of unity and they are 1, i, -1 and $-i$. (See NTH ROOT OF UNITY.)

fractal A set whose *fractal dimension exceeds its topological *dimension. Fractals are sets with infinitely complex structure and usually possess some measure of self-similarity, whereby any part of the set contains within it a scaled-down version of the whole set. Examples are the *Cantor set, the *Koch curve, and the *Sierpinski triangle.

(((●))) SEE WEB LINKS
• Examples of colourful fractal images.

fractal dimension One of the many extensions (see HAUSDORFF DIMENSION) of the notion of dimension to objects for which the traditional concept of dimension is not appropriate. The fractal dimension may have a non-integer value. The *Koch curve has dimension $\log_3 4 \approx 1.26$. The *Cantor set has dimension $\log_3 2 \approx 0.63$. Fractal dimension has found applications analysing chaotic or noisy processes (see CHAOS).

fraction The fraction a/b, where a and b are positive integers, was historically obtained by dividing a unit length into b parts and taking a of these parts. The number a is the numerator and the number b is the denominator. It is a proper fraction if $a < b$ and an improper fraction if $a > b$. Any fraction can be expressed as $c + d/e$, where c is an *integer and d/e is a proper fraction, and in this form it is called a mixed number. For example, $3\frac{1}{2}$ is a mixed fraction (equal to 7/2). See also

COMPOUND FRACTION, CONTINUED FRACTION, SIMPLE FRACTION.

fractional part For any real number x, its fractional part is equal to $x - [x]$, where $[x]$ is the *integer part of x. It may be denoted by $\{x\}$. The fractional part r of any real number always satisfies $0 \leq r < 1$.

frame of reference In mechanics, a means by which an observer specifies position and time to describe events and the motion of bodies around them. In some circumstances, it may be useful to consider two or more different frames of reference, each with its own observer. One frame of reference, its origin and axes, may be moving relative to another.

A frame of reference in which *Newton's laws of motion hold is called an inertial frame (of reference). Any frame of reference that is at rest or moving with constant velocity relative to an inertial frame is an inertial frame. A frame of reference that is accelerating or rotating with respect to an inertial frame is not an inertial frame.

A frame of reference fixed on the Earth is not an inertial frame because of the rotation of the Earth. However, such a frame of reference may be assumed to be an inertial frame in problems where the rotation of the Earth has little effect. *See* FICTITIOUS FORCE, GALILEAN RELATIVITY, SPECIAL RELATIVITY.

framework (in mechanics) *See* LIGHT FRAMEWORK.

Fredholm alternative A simple form of the Fredholm alternative states that for a square matrix $\mathbf{A}$

$$\text{either } \mathbf{v} = \mathbf{A}\mathbf{v} + \mathbf{b} \text{ has a solution } \mathbf{v} \text{ for all } \mathbf{b}$$
$$\text{or } \mathbf{v} = \mathbf{A}\mathbf{v} \text{ has a non} - \text{zero solution.}$$

The former holds when $\mathbf{A} - \mathbf{I}$ is *invertible and the latter when it is *singular. The more general Fredholm alternative is an important result in *functional analysis, where $\mathbb{R}^n$ is replaced by an infinite-dimensional *Hilbert space and $\mathbf{A}$ is replaced by a *self-adjoint *compact operator.

Fredholm integral equation The *integral equations

$$\varphi(x) = \int_a^b K(x,t)f(t) \ dt$$

and

$$f(x) = \int_a^b K(x,t)f(t) \ dt + \varphi(x),$$

are respectively known as Fredholm equations of the first and second kind. Here φ and K are given and a solution f is sought.

Equations of the first kind often have no solution, or if a solution exists, it may not be unique or depend continuously on $\varphi(x)$. Equations of the second kind can be shown to have a unique solution when K a *continuous function with a suitably small norm.

Freedman, Michael Hartley (1951–) American mathematician awarded the *Fields Medal in 1986 for his work on the 4-dimensional Poincaré conjecture (*see* MILLENNIUM PRIZE PROBLEMS).

free group Given a set S, the free group $F(S)$ is the set of *words using elements x of S and their formal inverses x^{-1} with group operation being *concatenation. Two distinct words are equal if and only if this involves replacing a product xx^{-1} or $x^{-1}x$ with the identity e or vice versa, that is, there are no non-trivial *relations. Every *group is a *quotient group of a free group. *See* PRESENTATION.

freely hinged A synonym for SMOOTHLY HINGED.

free module A *module with a *basis. Unlike *vector spaces, modules typically do not have bases. A free module over a ring R is *isomorphic to R^n for some n, and all bases of R^n have n elements. Note $\mathbb{Z}_2$ is free as a $\mathbb{Z}_2$-module, as $\{1\}$ is a basis, but not free as a $\mathbb{Z}$-module, as 1 is a *torsion element.

Frege, (Friedrich Ludwig) Gottlob

(1848–1925) German mathematician and philosopher, founder of the subject of mathematical logic. In his works of 1879 and 1884, he developed fundamental ideas, invented the standard notation of *quantifiers and variables, and studied the foundations of arithmetic. Not widely recognized at the time, his work was disseminated primarily by *Peano and *Russell.

frequency (in mechanics) When *oscillations, or cycles, occur with *period T, the frequency is equal to $1/T$. The frequency is equal to the number of oscillations or cycles that take place per unit time.

Frequency has the dimensions T^{-1}, and the *SI unit of frequency is the *hertz.

frequency (in statistics) The number of times that a particular value occurs as an observation. In *grouped data, the frequency corresponding to a group is the number of observations that lie in that group. If numerical data are grouped by means of *class intervals, the frequency corresponding to a class interval is the number of observations in that interval. *See also* RELATIVE FREQUENCY.

frequency analysis In *cryptography, a means of decrypting *ciphers first developed by Arab mathematicians in the 9th century. As letters in a text have certain frequencies (for example E and T are the two most common letters in English), their encrypted forms will appear as commonly in a suitably long piece of *ciphertext, which will restrict enormously the possible ciphers to consider. Modern cryptographic methods, such as *RSA, are still essentially ciphers but work on an 'alphabet' of size 2^{4096},

rather than 26, and so are not susceptible to frequency analysis.

frequency distribution For nominal or discrete data, the information consisting of the possible values and the corresponding *frequencies is called the frequency distribution. For *grouped data, it gives the information consisting of the groups and the corresponding frequencies. It may be presented in a table or in a diagram such as a *bar chart, *histogram or *stem-and-leaf plot.

frequentist inference *See* BAYESIAN INFERENCE, HYPOTHESIS TESTING.

Freudenthal, Hans (1905–90) German mathematician who became one of the most important figures in *mathematics education. He founded the Institute for the Development of Mathematical Education in Utrecht in 1971 and was its first director. It was renamed the Freudenthal Institute in 1991 after his death.

Frey's curve Frey's curve was an important link between *Fermat's Last Theorem and proving the *modularity theorem. If there were a counterexample to Fermat's Last Theorem, $X^p + Y^p = Z^p$, where p is an odd prime, then the *elliptic curve with equation

$$y^2 = x(x - X^p)(x + Y^p)$$

would not be modular. So if all elliptic curves are modular, then no such X, Y, Z can exist, which thus proves Fermat's Last Theorem.

friction (frictional force) For two bodies in contact, say the frictional force and the *normal reaction have magnitudes F and N respectively (*see* CONTACT FORCE). The coefficient of static friction μ_s is the ratio F/N in the limiting case when the bodies are just about to move relative to each other. Thus if the bodies are at rest relative to each other, $F \leq \mu_s N$. The coefficient of kinetic friction μ_k is the ratio F/N when the bodies are sliding, in contact with each other. These coefficients of friction depend on the materials of which the bodies are made. Normally, μ_k is somewhat less than μ_s. *See also* ANGLE OF FRICTION.

frieze group A frieze pattern is a planar pattern that is repeating in one direction. As such they are common in decorations and architecture. Formally a frieze pattern is such that its *group of *symmetries—called a frieze group—has a *subgroup of *translations that is *isomorphic to the integers.

The frieze group of the pattern in the figure is generated by x, a translation two tiles to the right, and y, a half-turn, and is isomorphic to

$$\langle x, y \mid y^2 = e, yx = x^{-1}y \rangle,$$

and its subgroup of translations is $\langle x \rangle$. It can be shown that, up to isomorphism, there are seven frieze groups. *Compare* CRYSTALLOGRAPHIC GROUPS.

Frobenius, (Ferdinand) Georg (1849–1917) German mathematician known especially for his work in *number theory and *algebra, particularly in the *representation theory and *character theory of *groups.

Frobenius endomorphism For a *field F with prime *characteristic p, the Frobenius *endomorphism is the map φ: $F \rightarrow F$ defined by $\varphi(r) = r^p$. Note φ is clearly multiplicative and is additive as the *binomial coefficient $\binom{p}{k}$ is divisible by p for $0 < k < p$. By *Fermat's Little Theorem, φ fixes the *prime subfield. If F

is finite, then φ is an *automorphism and generates the *Galois group of F over the prime subfield.

Frobenius normal form A synonym for RATIONAL CANONICAL FORM.

from above, from the right *See* RIGHT-HAND LIMIT.

from below, from the left *See* LEFT-HAND LIMIT.

frontier A synonym for boundary (of a set).

FRS Abbreviation for 'Fellow of the Royal Society'. The Royal Society is the UK's academy of sciences, founded in 1660.

frustum (frusta) A frustum of a right-circular *cone is the part between two *parallel *planes *perpendicular to the axis. Suppose that the planes are a distance h apart and that the circles that form the top and bottom of the frustum have radii a and b. Then the volume of the frustum equals $\frac{1}{3}\pi h(a^2 + ab + b^2)$.

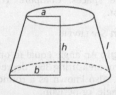

A frustum of a cone

Let l be the slant height of the frustum; that is, the length of the part of a generator between the top and bottom of the frustum. Then the area of the curved surface of the frustum equals $\pi(a + b)l$.

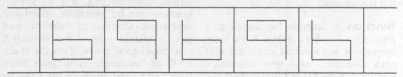

An example of a frieze pattern

Fry, Hannah (1984–) Mathematician and popularizer of mathematics. She has a long-running radio programme with the geneticist Adam Rutherford and has written several books. Her work often relates to what mathematics and statistics can say about societal behaviour. In 2019 she gave the *Royal Institution Christmas lectures, and in 2020 she received the *LMS *Zeeman medal for her public engagement on mathematics.

(🌐 SEE WEB LINKS)
• Hannah Fry's official website.

F-test A test that uses the *F-distribution.

Fubini's Theorem For a Lebesgue integrable function $f:\mathbb{R}^2 \to \mathbb{R}$

$$\int_{\mathbb{R}^2} f(x,y) \, dA = \int_{\mathbb{R}} \left(\int_{\mathbb{R}} f(x,y) \, dx \right) dy$$
$$= \int_{\mathbb{R}} \left(\int_{\mathbb{R}} f(x,y) \, dy \right) dx.$$

This result generalizes to products of *measure spaces. *Compare* TONELLI'S THEOREM.

fulcrum *See* LEVER.

full angle An angle equal to one complete turn, measuring 360 *degrees or 2π *radians. Also known as a round angle, whole angle, or perigon.

full measure In a *measure space, a set whose complement is a *null set.

full rank A *matrix has full rank if its *rank equals the smaller of the number of *rows and the number of *columns. A square matrix has full rank if and only if it is *invertible.

function A function (or mapping) f from S to T, where S and T are non-empty sets, is a rule that associates with each element s of S (the domain) a unique element $f(s)$ of T (the codomain). s is the argument or input, and $f(s)$ is the image or output. The notation $f: S \to T$, read as "f from S to T", is used. The subset of T consisting of those elements that are images, that is $\{f(s) \mid s \varepsilon S\}$, is the image or range of f. If $f(s) = t$, it is said that f maps s to t, written $s \mapsto t$. *See* BIJECTION, GRAPH (OF A FUNCTION), ONE-TO-ONE MAPPING, ONTO MAPPING, REAL FUNCTION, MORPHISM.

functional The term, 'functional' is variously used, typically referring to a scalar-valued function on a set. In *calculus of variations, the aim is to minimize a functional which might represent length, area, work, etc.. In *linear algebra and *functional analysis, the term is used for a *linear map $\varphi:V \to F$ from a *vector space to its *base field, and so an element of the *dual space.

functional analysis The branch of mathematics studying *function spaces. It is largely concerned with the study of *complete *normed vector spaces with important applications in the study of *quantum theory and *differential equations. *See* BAIRE CATEGORY THEOREM, BANACH SPACE, HILBERT SPACE.

functionally separable Two subsets A and B of a *topological space X are functionally separable if there exists a continuous function $f : X \to [0, 1]$ such that $f(a) = 0$ for any $a \in A$ and $f(b) = 1$ for any $b \in B$. *See* URYSOHN'S LEMMA.

function of a function *See* COMPOSITION.

function space A general term relating to a space of *functions between two given *sets X and Y. If the *codomain has further structure (e.g. if $Y = \mathbb{R}$), then functions can be *pointwise added and multiplied, forming an *algebra, and given a *partial order. If the *domain X is a *topological space, then the space $C^*(X)$ of *bounded, *continuous functions real-valued functions on X can be made into a *normed vector space with

$\|f\| = \sup_{x \in X} |f(x)|$. For *vector spaces V and W, then $\text{Hom}(V,W)$, denoting the space of *linear maps from V to W, is a vector space itself. *See also* C^0, C^1, C^∞, C^ω, L^p.

functor A covariant functor F, between two *categories C and D, is an assignment to each object X of C an object $F(X)$ of D and to each morphism $f:X \to Y$ of C a morphism $F(f):F(X) \to F(Y)$ satisfying $F(1_X) = 1_{F(X)}$ and $F(g \circ f) = F(g) \circ F(f)$. The functor is contravariant if, instead, $F(g \circ f) = F(f) \circ F(g)$. Assigning every real *vector space to its *dual space, and every *linear map to its *dual map, is a contravariant functor on the category of real vector spaces.

fundamental group A topological invariant important in *algebraic topology and the first of the homotopy groups. For a *path-connected space X with base point x_0, the fundamental *group $\pi_1(X)$ is the set of *equivalence classes of continuous loops $\gamma:[0,1] \to X$ such that $\gamma(0) = \gamma(1) = x_0$, and where two loops are equivalent if they are *homotopic. The product $\gamma_{1*}\gamma_2$ of two loops is the loop γ_2 followed by γ_1 and the *inverse of γ is the loop in reverse. A *continuous function $f:X \to Y$ induces a *homomorphism $f_*:\pi_1(X) \to \pi_1(Y)$ by sending γ to $f(\gamma)$. The circle's fundamental group is $\mathbb{Z}$; the knot group of a *knot is the fundamental group of its *complement.

Fundamental Theorem of Algebra Every *polynomial equation

$$f(z) = a_n z^n + a_{n-1} z^{n-1} + \ldots + a_1 z + a_0 = 0,$$

where the a_i are real or complex numbers and $a_n \neq 0$, has a root in the set of *complex numbers.

It follows that there exist complex numbers $\alpha_1, \alpha_2, \ldots, \alpha_n$ (not necessarily distinct) such that

$$f(z) = a_n(z - \alpha_1)(z - \alpha_2) \ldots (z - \alpha_n).$$

Fundamental Theorem of Arithmetic Any positive integer ($\neq 1$) can be expressed as a product of *primes. This expression is unique except for the order in which the primes occur.

Thus, any positive integer $n(\neq 1)$ can be written uniquely as:

$$p_1^{\alpha_1} p_2^{\alpha_2}, \ldots, p_r^{\alpha_r}$$

where $p_1, p_2, \ldots, p_r$ are primes, satisfying $p_1 < p_2 < \ldots < p_r$, and $\alpha_1, \alpha_2, \ldots, \alpha_r$ are positive integers. This is the prime *decomposition of n, for example $360 = 2^3 \times 3^2 \times 5$. Note this decomposition would not be unique if 1 were considered a prime number. Prime factorization of large integers is computationally difficult and so commonly used in *cryptography. *See* UNIQUE FACTORIZATION DOMAIN.

Fundamental Theorem of Calculus This theorem makes explicit how *integration is the inverse of *differentiation.

If f is a *continuous function on $[a,b]$ and F is an antiderivative of f, so that $F'(x) = f(x)$ for all x in $[a,b]$, then

$$\int_a^b f(x) \, dx = F(b) - F(a).$$

It follows that

$$\frac{d}{dx} \int_a^x f(t) \, dt = f(x).$$

Fundamental Theorem of Game Theory (Minimax Theorem) Due to *Von Neumann, the theorem states:

in a matrix game, with $E(\mathbf{x},\mathbf{y})$ denoting the expectation, where $\mathbf{x}$ and $\mathbf{y}$ are mixed strategies for the two players, then

$$\max_{\mathbf{x}} \min_{\mathbf{y}} E(\mathbf{x}, \mathbf{y}) = \min_{\mathbf{y}} \max_{\mathbf{x}} E(\mathbf{x}, \mathbf{y}).$$

By using a maximin strategy (*see* CONSERVATIVE STRATEGY), one player, R, ensures that the expectation is at least as large as the left-hand side of the equation. Similarly, by using a minimax strategy, the other player, C, ensures that the expectation is less than or equal to the right-hand side. Such strategies may be called optimal strategies for R and C. Since, by the theorem, the two sides of the equation are equal, then if R and C use optimal strategies the expectation is equal to the common value, which is called the value of the game.

For example, consider the game given by the matrix

$$\begin{bmatrix} 4 & 2 \\ 3 & 4 \end{bmatrix}.$$

if $\mathbf{x}^* = (\frac{1}{3}, \frac{2}{3})$, it can be shown that $E(\mathbf{x}^*, \mathbf{y}) \geq 10/3$ for all $\mathbf{y}$. Also, if $\mathbf{y}^* = (\frac{2}{3}, \frac{1}{3})$, then $E(\mathbf{x}, \mathbf{y}^*) \leq 10/3$ for all $\mathbf{x}$. It follows that the value of the game is 10/3, and $\mathbf{x}^*$ and $\mathbf{y}^*$ are optimal strategies for the two players.

fuzzy set theory In standard *set theory, an *element either is or is not a member of a particular *set. However, in some instances, for example in pattern recognition or decision-making, it may not be known whether an element is in the set. Fuzzy set theory blurs this distinction and replaces this binary option with the interval [0,1] with, possibly, the value representing a *probability that an element is a member of a particular set.

g See GRAVITY.

G Abbreviation for *giga-.

G See GRAVITATIONAL CONSTANT.

Gabriel's horn The *surface of revolution formed by rotating the curve $y = 1/x$ for $x > 1$ about the x-axis, which has seemingly paradoxical properties. The *volume within the surface is finite, but the *surface area is infinite.

Galilean relativity Two observers O and O' moving in 3-dimensional space will, according to *Galileo's theory of relativity, have coordinate systems related by

$$\begin{bmatrix} t \\ \mathbf{r} \end{bmatrix} = \begin{bmatrix} 1 & 0 \\ \mathbf{v} & \mathbf{M} \end{bmatrix} \begin{bmatrix} t' \\ \mathbf{r}' \end{bmatrix} + \mathbf{c}$$

where t, t' and $\mathbf{r}$, $\mathbf{r}'$ are the time and space coordinates for O, O', respectively, $\mathbf{v}$ is the *velocity of O with respect to O', $\mathbf{M}$ is a 3×3 *orthogonal, *determinant 1 matrix, and $\mathbf{c}$ is a constant *column vector. In Galileo's theory, observers agree on the *simultaneity of events, the time between events (see TIME DILATION), and the distance between simultaneous events (see LORENTZ-FITZGERALD CONTRACTIONS). Compare LORENTZ GROUP.

Galileo Galilei (1564–1642) Italian mathematician, astronomer, and physicist who established the method of studying dynamics by a combination of theory and experiment. He formulated and verified by experiment the law $s = \frac{1}{2}at^2$ of constant *acceleration for falling bodies and derived the parabolic path of a *projectile. He developed the telescope and was the first to use it to make significant and outstanding astronomical observations. In later life, his support for the Copernican theory that the planets travel round the Sun resulted in conflict with the Church and consequent trial and house arrest.

Galois, Évariste (1811–32) French mathematician who had made major contributions to the theory of equations before he died at the age of 20, shot in a duel. His work developed the necessary *group theory to deal with the question of whether a *polynomial equation can be *solved by radicals. He spent the night before the duel writing a letter containing notes of his discoveries, but his work would only be appreciated decades later thanks to *Liouville and *Jordan.

Galois correspondence Let K denote the *splitting field of a polynomial $f(x)$ over $\mathbb{Q}$ and let G denote the *Galois group of K: $\mathbb{Q}$. Then the degree of K: $\mathbb{Q}$ equals the order of G. Further for every *subgroup H of G we can define its fixed field

$$H^\dagger = \{k \in K | g(k) = k \quad \text{for all} \quad g \in H\},$$

which is a *subfield of K. And for every subfield F of K,

$$F^* = \text{Gal}(K : F) = \{g \in G | g(f) = f$$
$$\text{for all} \quad f \in F\},$$

is a subgroup of G. Then * and † are *inverses of one another. Further N is a *normal subgroup of G if and only if $N^\dagger$ is the splitting field of some polynomial. (With further technical additions the correspondence can be extended to base fields other than $\mathbb{Q}$.)

Galois field A finite *field. The order of a finite field is necessarily p^n, where p is a prime, and for each such order there is a finite field which is unique up to *isomorphism. The Galois field of order q is denoted $\mathbb{F}_q$ and $\mathbb{F}_p$ is $\mathbb{Z}_p$.

Galois group The Galois *group Gal(K: F) of a *field extension $F \subseteq K$ is the group Aut$_F(K)$ of field *automorphisms of K, which pointwise fix the base field F; the group operation is *composition. The Galois group Gal($\mathbb{C}$:$\mathbb{R}$) has order 2 and is generated by complex *conjugation.

If f is a *polynomial over F, then the Galois group of f is the Galois group of its *splitting field. The Galois group of f acts (*see* ACTION) on the set of roots of f. A polynomial f over $\mathbb{Q}$ is *solvable by radicals if and only if its Galois group is *solvable. The Galois group of $x^5 - x - 1$ is the *symmetric group S_5, which is not solvable. *See* GALOIS CORRESPONDENCE.

Galton, Francis (1822–1911) British explorer and anthropologist, a cousin of Charles Darwin. His primary interest was eugenics. He made contributions to *statistics in the areas of *regression and *correlation.

gambler's ruin *See* RANDOM WALK.

⊕ SEE WEB LINKS

• A Java applet showing the gambler's ruin problem.

game (game theory) An attempt to represent and analyse mathematically some conflict situation in which the outcome depends on the choices made by the opponents. The applications of game theory are not primarily concerned with recreational activities. Games may be used to investigate problems in business, personal relationships, military manoeuvres, and other areas involving decision-making. One particular kind of game for which the theory has been well developed is the *matrix game.

gamma distribution If X is a random variable with pdf given by $f(x) = \dfrac{\lambda^v x^{v-1} e^{-\lambda x}}{\Gamma(v)}$ where Γ is a *gamma function and λ, v, x are all positive, then we say that X has a gamma distribution with parameters λ, v. When $v = 1$, $f(x)$ reduces to $\lambda e^{-\lambda x}$ which is the *exponential distribution.

gamma function The function defined by $\Gamma(x) = \int_0^\infty t^{x-1} e^{-t} dt$ for $x > 0$. *Integration by parts yields $\Gamma(x + 1) = x\Gamma(x)$, and $\Gamma(1) = 1$ so if n is a positive integer $\Gamma(n) = (n-1)!$.

Gantt charts *See* CASCADE CHARTS.

Gardner, Martin (1914–2010) Prolific American writer of popular and recreational mathematics and science. He was an authority on the work of Lewis Carroll and helped popularize the art of M. C. Escher. Through his column *Mathematical Games* he did much to introduce the wider public and, commonly, mathematicians to a diverse array of mathematical topics.

Gauss, Carl Friedrich (1777–1855) German mathematician and astronomer, arguably the greatest pure mathematician of all time. He also made enormous contributions to other parts of mathematics, physics, and astronomy. He was highly talented as a child. At the age of 19, made the new discovery that a 17-sided regular polygon could be constructed with ruler and compasses. In 1799, in his doctoral thesis, he proved the *Fundamental Theorem of Algebra. At the age of 24, he published his *Disquisitiones arithmeticae*, a book that was to have a profound influence on *number theory. In this, he proved the *theorem of *quadratic reciprocity. In later work, he developed the theory of curved surfaces using methods now known as *differential geometry. His work on complex functions was fundamental, but, like his discovery of *non-Euclidean geometry, it was not published at the time. He introduced what is now known to statisticians as the *Gaussian distribution. His memoir on potential theory was just one of his contributions to applied mathematics. In astronomy, his great powers of mental calculation allowed him to calculate the orbits of comets and asteroids from limited observational data. He was the first mathematician to consider the behaviour of *knots in a formal mathematical sense.

Gauss error function See ERROR FUNCTION.

Gauss flux theorem If a *closed surface R contains matter of total *mass M and if $\mathbf{F}$ denotes the *gravitational field, then the gravitational flux equals

$$\iint_R \mathbf{F} \cdot \mathrm{d}S = -4\pi GM,$$

where G is the *gravitational constant, S denotes *surface area and $\mathbf{n}$ is the outward-pointing *unit *normal. An equivalent differential version reads

$$\mathrm{div}\mathbf{F} = -4\pi G\rho$$

where ρ denotes *density. There are similar theorems relating *electric and *magnetic fields, the differential forms of which appear as two of *Maxwell's equations.

Gauss-Bonnet theorem (for closed surfaces) The theorem states, for a smooth, *orientable, *closed surface $\int X$ with *Euler characteristic $\chi(X)$ that $\int_X K\mathrm{d}A = 2\pi\chi(X)$, where K is the *Gaussian curvature. The result is surprising, as K relates to the *geometry of the surface and χ to its *topology.

Gaussian curvature The curvature of a *surface. If u and v are local coordinates on a *smooth surface, with *first and *second fundamental forms $E\mathrm{d}u^2 + 2F\mathrm{d}u\mathrm{d}v + G\mathrm{d}v^2$ and $L\mathrm{d}u^2 + 2M\mathrm{d}u\mathrm{d}v + N\mathrm{d}v^2$, then the Gaussian curvature is $K = (LN - M^2)/(EG - F^2)$. For a *sphere of radius r, then $K = 1/r^2$, for a plane or cylinder $K = 0$, for the *hyperbolic plane $K = -1$; more generally K varies on a surface. If U is a *neighbourhood of a point p in the surface and $\mathbf{n}$ denotes the *Gauss map, then $|K(p)|$ is the limit of area$(\mathbf{n}(U))/$area(U) as the area of U tends to 0. See also GAUSS-BONNET THEOREM, THEOREMA EGREGIUM.

Gaussian distribution A synonym for NORMAL DISTRIBUTION.

Gaussian elimination A particular systematic procedure for solving a set of *linear equations in several unknowns. This is normally carried out by applying *elementary row operations to the augmented matrix

$$\begin{bmatrix} a_{11} & a_{12} & \cdots & a_{1n} & b_1 \\ a_{21} & a_{22} & \cdots & a_{2n} & b_2 \\ \vdots & \vdots & \ddots & \vdots & \vdots \\ a_{m1} & a_{m2} & \cdots & a_{mn} & b_m \end{bmatrix}$$

to transform it to *echelon form. The method is to divide the first row by a_{11} and then subtract suitable multiples of the first row from the subsequent rows, to obtain a matrix of the form

$$\begin{bmatrix} 1 & a'_{12} & \cdots & a'_{1n} & b'_1 \\ 0 & a'_{22} & \cdots & a'_{2n} & b'_2 \\ \vdots & \vdots & \ddots & \vdots & \vdots \\ 0 & a'_{m2} & \cdots & a'_{mn} & b'_m \end{bmatrix}.$$

(If $a_{11} = 0$, it is necessary to interchange two rows first.) The first row now remains untouched, and the process is repeated with the remaining rows, dividing the second row by a'_{22} to produce a 1 and subtracting suitable multiples of the new second row from the subsequent rows to produce zeros below that

1. The method continues in the same way. The essential point is that the corresponding set of equations at any stage has the same solution set as the original. (*See also* GAUSS-JORDAN ELIMINATION, SIMULTANEOUS LINEAR EQUATIONS.)

Gaussian function The function e^{-x^2} which has the property $\int_{-\infty}^{\infty} e^{-x^2} dx = \sqrt{\pi}$ which is the function underlying the *normal distribution.

Gaussian integer A *complex number whose real and imaginary parts are both integers, so $z = a + ib$ is a Gaussian integer if $a, b \in \mathbb{Z}$. The set of Gaussian integers is denoted $\mathbb{Z}[i]$. They form a *Euclidean domain with Euclidean function $d(a + ib) = a^2 + b^2$.

Gaussian quadrature An approach in *numerical analysis to the *approximation of *integrals. The aim is to approximate integrals of the form

$$I(f) = \int_a^b f(x)w(x) \ dx$$

where f, w are *continuous functions and the weight function $w(x)$ is positive. Then

$$\langle f, g \rangle = \int_a^b f(x)g(x)w(x) \ dx$$

defines an *inner product on the space of continuous functions, and an orthonormal sequence $p_n(x)$ of polynomials may be constructed such that $p_n(x)$ has degree n. Gauss then showed that $p_{n+1}(x)$ has $n + 1$ distinct real roots $x_0, \ldots, x_n$ in (a, b) and that there are unique solutions $W_0, \ldots, W_n$ to the equations

$$\int_a^b x^k w(x) \ dx = \sum_{i=0}^n W_i x_i^k$$

for $0 \leq k \leq 2n + 1$.

It is then the case that

$$Q_n(f) = \sum_{i=0}^n W_i f(x_i)$$

exactly equals $I(f)$ for polynomials of degree $2n + 1$ or less, and more generally for continuous functions $Q_n(f)$ tends to $I(f)$ as $n \rightarrow \infty$.

Gauss-Jordan elimination An extension of the method of *Gaussian elimination. At the stage when the i-th row has been divided by a suitable value to obtain a 1, suitable multiples of this row are subtracted, not only from subsequent rows but also from preceding rows to produce zeros both below and above the 1. The result of this systematic method is that the *augmented matrix is transformed into *reduced echelon form. However, as a method for solving *simultaneous linear equations, Gauss-Jordan elimination in fact requires more work than Gaussian elimination followed by *backward substitution and is less *numerically stable.

Gauss' Lemma The product of two *primitive polynomials is primitive. Consequently, a primitive polynomial is *irreducible over the *integers if and only if it is irreducible over the *rational numbers. *See also* CONTENT.

Gauss map The unit normal map **n**: $X \rightarrow S^2$ from a smooth oriented *surface X to the unit sphere S^2, which sends a point x to the choice of unit normal $\mathbf{n}(x)$ at x. *See* GAUSSIAN CURVATURE.

Gauss-Markov Theorem In a *linear regression model in which the errors have zero *mean, are uncorrelated, and have equal *variances, the best linear unbiased *estimators of the coefficients are the *least squares estimators. Here, 'best' means that it has minimum variance amongst all linear unbiased estimators.

Gauss-Seidel iterative method A technique for solving a set of n linear equations in n unknowns. If the system is summarized by $\mathbf{Ax} = \mathbf{b}$, then taking initial values as $x_i^{(1)} = \frac{b_i}{a_{ii}}$,

it uses the iterative relation
$$x_i^{(k)} = \frac{b_i - \sum_{j<i} a_{ij}x_j^{(k)} - \sum_{j>i} a_{ij}x_j^{(k-1)}}{a_{ii}},$$ so it
uses the new values immediately they are available, unlike the *Jacobi method.

gcd An abbreviation for GREATEST COMMON DIVISOR.

Gelfond-Schneider theorem If a and b are *algebraic numbers, with $a \neq 0,1$ and b *irrational, then a^b is *transcendental.

generalization The act of appreciating a particular result or problem as a specific case of a broader family of results or problems. Once a mathematical theorem is proved, such as *Fermat's Little Theorem, mathematicians will look to extend the result—as *Euler did with the *totient function. It can also help to place a particular unsolved problem in a broader setting which makes the nature of the problem more apparent. *Abstraction is a form of generalization, usually involving viewing the problem in a new and different light.

generalized coordinates In *mechanics, *coordinates $q_1, \ldots, q_n$ which uniquely specify the configuration of a system. Here n is the number of *degrees of freedom of the system.

generalized eigenvector A non-zero *vector $\mathbf{v}$ is a generalized eigenvector of a matrix $\mathbf{A}$, with generalized eigenvalue λ, if $(\mathbf{A} - \lambda\mathbf{I})^m\mathbf{v} = \mathbf{0}$ for some m. If a matrix is in *Jordan normal form, then the *basis used consists of generalized eigenvectors. *Compare* EIGENVALUE, EIGENVECTOR.

generalized function A synonym for DISTRIBUTION (Schwartz distribution).

generalized maximum likelihood ratio test statistic The ratio of the *maximum likelihoods of the observed value under the parameters for the null and alternative hypotheses.

generalized mean value theorem A synonym for CAUCHY'S MEAN VALUE THEOREM

general linear group *See* MATRIX GROUPS.

general position Points in a *projective space which may be used to assign *homogeneous coordinates to the space. Specifically $n + 2$ points in an n-dimensional projective space are in general position if no $n + 1$ of them lie in an $(n-1)$-dimensional subspace. For example, four points in a projective plane, no three of which are *collinear; points which might then be assigned homogeneous coordinates [1:0:0], [0:1:0], [0:0:1], [1:1:1].

general relativity *See* RELATIVITY THEORY.

general solution A description, often involving arbitrary constants or functions, of all solutions to a problem. This might be a *differential equation, a system of linear equations, or a *recurrence relation.

general topology (point-set topology) The branch of *topology concerned with foundational issues such as *continuity, *convergence, *metrizability, *dimension, and *separation axioms. *See also* ALGEBRAIC TOPOLOGY, COMBINATORIAL TOPOLOGY, DIFFERENTIAL TOPOLOGY, GEOMETRIC TOPOLOGY.

generating function The power series $G(x)$, where

$$G(x) = g_0 + g_1x + g_2x^2 + g_3x^3 + \ldots,$$

is the generating function for the infinite sequence $g_0, g_1, g_2, g_3, \ldots$. (Notice that it is convenient here to start the sequence with a term with subscript 0.) Such power series can be manipulated algebraically, and it can be shown, for example, that

$$\frac{1}{1-x} = 1 + x + x^2 + x^3 + \ldots,$$

$$\frac{1}{(1-x)^2} = 1 + 2x + 3x^2 + 4x^3 + \ldots.$$

Hence, $1/(1-x)$ and $1/(1-x)^2$ are the generating functions for the sequences 1, 1, 1, 1,... and 1, 2, 3, 4,..., respectively.

The Fibonacci sequence F_0, F_1, F_2,... is given by $F_0 = 1$, $F_1 = 1$, and $F_{n+2} = F_{n+1} + F_n$. It can be shown that the generating function for this sequence is $1/(1 - x - x^2)$.

The use of generating functions enables sequences to be handled concisely and algebraically. A *difference equation for a sequence can lead to an equation for the corresponding generating function, and the use of *partial fractions, for example, may then lead to a formula for the n-th term of the sequence.

*Probability and *moment generating functions are very powerful tools in *statistics.

generator (group theory) A subset S of a *group G generates G if the smallest subgroup of G containing S is G itself; the elements of S may then be referred to as generators of G. Typically, we will wish S to be minimal, in the sense that no *proper subset of S generates G. A *cyclic group is a group which is generated by a single element. *See* PRESENTATION, RELATION (group theory), SUBGROUP (generated by a set).

generator *See* CONE, CYLINDER.

genus The maximum number of times a *surface can be cut along simple closed curves without the surface separating into disconnected parts. It is the same as the number of handles on the surface.

geodesic A curve on a *surface, joining two given points, that is the shortest curve between the two points. On a *sphere, a geodesic is an arc of a great circle through the two given points.

For a curve $\gamma(s)$, parameterized by *arc length s, this is equivalent to the acceleration vector $\ddot{y}(s)$ being *normal to the surface. A curve $\gamma(s) = \mathbf{r}(u(s),v(s))$ in a *parameterized surface $\mathbf{r}(u,v)$ is a geodesic if it satisfies the equations

$$\frac{d}{ds}(E\dot{u} + F\dot{v}) = \frac{1}{2}(E_u\dot{u}^2 + 2F_u\dot{u}\dot{v} + G_u\dot{v}^2),$$

$$\frac{d}{ds}(F\dot{u} + G\dot{v}) = \frac{1}{2}(E_v\dot{u}2 + 2F_v\dot{u}\dot{v} + G_v\dot{v}2),$$

where $E, 2F, G$ are the coefficients of the *first fundamental form. *See also* CHRISTOFFEL SYMBOLS.

geodesic polar coordinates When *polar coordinates (r, θ) are used to parameterize the *plane, centred on the origin, then the *first fundamental form is

$$ds^2 = dr^2 + r^2 d\theta^2.$$

More generally, at a point on a surface, there is a unique locally defined *geodesic in each given *tangent direction. Using geodesic polars, so that coordinates (r, θ) are assigned to the point distance r along the geodesic in direction θ,

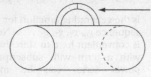

One cut can be made in the handle and still the whole of the surface is connected so the genus = 1.

A surface of genus 1

the first fundamental form has $E = 1$, $F = 0$, as above.

geometric distribution The discrete probability *distribution for the number of experiments required to achieve the first success in a sequence of independent experiments, all with the same probability p of success. The *probability mass function is given by $\Pr(X = r) = p(1-p)^{r-1}$, for $r = 1, 2, \ldots$. It has *mean $1/p$ and *variance $(1-p)/p^2$.

geometric mean *See* MEAN.

geometric multiplicity The geometric multiplicity of an *eigenvalue of a square matrix is the *dimension of the corresponding *eigenspace. The geometric multiplicity never exceeds the *algebraic multiplicity. *See* DIAGONALIZABLE.

geometric sequence (geometric progression) A real or complex, finite or infinite *sequence $a_1, a_2, a_3, \ldots$ with a common ratio r, so that $a_2/a_1 = r$, $a_3/a_2 = r, \ldots$. The first term is usually denoted by a. For example, 3, 6, 12, 24, 48, ... is the geometric sequence with $a = 3$, $r = 2$. In such a geometric sequence, the nth term a_n is given by $a_n = ar^{n-1}$.

geometric series A real or complex, finite or infinite series $a_1 + a_2 + a_3 + \ldots$ in which the terms form a *geometric sequence. Thus the terms have a common ratio r with $a_k/a_{k-1} = r$ for all k. If the first term a_1 equals a, then $a_k = ar^{k-1}$. Let s_n be the sum of the first n terms, so that $s_n = a + ar + ar^2 + \ldots + ar^{n-1}$. Then s_n is given (when $r \neq 1$) by the formulae

$$s_n = \frac{a(1 - r^n)}{1 - r} = \frac{a(r^n - 1)}{r - 1}.$$

If the common ratio r satisfies $|r| < 1$, then $r^n \to 0$ and it can be seen that $s_n \to a/(1-r)$. The value $a/(1-r)$ is called the sum to infinity of the series $a + ar + ar^2 + \ldots$. In particular, for $|x| < 1$, the geometric series $1 + x + x^2 + \ldots$ has

sum to infinity equal to $1/(1-x)$. For example, putting $x = \frac{1}{2}$, the series $1 + \frac{1}{2} + \frac{1}{4} + \frac{1}{8} + \ldots$ as sum 2. If $|x| \geq 1$, then s_n does not tend to a limit and the series has no sum to infinity.

geometric topology The branch of *topology focusing on *manifolds and *embeddings of manifolds, and so encompassing *knot theory.

geometry The area of mathematics related to the study of points and figures, and their properties.

Germain, (Marie) Sophie (1776–1831) French mathematician who corresponded with *Lagrange, *Legendre, and *Gauss under the pseudonym Monsieur Le Blanc because of prejudice against women in mathematics at the time. She did pioneering work on the *modelling of elasticity and on *Fermat's Last Theorem.

Germain prime, Sophie A *prime number p such that $2p + 1$ is also prime. It is currently unknown whether there are infinitely many Sophie Germain primes.

Gershgorin's Theorem All the *eigenvalues of a *square complex matrix **A** lie within the circles centred on each entry in the leading diagonal a_{ii}, with radius $r_i = \sum_{j=1,\ j \neq i}^{n} |a_{ij}|$.

Gibbs, (Josiah) Willard (1839–1903) American mathematician and theoretical physicist who made significant contribution to thermodynamics and statistical mechanics, and in his *Vector Analysis* of 1881 introduced the *dot and *cross-products.

Gibbs' inequality The inequality $-\sum_1^n p_i \log p_i \leq -\sum_1^n p_i \log q_i$ where $(p_1, \ldots, p_n)$ and $(q_1, \ldots, q_n)$ are probability distributions, with equality if and only if $p_i = q_i$ for all i. The inequality is important in *information theory. *See also* ENTROPY (Shannon entropy).

giga- Prefix used with *SI units to denote multiplication by 10^9. Abbreviated to G. For binary giga, *see* KILO (binary).

Girard's theorem *See* SPHERICAL TRIANGLE.

given Something which is already known independently, or something which is to be used in the course of a proof. For example the epsilon-delta method of proof usually states 'Given $\varepsilon > 0$, there exists a δ', by which is meant 'For any ε you choose, no matter how small, I can find a value of δ for which...'.

glb An abbreviation for GREATEST LOWER BOUND.

glide reflection An *isometry of the plane which is the *composition of *reflection in a line followed by a *translation *parallel to that line. All *indirect isometries of the plane are glide reflections.

global maximum (global minimum) A global maximum (resp. minimum) for a *real function on a set X is a value $f(x_0)$ such that $f(x) \leq f(x_0)$ for all x (resp. $f(x) \geq f(x_0)$ for all x). The function $f(x) = x^3 - 3x$ on the interval $[0,2]$ has a global minimum of -2 achieved at $x = 1$ which, being an *interior point, is also a *stationary point by *Fermat's Theorem.

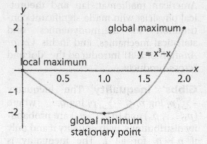

The extrema of $f(x)$

$f(2) = 2$ is a global maximum which is not a stationary point and $f(0) = 0$ is a local maximum which is not global and is not stationary. *See also* WEIERSTRASS THEOREM.

Gödel, Kurt (1906–78) Logician and mathematician who showed that the consistency of elementary arithmetic could not be proved from within the system itself. This result followed from his proof that any formal axiomatic system contains *undecidable propositions. It undermined the hopes of those who had been attempting to determine *axioms from which all mathematics could be deduced. Born in Brno, he was at the University of Vienna from 1930 until he emigrated to the United States in 1940.

Gödel numbering A Gödel numbering assigns to each distinct *word in a finite *alphabet a unique *natural number. As an example, if A,B,...,Z are counted as $1,2,...,26$, then the word NUMBER would be assigned the number $2^{14}3^{21}5^{13}7^2 11^5 13^{18}$. Using the *primes $2,3,5,...$ in order and as N is the 14th letter, U the 21st, etc. As *prime factorization is unique, Gödel numbering is *one-to-one. *See also* HALTING PROBLEM.

Gödel's Completeness Theorem The theorem which states that in mathematical *logic, if a formula is logically valid, then there is a finite formal proof of the formula based on the *axioms of the system—something that a computer could be programmed to do. This is important in its own right but is perhaps even more important as the precursor to *Gödel's Incompleteness Theorems.

Gödel's Incompleteness Theorems Taken together, the two incompleteness theorems say that it is not possible to find a set of axioms for arithmetic which are totally *adequate. The first theorem says no set of axioms for arithmetic can be *consistent and *complete. So, any formal system that proves certain basic arithmetic truths must contain an arithmetical statement that is true but which cannot be proved from the axioms. The second theorem is really just a tightening-up of a particular aspect of the first incompleteness theorem. It states that if a set of axioms A is consistent then the consistency of A cannot be proved by A.

The key implication of these incompleteness theorems is that you might be able to prove all true statements about numbers (or, equivalently, about any other branch of mathematics) within a system by going outside the system to define new rules or axioms, but if you do so then you only create a larger system which will have its own unprovable statements.

Goldbach, Christian (1690–1764) Mathematician born in Prussia, who later became professor in St Petersburg and tutor to the Tsar in Moscow. *Goldbach's conjecture, for which he is remembered, was proposed in 1742 in a letter to Euler.

Goldbach's conjecture The conjecture that every even integer greater than 2 is the sum of two *primes. Goldbach's conjecture remains one of the most famous unsolved problems in *number theory. The conjecture was first made in 1742 in a letter from Chirstian Goldbach to *Euler. Computer calculations have verified the conjecture up to 10^{18}. In 1937, Vinogradov showed that all sufficiently large odd numbers can be written as the sum of three primes. In 1973, Chen showed that every sufficiently large even number can be written either as the sum of two primes or as the sum of a prime and the product of two primes.

golden ratio (golden rectangle, golden section) A line segment is divided in golden section if the ratio of the whole length to the larger part is equal to

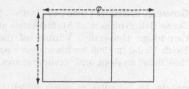

A golden rectangle

the ratio of the larger part to the smaller part. This definition implies that if the smaller part has unit length and the larger part has length φ, then $(\varphi + 1)/\varphi = \varphi/1$. It follows that $\varphi^2 - \varphi - 1 = 0$, which gives $\varphi = \frac{1}{2}(1 + \sqrt{5}) = 1.6180$ to 4 decimal places. This number φ is the golden ratio. A golden rectangle whose sides are in this ratio has throughout history been considered to have a particularly pleasing shape. It has the property that the removal of a square from one end of it leaves a rectangle that has the same shape.

(()) SEE WEB LINKS

• Shows the golden ratio in biology and the relationship to Fibonacci numbers.

goodness-of-fit test A synonym for CHI-SQUARED TEST.

googol A fanciful name for the number 10^{100}, written in decimal notation as a 1 followed by 100 zeros.

googolplex The number equal to the googolth power of 10, written in decimal notation as a 1 followed by a *googol of zeros.

Gosset, William Sealy (1876–1937) British industrial scientist and statistician best known for his discovery of the *t-distribution. His statistical work was motivated by his research for the brewery firm that he was with all his life. The most important of his papers, which were published under the pseudonym 'Student', appeared in 1908.

Gowers, William Timothy (1963–)
Rouse Ball Professor of Mathematics at
Cambridge University. Winner of the
Fields Medal in 1998 for his research on
*functional analysis and *combinatorics.

grade An *angular measure which is
one-hundredth of a *right angle.

gradient (of a curve) The gradient of a
curve at a point P may be defined as
equal to the gradient of the *tangent to
the curve at P. This definition presup-
poses an intuitive idea of what it means
for a line to touch a curve. At a more
advanced level, it is preferable to define
the gradient of a curve by the methods of
*differential calculus. In the case of a
graph $y = f(x)$, the gradient is equal to
$f'(x)$, the value of the derivative. The
tangent at P can then be defined as the
line through P whose gradient equals the
gradient of the curve.

gradient (of a straight line) In *coordinate geometry, suppose that A and B are
two points on a given straight line, and
let M be the point where the line through
A parallel to the x-axis meets the line
through B parallel to the y-axis. Then
the gradient of the straight line is equal
to MB/AM. (Notice that here MB is the
measure of $\overrightarrow{MB}$ where the line through
M and B has positive direction upwards.
In other words, MB equals the length
$|MB|$ if B is above M, and equals $-|MB|$
if B is below M. Similarly, $AM = -|AM|$ if
M is to the right of A, and $AM = -|AM|$ if

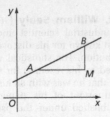

B is above A

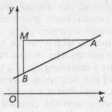

B is below A

M is to the left of A. Two cases are illustrated in the figures.)

The gradient of the line through A and
B may be denoted by m_{AB}, and, if A and
B have coordinates (x_1, y_1) and (x_2, y_2),
with $x_1 \neq x_2$, then

$$m_{AB} = \frac{y_2 - y_1}{x_2 - x_1}.$$

Though defined in terms of two points A
and B on the line, the gradient of the line
is independent of the choice of A and B.
The line in the figures has gradient $\frac{1}{2}$.

Alternatively, the gradient may be
defined as equal to $\tan\theta$, where either
direction of the line makes an angle θ
with the positive x-axis. (The different
possible values for θ give the same
value for $\tan\theta$.) If the line through A
and B is vertical, that is, parallel to the
y-axis, it is customary to say that the
gradient is infinite. The following properties hold:

(i)   Points A, B and C are collinear if
      and only if $m_{AB} = m_{AC}$. (This
      includes the case when m_{AB} and
      m_{AC} are both infinite.)

(ii) The lines with gradients m_1 and
      m_2 are parallel if and only if $m_1 =$
      m_2. (This includes m_1 and m_2
      both infinite.)

(iii) The lines with gradients m_1 and
      m_2 are perpendicular if and only
      if $m_1 m_2 = -1$. (This must be
      reckoned to include the cases
      when $m_1 = 0$ and m_2 is infinite
      and vice versa.)

gradient (grad) The vector obtained by applying the differential operator del, $\nabla = \mathbf{i}\frac{\partial}{\partial x} + \mathbf{j}\frac{\partial}{\partial y} + \mathbf{k}\frac{\partial}{\partial z}$, to a scalar function of position $\phi(\mathbf{r})$. This gives the gradient of ϕ as $\mathrm{grad}\phi = \nabla\phi = \mathbf{i}\frac{\partial\phi}{\partial x} + \mathbf{j}\frac{\partial\phi}{\partial y} + \mathbf{k}\frac{\partial\phi}{\partial z}$. The gradient of ϕ points in the direction in which ϕ is increasing fastest. *See also* CURL, DIVERGENCE.

Graeco-Latin square The notion of a *Latin square can be extended to involve two sets of symbols. Suppose that one set of symbols consists of Roman letters and the other of Greek letters. A Graeco-Latin square is a square array in which each position contains one Roman letter and one Greek letter, such that the Roman letters form a Latin square, the Greek letters form a Latin square, and each Roman letter occurs with each Greek letter exactly once. An example of a 3×3 Graeco-Latin square is the following:

$$\begin{array}{ccc} A\alpha & B\beta & C\gamma \\ B\gamma & C\alpha & A\beta \\ C\beta & A\gamma & B\alpha \end{array}$$

Such squares are used in the *design of experiments.

gram In *SI units, it is the *kilogram that is the base unit for measuring mass. A gram is one-thousandth of a kilogram.

Gram–Schmidt method If $(\mathbf{b}_1, \mathbf{b}_2, \ldots, \mathbf{b}_n)$ is a set of vectors forming a *basis then an *orthonormal basis $(\mathbf{u}_1, \mathbf{u}_2, \ldots, \mathbf{u}_n)$ can be constructed by

$$\tilde{\mathbf{u}}_j = \mathbf{b}_j - \sum_{k=1}^{j-1}\left(\frac{\tilde{\mathbf{u}}_j^{\mathrm{T}}\mathbf{b}_k}{\tilde{\mathbf{u}}_j^{\mathrm{T}}\tilde{\mathbf{u}}_j}\right)\tilde{\mathbf{u}}_j, \mathbf{u}_j = \frac{\tilde{\mathbf{u}}_j}{|\tilde{\mathbf{u}}_j|}$$

graph A number of vertices (or points or nodes), some of which are joined by edges. The edge joining the vertex U and the vertex V may be denoted by (U, V) or (V, U). The vertex-set, that is, the set of vertices, of a graph G may be denoted by

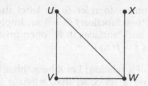

A graph

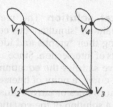

A *multigraph

$V(G)$ and the edge-set by $E(G)$. For example, the graph shown here on the left has $V(G) = \{U, V, W, X\}$ and $E(G) = \{(U, V), (U, W), (V, W), (W, X)\}$.

In general, a graph may have more than one edge joining a pair of vertices; when this occurs, these edges are called multiple edges. Also, a graph may have loops—a loop is an edge that joins a vertex to itself. In the other graph shown, there are 2 edges joining V_1 and V_3 and 3 edges joining V_2 and V_3; the graph also has three loops. *See* MULTIGRAPH.

Normally, $V(G)$ and $E(G)$ are finite, but if this is not so, the result may also be called a graph, though some prefer to call this an *infinite graph.

graph (of a function or mapping) For a *function $f:S\to T$ the graph of f is the subset $\{(s, f(s)) \mid s \in S\}$ of the *Cartesian product $S \times T$. Note that for each $s \in S$ there is a unique $t \in T$ such that (s, t) is in the graph; some authors define functions as such subsets of the Cartesian product of the *domain and *codomain.

It is common to refer to or label the graph of *real function $f : \mathbb{R} \to \mathbb{R}$ as simply $y = f(x)$, and *surfaces in $\mathbb{R}^3$ often arise as graphs $z = f(x,y)$.

graph (of a relation) Let R be a *binary relation on a set S, so that, when a is related to b, this is written aRb. The graph of R is the corresponding subset of the *Cartesian product $S \times S$, namely the set of all pairs (a, b) such that aRb.

graphical solution The method of solving pairs of *simultaneous equations by plotting their *graphs and identifying the points of intersection. Since any point on a curve satisfies the equation producing that curve, a point of intersection of two or more lines or curves must necessarily be a solution of the equations.

graph metric The vertex set of an (undirected) *graph and the distances between vertices form a *metric space if and only if the graph is *connected. Such a metric space is known as a graph metric.

graph paper Paper printed with intersecting lines for drawing graphs or diagrams. The simplest form has equally spaced perpendicular lines, but one or both axis may use logarithmic scales, or scales derived from probability distributions.

graph theory *See* GRAPH.

Grassmannian The Grassmannian $G(k,n)$, named after Hermann Grassmann (1809–77), is a smooth *manifold parameterizing all the k-dimensional *subspaces of $\mathbb{R}^n$. When $k = 1$, this is $(n-1)$-dimensional real *projective space. When $k = 2$, a subspace can be determined by a *basis $\mathbf{v}_1, \mathbf{v}_2$. Unfortunately, there are many different bases for the same subspace. However, if $\wedge$ denotes the *exterior product, then

$$(a\mathbf{v}_1 + b\mathbf{v}_2) \wedge (c\mathbf{v}_1 + d\mathbf{v}_2)$$
$$= (ad - bc)\mathbf{v}_1 \wedge \mathbf{v}_2,$$

and so any two bases give the same product in $\wedge^2 \mathbb{R}^n$ up to a scalar multiple. This means that $G(2,n)$ can be considered as a submanifold of the projective space $\mathbb{P}(\wedge^2 \mathbb{R}^n)$ and more generally $G(k,n) \subseteq \mathbb{P}(\wedge^k \mathbb{R}^n)$. The *dimension of $G(k,n)$ equals $k(n-k)$.

gravitational constant The constant of proportionality, denoted by G, that occurs in the *inverse square law of gravitation. Its value is dependent on the assumption that the *gravitational mass and the *inertial mass of a particle to have the same value. The dimensions of G are $L^3 M^{-1} T^{-2}$, and its value is $6.672 \times 10^{-11} Nm^2 kg^{-2}$.

gravitational force The force of attraction that exists between any two bodies, described by the *inverse square law of gravitation. *See also* GRAVITY.

gravitational mass The quantity associated with a body that arises in the *inverse square law of gravitation. *Compare* INERTIAL MASS, MASS, REST MASS.

gravitational potential *See* POTENTIAL.

gravitational potential energy The *potential energy associated with the gravitational force. When $\mathbf{F} = -\dfrac{GMm}{r^3}\mathbf{r}$, as in the *inverse square law of gravitation, it can be shown that the gravitational potential energy $V = -GMm/r +$ constant. When $\mathbf{F} = -mg\mathbf{k}$, as may be assumed near the Earth's surface, $V = mgz +$ constant.

gravity Near the Earth's surface, a body experiences the gravitational force between the body and the Earth, which may be taken to be constant. The resulting acceleration due to gravity is $-g\mathbf{k}$, where $\mathbf{k}$ is a unit vector directed vertically upwards from the Earth's surface, assumed to be a horizontal plane. The constant g, which is the magnitude of the acceleration due to gravity, is equal to GM/R^2, where G is the *gravitational constant, M is the mass of the Earth and R is the radius of the Earth. Near the Earth's surface, the value of g may be taken as 9.81 ms^{-2}, though it varies between 9.78 ms^{-2} at the equator and 9.83 ms^{-2} at one of the poles.

great circle A circle on the surface of a sphere with its centre at the centre of the sphere; otherwise, the circle is called a small circle. The shortest distance between two points on a sphere is along an arc of a great circle through the two points. This great circle is unique unless the two points are *antipodal. *See also* SPHERICAL TRIANGLE.

greatest common divisor (highest common factor) For two non-zero integers a and b, any integer that is a divisor of both is a *common divisor. Of all the common divisors, the greatest is the greatest common divisor (or gcd), denoted by (a,b) or gcd(a,b). The gcd of a and b has the property of being divisible by every other common divisor of a and b. *Bézout's lemma states that there are integers s and t such that the gcd can be expressed as $sa + tb$. If the prime decompositions of a and b are known, the gcd is easily found: for example, if $a = 168 = 2^3 \times 3 \times 7$ and $b = 180 = 2^2 \times 3^2 \times 5$, then the gcd is $2^2 \times 3 = 12$. Otherwise, the gcd can be found by the *Euclidean Algorithm, which can also be used to find s and t to express the gcd as $sa + tb$. Similarly, any finite set of non-zero integers $a_1, a_2, \dots, a_n$ has a gcd, denoted by $(a_1, a_2 \dots, a_n)$, and there are integers $s_1, s_2, \dots, s_n$ such that this can be expressed as $s_1 a_1 + s_2 a_2 + \dots + s_n a_n$.

greatest integer function A synonym for FLOOR and INTEGER PART.

greatest lower bound A synonym for INFIMUM.

greatest value *See* GLOBAL MAXIMUM.

greedy algorithm An umbrella term for any algorithm that makes the best choice at each term judged solely on the choices at that term and so may fail to be optimal overall. Such an algorithm may fail for the *travelling salesman problem but does work to find *Egyptian fractions, *Huffman codes, and *minimum cost spanning trees.

Green, George (1793–1841) British mathematician who developed the mathematical theory of electricity and magnetism. In his essay of 1828, following *Poisson, he used the notion of potential and proved the result now known as Green's Theorem, which has wide applications in the subject. He had worked as a baker and was self-taught in mathematics; he published other notable mathematical papers before beginning to study for a degree at Cambridge at the age of 40.

Green-Tao theorem *See* ERDŐS CONJECTURE.

Green's function For the inhomogeneous *ODE

$$\frac{d^2y}{dx^2} + p(x)\frac{dy}{dx} + q(x)y = r(x),$$
$$y(0) = y'(0) = 0,$$

the solution $y(x)$ can be written as

$$y(x) = \int_0^x G(x,t)r(t) \ dt,$$

for some function $G(x,t)$ known as Green's function. If the above solution is written as $G(r)$ and the ODE as $Ly = r$, so that $LG(r) = r$, then G acts as a *right inverse for *differential operator L.

Green's theorem The theorem that

$$\int_C P dx + Q dy = \iint_D \left(\frac{\partial Q}{\partial x} - \frac{\partial P}{\partial y}\right) dA$$

for differentiable functions $P(x,y)$ and $Q(x,y)$ defined in a region D of the plane bounded by a *simple, *closed, *positively oriented curve C. This is a special case of *Stokes' theorem applied to

$\mathbf{F} = (P,Q)$ and is closely connected to *Cauchy's integral theorem via the *Cauchy-Riemann equations.

Gregory, James (1638–75) Scottish mathematician who studied in Italy before returning to hold chairs at St Andrews and Edinburgh. He obtained infinite *series for certain *trigonometric functions such as $\tan^{-1}x$, and was one of the first to appreciate the difference between convergent and divergent series. A predecessor of *Newton, in 1668 he proved a simple version of the *Fundamental Theorem of Calculus and knew of *Taylor series 40 years before Taylor's publication. He died at the age of 36.

Gregory–Newton forward differ-ence formula Let $x_0, x_1, x_2, \ldots, x_n$ be equally spaced values, so that $x_i = x_0 + ih$, for $i = 1, 2, \ldots, n$. Suppose that the values $f_0, f_1, f_2, \ldots, f_n$ are known, where $f_i = f(x_i)$, for some function f. The Greg-ory–Newton forward difference formula is a formula involving *finite differences that gives an approximation for $f(x)$, where $x = x_0 + \theta h$, and $0 < \theta < 1$. It states that

$$f(x) \approx f_0 + \theta\Delta f_0 + \frac{\theta(\theta - 1)}{2!}\Delta^2 f_0$$
$$+ \frac{\theta(\theta - 1)(\theta - 2)}{3!}\Delta^3 f_0 + \ldots,$$

the series terminating at some stage. The approximation $f(x) \approx f_0 + \theta\Delta f_0$ gives the result of linear *interpolation. Terminating the series after one more term provides an example of quadratic interpolation.

Grelling's paradox Certain adjectives describe themselves and others do not. For example, 'short' describes itself, and so does 'polysyllabic'. But 'long' does not describe itself, 'monosyllabic' does not, and nor does 'green'. The word 'hetero-logical' means 'not describing itself'. The German mathematician K. Grelling pointed out what is known as Grelling's

paradox that results from considering whether 'heterological' describes itself or not. The paradox has some similarities with *Russell's paradox.

Grothendieck, Alexander (1928–2014) One of the great mathematicians of the 20th century, particularly influen-tial in the areas of *algebraic geometry and *category theory, winning the *Fields Medal in 1966. He introduced the notion of a scheme, a generalization of a *variety, generalized the *Riemann-Roch theorem, and made significant progress on the *Weil conjectures.

group An operation on a set is worth considering only if it has properties likely to lead to interesting and useful results. Certain basic properties recur in differ-ent parts of mathematics and, if these are recognized, use can be made of the simi-larities that exist in the different situ-ations. One such set of basic properties is specified in the definition of a group. The following, then, are all examples of groups: the set of real numbers with add-ition, the set of non-zero real numbers with multiplication, the set of 2×2 real matrices with matrix addition, the set of vectors in 3-dimensional space with vec-tor addition, the set of all bijections from a set S onto itself with composition, the four numbers $1, i, -1, -i$ with multiplication. The definition is as follows: a group is a set G *closed under an operation $\circ$ such that

(i)   for all a, b and c in G, $a \circ (b \circ c) = (a \circ b) \circ c$,
(ii) there is an identity element e in G such that $a \circ e = e \circ a = a$ for all a in G,
(iii) for each a in G, there is an inverse element a' in G such that $a \circ a' = a' \circ a = e$.

The group may be denoted by $(G, \circ)$, when it is necessary to specify the operation, but it may be called simply the group G when the intended operation is clear.

group action The group action of a group G on a non-empty set X is a *homomorphism h from G to the *symmetric group of X. Examples include the group of *invertible $n \times n$ real *matrices acting on column vectors in $\mathbb{R}^n$ or a group acting on itself by *premultiplication. *See also* LINEAR ACTION, REPRESENTATION.

group algebra Given a finite *group $G = \{g_1, \ldots, g_n\}$ and a *field F, the group algebra FG consists of the formal *linear combinations

$$\sum_{k=1}^{n} f_i g_i \quad \text{where } f_1, \ldots, f_n \in F.$$

Such linear combinations then can be added and multiplied naturally, giving FG the structure of an *algebra. *See* REPRESENTATION THEORY.

grouped data A set of *data is said to be grouped when certain groups or categories are defined and the observations in each group are counted to give the frequencies. For numerical data, groups are often defined by means of *class intervals.

group table A synonym for CAYLEY TABLE.

h Abbreviation for *hecto-.

ℍ Denoting the *quaternions, the letter 'H' being used to acknowledge *Hamilton.

Haar measure For a *locally compact *topological group G there exists a non-trivial *measure μ on the *Borel sets such that

- μ is invariant meaning $\mu(gS) = \mu(S)$ for a Borel set S and $g \in G$.
- μ is countably additive meaning $\mu(\bigcup_1^\infty S_k) = \sum_1^\infty \mu(S_k)$ for *disjoint S_k.
- $\mu(S)$ is finite for any *compact Borel set S.
- μ is unique up to a multiplication by a positive scalar.

Hadamard, Jacques (1865–1963) French mathematician, known for his proof in 1896 of the *prime number theorem. He also worked, amongst other things, on the *calculus of variations and the beginnings of *functional analysis.

Hadamard's inequality For an $n \times n$ *matrix $\mathbf{A}$ with columns $\mathbf{v}_1, \mathbf{v}_2, \ldots, \mathbf{v}_n$ it follows that $|\det \mathbf{A}| \le |\mathbf{v}_1|\,|\mathbf{v}_2| \times \cdots \times |\mathbf{v}_n|$, where det denotes *determinant. If $\mathbf{A}$ is *invertible, then equality holds if and only if the columns are mutually *perpendicular.

Hairy Ball Theorem Informally, it is impossible to comb flat the hair on a ball without creating a tuft or cowlick. More formally, the theorem says that a *tangent vector field on a *sphere must have a *singularity. This is a consequence of the *Poincaré-Hopf Theorem, as the sphere has *Euler characteristic 2.

half-angle formula Trigonometric formulae expressing a function of a half-angle in terms of the full angle or expressing a function of an angle in terms of half-angles.

$$\sin\frac{x}{2} = \pm\sqrt{\frac{1-\cos x}{2}}, \cos\frac{x}{2} = \pm\sqrt{\frac{1+\cos x}{2}},$$
$$\text{and if } t = \tan\frac{x}{2}, \sin x = \frac{2t}{1+t^2},$$
$$\cos x = \frac{1-t^2}{1+t^2}, \tan x = \frac{2t}{1-t^2}.$$

half-closed An *interval which includes one end but not the other, so in the form $[a, b)$ or $(a, b]$.

half-life *See* EXPONENTIAL DECAY.

half-line The part of a line extending from a point indefinitely in one *direction. Also called a ray.

half-open A synonym for HALF-CLOSED.

half-plane One of the two regions into which a line divides a plane. The half-plane is said to be closed or open depending on whether it includes the dividing line or not.

half-space One of the two regions into which a plane divides 3-dimensional space. The half-space is said to be closed or open depending on whether it includes the dividing plane or not.

half-turn symmetry *See* SYMMETRICAL ABOUT A POINT.

Halley, Edmond (1656–1742) Published a catalogue of the stars in the southern hemisphere in 1687 and studied planetary orbits correctly predicting the return of Halley's comet in 1758. Published the Breslau Table of Mortality in 1693 which was the foundation of actuarial science for calculating life insurance premiums and annuities.

Halley's method A method of solving an equation in one variable $f(x) = 0$ with a first approximation to the solution x_0 using the iteration

$$x_{n+1} = x_n - \frac{f(x_n)}{f(x_n) - \left(\frac{f(x_n)f''(x_n)}{f'(x_n)}\right)}$$

$$n = 0, 1, 2, \ldots$$

This method converges faster than *Newton's method, but more computation is required at each iteration, so neither method can claim to be more efficient.

Hall's Theorem (Hall's Marriage Theorem) Gives conditions for when a part of a *bipartite graph has a *matching. For a bipartite graph G with parts A and B, there is a matching for every element of A if and only if for every $W \subseteq A$ it is the case that $|W| \le |N_G(W)|$, where $N_G(W)$ denotes the *neighbourhood of W.

Halmos, Paul Richard (1916–2006) Hungarian mathematician who made important contributions to *functional analysis, *measure theory *ergodic theory, and operator theory but is perhaps best known for his great ability to communicate mathematics in a textbook style.

Halting problem An *undecidable problem of *computability. As there are *denumerably many programs (or computable functions), such programs can be counted and assigned a *Gödel number. The halting problem asks whether the ith program terminates when run

with the ith input, but no *algorithm can decide this.

Hamel basis A Hamel basis, or algebraic basis, is a subset of a *vector space such that every vector can be uniquely written as a finite *linear combination of the *basis. Thus, this agrees with the usual notion of basis in a finite-dimensional space. Assuming the *axiom of choice, every vector space has a Hamel basis, but in an infinite-dimensional *Banach space a Hamel basis is necessarily *uncountable. *Compare* SCHAUDER BASIS.

Hamilton, William Rowan (1805–65) Ireland's greatest mathematician, whose work in general *mechanics facilitated later advances in *quantum theory. He is perhaps best known in pure mathematics for his algebraic theory of complex numbers, the invention of *quaternions, and the exploitation of non-commutative algebra. A child prodigy, he could, it is claimed, speak 13 languages at the age of 13. He became Professor of Astronomy at Dublin and Royal Astronomer of Ireland at the age of 22.

Hamiltonian *See* HAMILTONIAN MECHANICS.

Hamiltonian graph In graph theory, one area of study concerns the possibility of travelling around a *graph, going along edges in such a way as to visit every vertex exactly once. A Hamiltonian cycle is a *cycle that contains every vertex, and a graph is called *Hamiltonian if it has a Hamiltonian cycle. The term arises from Hamilton's interest in the existence of such cycles in the graph formed from the *dodecahedron's vertices and edges. *Compare* EULERIAN GRAPH.

Hamiltonian mechanics A rewriting of *Lagrange's equations which foreshadowed advances in *statistical mechanics and *quantum theory. In the notation of Lagrange's equations, we set $p_i = \partial L / \partial \dot{q}_i$ and introduce the Hamiltonian

$$H(\mathbf{q}, \mathbf{p}, t) = \sum_{i=1}^{n} p_i q_i - L(\mathbf{q}, \mathbf{p}, t).$$

Hamilton's equations then have the form

$$\frac{\mathrm{d}q_i}{\mathrm{d}t} = \frac{\partial H}{\partial p_i}, \quad \frac{\mathrm{d}p_i}{\mathrm{d}t} = -\frac{\partial H}{\partial q_i}.$$

Hamming distance The number of digits differing between two *binary codewords of the same length. This defines a *metric on the set of such codewords. If codewords are used that are at least distance 3 apart, in the event of a single transmission error, the error can be corrected by using the nearest codeword.

handshaking lemma The result that, in any *graph, the sum of the *degrees of all the vertices is even. It follows from the observation that the sum of the degrees of all the vertices of a graph is equal to twice the number of edges. Consequently, in any graph, the number of vertices of odd degree is even.

hardware See COMPUTER.

Hardy, Godfrey Harold (1877–1947) British mathematician, a leading figure in mathematics in Cambridge. Often in collaboration with J. E. *Littlewood, he published many papers on *prime numbers, other areas of *number theory, and mathematical *analysis.

Hardy–Weinberg ratio The ratio of genotype frequencies that evolve when mating in a population is random. In the case where only two characteristics A and B are involved, with proportions p and $1-p$ in the population, then the ratio of the three possible pairs of genes can be derived from simple combinatorial probabilities. AA, AB, and BB will occur in the ratio $p^2 : 2p(1-p) : (1-p)^2$ after a single generation.

harmonic analysis The area of mathematics relating to the study of functions by expressing them as the sum or integral of simple *trigonometric functions. See FOURIER SERIES.

harmonic conjugate The *real and *imaginary parts of a *holomorphic function are *harmonic functions. Given a harmonic function $u(x,y)$, a harmonic conjugate for u is a function $v(x,y)$ such that $u + iv$ is holomorphic.

harmonic function A solution of *Laplace's equation.

harmonic mean See MEAN.

harmonic number A number of the form $H_n = 1 + \frac{1}{2} + \frac{1}{3} + \dots + \frac{1}{n}$, where n is a positive integer. As n becomes large, $H_n - \ln n$ converges to *Euler's constant.

harmonic range When the *cross ratio equals -1, the four points in a projective line (see PROJECTIVE SPACE) are said to form a harmonic range.

harmonic sequence (harmonic progression) A sequence $a_1, a_2, a_3, \dots$ such that $1/a_1, 1/a_2, 1/a_3, \dots$ is an *arithmetic sequence. The most commonly occurring harmonic sequence is the sequence $1, \frac{1}{2}, \frac{1}{3}, \frac{1}{4}, \dots$.

harmonic series A series $a_1 + a_2 + a_3 + \dots$ in which $a_1, a_2, a_3, \dots$ is a *harmonic sequence. Often the term refers to the particular series $1 + \frac{1}{2} + \frac{1}{3} + \frac{1}{4} \dots$, where the n-th term a_n equals $1/n$. For this series, $a_n \to 0$. However, the series does not have a sum to infinity since, if s_n is the sum of the first n terms, then $s_n \to \infty$. For large values of n, $s_n \approx \ln n + \gamma$, where γ is *Euler's constant.

hash function Any algorithm that maps *data of variable length to a hash value which serves as a safeguard against accidental corruption and provides a means for efficient searching of the data. *Checksums are a very simple example, but more sophisticated hash functions may be *one-way functions and so provide some information security.

Hausdorff, Felix (1868–1942) German mathematician who made seminal contributions in *set theory, *topology, and the study of *metric spaces. He committed suicide with his family when they were to be sent to concentration camps during the Second World War.

Hausdorff dimension Hausdorff dimension is a form of *fractal dimension. Hausdorff measure $\mathcal{H}^s$ is defined for each $s>0$ positive; when s is an integer, $\mathcal{H}^s$ is a scalar multiple of *Lebesgue measure $\mathcal{L}^s$. For any *metric space X, there is a real number d such that $\mathcal{H}^s(X) = \infty$ for $s < d$ and $\mathcal{H}^s(X) = 0$ for $s > d$. The number d is the Hausdorff dimension of X. For a *compact metric space, it is always at least the topological *dimension.

Hausdorff space A *topological space in which distinct points have *disjoint *neighbourhoods. This is T_2 from the *separation axioms.

Hawking, Stephen William (1942–2018) British mathematician and theoretical physicist who was Lucasian Professor of Mathematics at Cambridge University. He suffered from motor neurone disease for most of his adult life. His main contributions relate to the understanding of the nature of black holes: together with Roger *Penrose, he mathematically proved that, in a cosmos obeying the laws of *general relativity and standard cosmological models, singularities would emerge and that the universe necessarily began with a singularity; he also theorized the 'Hawking radiation' emitted by black holes.

Hawthorne effect The effect where participants in an experiment may respond differently than they normally would because they know they are part of an experiment.

hcf An abbreviation for HIGHEST COMMON FACTOR.

heat A form of *energy and the study of thermodynamics. The heat in a medium, with density ρ, specific heat capacity c, at temperature T, occupying a region R, equals

$$\iiint_R \rho c T \, dV.$$

In that it is energy, the unit of heat is the *joule.

heat equation The heat equation states that the *temperature $T(x,t)$ in a uniform, solid medium satisfies the *parabolic *PDE

$$\frac{\partial T}{\partial t} = \kappa \frac{\partial^2 T}{\partial x^2},$$

where κ denotes the thermal diffusivity of the medium. In more than one spatial dimension the second derivative is replaced by the *Laplacian $\nabla^2 T$, and if κ is not constant, then the RHS is replaced with $\text{div}(\kappa \nabla T)$. The equation models diffusion more generally, for example that of particles, and so is often referred to as the diffusion equation.

Heaviside function The function H, named after Oliver Heaviside, which takes value 1 for positive values and 0 for negative values. As a *distribution, its derivative is the *Dirac delta function.

hectare A metric unit of *area, of a square with side 100 metres, so 1 hectare = 10 000m².

hecto- Prefix used with *SI units to denote multiplication by 10^2. Abbreviated as h.

-hedron Ending denoting a solid geometrical object, as in *polyhedron and *dodecahedron.

height (of a triangle) *See* BASE (of a triangle).

Heine-Borel theorem The *compact subsets of $\mathbb{R}^n$ are the *closed and *bounded subsets. Generally, in a *metric space, compact subsets are closed and bounded, but the converse does not hold.

Heisenberg, Werner Karl (1901–76) German mathematician and theoretical physicist who was the first to work on what developed into *quantum theory for which he was awarded the Nobel Prize for Physics in 1932. He is best known for *Heisenberg's uncertainty principle.

Heisenberg's uncertainty principle The principle in *quantum theory stated by Werner Heisenberg in 1927 which says that it is not possible to simultaneously determine the *position and *momentum of a particle. Mathematically, this means that the momentum *observable P and the position observable X do not commute or, more precisely $\Delta_\psi(P)\Delta_\psi(X) \geq \hbar/2$, where Δ_ψ denotes *dispersion of a *wave function ψ, and $\hbar$ is the reduced *Planck's constant.

helicoid A helicoid is a *ruled surface which can be parametrized by

$$x(r,\theta) = r\cos(a\theta), \quad y(r,\theta) = r\sin(a\theta),$$
$$z(r,\theta) = \theta.$$

A curve of constant r is a *helix. Fixing $\theta = 0$ gives the x-axis; the helicoid is generated moving the x-axis up and down the z-axis, rotating the generating line at a constant *angular speed a. The helicoid is, aside from the plane, the only ruled surface which is also a *minimal surface.

helix A curve on the surface of a (right-circular) cylinder that cuts the generators of the cylinder at a constant angle. Thus, it is 'like a spiral staircase'. The cylinder's *geodesics are the helices, *meridians, and *latitudes.

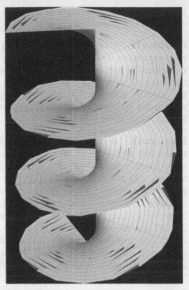

A helicoid

• An animation of a helix.

hemi- Prefix denoting half. See SEMI-.

hemisphere One half of a *sphere cut off by a plane through its centre.

hendecagon An eleven-sided *polygon.

hepta- Prefix denoting seven. *Compare* SEPT-.

heptagon A seven-sided *polygon.

hereditary property A property of a space is hereditary if all subspaces must necessarily have the property. In *topology, a property necessarily inherited by *closed subspaces is termed weakly hereditary, e.g. *normality.

Hermite, Charles (1822–1901) French mathematician who worked in *algebra and *analysis. In 1873, he proved that e is *transcendental. Also notable is his proof that the general *quintic equation can be solved using *elliptic functions.

Hermite polynomials *Orthogonal polynomials $H_n(x)$, for $n \geq 0$, defined by

$$H_n(x) = (-1)^n e^{x^2} \frac{d^n}{dx^n}(e^{-x^2}).$$

The polynomials are orthogonal with weight function $w(x) = \exp(-x^2)$ on $\mathbb{R}$. The polynomials arise variously, including in modelling the harmonic oscillator with *Schrödinger's equation.

Hermitian A complex *matrix M such that $M = M^*$ where $M^* = \bar{M}^T$ is the *conjugate *transpose of M. Hermitian matrices are the complex equivalent of real *symmetric matrices, to which a version of the *spectral theorem applies. Hermitian linear maps are the *self-adjoint linear maps of complex inner product spaces. In *quantum theory, *observables are represented by Hermitian maps and matrices.

Hermitian conjugate (Hermitian adjoint) The Hermitian conjugate of a complex *matrix M is $\bar{M}^T$, the *conjugate *transpose of M. It is often denoted M^*.

Hero (Heron) of Alexandria (1st century AD) Greek scientist whose interests included optics, mechanical inventions, and practical mathematics. A long-lost book, the *Metrica*, rediscovered in 1896, contains examples of *mensuration showing, for example, how to work out the areas of regular *polygons and the volumes of different *polyhedra. It also includes the earliest known proof of *Hero's formula.

Hero's formula *See* TRIANGLE.

Hero's method An iterative method of approximating the *square root of a number. If $\sqrt{k}$ is required, and x_0 is an initial approximation, then $x_{n+1} = \dfrac{\left(x_n + \frac{k}{x_n}\right)}{2}, \quad n = 0,1,2,\ldots$ will

*converge to the square root of k. For example, to calculate the square root of 5, using a first approximation of 2, will give $x_1 = \dfrac{(2+2.5)}{2} = 2.25$, $x_2 = 2.23611111\ldots$, $x_3 = 2.236067978\ldots$, $x_4 = 2.236067978\ldots$ and $\sqrt{5} = 2.236067978\ldots$. So this method has found the square root to 8 *dp after only three iterations.

hertz The *SI unit of *frequency, abbreviated to 'Hz', equivalent to s^{-1}. It is commonly used to measure the frequency of cycles or oscillations.

Hessian The Hessian of a *differentiable function $f(x,y)$ of two variables is the *matrix

$$H = \begin{pmatrix} f_{xx} & f_{xy} \\ f_{yx} & f_{yy} \end{pmatrix}.$$

If the *partial derivatives are *continuous functions, then the *mixed derivatives are equal and so H is *symmetric. The Hessian appears in the *Taylor series of f as

$$f(x+h, y+k) = f(x,y) + (hf_x + kf_y)$$
$$+ \frac{1}{2}(h^2 f_{xx} + 2hk f_{xy} + k^2 f_{yy}) + \ldots$$
$$= f(x,y) + (hf_x + kf_y) + \frac{1}{2}(h,k)H\begin{pmatrix} h \\ k \end{pmatrix} + \ldots$$

and is used in the classification of *stationary points which are non-degenerate; that is $\det H \neq 0$.

heteroscedastic Essentially relating to different *variances in some sense. A number of variables may have different variances. In a *bivariate or *multivariate context the variance of the independent variable may change as the dependent variable changes value, for example, if variability is proportional to the size of the variable.

heuristic Problem-solving based on experience of working with other problems which share some characteristics of the current problem but for which no *algorithm is known. Good heuristics can reduce the time needed to solve problems by recognizing which possible approaches are unlikely to be successful. George *Polya brought the notion of heuristics to a wide audience through his book *How To Solve It*.

hexa- Prefix denoting six. *Compare* SEX-.

hexadecimal representation The representation of a number to *base 16. In this system, 0–9 and A–F are used as the digits, with A–F representing the numbers 10–15. For example, the hexadecimal representation of 712 is

$$712 = 2 \times 16^2 + 12 \times 16 + 8 = 2C8_{16}.$$

Hexadecimal notation is important in computing; it translates easily into binary notation but is more concise and easier to read.

hexagon A six-sided *polygon.

hexagram The plane figure shown formed by extending the six sides of a regular hexagon to meet one another.

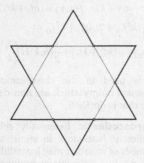

A hexagram

hexahedron A solid figure with six plane faces. The regular hexahedron is a *cube. *See* PLATONIC SOLIDS.

higher arithmetic A synonym for NUMBER THEORY.

higher derivative If the function f is *differentiable on an *interval, its

*derivative f' is defined. If f' is also differentiable, then the derivative of this, denoted by f'', is the second derivative function of f.

Similarly, if f'' is differentiable, then $f'''(x)$ or d^3f/dx^3, the third derivative of f at x, can be formed, and so on. The nth derivative of f at x is denoted by $f^{(n)}(x)$ or d^nf/dx^n. The nth derivatives, for $n \geq 2$, are called the higher derivatives of f. When $y = f(x)$, the higher derivatives may be denoted by $d^2y/dx^2, \ldots, d^ny/dx^n$ or $y'', y''', \ldots, y^{(n)}$.

higher-order partial derivative Given a function f of n variables $x_1, x_2, \ldots, x_n$, the *partial derivative $\partial f/\partial x_i$, where $1 \leq i \leq n$, may also be reckoned to be a function of $x_1, x_2, \ldots, x_n$. So the partial derivatives of $\partial f/\partial x_i$ can be considered. Thus

$$\frac{\partial}{\partial x_i}\left(\frac{\partial f}{\partial x_i}\right) \text{ and } \frac{\partial}{\partial x_j}\left(\frac{\partial f}{\partial x_i}\right) \text{ (for } j \neq i)$$

can be formed, and these are denoted, respectively, by

$$\frac{\partial^2 f}{\partial x_i^2} \text{ and } \frac{\partial^2 f}{\partial x_j \partial x_i}.$$

These are the second-order partial derivatives. When $j \neq i$,

$$\frac{\partial^2 f}{\partial x_i \partial x_j} \text{ and } \frac{\partial^2 f}{\partial x_j \partial x_i}$$

are different by definition, but the two are equal for most 'straightforward' functions f—it is sufficient that either *mixed derivative be continuous. Similarly, third-order partial derivatives such as

$$\frac{\partial^3 f}{\partial x_1^3}, \quad \frac{\partial^3 f}{\partial x_1 \partial x_2^2}, \quad \frac{\partial^3 f}{\partial x_1 \partial x_2 \partial x_3}, \quad \frac{\partial^3 f}{\partial x_2 \partial x_3 \partial x_1},$$

can be defined, and so on. Then the nth-order partial derivatives, where $n \geq 2$, are called the higher-order partial derivatives.

When f is a function of two variables x and y, and the partial derivatives are denoted by f_x and f_y, then f_{xx}, f_{xy}, f_{yx}, f_{yy} are used to denote

$$\frac{\partial^2 f}{\partial x^2}, \quad \frac{\partial^2 f}{\partial y \partial x}, \quad \frac{\partial^2 f}{\partial x \partial y}, \quad \frac{\partial^2 f}{\partial y^2}$$

respectively, noting particularly that f_{xy} means $(f_x)_y$ and f_{yx} means $(f_y)_x$. Alternatively, these partial derivatives f_x and f_y of $f(x,y)$ are denoted by f_1 and f_2, with similar notations for the higher-order partial derivatives.

highest common factor (hcf) A synonym for GREATEST COMMON DIVISOR.

Hilbert, David (1862–1943) German mathematician who was one of the founding fathers of 20th-century pure mathematics and in many ways the originator of the *formalist school of mathematics which has been dominant ever since. Born at Königsberg (Kaliningrad), he became professor at Göttingen in 1895, where he remained for the rest of his life. One of his fundamental contributions to formalism was his *Grundlagen der Geometrie* (Foundations of Geometry), published in 1899, which served to put geometry on a proper axiomatic basis, unlike the rather more intuitive 'axiomatization' of *Euclid. He made major contributions to mathematical *analysis and many other areas. At the International Congress of Mathematics in 1900, he opened the new century by posing his famous list of 23 problems, which have since influenced much research in mathematics.

(⊕) SEE WEB LINKS

• The 23 mathematical problems of Hilbert.

Hilbert's basis theorem If a *ring R is *Noetherian, then the *polynomial ring $R[x]$ is Noetherian.

Hilbert space An *inner product space which is a *complete metric space with respect to the *norm. A *Banach space is a Hilbert space if and only if the *parallelogram law holds. *See* RIESZ REPRESENTATION THEOREM.

Hilbert's paradox (Hilbert's hotel paradox) A paradox illuminating the nature of infinite sets: If a hotel with infinitely many rooms is full and another guest arrives, that guest can be accommodated by each existing guest moving from their current room to the room with the next highest number, leaving Room 1 free for the new arrival. Indeed, if an infinite number of extra guests arrived, they could be accommodated efficiently by each existing guest moving to the room whose number is twice their existing room number, leaving an infinite number of odd-numbered rooms available for the new arrivals.

Hilbert's programme In the early 20th century, *Hilbert's *formalist school of mathematics sought to resolve current issues with the *foundations of mathematics by grounding all areas of mathematics with finite, *complete sets of *axioms. This was essentially proven impossible by *Gödel's Incompleteness Theorems, though *proof theory might be seen as a natural extension of the programme's aims.

Hilbert's tenth problem One of *Hilbert's 23 problems: the problem posed was on the existence of an *algorithm to determine whether or not a given *Diophantine equation has solutions. It was proved by Y. Matijasevich in 1970 that no such algorithm exists.

Hindu-Arabic number system *See* NUMBER SYSTEMS.

histogram A diagram representing the *frequency distribution of continuous *data grouped by means of *class intervals. It consists of a sequence of rectangles, each of which has as its base one of the class intervals and is of a height taken so that the area is proportional to the frequency. If the class

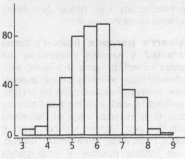

A histogram

intervals are of equal lengths, then the heights of the rectangles are proportional to the frequencies.

The figure shows a histogram of a sample of 500 observations. *Compare* BAR CHART.

history of mathematics The history of mathematics is a rich area of study. Its topics include the origins of theory (e.g. the long development of calculus), influences on mathematicians' thinking, and the varying emphases at different times (e.g. the rise of *abstraction), often linking with the *philosophy of mathematics. An appreciation of the history of mathematics helps contextualize its results and makes clearer the benefits of stronger theorems and different emphases. It also shows the colossal human endeavour that has been involved and how current appreciation of some relatively straightforward topics (e.g. *complex numbers, *limits) was often generations in the making. For a brief historical timeline of a selection of significant results, *see* APPENDIX 24.

Hölder's inequality Hölder's *inequality on $\mathbb{R}^n$ states that

$$|\mathbf{x} \cdot \mathbf{y}| \leq \|\mathbf{x}\|_p \|\mathbf{y}\|_q$$

where p, $q \geq 1$ and $1/p + 1/q = 1$ (*see* P-NORMS). When $p = q = 2$, this gives the *Cauchy-Schwarz inequality.

holomorphic A complex-valued function f defined on an open subset (*see* OPEN SET) $U \subseteq \mathbb{C}$ is said to be holomorphic if it is differentiable in the sense that the limit

$$f'(z) = \lim_{h \to 0} \frac{f(z+h) - f(z)}{h}$$

exists for all z in U. Unlike with real differentiable functions, a holomorphic function is then infinitely differentiable and in fact *analytic. *See* COMPLEX ANALYSIS, CAUCHY-RIEMANN EQUATIONS, LIOUVILLE'S THEOREM.

holonomic In a mechanical system, with generalized coordinates $q_1, q_2, \ldots, q_n$, a holonomic constraint is one involving only $q_1, q_2, \ldots, q_n$ and time t. For example, such a constraint would be $z = r^2$ for a particle at (r, θ, z) moving on the surface of a *paraboloid. *See* LAGRANGE'S EQUATIONS.

Hom For two *vector spaces V, W over a *field F, then $\text{Hom}(V, W)$ denotes the space of *linear maps from V to W. Its *dimension equals $\dim V \times \dim W$. Note that $\text{Hom}(V, V)$ is the *endomorphism ring $\text{End}(V)$ and that $\text{Hom}(V, F)$ is the *dual space of V.

The notation is more generally used; for example $\text{Hom}(G, H)$, the additive group of homomorphisms where G, H are *abelian groups, or to denote Hom *functors in *category theory.

homeomorphism A *continuous function between *topological spaces that has a continuous *inverse function. It is a mapping which preserves all the topological properties of the space, and therefore two homeomorphic spaces are equivalent from a topological perspective. *See* CLASSIFICATION THEOREM FOR SURFACES. *Compare* ISOTOPY.

homogeneous (Markov chains) A *Markov chain $X_1, X_2, X_3, \ldots$ is homogeneous if the transition matrix from X_n to

X_{n+1} is constant, that is, independent of n.

homogeneous (statistics) Literally meaning 'of the same kind or nature'. A homogeneous population is one where the members are similar in nature. If this is not the case, then it may be more appropriate to use some form of *representative sampling, such as quota or *stratified.

homogeneous coordinates Coordinates used in *projective spaces. In the *real projective plane $\mathbb{P}^2 = (\mathbb{R}^3 - \{0\})/\mathbb{R}^*$ the vectors (x,y,z) and (cx,cy,cz), where c is non-zero, represent the same point. Then $[x{:}y{:}z]$ are the homogeneous coordinates for this point, so that $[x{:}y{:}z] = [cx{:}cy{:}cz]$. Points of the complex projective line have coordinates of the form $[1{:}z]$ or $[0{:}1]$, so that $[0{:}1]$ equates with ∞ in the *extended complex plane.

homogeneous first-order differential equation A first-order *differential equation $dy/dx = f(x,y)$ in which the function f, of two variables, has the property that $f(kx,ky) = f(x,y)$ for all k. Examples of such functions are

$$\frac{x^2 + 3y^2}{2x^2 - 5xy}, \quad 1 + e^{x/y}, \quad \frac{x}{\sqrt{x^2 + y^2}}.$$

Any such function f can be written as a function of one variable v, where $v = y/x$. The method of solving homogeneous first-order differential equations is to let $y = vx$ so that $dy/dx = x(dv/dx) + v$. The differential equation for v as a function of x that is obtained is always *separable.

homogeneous linear differential equation *See* LINEAR DIFFERENTIAL EQUATION WITH CONSTANT COEFFICIENTS.

homogeneous polynomial A *polynomial where all terms are of the same total degree. For example, $x^3 + 5x^2y - y^3$

is a degree 3 homogeneous polynomial. *Compare* FORM.

homogeneous set of linear equations A set of m linear equations in n unknowns $x_1, x_2, \ldots, x_n$ that has the form

$$a_{11}x_1 + a_{12}x_2 + \ldots + a_{1n}x_n = 0,$$
$$a_{21}x_1 + a_{22}x_2 + \ldots + a_{2n}x_n = 0,$$
$$\vdots$$
$$a_{m1}x_1 + a_{m2}x_2 + \ldots + a_{mn}x_n = 0.$$

Here, unlike the inhomogeneous case, the numbers on the right-hand sides of the equations are all zero. In *matrix notation, this set of equations can be written $\mathbf{Ax=0}$, where the unknowns form a *column vector $\mathbf{x}$. Thus $\mathbf{A}$ is the $m \times n$ matrix $[a_{ij}]$, and

$$\mathbf{x} = \begin{bmatrix} x_1 \\ \vdots \\ x_n \end{bmatrix}.$$

There is always the trivial solution $\mathbf{x} = \mathbf{0}$; in fact, the solutions form a *vector space. If $m = n$, so that $\mathbf{A}$ is square then the set of equations has non-trivial solutions if and only if det $\mathbf{A} = 0$. If $m < n$, there will necessarily be non-trivial solutions.

homomorphism A *function between two similar algebraic structures which preserves the relational properties of elements in the two structures. So if f is a homomorphism, and $*$ and $\circ$ are corresponding *binary operations in the two structures, $f(x * y) = f(x) \circ f(y)$.

homoscedastic Essentially relating to same *variance in some sense. So a number of variables may have the same variance. In a *bivariate or *multivariate context the variance of the independent variable may remain the same as the dependent variable changes value.

homotopy (homotopic) A *continuous transformation of one *function f: $X \to Y$ between two *topological spaces (or *metric spaces) into another function

$g:X \rightarrow Y$. Formally, a homotopy is a *continuous function $H:X \times [0,1] \rightarrow Y$ such that $H(x,0) = f(x)$ and $H(x,1) = g(x)$, and the functions f,g are said to be homotopic, which is written $f \simeq g$.

homotopy equivalence Two *topological spaces (or *metric spaces) X, Y are homotopy equivalent if there exist *continuous functions $f:X \rightarrow Y$ and $g:Y \rightarrow X$ such that $g \circ f$ is *homotopic to the *identity map on X, and $f \circ g$ is homotopic to the identity map on Y. This is an *equivalence relation, weaker than *homeomorphic; for example, all convex subsets (see CONVEX SET) of $\mathbb{R}^n$ are homotopy equivalent to a *point. Homotopy equivalent spaces have *isomorphic *fundamental groups.

Hooke's law The law that says that the *tension in a spring or a stretched elastic string is proportional to the *extension. Suppose that a spring or elastic string has natural length l and actual length x. Then the tension $T = (\lambda/l)(x-l)$, where λ is the *modulus of elasticity, or $T = k(x-l)$, where k is the *stiffness.

Consider the vertical motion of a particle (mass m) suspended from a fixed support by a spring (stiffness k, natural length l). Let x be the length of the spring at time t. By using Hooke's law, the equation of motion $m\ddot{x} = mg - k(x - l)$ is obtained. The equilibrium position is given by $0 = mg - k(x-l)$; that is, $x = l + mg/k$. Letting $x = l + mg/k + X$ gives $\ddot{X} + \omega^2 X = 0$, where $\omega^2 = k/m$. Thus the particle performs *simple harmonic motion about the equilibrium.

(⊕) SEE WEB LINKS

• An interactive illustration of Hooke's law.

Horner's rule (Horner's method) A polynomial $f(x) = a_0 + a_1 x + a_2 x^2 + \dots + a_n x^n$ can be written as

$$f(x) = a_0 + x(a_1 + x(a_2 + x(a_3 + \dots \\ x(a_{n-1} + a_n x) \dots))).$$

Requiring n multiplications and n additions, this rearrangement is optimal. The method was widely used in the evaluation of polynomials before the advent of *computers. When combined with *Newton-Raphson to find roots, it is sometimes called the Horner-Newton method.

Householder transformation (Householder reflection) Named after Alston Householder (1904–93), a Householder transformation is a reflection in a *hyperplane in $\mathbb{R}^n$. A hyperplane can be characterized by a *unit normal $\mathbf{u}$, in which case the Householder transformation is

$$\mathbf{x} \mapsto \mathbf{x} - 2(\mathbf{x} \cdot \mathbf{u})\mathbf{u}.$$

Householder transformations play a key role in finding the *QR decomposition.

Hoyles, Dame Celia (1946–) British mathematics educationalist who was appointed as the government's Chief Adviser for Mathematics following the publication of Adrian *Smith's inquiry into the state of post-14 mathematics education in the UK, and subsequently the Director of the *National Centre for Excellence in the Teaching of Mathematics (NCETM). In 2004 she was chosen as the first recipient of the Hans Freudenthal medal for her research.

Huffman coding Huffman's coding method provides an optimal *prefix code, which is optimal in the sense that the average *codeword length is minimal. An example is given in the table. The method begins by combining the least likely source words—here beginning with N and S—and effectively reducing the number of sourcewords at each stage. So with 5 source words there are necessarily 4 combination stages (C1-C4). The codeword for each source

Source	Prob	C1	Prob	C2	Prob	C3	Prob	C4	Prob	Code
North	1/6	↘0	1/3	→	1/3	→	1/3	↘0	1	00
South	1/6	↗1								01
East	1/6	→	1/6	↘0	1/3	↘0	2/3	↗1		100
West	1/6	→	1/6	↗1						101
Stop	1/3	→	1/3	→	1/3	↗1				11

word is then read back from the right. The average codeword length is

$$\frac{2}{6}+\frac{2}{6}+\frac{3}{6}+\frac{3}{6}+\frac{2}{3}=2\frac{1}{3},$$

which cannot be bettered.

Huygens, Christiaan (1629–95) Dutch mathematician, astronomer, and physicist. In mathematics, he is remembered for his work on pendulum clocks and his contributions in the field of dynamics. These concerned, for example, the period of oscillation of a simple pendulum and matters such as the *centrifugal force in uniform *circular motion.

hydrodynamics The area of mechanics which studies liquids in motion.

hydrostatics The area of mechanics which studies liquids at rest.

Hypatia (370–415) Greek philosopher who was the head of the Neoplatonist school in Alexandria. It is known that she was consulted on scientific matters and that she wrote commentaries on works of *Diophantus and *Apollonius. Her brutal murder by a fanatical mob is often taken to mark the beginning of Alexandria's decline as the outstanding centre of learning.

hyperbola A *conic with *eccentricity greater than 1. Thus it is the locus of all points P such that the distance from P to a fixed point F_1 (the focus) is equal to e (>1) times the distance from P to a fixed line l_1 (the directrix). It turns out that there is another point F_2 and another

line l_2 such that the same set of points would be obtained with these as focus and directrix. The hyperbola is also the conic section that results when a plane cuts a (double) cone in such a way that a section in two separate parts is obtained.

The line through F_1 and F_2 is the transverse axis, and the points V_1 and V_2 where it cuts the hyperbola are the vertices. The length $|V_1V_2|$ is the length of the transverse axis and is usually taken to be $2a$. The midpoint of V_1V_2 is the centre of the hyperbola. The line through the centre perpendicular to the transverse axis is the conjugate axis. It is usual to introduce b (>0) defined by $b^2=a^2(e^2-1)$, so that $e^2=1+b^2/a^2$. The two separate parts of the hyperbola are the two branches.

If a coordinate system is taken with origin at the centre of the hyperbola, and x-axis along the transverse axis, the foci have coordinates $(ae, 0)$ and $(-ae, 0)$, the directrices have equations $x=a/e$ and

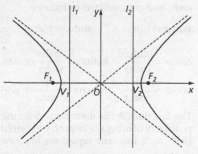

A hyperbola in normal form

$x = -a/e$, and the hyperbola's equation has the *normal form

$$\frac{x^2}{a^2} - \frac{y^2}{b^2} = 1.$$

The hyperbola can be *parametrized by $x = a\sec\theta$, $y = b\tan\theta$ ($0 \le \theta < 2\pi$, $\theta \ne \pi/2$, $3\pi/2$) or alternatively by $x = a\cosh t$, $y = b\sinh t$ ($t\epsilon\mathbb{R}$) (see HYPERBOLIC FUNCTION), but this gives only one branch of the hyperbola.

The hyperbola, in normal form, has two *asymptotes, $y = (b/a)x$ and $y = (-b/a)x$. The shape of the hyperbola is determined by the eccentricity. The particular value $e = \sqrt{2}$ is important for this gives $b = a$. Then the asymptotes are perpendicular, and the hyperbola is called a *rectangular hyperbola.

hyperbolic cylinder A *cylinder in which the fixed curve is a *hyperbola and the fixed line to which the generators are parallel is perpendicular to the plane of the hyperbola. It is a *quadric and in a suitable coordinate system has equation

$$\frac{x^2}{a^2} - \frac{y^2}{b^2} = 1.$$

hyperbolic function Any of the functions hyperbolic sine (sinh), hyperbolic cosine (cosh), and hyperbolic tangent (tanh), and their reciprocals cosech, sech, and coth, defined as follows:

$$\cosh x = \tfrac{1}{2}(e^x + e^{-x}), \quad \sinh x = \tfrac{1}{2}(e^x - e^{-x}),$$

$$\tanh x = \frac{\sinh x}{\cosh x}, \quad \coth x = \frac{\cosh x}{\sinh x} \, (x \ne 0),$$

$$\text{sech} x = \frac{1}{\cosh x}, \quad \text{cosech} x = \frac{1}{\sinh x} \, (x \ne 0).$$

The functions derive their name from the possibility of using $x = a\cosh t$, $y = b\sinh t$ ($t\epsilon\mathbb{R}$) as *parametric equations for (one

branch of) a *hyperbola. (The pronunciation of these functions causes difficulty. For instance, tanh may be pronounced as 'tansh' or 'than' (with the 'th' as in 'thing'); and sinh may be pronounced as 'shine' or 'sinch'. Many of the formulae satisfied by the hyperbolic functions are similar to corresponding formulae for the trigonometric functions, but some changes of sign must be noted (see OSBORNE'S RULE). For example:

$$\cosh^2 x = 1 + \sinh^2 x,$$
$$\text{sech}^2 x = 1 - \tanh^2 x,$$
$$\sinh(x+y) = \sinh x \cosh y + \cosh x \sinh y,$$
$$\cosh(x+y) = \cosh x \cosh y + \sinh x \sinh y,$$
$$\sinh 2x = 2 \sinh x \cosh x,$$
$$\cosh 2x = \cosh^2 x + \sinh^2 x.$$

Since $\cosh(-x) = \cosh x$ and $\sinh(-x) = -\sinh x$, cosh is an *even function and sinh is an *odd function. The graphs of cosh x and sinh x are shown below.

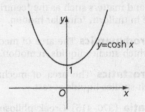

Graph of hyperbolic cosine

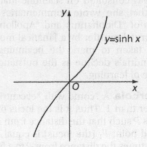

Graph of hyperbolic sine

The graphs of the other hyperbolic functions are:

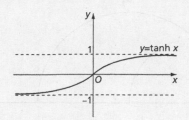

Graph of hyperbolic tangent

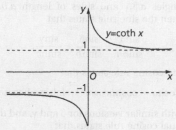

Graph of hyperbolic cotangent

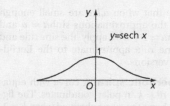

Graph of hyperbolic secant

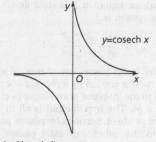

Graph of hyperbolic cosecant

The following derivatives are readily established:

$$\frac{d}{dx}(\cosh x) = \sinh x, \frac{d}{dx}(\sinh x) = \cosh x,$$

$$\frac{d}{dx}(\tanh\ x) = \operatorname{sech}^2 x.$$

See also INVERSE HYPERBOLIC FUNCTION.

hyperbolic geometry *See* HYPERBOLIC PLANE, NON-EUCLIDEAN GEOMETRY.

hyperbolic paraboloid A *quadric whose equation in a suitable coordinate system is

$$\frac{x^2}{a^2} - \frac{y^2}{b^2} = \frac{2z}{c}.$$

The *yz*-plane and the *zx*-plane are planes of symmetry. Sections by planes parallel to the *xy*-plane are *hyperbolas, the section by the *xy*-plane itself being a pair of straight lines. Sections by planes parallel to the other axial planes are *parabolas. Planes through the *z*-axis cut the paraboloid in parabolas with vertex at the origin. The origin is a *saddle-point.

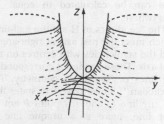

Hyperbolic paraboloid in normal form

hyperbolic partial differential equation *See* PARABOLIC PARTIAL DIFFERENTIAL EQUATION.

hyperbolic plane Hyperbolic geometry is an example of *non-Euclidean geometry. The following model for

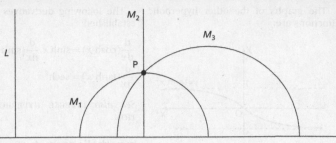

3 parallels through P not meeting L

the hyperbolic plane is due to *Poincaré. The set of points comprises the upper half of the *complex plane $H = \{z \mid \mathrm{Im}z > 0\}$, with *angles measured as usual, but the hyperbolic distance between two points a,b in H is given by the formula

$$d_H(z, w) = 2\tanh^{-1}\left|\frac{z - w}{z - \overline{w}}\right|.$$

Note that as w moves towards the real axis, this distance tends to infinity. The *isometries of H are the *Möbius transformations $z \mapsto (az + b)/(cz + d)$, where a,b,c,d are real and $ad - bc = 1$. The *first fundamental form has $E = G = y^{-2}$ and $F = 0$, from which the *Gaussian curvature can be calculated to equal -1 everywhere.

The *geodesics of H are the half-lines which meet the real axis at right angles and the semicircles with centres on the real axis. These geodesics correspond to the lines of *Euclidean geometry. The Euclidean *parallel postulate states that for any line L, and any point P not on that line, there is a unique line m through P which does not meet L, a so-called *parallel. From the figure, we see there infinitely many such parallels in H. There are many other unusual aspects: *similar triangles in H are *congruent; the area of a triangle equals $\pi - \alpha - \beta - \gamma$, where α, β, γ are the triangle's angles; there need not be a circle through three non-collinear points.

Given a hyperbolic triangle, with angles α, β, γ and sides of length a,b,c, then the sine rule states that

$$\frac{\sin\alpha}{\sinh a} = \frac{\sin\beta}{\sinh b} = \frac{\sin\gamma}{\sinh c},$$

the cosine rule states that

$$\cosh a = \cosh b \cosh c - \sinh b \sinh c \cos\alpha,$$

with similar versions for β and γ, and the dual cosine rule states that

$$\cos\alpha = -\cos\beta\cos\gamma + \sin\beta\sin\gamma\cosh a.$$

Note that when a,b,c are small enough that the approximations $\sinh a \approx a$ and $\cosh a \approx 1 + a^2/2$ apply, the sine rule and cosine rule approximate to the Euclidean versions.

hyperbolic spiral A curve with equation $r\theta = k$ in polar coordinates. The figure shows the curve obtained when $k = 1$.

hyperboloid of one sheet A *quadric whose equation in a suitable coordinate system is

$$\frac{x^2}{a^2} + \frac{y^2}{b^2} - \frac{z^2}{c^2} = 1.$$

The *axial planes are planes of symmetry. The section by a plane $z = k$ parallel to the xy-plane is an ellipse (a circle if $a = b$). The hyperboloid is all in one piece or sheet. Sections by planes parallel to the other two axial planes are hyperbolas.

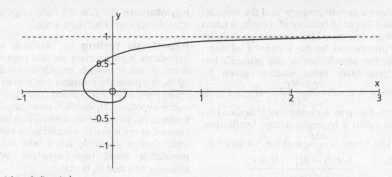

A hyperbolic spiral

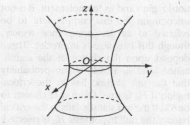

Hyperboloid of one sheet in normal form

hyperboloid of two sheets A *quadric whose equation in a suitable coordinate system is

$$\frac{x^2}{a^2} + \frac{y^2}{b^2} - \frac{z^2}{c^2} = -1.$$

The *axial planes are planes of symmetry. The section by a plane $z = k$ parallel to the xy-plane is, when non-empty, an *ellipse (a circle if $a = b$). When $-c < k < c$, the plane $z = k$ has no points of intersection with the hyperboloid, and the hyperboloid is thus in two pieces or sheets. Sections by planes parallel to the other two axial planes are *hyperbolas.

hypercube The generalization in n dimensions of a *square in two dimensions and a *cube in three dimensions.

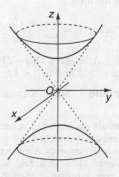

Hyperboloid of two sheets in normal form

In the plane, the four points with coordinates $(\pm 1, \pm 1)$ are the vertices of a square. In 3-dimensional space, the eight points $(\pm 1, \pm 1, \pm 1)$ are the vertices of a cube. So, in *n-dimensional space, the 2^n points with coordinates $(x_1, x_2, \ldots, x_n)$, where each $x_i = \pm 1$, are the vertices of a hypercube. Two vertices are joined by an edge if they differ in exactly one of their coordinates, and so there are $n2^{n-1}$ edges. The vertices and edges of a hypercube in n dimensions form the vertices and edges of the graph known as the *n-cube.

hypergeometric distribution Suppose that, in a set of N items, M items

have a certain property and the remainder do not. A *sample of n items is taken from the set. The discrete probability *distribution for the number r of items in the sample having the property has *probability mass function given by
$$\Pr(X = r) = \frac{{}^{M}C_r \, {}^{N-M}C_{n-r}}{{}^{N}C_n}$$
where r runs from 0 up to the smaller of M and n. This is called a hypergeometric distribution. The *mean is $\frac{nM}{N}$, and the *variance is
$$\frac{nM(N - M)}{N^2(N - 1)} \, (N - n).$$

hypergeometric series A *power series of the form
$$F(a, b; c; x) = \sum_{k=0}^{\infty} \frac{(a)_n (b)_n}{(c)_n} \frac{x^n}{n!}$$
where $(a)_n = a(a+1)(a+2)\ldots(a+n-1)$ for $n \geq 1$ and $(a)_0 = 1$.

hyperplane *See* N-DIMENSIONAL SPACE.

hyperreals An extended version of the *real numbers that includes *infinite and *infinitesimal elements. A hyperreal is an *equivalence class of a real sequence, with a real number x identified with the constant sequence $\{x\}$, while the sequence $\{n\}$ is an infinite hyperreal with its *reciprocal $\{1/n\}$ being an infinitesimal. When algebraic operations are defined *componentwise, the *ring of sequences has *zero-divisors; to avoid this, two sequences are made equivalent if they agree for almost all components (in a technical sense) so that $\{a\}\{b\} = \{0\}$ if and only if $\{a\} = \{0\}$ or $\{b\} = \{0\}$. The hyperreals then form a field. The study of the hyperreals is called non-standard analysis, though this term is sometimes more broadly used.

hypocycloid The curve traced out by a point on the circumference of a circle rolling round the inside of a fixed circle. A special case is the *astroid.

hypotenuse The side of a right-angled triangle opposite the *right angle.

hypothesis testing In *statistics, a hypothesis is an assertion about a population. A null hypothesis, usually denoted by H_0, is a particular assertion that is to be either rejected or not rejected. The alternative hypothesis H_1 specifies some alternative to H_0. To decide whether H_0 is to be rejected or not rejected, a significance test tests whether a sample taken from the population could have occurred by chance, given that H_0 is true.

From the sample, the *test statistic is calculated. The test partitions the set of possible values into the *critical (or rejection) region and its *complement. It is not uncommon for this complement to be referred to as the 'acceptance region', though this language is imprecise. These depend upon the choice of the significance level α, which is the probability that the test statistic lies in the critical region if H_0 is true. Often α is chosen to be 5%. If the test statistic falls in the critical region, the null hypothesis H_0 is rejected; otherwise, the conclusion is that there is no evidence for rejecting H_0 and H_0 is not rejected.

A Type I error, or false positive, occurs if H_0 is rejected when it is in fact true. The probability of a Type I error is α, so this depends on the choice of significance level. A Type II error, or false negative, occurs if H_0 is not rejected when it is in fact false. If the probability of a Type II error is β, the *power of the test is defined as $1-\beta$. This depends upon the choice of alternative hypothesis. The null hypothesis H_0 normally specifies that a certain parameter has a certain value. If the alternative hypothesis H_1 says that the parameter is not equal to this value, then the test is said to be two-tailed (or two-sided). If H_1 says that the parameter is greater than this value (or less than this value), then the test is one-tailed (or one-sided). *See also* P-VALUE.

i See COMPLEX NUMBER, QUATERNION.

i (*j*) A unit vector, usually in the direction of the *x*-axis.

I The Roman *numeral for 1.

I Denotes the *identity matrix or *identity function.

ICM A synonym for INTERNATIONAL CONGRESS OF MATHEMATICIANS.

icosahedron (icosahedra) A *polyhedron with 20 faces. The regular icosahedron is one of the *Platonic solids, and its *faces are *equilateral triangles. It has 12 *vertices and 30 *edges.

icosidodecahedron One of the *Archimedean solids, with 12 pentagonal *faces and 20 triangular faces. It can be formed by cutting off the corners of a *dodecahedron to obtain a polyhedron whose vertices lie at the midpoints of the edges of the original dodecahedron.

ideal A *subring *I* of a *ring *R* such that for every *a* in *R* and every *x* in *I* both *ax* and *xa* are in *I*. Every ideal of $\mathbb{Z}$ is of the form $n\mathbb{Z}$ for some *n* and so *principal. Ideals are those subrings for which a *well-defined *quotient ring *R/I* can be defined. See MAXIMAL IDEAL, PRIME IDEAL.

ideal element An element added to a mathematical structure with the purpose of eliminating anomalies, special cases, and exceptions. For example, by including infinity in the *extended complex plane, circles and lines can together be treated as *circlines.

ideal point A synonym for POINT AT INFINITY.

idempotent An element *x* in a *ring where $x^2 = x$. In a *Boolean ring all elements are idempotent. In the *endomorphism ring of a *vector space, the idempotent elements are the *projections.

identically distributed A number of *random variables are identically distributed if each has the same *distribution.

identification space A synonym for QUOTIENT SPACE.

identity As opposed to an *equation, an identity holds true for a range of values, such as $x^2 - y^2 = (x+y)(x-y)$ for real *x* and *y*. Sometimes the symbol $\equiv$ is used instead of = to indicate that a statement is an identity.

identity (identity element) An element *e* is an identity element (or neutral element) for a *binary operation ∘ on a set *S* if, for all *a* in *S*, $a \circ e = e \circ a = a$. In a *group, the identity element is commonly denoted *e*. In a *ring, the additive identity is denoted 0 and referred to as 'zero' and the multiplicative identity (if it exists) is denoted 1 and referred to as 'one'. Even though the numerical symbols '0' and '1' are used, 0 may denote a zero *function or 1 might represent the *identity matrix, for example.

identity function (identity mapping) The identity function $i_S: S \to S$, on a set *S*, is defined by $i_S(s) = s$ for all *s* in *S*. Identity functions have the property that if $f: S \to T$ is a function, then $f \circ i_S = f$ and $i_T \circ f = f$.

identity matrix The $n \times n$ identity matrix $\mathbf{I}$, or $\mathbf{I}_n$, is the $n \times n$ matrix

$$\begin{bmatrix} 1 & 0 & 0 & \cdots & 0 \\ 0 & 1 & 0 & \cdots & 0 \\ 0 & 0 & 1 & \cdots & 0 \\ \vdots & \vdots & \vdots & \ddots & \vdots \\ 0 & 0 & 0 & \cdots & 1 \end{bmatrix},$$

with each diagonal entry equal to 1 and all other entries equal to 0. It is the *identity element for the set of $n \times n$ matrices with multiplication. See also KRONECKER DELTA.

identity theorem A *holomorphic function $f: U \to \mathbb{C}$ on a *connected, open subset (see OPEN SET) with a zero set $f^{-1}(0)$ that has a *limit point in U, is identically zero in U. This theorem is in stark contrast to real analysis and demonstrates how rigid holomorphic functions are in comparison with *smooth real functions.

if and only if See CONDITION, NECESSARY AND SUFFICIENT.

iff Abbreviation for *if and only if.

iid An abbreviation for INDEPENDENT and IDENTICALLY DISTRIBUTED.

ill-conditioned A problem is said to be ill-conditioned if a small change in the input data (or coefficients in an equation) gives rise to large changes in the output data (or solutions to the equation).

Im Abbreviation and symbol for the *imaginary part of a *complex number.

IMA Abbreviation for the Institute of Mathematics and its Applications, one the UK's learned societies for mathematics alongside the *LMS and the *RSS. It was founded in 1963, its first president being James *Lighthill.

image See FUNCTION.

imaginary axis The y-axis in the *complex plane. Its points represent *pure imaginary numbers.

imaginary number A synonym for PURE IMAGINARY.

imaginary part A *complex number z can be written $x + yi$, where x and y are real, and then y is the imaginary part. It is denoted by $\mathrm{Im}z$ or $\Im z$.

implication If p and q are statements, the statement 'p implies q' or 'if p then q' is an implication and is denoted by $p \Rightarrow q$. It is reckoned to be false only in the case when p is true and q is false. So its *truth table is as follows:

p	q	$p \Rightarrow q$
T	T	T
T	F	F
F	T	T
F	F	T

p is said to be *sufficient for q and q to be *necessary for p.

implicit Where the *dependent variable is not explicitly given as a function of the *independent variable. For example, $3x + 5y - 1 = 0$ is an implicit form while the equivalent equation $y = -0.6x + 0.2$ is explicit.

implicit differentiation When x and y satisfy a single equation, it may be possible to regard this as defining y as a function of x with a suitable domain (see IMPLICIT FUNCTION THEOREM). Implicit differentiation involves applying the *chain rule to obtain the *derivative of y in such cases. For example, if $xy^2 + x^2y^3 - 1 = 0$, then

$$2xy\frac{dy}{dx} + y^2 + 3x^2y^2\frac{dy}{dx} + 2xy^3 = 0$$

and so, if $2xy + 3x^2y^2 \neq 0$,

$$\frac{dy}{dx} = -\frac{y^2 + 2xy^3}{2xy + 3x^2y^2}.$$

$$\int_1^\infty \frac{1}{x^2}\,dx = 1.$$

implicit function theorem The equation $f(x,y) = x^2 + y^2 - 1 = 0$ defines y implicitly in terms of x. At any point of the circle, except $(\pm1,0)$, the curve can be locally defined as a graph, either $y = g(x) = \sqrt{1 - x^2}$ or $y = g(x) = -\sqrt{1 - x^2}$. But this is not possible when $x = \pm1$, as the derivative dx/dy is zero. More generally, when a *dependent variable y is implicitly related to *independent variables $x_1, x_2, \ldots, x_k$ by $f(x_1, x_2, \ldots, x_k, y) = 0$, then y can be written as a graph $y = g(x_1, x_2, \ldots, x_k)$, local to a point p, if $\partial f/\partial y(p) \neq 0$. The theorem more generally applies when m dependent variables y_i are implicitly defined in terms of n independent variables x_i by m equations f_i, in which case the *Jacobian $(\partial f_i/\partial y_j)$ must be *invertible.

imply To enable a conclusion to be validly inferred. For example, n being divisible by 2 implies that n is even.

improper fraction *See* FRACTION.

improper integrals *Riemann integration deals only with *bounded functions on bounded *intervals. But other integrals may be considered as improper integrals. The first kind is one in which the interval of integration is infinite as, for example, in

$$\int_a^\infty f(x)\,dx.$$

This integral exists as an improper integral and has value l, if the the integral from a to X tends to a limit l as $X \to \infty$. For example,

$$\int_1^X \frac{1}{x^2}\,dx = 1 - \frac{1}{X}$$

and, as $X \to \infty$, the right-hand side tends to 1. So, as an improper Riemann integral

Similar definitions can be given for improper integrals from $-\infty$ to a and from $-\infty$ to ∞.

A second kind of improper integral is one in which the function becomes infinite at some point. For example, in

$$\int_0^1 \frac{1}{\sqrt{x}}\,dx.$$

The function $1/\sqrt{x}$ is not bounded on the interval $(0, 1]$, and so is not Riemann integrable. However, the function is bounded on the interval $[\delta, 1]$, where $0 < \delta < 1$, and

$$\int_\delta^1 \frac{1}{\sqrt{x}}\,dx = 2 - 2\sqrt{\delta}.$$

As $\delta \to 0$, the right-hand side tends to the limit 2. So the integral above, from 0 to 1, exists as an improper integral with value 2.

impulse The impulse $\mathbf{J}$ associated with a force $\mathbf{F}$ acting during the time interval from $t = t_1$ to $t = t_2$ is the definite integral of the force, with respect to time, over the interval. Thus $\mathbf{J}$ is the vector quantity given by

$$\mathbf{J} = \int_{t_1}^{t_2} \mathbf{F}\,dt.$$

Suppose that $\mathbf{F}$ is the total force acting on a particle of mass m. From Newton's second law of motion in the form $\mathbf{F} = (d/dt)(m\mathbf{v})$, it follows that the impulse $\mathbf{J}$ is equal to the change in the *linear momentum of the particle.

If the force $\mathbf{F}$ is constant, then $\mathbf{J} = \mathbf{F}(t_2 - t_1)$. This can be stated as 'impulse = force × time'.

The concept of impulse is relevant in problems involving *collisions.

Impulse has the *dimensions MLT^{-1}, and the *SI unit of measurement is the kilogram metre per second, abbreviated to kgms^{-1}.

incentre The centre of the *incircle of a triangle. It is the point at which the three internal *bisectors of the angles of the triangle are *concurrent.

incidence (incident) Two geometric figures are said to be incident if they have *non-empty *intersection.

incircle The circle that lies inside a triangle and *touches the three sides. Its centre is the *incentre of the triangle.

inclination The acute angle between the horizontal and a given line (*see* DEPRESSION) measured positively in the direction of the upwards vertical. θ is the angle of inclination of the top of the cliff from the boat in the figure:

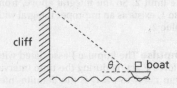

θ is the angle of inclination

inclined plane A plane that is not horizontal. Its angle of inclination is the angle that a line of greatest slope makes with the horizontal.

include *See* SUBSET.

inclusion Given a set A and $B \subseteq A$, then inclusion is the map $\iota:B \to A$ defined by $\iota(b) = b$ for all b in B. Note that ι is the *restriction of the *identity map on A to B.

inclusion–exclusion principle The number of elements in the union of 3 sets is given by

$$|A \cup B \cup C| = |A| + |B| + |C| - |A \cap B| \\ -|A \cap C| - |B \cap C| + |A \cap B \cap C|$$

where $|X|$ denotes the *cardinality of the set X. This may be written

$$|A \cup B \cup C| = \alpha_1 - \alpha_2 + \alpha_3$$

where $\alpha_1 = |A| + |B| + |C|$, $\alpha_2 = |A \cap B| + |A \cap C| + |B \cap C|$ and $\alpha_3 = |A \cap B \cap C|$. Now suppose that there are n sets instead of 3. Then the following, known as the inclusion–exclusion principle, holds:

$$|A_1 \cup A_2 \cup \ldots \cup A_n| = \alpha_1 - \alpha_2 + \alpha_3 - \ldots \\ + (-1)^{n-1}\alpha_n$$

where α_i is the sum of the cardinalities of the intersections of the sets taken i at a time.

inclusive disjunction If p and q are statements, then the statement 'p or q', denoted by $p \vee q$, is the disjunction of p and q. For example, if p is 'It is raining' and q is 'It is Monday' then $p \vee q$ is 'It is raining or it is Monday'. To be quite clear, notice that $p \vee q$ means 'p or q or both': the disjunction of p and q is true when at least one of the statements p and q is true. The *truth table is:

p	q	$p \vee q$
T	T	T
T	F	T
F	T	T
F	F	F

This is sometimes explicitly known as 'inclusive disjunction' to distinguish it from *exclusive disjunction (or xor) which requires exactly one of p, q to be true.

incommensurable Not *commensurable. Often used as a synonym for *irrational.

incompatible A synonym for INCONSISTENT.

incompressible In *fluid dynamics, a flow **u** is incompressible if the *density of material does not change as the material moves with the flow. This is equivalent to the *divergence div**u** equalling 0.

inconsistent A set of *equations is inconsistent if it has no *solution.

increasing function A *real function f is increasing on its *domain S if $f(x_1) \leq f(x_2)$ whenever x_1 and x_2 are in S with $x_1 \leq x_2$. Also, f is strictly increasing if $f(x_1) < f(x_2)$ whenever $x_1 < x_2$.

increasing sequence A sequence a_1, a_2, a_3, ... is increasing if $a_m \leq a_n$ whenever $m \leq n$, and strictly increasing if $a_m < a_n$ whenever $m < n$. This is equivalent to the function $i \mapsto a_i$ being increasing or strictly increasing. *See also* NESTED SETS.

increment The amount by which a certain quantity or variable is increased. An increment can normally be positive or negative, or zero, and is often required to be in some sense 'small'. If the variable is x, the increment may be denoted by Δx or δx.

indefinite integral *See* INTEGRAL.

independent In *logic, a set of statements is independent when the truth of any of the statements cannot be deduced from knowledge of the others.

independent events Events A and B are independent if the occurrence of either does not affect the *probability of the other occurring. Otherwise, A and B are dependent. For independent events, the probability that they both occur is given by the product law $\Pr(A \cap B) = \Pr(A) \Pr(B)$. This is equivalent to the *conditional probability $\Pr(A \mid B)$ equalling $\Pr(A)$, provided $\Pr(B) \neq 0$.

For example, when an unbiased coin is tossed twice, the probability of obtaining 'heads' on the first toss is $\frac{1}{2}$, and the probability of obtaining 'heads' on the second toss is $\frac{1}{2}$. These two events are independent. So the probability of obtaining 'heads' on both tosses is equal to $\frac{1}{2} \times \frac{1}{2} = \frac{1}{4}$.

independent random variables Two *discrete random variables X and Y are independent if $X = x$ and $Y = y$ are independent events for all x and y; that is, $\Pr(X = x, Y = y) = \Pr(X = x) \Pr(Y = y)$. Two continuous random variables X and Y, with *joint probability density function $f(x, y)$, are independent if $f(x, y) = f_1(x) f_2(y)$, where f_1 and f_2 are the *marginal probability density functions.

independent variable (in regression) A synonym for EXPLANATORY VARIABLE. *Compare* DEPENDENT VARIABLE (STATISTICS).

independent variable *See* DIFFERENTIAL EQUATION, FUNCTION.

indeterminate A *variable, in particular a formal variable, such as in the *polynomial ring $F[x]$ or the formal *power series ring $F[[x]]$ over a *field F. Such polynomials might be formally differentiated, but without reference to limits, and such power series multiplied, but without reference to *convergence.

indeterminate equations One or more *equations which have more than one *solution.

indeterminate form Suppose that $f(x) \to 0$ and $g(x) \to 0$ as $x \to a$. Then the limit of the quotient $f(x)/g(x)$ as $x \to a$ is said to give an indeterminate form, sometimes denoted by $0/0$. It may be that a specific limit $f(x)/g(x)$ can nevertheless be found by some method such as *L'Hôpital's rule.

Other indeterminate forms are ∞/∞, $\infty - \infty$, $0 \times \infty$, 0^0, 1^∞, ∞^0. In general, these forms cannot be given definite values but in some contexts have an understood value; for example setting $x = 0$ in the polynomial $f(x) = \sum_0^n a_k x^k$ to give $f(0) = a_0$ assigns 0^0 the value 1.

index (indices) Suppose that a is a real number. When the product $a \times a \times a \times a \times a$ is written as a^5, the number 5 is called the index. When the index is a positive integer p, then a^p means $a \times a \times \ldots \times a$, where there are p occurrences of a. It can then be shown that

(i) $a^p \times a^q = a^{p+q}$

(ii) $a^p/a^q = a^{p-q}$ $(a \neq 0)$,

(iii) $(a^p)^q = a^{pq}$,

(iv) $(ab)^p = a^p b^p$,

where, in (ii), for the moment, it is required that $p > q$. The meaning of a^p can be extended for a general integer p by the following definitions:

(v) $a^0 = 1$,

(vi) $a^{-p} = 1/a^p$ $(a \neq 0)$,

noting (i)-(vi) still hold. And if $a > 0$, m, n are integers with $n > 0$, we define

(vii) $a^{m/n} = \sqrt[n]{a^m}$

For $a > 0$ and x a real number, we define

$$a^x = \exp(x \ln a)$$

and again rules (i)-(vii) hold. The same notation is used in other contexts; for example, to define z^p, where z is a *complex number, to define $\mathbf{A}^p$, where $\mathbf{A}$ is a *square matrix, or to define g^p, where g is an element of a multiplicative *group. In such cases, some of the above rules may hold and others may not.

index (group theory) The index of a *subgroup H in a *group G, denoted $|G : H|$, is the number of (left or right) *cosets of H. If G is finite, then $|G : H| = |G|/|H|$ by *Lagrange's theorem.

index (permutations) The index $\text{ind}(\sigma)$ of a *permutation σ of a finite set is the least

number k such that σ can be written as a product of k *transpositions (or 2-cycles). Given two permutations σ and τ, then

$$\text{ind}(\sigma\tau) \leq \text{ind}(\sigma) + \text{ind}(\tau).$$

index (vector fields) For a *vector field $\mathbf{v}$: $\mathbb{R}^2 \rightarrow \mathbb{R}^2$, a singularity is a point p such that $\mathbf{v}(p) = \mathbf{0}$. On a *simple, *closed, oriented (*see* ORIENTATION) curve C about p (which contains no other singularities), the direction of $\mathbf{v}$ wraps around the unit circle an integer number of times; that integer is the index of the singularity.

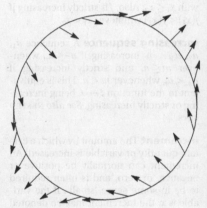

Singularity with index −1

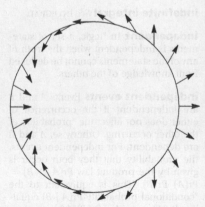

Singularity with index 2

For the first figure $\mathbf{v}(x,y) = (y,x)$, and $(0,0)$ has an index of -1; note how the vector field goes anticlockwise once as we move clockwise around the circle. For the second field $\mathbf{v}(x,y) = (x^2-y^2, 2xy)$, then $(0,0)$ has index 2. This definition then generalizes to tangent vector fields on *surfaces (*see* POINCARÉ-HOPF THEOREM) and to higher dimensions (*see* DEGREE (of a map)).

index (statistics) A figure used to show the variation in some quantity over a period of time, usually standardized relative to some base value. The index is given as a percentage with the base value equal to 100%. For example, the retail price index and consumer price index are used to measure changes in the cost of household items and inflation. Indices are often calculated by using a weighted mean of a number of constituent parts.

index set A set S may consist of elements, each of which corresponds to an element of a set I. Then I is the index set, and S is said to be indexed by I. For example, the set S may consist of elements a_i, where $i \in I$. This is written $S = \{a_i \mid i \in I\}$. A *denumerable set can be indexed by the set of *natural numbers.

indicator function (characteristic function) The indicator function of a subset $S \subseteq X$ is the function $\mathbf{1}_S : X \rightarrow \mathbb{R}$ such that $\mathbf{1}_S(x) = 1$ if $x \in S$ and $\mathbf{1}_S(x) = 0$ otherwise. Such functions can be useful in probability as *random variables indicating whether a certain event occurred.

indifference curve A *contour line for a *utility function. Since all points give the same value of the utility function, there is no reason to prefer one of the points on the indifference curve to any other.

indirect isometry *See* DIRECT ISOMETRY.

indirect proof A theorem of the form $p \Rightarrow q$ can be proved by establishing instead its *contrapositive, by supposing $\neg q$ and showing that $\neg p$ follows. Such a method is called an indirect proof. Another example of an indirect proof is the method of *proof by contradiction.

indirect proportion *See* PROPORTION.

indivisible by A synonym for not DIVISIBLE by.

induce (induced) A structure on a mathematical object may naturally introduce a structure to subsets or related sets; that inherited structure is commonly referred to as 'induced'. Likewise a function between sets might naturally give rise to or 'induce' other functions between related sets.

For example, a *metric on a set induces a metric on each subset and the *topology on a set induces the *subspace topology on a subset; a function between sets induces a function between their *power sets; a *continuous function between two spaces induces a *homomorphism between their *fundamental groups.

induction *See* MATHEMATICAL INDUCTION.

inequality The following symbols have the meanings shown:

$\neq$ is not equal to,
$<$ is less than,
$\leq$ is less than or equal to,
$>$ is greater than,
$\geq$ is greater than or equal to.

An inequality is a statement of one of the forms: $a \neq b$, $a < b$, $a \leq b$, $a > b$ or $a \geq b$, where a and b are suitably comparable expressions. Inequalities involving $<$ and $>$ are called 'strict', and those involving $\leq$ and $\geq$ are termed 'weak'.

Given an inequality involving one real variable x, the problem may be to find the *solutions, that is, the values of x that satisfy the inequality, in an explicit form. Often, the set of solutions, known as the *solution set, may be given as an *interval or as a *union of intervals. For example, the solution set of the inequality

$$x - 2 > \frac{6}{x+3}$$

is the set $(-4, -3) \bigcup (3, \infty)$.

(Some authors use 'inequation' for a statement which holds only for some values of the variables involved. For them, an inequation has a solution set but an inequality does not; an inequality is comparable to an *identity such as $x^2 + 2 > 2x$ which holds for all real x)

inertia The resistance of a *body to any change in its *velocity. The 'principle of inertia' is synonymous with Newton's first law of motion.

inertial frame of reference See FRAME OF REFERENCE.

inertial mass The inertial mass of a particle is the constant of proportionality that occurs in the statement of *Newton's second law of motion, linking the *acceleration of the particle with the total *force acting on the particle. Thus $ma = F$, where m is the inertial mass of the particle, a is the acceleration, and F is the total force acting on the particle. See also GRAVITATIONAL MASS, MASS, REST MASS.

inertia matrix (inertia tensor) The inertia matrix (or inertia tensor) I_P of a *rigid body (associated with a given point P) arises in the calculation of the *kinetic energy and the *angular momentum of the body in general motion. It is the *symmetric matrix

$$I_P = \begin{bmatrix} I_{xx} & I_{xy} & I_{xz} \\ I_{xy} & I_{yy} & I_{yz} \\ I_{xz} & I_{yz} & I_{zz} \end{bmatrix}$$

(see MOMENT OF INERTIA, PRODUCT OF INERTIA). By the *spectral theorem,

there is an *orthogonal set of axes—the principal axes—with respect to which the inertia matrix is *diagonal with diagonal entries being the principal moments of inertia.

inextensible string A string that has constant *length, no matter how great the *tension in the string.

inf Abbreviation for *infimum.

infeasible A *constrained optimization problem for which the constraints cannot be satisfied simultaneously.

inference The process of coming to some conclusion about a *population based on a *sample; or the conclusion about a population reached by such a process. The subject of statistical inference is concerned with the methods that can be used and the theory behind them. See BAYESIAN INFERENCE.

infimum (infima) See BOUND.

infinite Not *finite. Having a *cardinality greater than any *natural number.

infinite product From an infinite sequence $a_1, a_2, a_3, \ldots$, an infinite product $a_1 \, a_2 \, a_3 \ldots$ can be formed and denoted by

$$\prod_{r=1}^{\infty} a_r.$$

Let P_n be the nth *partial product, so that

$$P_n = \prod_{r=1}^{n} a_r.$$

If P_n tends to a limit P as $n \to \infty$, then P is the value of the infinite product. For example,

$$\prod_{r=2}^{\infty} \left(1 - \frac{1}{r^2}\right)$$

has the value $\frac{1}{2}$, since it can be shown that $P_n = (n+1)/2n$ and $P_n \to \frac{1}{2}$.

infinite sequence *See* SEQUENCE.

infinite series *See* SERIES.

infinite set Not *finite. A set whose elements cannot be put into *one-to-one correspondence with a bounded sequence of the *natural numbers. Equivalently a set which can be put in *one-to-one correspondence with a *proper subset of itself.

infinitesimal A variable whose *limit is zero, usually involving increments of functions, and of particular use in defining *differentiation. The ratio $\frac{\delta y}{\delta x}$ tends to the differential $\frac{dy}{dx}$ as $\delta x \to 0$. *See also* HYPERREALS.

infinity For the *real numbers, a value greater than any finite number, denoted by ∞. It is natural to introduce notions of tending to ∞ and $-\infty$, though these are not themselves real numbers. For the *complex numbers, it is natural to introduce a notion of a single infinity ∞ to make the *extended complex plane.

inflection (inflexion) *See* POINT OF INFLEXION.

information The information associated with an *event E is $-\log_2 \Pr(E)$, and the unit of information is a *bit. So the information in a fair coin toss is 1 bit and in a fair die roll is $\log_2 6$ bits (which is between 2 and 3 bits). Note that the information relating to two *independent events is equal to the sum of the separate events' information.

information theory The area of mathematics concerned with the transmission and processing of information, especially concerned with methods of coding, decoding, storage, and retrieval and evaluating likelihoods of degree of accuracy in the process. *See* ENTROPY, SHANNON'S THEOREM.

inhomogeneous Not *homogeneous.

initial conditions In applied mathematics, when *differential equations with respect to *time are to be solved, the known information about the system at a specified time, usually taken to be $t = 0$, may be called the initial conditions.

initialize Set parameters or variables at the start of an *algorithm.

initial line The axis used in *polar coordinates (from which the *angle coordinate is measured).

initial value The value of any quantity at a specified time reckoned as the starting-point, usually taken to be $t = 0$.

initial value problem Usually a *differential or *partial differential equation together with specified *initial conditions.

injection (injective mapping) A synonym for ONE-TO-ONE MAPPING.

inner product (inner product space) An inner product is a generalization of the *scalar product. An inner product on $\mathbb{R}^n$ is usually denoted $\langle \mathbf{u}, \mathbf{v} \rangle$ rather than $\mathbf{u} \cdot \mathbf{v}$ and satisfies:

(i) $\langle (a\mathbf{u} + b\mathbf{v}), \mathbf{w} \rangle = a\langle \mathbf{u}, \mathbf{w} \rangle + b\langle \mathbf{v}, \mathbf{w} \rangle$.

(ii) $\langle \mathbf{u}, \mathbf{v} \rangle = \overline{\langle \mathbf{v}, \mathbf{u} \rangle}$.

(iii) $\|\mathbf{v}\|^2 \geq 0$ and if $\langle \mathbf{v}, \mathbf{v} \rangle = 0$, then $\mathbf{v} = \mathbf{0}$.

The *norm (or *length) $\|\mathbf{v}\|$ of a vector then equals $\sqrt{\langle \mathbf{v}, \mathbf{v} \rangle}$. The *distance between points $\mathbf{v}$ and $\mathbf{w}$ is then defined as $\|\mathbf{v} - \mathbf{w}\|$ and the *angle between $\mathbf{v}$ and $\mathbf{w}$ equals

$$\cos^{-1}\left(\frac{\langle \mathbf{v}, \mathbf{w} \rangle}{\|\mathbf{v}\| \|\mathbf{w}\|} \right),$$

(*see* CAUCHY-SCHWARZ INEQUALITY). A real inner product space is a real *vector space with an inner product. A complex inner product space is a complex vector space with an inner product that satisfies instead

(ii)′ $\langle \mathbf{u}, \mathbf{v} \rangle = \overline{\langle \mathbf{v}, \mathbf{u} \rangle}$

to ensure that $\langle \mathbf{v}, \mathbf{v} \rangle$ is real. *See also* HILBERT SPACE.

input *See* FUNCTION.

inscribe Construct a geometric figure inside another so they have points in common but the inscribed figure does not have any part of it outside the other.

inscribed circle (of a triangle) A synonym for INCIRCLE.

insoluble (insolvable) Without a solution. So $x^2 + 1 = 0$ is insoluble within the set of *real numbers, though it can be solved within the set of *complex numbers.

instance A particular case, often derived from a general expression by substituting values for one or more parameters.

instantaneous Occurring at a single point in time, with zero duration.

instantaneous code A synonym for PREFIX CODE.

instrumental understanding In *mathematics education, in contrast to *relational understanding, the term refers to sufficient understanding of a mathematical rule to accurately implement that rule without necessarily being able to appreciate why the rule works or understand it in any broader context.

integer One of the 'whole' numbers: ...,-3,-2,-1, 0, 1, 2, 3, ... The set of all integers is often denoted by $\mathbb{Z}$. With the normal addition and multiplication, $\mathbb{Z}$ forms an *integral domain. The use of the letter zed comes from the German word *Zahlen* meaning 'numbers'.

integer factorization (in cryptography) While the *Fundamental Theorem of Arithmetic states that any number can be written uniquely as a product of prime factors, there is no known efficient factorization *algorithm for very large numbers. The difficulty of factorizing large composite integers, especially *semiprimes, lies at the heart

of many modern secure systems of data transmission such as *RSA.

integer part For any real number x, there is a unique integer n such that $n \leq x < n + 1$. This integer n is the integer part of x and is denoted by $[x]$. For example, $\left[\frac{9}{4}\right] = 2$ and $[\pi] = 3$, but notice that $\left[-\frac{9}{4}\right] = -3$. It is equivalent to *floor.

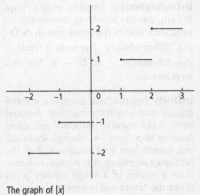

The graph of $[x]$

See CEILING, FRACTIONAL PART.

integer programming The restriction of *linear programming to cases when one or more variables can only take integer values.

integrable A function that has an *integral within a given theory of integration, e.g. *Riemann integrable or *Lebesgue integrable.

integrable system A *dynamical system which is determined by its initial conditions and such that the *phase space can be partitioned into disjoint evolutions of the system (called integral curves or leaves). The motion of a *rigid body about its *centre of mass is integrable whilst the *three-body problem is generally not integrable. Integrable systems have conserved invariants and do not have chaotic behaviour (*see* CHAOS).

integral Let f be a *bounded function defined on the closed interval $[a,b]$. Take points $x_0, x_1, x_2, \ldots, x_n$ such that $a = x_0 < x_1 < x_2 < \ldots < x_{n-1} < x_n = b$, and in each subinterval $[x_i, x_{i+1}]$ take a point c_i. Form the *Riemann sum

$$\sum_{i=0}^{n-1} f(c_i)(x_{i+1} - x_i).$$

Geometrically, this gives the sum of the areas of n rectangles, and is an approximation to the area under the curve $y = f(x)$ between $x = a$ and $x = b$.

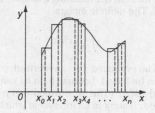

Area represented by a Riemann sum

The (Riemann) integral of f over $[a, b]$ is defined to be the limit I, if it exists, of such a Riemann sum as n, the number of points, increases and the maximal length of the subintervals tends to zero. The value of I is denoted by

$$\int_a^b f(x) \, dx.$$

The intention is that the value of the integral is equal to what is intuitively understood to be the signed area under the curve $y = f(x)$. (*See* DARBOUX INTEGRAL.) Such a limit does not always exist, but it can be proved that it does if f is a *continuous function on $[a,b]$. An example where the limit does not exist is the *Dirichlet function.

If f is continuous on $[a,b]$ and F is defined by

$$F(x) = \int_a^x f(t) \, dt,$$

then $F'(x) = f(x)$ for all x in $[a, b]$, so that F is an *antiderivative of f. Moreover, if an antiderivative ϕ of f is known, the *Fundamental Theorem of Calculus evaluates the integral

$$\int_a^b f(t) \, dt$$

as $\phi(b) - \phi(a)$. Of the two integrals

$$\int_a^b f(x) \, dx \text{ and } \int f(x) \, dx,$$

the first, with limits, is called a definite integral; the second, which denotes an antiderivative of f, is an indefinite integral and is defined only up to addition by arbitrary constant. *See also* LEBESGUE MEASURE, LINE INTEGRAL, MULTIPLE INTEGRAL, SURFACE INTEGRAL.

integral calculus The study, within *calculus, of accumulated changes. The *integral is defined by a limiting process based on an intuitive idea of the *area under a curve; the fundamental discovery was the link that exists between this and the *differential calculus.

integral domain A commutative *ring R with identity, with the additional property that

(ix) For all a and b in R, $ab = 0$ only if $a = 0$ or $b = 0$.

(The axiom numbering follows on from that used for *ring.) Thus an integral domain is a commutative ring with identity with no *zero-divisors. The natural example is the set $\mathbb{Z}$ of all integers with the usual addition and multiplication. Any *field is an integral domain. Further examples of integral domains (these are not fields) are: the set $\mathbb{Z}[\sqrt{2}]$ of all real numbers of the form $a + b\sqrt{2}$, where a and b are integers, and the set $\mathbb{R}[x]$ of all *polynomials in a variable x, with real coefficients, each with the normal addition and multiplication.

integral equation An equation that involves *integration of functions. In particular, the information given in an *initial-value problem such as $y'(x) = y$ with $y(0) = 1$ is equivalent to the integral equation

$$y(x) = 1 + \int_{t=0}^{t=x} y(t)\,dt.$$

This rewriting can be useful as the integral equation lends itself better to treatment by theorems of *analysis, particularly to find approximate solutions.

integral test Let $f(x)$ be a non-negative, *continuous, *decreasing function defined for $x \geq N$, where N is an integer. Then, the *series $\sum_{n=N}^{\infty} f(n)$ *converges if and only if the integral $\int_N^{\infty} f(x)\,dx$ converges. For example, setting $N = 1$ and $f(x) = 1/x$ shows that the *harmonic series diverges as $\int_1^X \frac{dx}{x} = \ln X$ tends to infinity as X tends to infinity.

integral transform A relationship between two or more functions can be expressed in the form $f(x) = \int K(x, y) F(y)\,dy$ then $f(x)$ is the integral transform of $F(x)$, and $K(x,y)$ is the kernel of the transform. If $F(x)$ can be found from $f(x)$ then the transform can be inverted. Integral transforms, such as the *Fourier transform and the *Laplace transform, are useful in transforming *differential equations in $F(x)$ into *algebraic equations in $f(x)$ which can then be solved, and the solution $f(x)$ inverted to find $F(x)$, the solution to the original problem.

integrand The expression $f(x)$ in either of the integrals

$$\int_a^b f(x)\,dx, \quad \int f(x)\,dx.$$

integrating factor See LINEAR FIRST-ORDER DIFFERENTIAL EQUATION.

integration The process of finding an *antiderivative of a given function f. 'Integrate f' means 'find an antiderivative of f'. Such an antiderivative may be called an indefinite integral of f and be denoted by

$$\int f(x)\,dx.$$

Such antiderivatives are only defined up to addition of an arbitrary constant.

The term 'integration' is also used for any method of evaluating a definite integral. The definite integral

$$\int_a^b f(x)\,dx$$

can be evaluated if an antiderivative ϕ of f can be found, because then its value is $\phi(b) - \phi(a)$. (This is provided that a and b both belong to an interval in which f is *continuous.) However, for many functions f, it can be shown there is no antiderivative expressible in terms of elementary functions, and other methods of evaluation have to be employed such as *numerical integration.

What ways are there, then, of finding an antiderivative? If the given function can be recognized as the derivative of a familiar function, an antiderivative is immediately known. For some standard integrals, see APPENDIX 8; more extensive tables of integrals are available. Certain techniques of integration may also be tried, among which are the following:

Change of variable/Substitution If it is possible to find a suitable function g such that the *integrand can be written as $f(g(x))\, g'(x)$, it may be possible to find an indefinite integral using the change of variable $u = g(x)$; this is because

$$\int f(g(x))g'(x)\ dx = \int f(u)\ du,$$

a rule derived from the *chain rule for differentiation. For example, in the integral

$$\int 2x(x^2 + 1)^8\ dx,$$

let $u = g(x) = x^2 + 1$. Then $g'(x) = 2x$ (this can be written '$du = 2x\ dx$'), and, using the rule above with $f(u) = u^8$, the integral equals

$$\int (x^2 + 1)^8 2x\ dx = \int u^8\ du$$
$$= \tfrac{1}{9} u^9 = \tfrac{1}{9}(x^2 + 1)^9.$$

Commonly, it may seem more natural to treat x as a function of u and 'substitute' $x = g(u)$ according to

$$\int f(x)\ dx = \int f(g(u))g'(u)\ du.$$

Note in the following example of a definite integral that it is necessary to change from x-limits to u-limits.

Let $x = g(u) = \tan u$. Then $g'(u) = \sec^2 u$ (this can be written '$dx = \sec^2 u\ du'$), and, recalling $1 + \tan^2 u = \sec^2 u$, the integral becomes

$$\int_{x=0}^{x=\infty} \frac{dx}{(1 + x^2)^{\frac{3}{2}}} = \int_{u=0}^{u=\frac{\pi}{2}} \frac{\sec^2 u\ du}{(1 + \tan^2 u)^{\frac{3}{2}}}$$
$$= \int_{u=0}^{u=\frac{\pi}{2}} \cos u\ du = \sin \frac{\pi}{2} - \sin 0 = 1.$$

Integration by parts The rule for integration by parts,

$$\int f(x)g'(x)dx = f(x)g(x) - \int g(x)f'(x)dx,$$

is derived from the rule for differentiating a product $f(x)g(x)$, and is useful when the integral on the right-hand side is

easier to find than the integral on the left. For example, in the integral

$$\int x \cos x\, dx,$$

let $f(x) = x$ and $g'(x) = \cos x$. Then $g(x)$ can be taken as $\sin x$ and $f'(x) = 1$, so the method gives

$$\int x \cos x\ dx = x \sin x - \int \sin x\, 1\, dx$$
$$= x \sin x + \cos x.$$

See also INTEGRABLE, LEBESGUE INTEGRAL, PARTIAL FRACTIONS, REDUCTION FORMULA, SEPARABLE FIRST-ORDER DIFFERENTIAL EQUATIONS.

integration by parts *See* INTEGRATION.

interacting variables Sometimes a second variable interacts with one of the variables in an experiment. If the experiment fails to take account of this, the reported results are unlikely to have any general applicability.

intercept *See* LINE (in two dimensions).

interest Effectively the price attached to the use of money belonging to someone else. It is expressed as a rate, usually as a percentage per year, though often how the interest is calculated can make an important difference. *Simple interest calculations involve interest being paid on the whole principle for the whole term, so if £5000 is borrowed at 8% pa, and the finance is for 3 years, the buyer will pay $5000 \times 8/100 \times 3 = £1200$ in interest charges and will have to pay $6200 \div 36 = £127.22$ per month for 3 years. An interest-bearing account will usually reinvest the interest (so the interest is added to the principal and subsequently interest will be paid on the larger amount), and this is known as *compound interest.

interior (of a curve) *See* JORDAN CURVE THEOREM.

interior angle (of a polygon) The angle between two adjacent sides of a *polygon that lies inside the polygon.

interior angle (with respect to a transversal of a pair of lines) *See* TRANSVERSAL.

interior of a set The set of all *interior points. A set is *open if and only if it equals its interior.

interior point A point in a *topological space (or *metric space) contained within an open subset of the space. When the *real line is being considered, the real number x is an interior point of the set S of real numbers if there is an open interval $(x-\delta, x+\delta)$, where $\delta > 0$, included in S.

intermediate value theorem The theorem states that if a real function f is *continuous on the closed interval $[a, b]$ and η is a real number between $f(a)$ and $f(b)$, then $f(c) = \eta$ for some c in (a, b).

The theorem relies on the *connectedness of the interval and naturally generalizes to real-valued functions on other connected spaces.

intermediate vertex (in a network) A vertex between the *source and the *sink.

internal bisector Name sometimes given to the *bisector of the *interior angle of a *polygon.

internal division Let AB be a *line segment; then the point E is the internal division of AB in the *ratio 1: k if

$$k\,\overrightarrow{AE} = \overrightarrow{EB}$$

where $\overrightarrow{AB}$ is the *vector from A to B. Alternatively E is the point $\frac{1}{1+k}$ of the way along from A to B. If A and B have position vectors $\mathbf{a}$ and $\mathbf{b}$, from some origin, then the position vector of E is

$$\mathbf{e} = \mathbf{a} + \frac{1}{k+1}(\mathbf{b} - \mathbf{a}) = \frac{k\mathbf{a} + \mathbf{b}}{k+1},$$

sometimes referred to as the section formula. *Compare* EXTERNAL DIVISION.

internal force When a system of *particles or a *rigid body is being considered as a whole, an internal force is a force exerted by one particle of the system on another or by one part of the rigid body on another. As far as the whole is concerned, the sum of the internal forces is zero. *Compare* EXTERNAL FORCE.

International Congress of Mathematicians (ICM) A large academic conference, usually occurring every 4 years, with invited talks from the world's foremost research mathematicians and at which the *Fields Medals are awarded. The first ICM was held in Zurich in 1897, and *Hilbert famously presented 10 of his 23 unsolved problems at the 1900 congress in Paris.

interpolation Suppose that the values $f(x_0), f(x_1), \ldots, f(x_n)$ of a certain function f are known for the particular values $x_0, x_1, \ldots, x_n$. A method of finding an approximation for $f(x)$, for a given value of x somewhere between these particular values, is called interpolation. If $x_0 < x < x_1$, the method known as linear interpolation gives

$$f(x) \approx f(x_0) + \frac{x - x_0}{x_1 - x_0}\left(f(x_1) - f(x_0)\right).$$

This is obtained by supposing, as an approximation, that between x_0 and x_1 the graph of the function is a straight line joining the points $(x_0, f(x_0))$ and $(x_1, f(x_1))$. More complicated methods of interpolation use the values of the function at more than two values. *See* LAGRANGE INTERPOLATION, SPLINE.

interquartile range A measure of *dispersion equal to the difference between the first and third *quartiles in a set of numerical *data.

intersect Two geometrical figures or curves are said to intersect if they have at least one point in common. *Compare* TOUCH.

intersection The intersection of sets A and B is the set consisting of all *elements that belong to both A and B, and it is denoted by $A \cap B$ (read as 'A intersect B' or 'A cap B'). The term 'intersection' is used for both the resulting set and the operation, a *binary operation on the set of all subsets of a set. The following properties hold:

(i) For all A, $A \cap A = A$ and $A \cap \emptyset = \emptyset$.
(ii) For all A and B, $A \cap B = B \cap A$; that is, $\cap$ is commutative.
(iii) For all A, B and C, $(A \cap B) \cap C = A \cap (B \cap C)$; that is, $\cap$ is associative.

Given a family of sets A_i with *index set I, then the intersection

$$\bigcap_{i \in I} A_i$$

consists of those elements x such that x is an element of A_i for all i in I. For the intersection of two events, *see* EVENT.

interval A subset I of $\mathbb{R}$ is an interval if whenever $x < y < z$ and $x \in I$, $z \in I$, then $y \in I$.

A *bounded interval on the real line is a subset of $\mathbb{R}$ defined in terms of endpoints a and b. Since each end-point may or may not belong to the subset, there are four types of bounded interval:

(i) the closed interval $\{x \mid x \in \mathbb{R}$ and $a \le x \le b\}$, denoted by $[a, b]$,
(ii) the open interval $\{x \mid x \in \mathbb{R}$ and $a < x < b\}$, denoted by (a, b),
(iii) the interval $\{x \mid x \in \mathbb{R}$ and $a \le x < b\}$, denoted by $[a, b)$,
(iv) the interval $\{x \mid x \in \mathbb{R}$ and $a < x \le b\}$, denoted by $(a, b]$.

There are also four types of unbounded interval:

(v) $\{x \mid x \in \mathbb{R}$ and $a \le x\}$, denoted by $[a, \infty)$,

(vi) $\{x \mid x \in \mathbb{R}$ and $a < x\}$, denoted by (a, ∞),
(vii) $\{x \mid x \in \mathbb{R}$ and $x \le a\}$, denoted by $(-\infty, a]$,
(viii) $\{x \mid x \in \mathbb{R}$ and $x < a\}$, denoted by $(-\infty, a)$.

Here ∞ (read as 'infinity') and $-\infty$ (read as 'minus infinity') are not real numbers, but the use of these symbols provides a convenient notation.

interval estimate *See* ESTIMATE.

interval scale A scale of measurement where differences in value are meaningful but not ratios of the values. Consequently the choice of zero on the scale is arbitrary. For example, temperature is an interval scale. On the *Celsius scale it is not meaningful to say that 10° is twice as hot as 5°.

intransitive relation A *binary relation which is not transitive for any three elements, i.e. $a \sim b$ and $b \sim c$ requires that a does not $\sim c$. Note that this is a different condition from the relation being *non-transitive which implies that transitivity does not hold for some combination of three elements.

intrinsic A *curve or *surface may be extrinsically studied while *embedded in *Euclidean space, its structure being inherited from the ambient (i.e. surrounding) space. Intrinsic properties of the curve or surface, such as *area or *curvature, are properties determined by its metric properties and so independent of a particular embedding and are preserved by *isometries. An intrinsic approach may be preferred when embeddings are difficult or even impossible; for example, the *hyperbolic plane cannot be isometrically embedded in $\mathbb{R}^3$.

intrinsic coordinates (intrinsic equation) A planar *curve may be defined by use of intrinsic properties such as *arc length s (from some fixed point of the curve) and the *angle ψ that the curve's *tangent makes (with a fixed direction). The *curvature κ then satisfies $\kappa = d\psi/ds$ and alternatively κ and s might be used to define the curve. For a curve in three dimensions, arc length, curvature, and *torsion can be used.

intuitionism A *constructivist philosophy of mathematics, its main proponent being *Brouwer during the 20th century, in stark contrast to *Hilbert's *formalism. Intuitionism, in particular, rejects any notion of an *actual infinity and the *principle of the excluded middle and hence rejects *proof by contradiction. *Compare* LOGICISM, PLATONISM, STRUCTURALISM.

invalid Not valid. A conclusion which is not a necessary consequence of the conditions, so a *counterexample is sufficient to show a proposition is invalid.

invariable A synonym for CONSTANT.

invariant A property or quantity that is not changed by one or more specified operations or transformations. For example, for a *conic with equation

$$Ax^2 + Bxy + Cy^2 + Dx + Ey + F = 0$$

the quantity $B^2 - 4AC$ is invariant under a *rotation of axes. The *distance between two points and the *angle between two lines are invariants under *translations and rotations of the plane, but distance is not invariant under *dilations, while angle is.

invariant set Given a function $f : S \to S$, a subset A of S is invariant if $f(A) \subseteq A$. Note that this is a much weaker condition than requiring that $f(a) = a$ for all $a \in A$. For example, if f reflects $\mathbb{R}^2$ in the x-axis, then the x-axis is invariant in the latter sense—every point is fixed—but any line parallel to the y-axis is an invariant set.

invariant subgroup A synonym for NORMAL SUBGROUP.

inverse correlation (negative correlation) *See* CORRELATION.

inverse element Suppose that, for the *binary operation $\circ$ on the set S, there is an *identity element e. An element a' is an inverse (or inverse element) of the element a if $a \circ a' = a' \circ a = e$. If the operation is called multiplication, the identity element may be denoted by 1. Then the inverse a' may be called a multiplicative inverse of a and be denoted by a^{-1}, so that $aa^{-1} = a^{-1} a = 1$ (or e). If the operation is addition, the identity element is denoted by 0, and the inverse a' may be called an additive inverse (or a negative) of a and be denoted by $-a$, so that $a + (-a) = (-a) + a = 0$. *See also* GROUP.

inverse function (inverse mapping) For a function f, its inverse function f^{-1} (or just inverse) is to be a function such that $y = f(x)$ if and only if $x = f^{-1}(y)$. If such a function exists, then it is unique and f is said to be invertible. Suppose that f has *domain S and *codomain T. To be invertible $f : S \to T$ needs to be onto so that every y in T equals $f(x)$ for some x in S. However, any inverse f^{-1}, with domain T and codomain S must uniquely assign such x to this y and so f must also be *one-to-one. The inverse is then defined by: for y in T, $f^{-1}(y)$ is the unique element x of S such that $f(x) = y$. Consequently, f is invertible if and only if it is a *bijection.

For a *real function defined on an interval I, if f is *strictly increasing on I or strictly decreasing on I, then f is one-to-one and so bijective onto its image.

When an inverse is required for a given function f, it may be necessary to restrict the domain and obtain instead the inverse function of this *restriction of f. For example, suppose that $f : \mathbb{R} \to \mathbb{R}$

is defined by $f(x) = x^2 - 4x + 5$. This function is not one-to-one but is strictly increasing for $x \geq 2$. Use f now to denote the function defined by $f(x) = x^2 - 4x + 5$ with domain $[2, \infty)$. The *image is $[1, \infty)$ and the function $f: [2, \infty) \to [1, \infty)$ has inverse $f^{-1}: [1, \infty) \to [2, \infty)$. A formula for f^{-1} can be found by setting $y = x^2 - 4x + 5$ and, remembering that $x \in [2, \infty)$, obtaining $x = 2 + \sqrt{y - 1}$. So, with a change of notation, $f^{-1}(x) = 2 + \sqrt{x - 1}$ for $x \geq 1$.

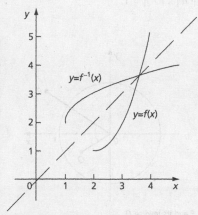

The graph of a function and its inverse

When the inverse function exists, the graphs $y = f(x)$ and $y = f^{-1}(x)$ are *reflections of each other in the line $y = x$. See INVERSE FUNCTION THEOREM, INVERSE HYPERBOLIC FUNCTION, INVERSE TRIGONOMETRIC FUNCTION, LEFT INVERSE, RIGHT INVERSE.

inverse function theorem A theorem which guarantees the existence of a local *inverse of a real function, which generalizes to multivariable functions. If the real continuously *differentiable function f has non-zero *derivative at a, then f has a continuously differentiable inverse in a *neighbourhood of a which satisfies:

$$(f^{-1})'(b) = \frac{1}{f'(a)}$$

where $f(a) = b$. Similarly, the multivariate version states that if $F: \mathbb{R}^n \to \mathbb{R}^n$ is continuously differentiable and has invertible *differential $dF_\mathbf{p}$ at a point $\mathbf{p}$, then F has a continuously differentiable inverse in a neighbourhood of $\mathbf{p}$ satisfying

$$d(F^{-1})_\mathbf{q} = (dF_\mathbf{p})^{-1},$$

where $\mathbf{q} = F(\mathbf{p})$.

The following result relating to *continuous real functions might also be referred to as the inverse function theorem. If $f:[a,b] \to [c,d]$ is continuous, strictly increasing, and satisfies $f(a) = c$, $f(b) = d$, then f has a strictly increasing continuous inverse.

inverse hyperbolic function Each of the *hyperbolic functions sinh and tanh is strictly increasing throughout the whole of its domain $\mathbb{R}$, so in each case an inverse function exists. In the case of cosh, the function has to be restricted to a suitable domain (*see* INVERSE FUNCTION), taken to be $[0, \infty)$. The domain of the inverse function is, in each case, the *image of the original function (after the restriction of the domain, in the case of cosh). The inverse functions obtained are: $\cosh^{-1}:[1, \infty) \to [0, \infty)$; $\sinh^{-1}: \mathbb{R} \to \mathbb{R}$; $\tanh^{-1}: (-1, 1) \to \mathbb{R}$. These functions are given by the formulae:

$$\cosh^{-1}x = \ln\left(x + \sqrt{x^2 - 1}\right) \text{ for } x \geq 1,$$

$$\sinh^{-1}x = \ln\left(x + \sqrt{x^2 + 1}\right) \text{ for all } x,$$

$$\tanh^{-1}x = \ln\sqrt{\frac{1 + x}{1 - x}} \text{ for } -1 < x < 1.$$

The following derivatives can be obtained:

$$\frac{d}{dx}(\cosh^{-1} x) = \frac{1}{\sqrt{x^2-1}} \quad (x \neq 1),$$

$$\frac{d}{dx}(\sinh^{-1} x) = \frac{1}{\sqrt{x^2+1}},$$

$$\frac{d}{dx}(\tanh^{-1} x) = \frac{1}{1-x^2}.$$

inverse image A synonym for PRE-IMAGE.

inverse points Given a circle with centre O and radius r, two points A and B are inverse points if O, A, and B are *collinear and $|OA||OB| = r^2$. Inversion—the map taking a point to its inverse point—has various theories associated with it; for example, inversion preserves *angles (though it reverses the sense of the angle) and also maps a *circle or *line not through O to a circle.

inversely proportional See PROPORTION.

inverse matrix An *inverse of a *square matrix $\mathbf{A}$ is a matrix $\mathbf{X}$ such that $\mathbf{AX} = \mathbf{I}$ and $\mathbf{XA} = \mathbf{I}$. (A matrix that is not square cannot have an inverse, though it may have a *left inverse or a *right inverse.) A square matrix $\mathbf{A}$ may or may not have an inverse, but if it has then that inverse is unique and $\mathbf{A}$ is said to be invertible. A matrix is invertible if and only if it is not *singular. Consequently, the term 'non-singular' is sometimes used for 'invertible'.

When $\det \mathbf{A} \neq 0$, the matrix $(1/\det\mathbf{A})$ adj$\mathbf{A}$ is the inverse of $\mathbf{A}$, where adj$\mathbf{A}$ is the *adjugate of $\mathbf{A}$. For example, the 2×2 matrix $\mathbf{A}$ below is invertible if $ad - bc \neq 0$, and its inverse $\mathbf{A}^{-1}$ is as shown:

$$\mathbf{A} = \begin{bmatrix} a & b \\ c & d \end{bmatrix}, \mathbf{A}^{-1} = \frac{1}{ad-bc}\begin{bmatrix} d & -b \\ -c & a \end{bmatrix}.$$

inverse of a complex number If z is a non-zero complex number and $z = x + yi$, the (*multiplicative) inverse of z, denoted by z^{-1} or $1/z$, is

$$\frac{x}{x^2+y^2} - \frac{y}{x^2+y^2}i.$$

When z is written in *polar form, so that $z = re^{i\theta} = r(\cos\theta + i\sin\theta)$, where $r \neq 0$, the inverse of z is $(1/r)e^{-i\theta} - (1/r)(\cos\theta - i\sin\theta)$. If z is represented by P in the *complex plane, then z^{-1} is represented by Q, where $\angle xOQ = -\angle xOP$ and $|OP| \times |OQ| = 1$.

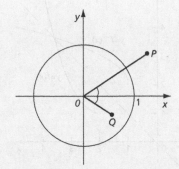

P and its inverse *Q*

inverse square law of gravitation
The following law, due to *Newton, that describes the *force of attraction that exists between two particles. Say particles P and S have gravitational masses m and M, respectively. Let $\mathbf{r} = \vec{SP}$ be the position vector of P relative to S and let $r = |\mathbf{r}|$. Then the force $\mathbf{F}$ experienced by P is given by

$$\mathbf{F} = -\frac{GMm}{r^3}\mathbf{r},$$

where G is the *gravitational constant. Since $\mathbf{r}/r$ is a unit vector, the magnitude of the gravitational force experienced by P is inversely proportional to the square of the distance between S and P. The force experienced by S is equal and opposite to that experienced by P.

inverse trigonometric function

Sine is not a *one-to-one function, and so not *invertible, but 'inverse sine' can be defined by choosing a set of *principal values. Thus, if $y = \sin^{-1} x$, then $x = \sin y$, but the converse need not hold. For each x in the interval $[-1,1]$ there is a unique y in the interval $[-\pi/2, \pi/2]$ such that $x = \sin y$ and this value of y is defined as $y = \sin^{-1} x$.

Similarly, for real x, $y = \tan^{-1} x$ if $x = \tan y$, and the unique value y is taken from the interval $(-\pi/2, \pi/2)$. Also, for x in the interval $[-1,1]$, $y = \cos^{-1} x$ if $x = \cos y$ and the unique value y is taken from the interval $[0, \pi]$.

The notation arcsin, arctan, and arccos for $\sin^{-1}$, $\tan^{-1}$, and $\cos^{-1}$ is also used. Their derivatives are:

$$\frac{d}{dx}(\sin^{-1}x) = \frac{1}{\sqrt{1-x^2}} \quad (x \neq \pm 1),$$

$$\frac{d}{dx}(\cos^{-1}x) = -\frac{1}{\sqrt{1-x^2}} \quad (x \neq \pm 1),$$

$$\frac{d}{dx}(\tan^{-1}x) = \frac{1}{1+x^2}.$$

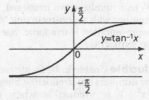

Graph of inverse tangent

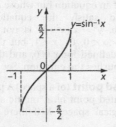

Graph of inverse sine

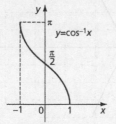

Graph of inverse cosine

invertible function See INVERSE FUNCTION.

invertible matrix See INVERSE MATRIX.

inviscid
Not viscous. The force exerted across a small surface element within an inviscid *fluid is $p\mathbf{n}dS$, where p denotes *pressure, $\mathbf{n}$ is the unit *normal, and dS is the surface area of the element. In a viscous fluid there would be tangential components to this *contact force due to *friction. See EULER'S EQUATIONS, NAVIER-STOKES EQUATIONS.

involution
A *function or *transformation which is its own inverse, that is to say applying it twice produces the *identity map. The *identity will also be an involution trivially, but more interesting ones are any reflection, rotation by 180°, multiplication by -1, taking *reciprocals, *complex conjugates, and taking *complements of sets.

irrational number
A real number that is not *rational. A famous proof, sometimes attributed to *Pythagoras, shows that $\sqrt{2}$ is irrational; the method can also be used to show that numbers

such as $\sqrt{3}$ and $\sqrt{7}$ are also irrational. It follows that numbers like $1 + \sqrt{2}$ and $1/(1 + \sqrt{2})$ are irrational. The proof that e is irrational is reasonably easy, and in 1761 Lambert showed that π is irrational. There are *uncountably many irrationals, compared with *denumerably many rationals, so in a sense 'most' real numbers are irrational. Any real number with a non-recurring *decimal representation is irrational. *See also* TRANSCENDENTAL.

irreducible Unable to be factorized. So *prime numbers are irreducible integers, and $x^2 + 1$ is irreducible when considered as a real polynomial though it is reducible to $(x + i) (x - i)$ as a complex polynomial where $i = \sqrt{-1}$. More generally, an element x of an *integral domain R is irreducible if x is non-zero, not a *unit, and whenever $x = yz$ where $y, z \in R$, then y or z is a unit. In a *unique factorization domain an element is irreducible if and only if it is a *prime element, but this is not the case generally in *rings.

irreducible (Markov chains) *See* COMMUNICATING CLASS.

irreducible fraction A *fraction m/n which cannot be reduced any further, that is, m and n are *coprime. Such a fraction is also said to be in lowest terms or in simplest form.

irreducible representation A *representation $\rho : G \to GL(V)$ of a *group G is irreducible if the only invariant subspaces of V are $\{0\}$ and V. A subspace W of V is invariant in this sense if $\rho(g)w \in W$ for all $w \in W$ and $g \in G$.

irrotational A *vector field is irrotational if it has zero *curl. *See* CONSERVATIVE FIELD, VORTICITY.

isoclines *See* TANGENT FIELDS.

isogon A *polygon with all *angles equal, so that all *regular polygons are isogons but not vice versa; for example, a rectangle is an isogon but not regular.

isolate To identify an *interval within which only one *root of an equation lies.

isolated point (of a curve) A *singular point which does not lie on the main curve of an equation but whose coordinates do satisfy the equation. For example, $y^2 = x^2(x^2 - 1)$ is not defined for $-1 < x < 0$, $0 < x < 1$, but the function is defined for $x = 0$, and the point $(0, 0)$ satisfies the equation. See figure.

isolated point (of a space) A point P is an isolated point of a *metric space (or *topological space) X if there exists a

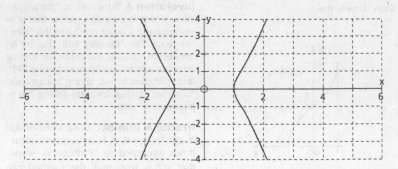

Curve with origin as an isolated point

*neighbourhood of P which does not contain another element in X which is distinct from the point P.

isolated singularity *See* SINGULAR POINT.

isometric graph paper Has three axes, equally spaced relative to one another, to enable 3-dimensional figures to be represented on a plane. The effect of the equal spacing is to produce a grid made up of equilateral triangles.

isometry If P and Q are points in the plane, $|PQ|$ denotes the distance between P and Q. An isometry is a *transformation of the plane that preserves the distance between points: it is a transformation with the property that, if P and Q are mapped to P' and Q', then $|P'Q'| = |PQ|$. Examples of isometries are *translations, *rotations, and *reflections. It can be shown that all the isometries of the plane can be obtained from translations, rotations, and reflections, by composition. Two figures are *congruent if there is an isometry that maps one onto the other.

isomorphic *See* ISOMORPHISM.

isomorphic graphs Two or more *graphs which have the same structure of *edges joining corresponding vertices. Depending on where the vertices are positioned these graphs may superficially look quite different.

isomorphism (of groups) Let $(G, \circ)$ and $(G', *)$ be *groups, so that $\circ$ is the operation on G and $*$ is the operation on G'. An isomorphism between $(G, \circ)$ and $(G', *)$ is a *one-to-one correspondence f from the set G to the set G' such that, for all a and b in G,

$$f(a \circ b) = f(a) * f(b).$$

This means that if f maps a to a' and b to b', then f maps $a \circ b$ to $a' * b'$. So an isomorphism is a *bijective *homomorphism.

f can be viewed as a relabelling of the *Cayley table of G to become the Cayley table of G'.

If there is an isomorphism between two groups, the two groups are isomorphic to each other. This is written $G \cong G'$. Two groups that are isomorphic to one another have essentially the same structure; the actual elements of one group may be quite different objects from the elements of the other, but the way in which they behave with respect to the operation is the same. For example, the group $1, i, -1, -i$ with multiplication is isomorphic to the group of elements 0, $1, 2, 3$ with addition modulo 4. *See* APPENDIX 17 for groups, of *order 15 or less, up to isomorphism.

isomorphism (of rings) Suppose that $(R, +, \times)$ and $(R', \oplus, \otimes)$ are *rings. An isomorphism between them is a *one-to-one onto mapping f from the set R to the set R' such that, for all a and b in R,

$$f(a + b) = f(a) \oplus f(b),$$
$$f(a \times b) = f(a) \otimes f(b).$$

If there is an isomorphism between two rings, the rings are isomorphic to each other and, as with isomorphic groups, the two have essentially the same structure.

isomorphism theorems A collection of theorems describing *isomorphisms for *groups (with similar versions for *rings, *modules, etc.) due to Emmy *Noether.

First isomorphism theorem Let f: $G \rightarrow H$ be a *homomorphism of groups. Then the *kernel of f is a *normal subgroup of G, the *image of f is a subgroup of H, and the *quotient group $G/\ker f$ is isomorphic to Im f via the map $g\ker f \mapsto f(g)$.

As examples:

$f: z \mapsto |z|$ is a homomorphism from $\mathbb{C}^*$ to $\mathbb{R}^*$ with kernel S^1 and image $(0, \infty)$ so that $\mathbb{C}^*/S^1$ is isomorphic to $(0, \infty)$.

$g{:}x \mapsto \ln x$ is a homomorphism from $(0,\infty)$ to $\mathbb{C}$ with kernel $\{1\}$ and image $\mathbb{R}$ so that $(0,\infty)$ is isomorphic to $\mathbb{R}$.

$h{:}x \mapsto e^{2\pi i x}$ is a homomorphism from $\mathbb{R}$ to $\mathbb{C}^*$ with kernel $\mathbb{Z}$ and image S^1 so that $\mathbb{R}/\mathbb{Z}$ is isomorphic to S^1.

Second isomorphism theorem Let G be a group, H be a subgroup of G, and N be a normal subgroup of G. Then $HN = \{hn : h \in H,\ n \in N\}$ is a subgroup of G, and $H \cap N$ is a normal subgroup of H. Further, the quotient groups $H/(H \cap N)$ and $(HN)/N$ are isomorphic. (This follows by applying the first theorem to the homomorphism $h \mapsto hN$.)

Third isomorphism theorem Let G be a group, K,N be normal subgroups such that $N \subseteq K \subseteq G$. Then K/N is a normal subgroup of G/N, and the quotient groups G/K and $(G/N)/(K/N)$ are isomorphic. (This follows by applying the first theorem to the homomorphism $g \mapsto gN(K/N)$.)

isoperimetric Two or more geometric figures with the same *perimeter.

isoperimetric inequality If p is the *perimeter of a *closed curve in a plane and the *area enclosed by the curve is A, then $p^2 \geq 4\pi A$, and equality is only achieved if the curve is a *circle. Turning around this condition, it says that for a fixed length of curve p the greatest area which can be enclosed is when the curve is a circle. The result can be generalized to surfaces in 3-dimensional space where the *sphere is the most efficient

shape at enclosing volume for a fixed *surface area, and to higher dimensions.

isoperimetric quotient (IQ) number of a closed curve $\text{IQ} = \dfrac{4\pi A}{p^2}$ where $A = $ area enclosed by the curve, with perimeter P. The circle maximizes the area which can be enclosed by isoperimetric curves, and the IQ gives the proportion of this maximum which is actually enclosed by the curve. IQ is always ≤ 1, and equality only occurs for the circle.

isosceles trapezium A *trapezium in which the two non-parallel sides have the same length.

isosceles triangle A triangle in which two sides have the same length. Sometimes the third side (which may be called the *base) is required to be not of the same length. The two angles opposite the sides of equal length (which may be called the base angles) are equal.

isotopy An isotopy, or ambient isotopy, of two *knots is a *knot equivalence.

iterated series A double or multiple series such as $\displaystyle\sum_{i=1}^{\infty} \sum_{j=1}^{\infty} a_{ij}$.

iteration A method uses iteration if it yields successive approximations to a required value by repetition of a certain procedure. Examples are *fixed-point iteration and *Newton's method for finding a root of an equation $f(x) = 0$.

j In the notation for *complex numbers, some authors, especially engineers, use j instead of i.

j ($\hat{\jmath}$) A *unit vector, usually in the direction of the y-axis or a vertical vector in the plane of a *projectile's motion.

J Symbol for *joule.

Jacobi, Carl Gustav Jacob (1804–51) German mathematician responsible for notable developments in the theory of *elliptic functions, a class of functions defined to invert *elliptic integrals. Applying them to number theory, he was able to reprove *Lagrange's Theorem that every integer is the sum of four perfect squares, showing further in how many different ways.

Jacobian (Jacobian determinant) For n functions f_i in n variables x_i, the Jacobian is the *determinant of the *Jacobian matrix. The *absolute value of the Jacobian determinant is a measure of how much *area or *volume is being stretched or contracted *locally*. So, for small changes we have

$$\delta f_1 \delta f_2 \ldots \delta f_n \approx \left| \frac{\partial(f_1, f_2, \ldots, f_n)}{\partial(x_1, x_2, \ldots, x_n)} \right| \delta x_1 \delta x_2 \ldots \delta x_n.$$

For example, with the usual *spherical polar coordinates r, θ, ϕ, the Jacobian determinant equals $r^2 \sin\theta$, showing that the same change in θ contributes to a larger element of volume near this equator ($\theta = \pi/2$) compared to the north pole ($\theta = 0$), but the longitude φ has no effect on the element of volume. *See* JACOBIAN MATRIX.

Jacobian matrix For m functions in n variables, $f_i(x_1, x_2, \ldots, x_n)$ where $1 \leq i \leq m$, the Jacobian matrix is the $m \times n$ matrix whose element in the i-th row and j-th column is the *partial derivative $\dfrac{\partial f_i}{\partial x_j}$.

This Jacobian is denoted $\dfrac{\partial(f_1, f_2, \ldots, f_m)}{\partial(x_1, x_2, \ldots, x_n)}$. Jacobians satisfy a chain rule, namely:

$$\frac{\partial(f_1, f_2, \ldots, f_m)}{\partial(u_1, u_2, \ldots, u_k)}$$
$$= \frac{\partial(f_1, f_2, \ldots, f_m)}{\partial(x_1, x_2, \ldots, x_n)} \frac{\partial(x_1, x_2, \ldots, x_n)}{\partial(u_1, u_2, \ldots, u_k)}$$

when each x_i is a function of $u_1, u_2, \ldots, u_k$. *See* DIFFERENTIAL (multivariate), JACOBIAN.

Jacobi's identity For a set with addition and product $(X, Y) \mapsto [X, Y]$ defined, the identity $[X, [Y, Z]] + [Y, [Z, X]] + [Z, [X, Y]] = 0$. The identity is satisfied by the *vector sum and *vector product, by the sum and commutator product $[X, Y] = XY - YX$ for $n \times n$ matrices, and in *Lie algebras.

Jacobi's iterative method A technique for solving a set of n *linear equations in n unknowns. If the system is summarized by $\mathbf{A}\mathbf{x} = \mathbf{b}$, then taking initial values as $x_i^{(1)} = \frac{b_i}{a_{ii}}$, it uses the iterative relation $x_i^{(k)} = \frac{b_i - \sum_{j \neq i} a_{ij} x_j^{(k-1)}}{a_{ii}}$, so it computes a complete set of new values together, unlike the *Gauss–Seidel method.

Jeffreys, Sir Harold (1891–1989) British mathematician and astronomer who made important contributions in *probability and *calculus as well as in our understanding of the solar system.

Jensen's inequality The inequality

$$f(tx_1 + (1-t)x_2) \leq tf(x_1) + (1-t)f(x_2)$$

for a convex (= *concave up) function f and $0 \leq t \leq 1$. This implies that the *chord connecting the points $(x_1, f(x_1))$ and $(x_2, f(x_2))$ lies above the graph $y = f(x)$ for the interval $x_1 \leq x \leq x_2$. In *probability, the inequality is commonly expressed as $f(E(X)) \leq E(f(X))$ where X is a *random variable.

jerk (jolt) The *derivative of *acceleration with respect to time, second derivative of *velocity, and third derivative of *displacement.

joint cumulative distribution function For two *random variables X and Y, the joint cumulative distribution function is the function F defined by $F(x, y) = \Pr(X \leq x, Y \leq y)$. The term applies also to the generalization of this to more than two random variables. For two variables, it may be called the *bivariate and, for more than two, the *multivariate cumulative distribution function.

joint distribution The joint distribution of two or more *random variables is concerned with the way in which the probability of their taking certain values, or values within certain intervals, varies. It may be given by the *joint cumulative distribution function. More commonly, the joint distribution of some number of discrete random variables is given by their *joint probability mass function, and that of some number of continuous random variables by their *joint probability density function. For two variables it may be called the *bivariate and, for more than two, the *multivariate distribution.

joint probability density function For two continuous *random variables X and Y, the joint probability density function is the function f such that

$$\Pr(a \leq X \leq b, c \leq Y \leq d)$$
$$= \int_a^b \int_c^d f(x, y) \, \mathrm{d}y \, \mathrm{d}x.$$

Then X and Y are *independent if and only if $f(x, y) = f_X(x)f_Y(y)$, where f_X and f_Y are the pdfs of X and Y. The term applies also to the generalization of this to more than two random variables. For two variables, it may be called the *bivariate and, for more than two, the *multivariate probability density function.

joint probability mass function For two discrete *random variables X and Y, the joint probability mass function is the function p such that $p(x_i, y_j) = \Pr(X = x_i, Y = y_j)$, for all i and j. The term applies also to the generalization of this to more than two random variables. For two variables, it may be called the *bivariate and, for more than two, the *multivariate probability mass function.

jolt A synonym for JERK.

Jordan, (Marie-Ennemond) Camille (1838–1922) French mathematician whose treatise on groups of permutations and the theory of equations drew attention to the work of *Galois. In a later work on analysis, there appears his formulation of what is now called the *Jordan Curve Theorem.

Jordan block A *block matrix such as

$$\begin{bmatrix} \lambda & 1 & 0 \\ 0 & \lambda & 1 \\ 0 & 0 & \lambda \end{bmatrix}$$

and more generally an $n \times n$ matrix with entries of λ on the diagonal, entries of 1 just above the diagonal, and 0 entries elsewhere. Often denoted $J(\lambda, n)$. The *characteristic polynomial and *minimal polynomial of $J(\lambda, n)$ both equal $(x - \lambda)^n$.

Jordan curve A *simple, *closed curve in the plane. Thus, a Jordan curve is a continuous plane curve that has no ends (or, in other words, it begins and ends at the same point) and does not intersect itself.

Jordan Curve Theorem The theorem states that a *Jordan curve divides the plane into two regions (more precisely, *connected components), the interior, which is *bounded, and the exterior of the curve. The result is intuitively obvious but the proof is not elementary.

Jordan-Hölder Theorem All *composition series for a finite group G have the same length and the same set of composition factors. For example, the three composition series for $C_{12} = \{e, g, \dots, g^{11}\}$ are

$$\{e\} \lhd \{e, g^6\} \lhd \{e, g^3, g^6, g^9\} \lhd C_{12},$$

$$\{e\} \lhd \{e, g^4, g^8\} \lhd \{e, g^2, g^4, g^6, g^8, g^{10}\} \lhd C_{12},$$

$$\{e\} \lhd \{e, g^6\} \lhd \{e, g^2, g^4, g^6, g^8, g^{10}\} \lhd C_{12}$$

which have composition factors C_2, C_2, C_3 in some order. But knowledge of the composition factors says relatively little about the structure of G. For example, all five groups of order 12 have composition factors C_2, C_2, C_3.

Jordan normal form A *block diagonal matrix where every block is a *Jordan block. Every complex *square matrix is *similar to a matrix in Jordan normal form. The *geometric multiplicity of an *eigenvalue λ equals the number of Jordan blocks involving λ; the *algebraic multiplicity of λ is the sum of those block's orders. The *characteristic polynomial of the matrix is the product of all the block's characteristic polynomials; the *minimal polynomial of the matrix is the product of the each eigenvalue's largest block's characteristic polynomials. Two such matrices are:

$$\begin{bmatrix} 1 & 0 & 0 & 0 & 0 & 0 \\ 0 & 1 & 0 & 0 & 0 & 0 \\ 0 & 0 & 2 & 1 & 0 & 0 \\ 0 & 0 & 0 & 2 & 1 & 0 \\ 0 & 0 & 0 & 0 & 2 & 1 \\ 0 & 0 & 0 & 0 & 0 & 2 \end{bmatrix}, \begin{bmatrix} 1 & 0 & 0 & 0 & 0 & 0 \\ 0 & 1 & 0 & 0 & 0 & 0 \\ 0 & 0 & 2 & 1 & 0 & 0 \\ 0 & 0 & 0 & 2 & 0 & 0 \\ 0 & 0 & 0 & 0 & 2 & 1 \\ 0 & 0 & 0 & 0 & 0 & 2 \end{bmatrix}.$$

Both matrices have characteristic polynomials $(x-1)^2(x-2)^4$; the minimal polynomials are $(x-1)(x-2)^4$ and $(x-1)(x-2)^2$ respectively.

Jordan's inequality For $0 < x < \pi/2$ then

$$\frac{2}{\pi} < \frac{\sin x}{x} < 1.$$

(Here the angle x is in *radians.) The inequality is often useful in estimating *contour integrals.

Josephus problem The story is told by the Jewish historian Josephus of a band of 41 rebels, who, preferring suicide to capture, decided to stand in a circle and kill every third remaining person until no one was left. Josephus, one of the number, quickly calculated where he and his friend should stand so that they would be the last two remaining and so avoid being a part of the suicide pact. In its most general form, the Josephus problem is to find the position of the one who survives when there are n people in a circle and every m-th remaining person is eliminated.

Joukovski, Nikolai Egorovich (Joukowski, Zhukovsky) (1847–1921) Russian mathematician and engineer, often labelled 'the father of Russian aviation', he was a pioneer in the early study of aerodynamics, hydrodynamics, and flight. He was the first to mathematically model aerodynamic *lift. *See* JOUKOVSKI MAP.

Joukovski map The Joukovski map, or transform, is the map $z = \zeta + \zeta^{-1}$ from the *complex plane to itself. The map has the property that a circle in the

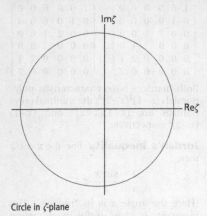

Circle in ζ-plane

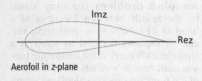

Aerofoil in z-plane

ζ-plane, not centred at zero, is mapped to an aerofoil shape in the z-plane as shown in the figures. The map was instrumental in the early study of flight.

joule The *SI unit of *work, abbreviated to 'J'. One joule is the work done when the point of application of a force of 1 *newton moves a distance of 1 metre in the direction of the force.

Julia set The boundary of the set of points z_0 in the *complex plane for which the *sequence $z_{n+1} = z_n^2 + c$ is bounded. Julia sets are almost always *fractals, and vary enormously in structure for different values of c. The *Mandelbrot set consists of those c whose Julia set is *connected; the *cardioid and different 'bulbs' of the Mandelbrot set classify the different possible structures of the Julia sets.

jump The absolute value of the difference between the *left-hand and *right-hand limits of a given function, i.e. $|(f(x_+) - f(x_-))|$.

jump discontinuity A point at which a function $f(x)$ has both *left-hand and *right-hand limits but the limits are not equal. For example, $f(x) = -1$ if $x < 0$ and $f(x) = 1$ if $x > 0$, f has a jump discontinuity at $x = 0$.

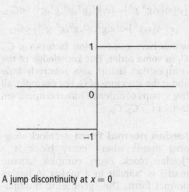

A jump discontinuity at $x = 0$

k Abbreviation for *kilo-.

k Abbreviation for *kilo (binary).

k (**_k_**) A unit vector, usually in the direction of the *z*-axis or a vertical vector when describing motion in three dimensions.

Karmarkar's algorithm A faster alternative to the *simplex method for solving large *linear programming problems. Karmarkar's algorithm runs in *polynomial time, whereas, in a worst-case scenario, the simplex method can take exponential time.

kelvin (symbol K) The temperature scale, and the unit of measurement of temperature, which takes $0°K = -273.15°C$ as absolute zero, the lowest possible temperature, at which all thermodynamic motion has ceased. Then $1°K$ equals $1°C$ as a temperature difference.

Kelvin, Lord (1824–1907) The British mathematician, physicist, and engineer, William Thomson, who made contributions to mathematics in the theories of electricity and magnetism and hydrodynamics and first identified *Stokes' Theorem.

Kendall's rank correlation coefficient *See* RANK CORRELATION.

Kepler, Johannes (1571–1630) German astronomer and mathematician, best known for *Kepler's laws of planetary motion. He proposed the first two laws in 1609 after observations had convinced him that the orbit of Mars was an

*ellipse, and the third law followed ten years later.

Kepler conjecture Now proven, the conjecture relates to optimal ways to pack spheres in three-dimensional space. The conjecture claims that the best ways are cubic close packing or hexagonal close packing, as shown in the figures. Each has a density of $\pi/(3\sqrt{2})$, which is approximately 74%.

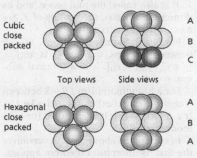

Optimal packings in 3d

Thomas Hales gave a computer-asssisted proof of the conjecture in 1998; at the time the proof consisted of 250 pages of mathematics and 3 gigabytes of data. By comparison, the optimal circle-packing arrangement in the plane was shown to be hexagonal packing in 1890 by Axel Thue; its density is $\pi/(\sqrt{12})$, which is approximately 90.7%.

Kepler's laws of planetary motion The following three laws about the motion of the planets round the Sun:

(i) The *orbit of each planet is an *ellipse with the Sun at one *focus.

(ii) The line joining a planet and the Sun sweeps out equal areas in equal times.

(iii) The squares of the times taken by the planets to complete an orbit are proportional to the cubes of the lengths of the major axes of their orbits.

These laws were formulated by Kepler, based on observations, but each can be deduced from *Newton's *inverse square law of gravitation.

(((•))) SEE WEB LINKS

• Animation which illustrates all three laws.

kernel For a *linear map $T:V \to W$ between *vector spaces, ker$T$, the kernel of T, is the subspace $\{v \in V \mid Tv = 0_W\}$ of V. It is also called the null space, and its dimension is called the *nullity of T. *See* RANK-NULLITY THEOREM.

For a *homomorphism $f:G \to H$ between *groups, the kernel is ker$f = \{g \in G \mid f(g) = e_H\}$. It is a *normal subgroup of G.

For a homomorphism $f:R \to S$ between *rings (or *modules), the kernel is ker$f = \{r \in R \mid f(r) = 0_S\}$. It is an *ideal (or submodule) of R.

In each of the above cases, a version of the first *isomorphism theorem applies.

ket Also known as *Dirac notation, the bra and ket notation highlights the difference between a *wave function being thought of as a vector or a *functional. So bra$(\phi) = \langle \phi \mid$ is a functional, in the *dual space, acting on states ψ by $\langle \phi \mid : \psi \mapsto \langle \phi, \psi \rangle$, whereas ket$(\phi) = \mid \phi \rangle$ represents a quantum state. Here $\langle , \rangle$ denotes the *inner product.

key In *cryptography, a key is a means of *encrypting the message to make the cryptogram or of decrypting the cryptogram to retrieve the message. As a simple example, a *cipher permutes the 26 letters of the alphabet and a key is a rule detailing this. This is an example of a private key, and a security issue with any such system is passing such keys between correspondents.

With public key cryptography, such as *RSA, the means of encrypting messages is in the public domain. This way a company allows a customer to send in an encrypted order. However, if knowledge of how to encrypt is public knowledge, then in principle so is knowledge of how to decrypt. For example, if one were given one half of an English-French dictionary, it would be possible, given enough time, to reconstruct the French-English half. For this reason encrypting with public keys has to have a computationally difficult inverse—they are *one-way functions—and public keys are frequently changed to avoid being decrypted.

kg Abbreviation and symbol for *kilogram.

Khwārizmī, Muhammad ibn Mūsā al- (about AD 800) Persian mathematician and astronomer. The word 'algorithm' is derived from his name. The word 'algebra' comes from the title of his work *Al-jabr wa'l muqābalah*.

kilo- Prefix used with *SI units to denote multiplication by 10^3. Abbreviated as k.

kilo- (binary) A kilobyte contains 1024 *bytes, rather than 1000. The figure of 1024, rather than 1000, is well suited to *binary, as it is 2^{10}. Similarly, mega- in binary refers to $2^{20} = 1,048,576$, giga- to $2^{30} = 1,073,741,824$, and tera- to $2^{40} = 1,099,511,627,776$. To avoid the confusion between decimal and binary versions, the prefixes kibi-, mebi-, gibi-, and tebi- have been introduced, for the binary versions, but are not widely used.

kilogram In *SI units, the base unit used for measuring *mass, abbreviated to 'kg'. One kilogram is approximately the mass of 1000cm³ of water, though from 2019 it has been precisely defined in terms of the *Planck constant.

kinematics The study of the motion of bodies without reference to the *forces causing the motion. It is concerned with such matters as the positions, the velocities and the accelerations of the bodies.

kinetic energy The *energy of a *body attributable to its motion. The kinetic energy is often denoted T. For a *particle of mass m with speed v, $T = \frac{1}{2}mv^2$. For a *rigid body rotating about a fixed axis with *angular speed ω, and with *moment of inertia I about the fixed axis, $T = \frac{1}{2}I\omega^2$. For a *rigid body in general motion, the kinetic energy is given by

$$T = \frac{1}{2}m|\mathbf{v}_G|^2 + \frac{1}{2}\boldsymbol{\omega}^T I_G \boldsymbol{\omega}$$

where m is the mass of the body, $\mathbf{v}_G$ is the velocity of the *centre of mass G, $\boldsymbol{\omega}$ is the *angular velocity, and I_G is the *inertia matrix at G.

Kinetic energy has the dimensions ML^2T^{-2}, and the SI unit of measurement is the *joule.

kinetic friction *See* FRICTION.

Kingman, Sir John Frank Charles (1939-) British mathematician who made important contributions to the theory and applications of *probability and stochastic analysis, including *operational research and population genetics, queuing theory, *measure theory, and *ergodic theory. He was president of the *LMS from 1990–2 and the first chairman of the Statistics Commission in 2000.

Kirchoff's Theorem A theorem which determines the number of *spanning trees in a *graph. Let $\mathbf{A}$ be the *adjacency matrix of a graph G with n vertices and let $\mathbf{L} = \mathbf{D} - \mathbf{A}$ where $\mathbf{D}$ is the *diagonal matrix whose entries are the *degrees of the vertices. Then the number of

spanning trees equals c_1/n, where c_1 is the coefficient of x in the *characteristic polynomial $\chi_L(x)$.

kite A *quadrilateral that has two adjacent sides of equal length and the other two sides of equal length. If the kite $ABCD$ has $AB = AD$ and $CB = CD$, the *diagonals AC and BD are *perpendicular, and AC *bisects BD.

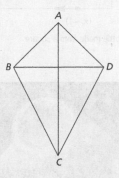

A kite

Klein, (Christian) Felix (1849-1925) German mathematician, who in his influential *Erlangen Program* launched a unification of *geometry, in which different geometries are classified by means of *group theory. In *topology, he rigorously defined what it is for a *surface to be *non-orientable and is remembered in the name of the *Klein bottle.

Klein bottle The Klein bottle is a closed *non-orientable (i.e. one-sided) *surface. Consider, in the plane, the square of all points with *Cartesian coordinates (x,y) such that $-1 \leq x \leq 1$ and $-1 \leq y \leq 1$. *Identifying each point $(x,1)$ on the top edge with the point $(x,-1)$ opposite, for all x, forms a *cylinder with two bounding circles. Identifying each point $(1, y)$ with $(-1,-y)$ glues these circles in reverse orientations. The Klein bottle is constructed by doing

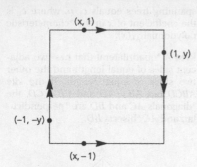

(x, 1)

(1, y)

(−1, −y)

(x, −1)

A Klein bottle made from a square

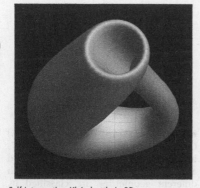

Self-intersecting Klein bottle in 3D

these two operations simultaneously. This is not practically possible in three dimensions without the Klein bottle intersecting itself but can be achieved in four dimensions. Note that the shaded area in the figure is a *Möbius strip.

(⊕) SEE WEB LINKS

• A fuller description and images of Klein bottles.

Klein four-group Essentially (i.e. up to *isomorphism), there are two *groups with four elements. One is *cyclic. The other is the Klein four-group, often

denoted V_4. It is *abelian and together with the *identity element has three commuting order two elements. It is isomorphic to the *product group $C_2 \times C_2$.

Kline, Morris (1908–92) Influential and prolific historian and popularizer of mathematics He also wrote many critiques on *mathematics education in schools and universities. Perhaps his most noted work is the comprehensive *Mathematical Thought from Ancient to Modern Times*.

knapsack problem The *bin packing problem with items of different sizes, where the items have an associated value, so as to maximize the total value. This is akin to a hiker packing a knapsack (or rucksack) with a limit on the weight they can carry, but it also models any resource allocation problem with constraints such as finances, time, or manpower. A method of solution is the *branch and bound method.

knot A *closed curve in three-dimensional space which does not intersect itself. The simplest non-trivial knot is the *trefoil. Any knot is topologically equivalent (i.e. *homeomorphic) to a circle, so the character of a knot is in how the circle has been *embedded in space.

knot equivalence Two mathematical *knots are equivalent or isotopic if one can be transformed into the other by a continuous deformation (*see* CONTINUOUS FUNCTION) of Euclidean space. The fundamental problem of knot theory is classifying knots up to equivalence. *See* CROSSING NUMBER.

knot polynomial A polynomial associated with a *knot, with the property that if two knots are *equivalent, then their polynomials are equal. Famous examples include the Alexander polynomial, introduced by James Alexander in

1923, and the Jones polynomial, introduced by Vaughan Jones in 1984.

knot theory In *topology, the study of *knots. It was first formally considered by *Gauss, while the physicist Peter Tait was the first to seek to classify knots up to *knot equivalence. Knot theory has applications in understanding the knotting of DNA and has been found more widely to have connnections with *statistical mechanics and even quantum computation. *See* CONNECTED SUM, CROSSING NUMBER.

Knuth, Donald (1938–) American computer scientist and mathematician, the creator of of the markup language *T$_E$X used for typesetting mathematics and the author of *The Art of Computer Programming*. Much of his work has related to *algorithms.

Koch curve Take an equilateral triangle as shown in the first diagram. Replace the middle third of each side by two sides of an equilateral triangle pointing outwards. This forms a six-pointed star, as shown in the second diagram. Repeat this construction to obtain the figure shown in the third diagram. The Koch curve, named after the Swedish mathematician Helge von Koch (1870–1924), is the curve obtained when this process is continued indefinitely. The interior of the curve has finite area, but the curve has infinite length. *See also* FRACTALS.

Kolmogorov, Andrei Nikolaevich (1903–87) Russian mathematician who made significant contributions in various areas of mathematics, including *topology and *dynamical systems. In *probability, his important contribution was to give the subject a rigorous foundation by using the language and notation of *measure theory. *See* PROBABILITY SPACE.

Kolmogorov–Smirnov test A *nonparametric test for testing the *null hypothesis that a given *sample has been selected from a *population with a specified *cumulative distribution function F. Let $x_1, x_2, \ldots, x_n$ be the sample values in ascending order, and let

$$z = \max_i \left| \frac{i}{n} - F(x_i) \right|.$$

If the null hypothesis is true, then z should be less than a value that can be obtained, for different significance levels, from tables.

Kolmogorov space A *topological space satisfying the T$_0$ *separation axiom.

Königsberg bridge problem *See* BRIDGES OF KÖNIGSBERG.

Kovalevskaya, Sofya (1850–91) Russian mathematician who made significant contributions in *analysis, *partial differential equations, and *dynamical systems, particularly in the study of

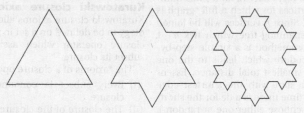

First three iterations for Koch curve

Saturn's rings. From 1871 she was taught privately by *Weierstrass in Berlin as the university would not permit her to attend classes. In 1874 she became the first woman to receive a doctorate in mathematics, and in 1884 she became a professor in Stockholm.

Kraft inequality There is an *instantaneous *binary code with n codewords of length $l_1, l_2, \ldots l_n$ if and only if $\sum_1^n 2^{-l_i} \leq 1$. This is most easily seen using a binary rooted *tree, as a codeword of length l precludes 2^{-l} of the remainder of the tree, codewords for which it would be a *prefix.

Kronecker, Leopold (1823–91) German mathematician responsible for significant contributions to *number theory and *algebra, classifying finitely generated *abelian groups. But he is more often remembered as the first to cast doubts on non-constructive existence proofs, over which he disputed with *Weierstrass and *Cantor.

Kronecker delta The two variable function δ_{ij} that takes the value 1 when $i = j$ and the value 0 otherwise. δ_{ij} is the row i, column j entry of an *identity matrix. *See* SIFTING PROPERTY.

Kronecker's Lemma If $\displaystyle\sum_{n=1}^{\infty} \frac{a_n}{n}$ converges then $\displaystyle\frac{1}{N} \sum_{n=1}^{N} a_n \to 0$ as $N \to \infty$.

Kruskal's algorithm (to solve the *minimum connector problem) This method is very effective for a small number of vertices for which a full *graph is available. Since n vertices will be joined by any *spanning tree which has $n - 1$ edges, this method is a simple step-by-step procedure which leads to the one with the smallest total distance. Essentially you start with the shortest edge (and any time there is a tie for the shortest edge, choose either one at random). At each stage add in another edge which

connects a new point to the connected tree being built up, without completing a cycle and adding the least distance to the total connected distance, i.e. the shortest edge which connects a new vertex without creating a cycle. Once all vertices have been connected, i.e. once $n - 1$ edges have been taken, the minimum connected path will have been found.

Kruskal–Wallis test A *non-parametric test for comparisons between more than two medians, using the relative rankings within the different samples.

k-simplex In *geometry, the k-simplex generalizes the notions of the *triangle and *tetrahedron from two and three dimensions to k-dimensional space. The k-simplex has $k + 1$ *vertices and is defined as the *convex hull of its vertices.

Kummer, Ernst (1810–93) German mathematician who made various contributions to mathematics but is perhaps now best remembered for his work in *abstract algebra. In 1847, Gabriel Lamé claimed to have proven *Fermat's Last Theorem, but his proof failed, as he incorrectly assumed *subrings of the *complex numbers to be *unique factorization domains. Kummer rectified this, introducing 'ideal numbers' (precursors to *ideals in *rings), and was able to prove Fermat's Last Theorem for all exponents which were 'regular' primes. In particular, this proved the theorem for all powers up to 100 except 37, 59, and 67.

Kuratowski closure axioms The Kuratowski closure axioms allow a *topology to be defined on a set in terms of a closure operator which assigns each subset its closure.

The *axioms of a closure operator are:

(i) Every subset is contained in its closure.

(ii) The closure of the closure of a subset equals the closure of the subset.

(iii) The closure of the union of two subsets is the union of their closures.

(iv) The closure of the empty set is empty.

Given a closure operator, *closed subsets can be defined as those sets which equal their closure, and the *complements of the closed subsets, i.e. the open subsets, satisfy the axioms of being a *topology.

Kuratowski's Theorem A graph is *planar if and only if it does not contain a *subdivision of either the *complete

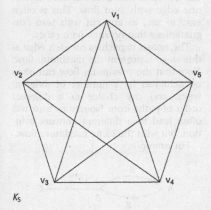

K_5

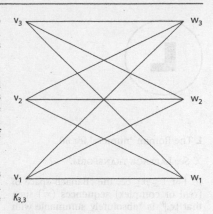

$K_{3,3}$

graph K_5 or of the *bipartite graph $K_{3,3}$ as a subgraph.

kurtosis Let $m_2 = E((X-\mu)^2)$ and $m_4 = E((X-\mu)^4)$, where $\mu = E(X)$, be the second and fourth moments about the mean. Then $\frac{m_4}{(m_2)^2}$ is the kurtosis of the distribution. For a normal distribution this has the value 3, which can be taken as a standard for comparison and hence a value of 3 is termed mesokurtic—from 'meso' meaning 'middle'. For values less than 3 the distribution is termed platykurtic, and for values greater than 3 it is leptokurtic.

L The Roman *numeral for 50.

$\mathcal{L}$ *See* LAPLACE TRANSFORM.

l^p For $1 \le p \le \infty$, the *Banach space of (real or complex) sequences (x_n) such that $|x_n|^p$ is *absolutely summable with the *p-norm $\|(x_n)\| = (\sum_1^\infty |x_n|^p)^{1/p}$. When $p = \infty$, this is understood as $\|(x_n)\| = \sup|x_n|$ for bounded sequences. When $p = 2$, l^2 is a *Hilbert space.

L^p For $1 \le p \le \infty$, the *Banach space of (real or complex) *measurable functions f such that $|f|^p$ is *Lebesgue integrable. The *norm is given by

$$\|f\| = \left(\int_\mathbb{R} |f|^p \, \mathrm{d}x \right)^{1/p}.$$

In L^p two functions that agree almost everywhere are considered equal; this is so that the norm property $\|f\| = 0$ implies $f = 0$ holds. When $p = 2$, then L^2 is the space of square-integrable functions and is a *Hilbert space.

labelling algorithm The algorithm for identifying the maximum flow possible across a *network. It is a step-by-step procedure whereby any initial flow can be increased until the maximum is found. The *max-flow/min-cut rule can tell you in advance what the maximum flow is, but the algorithm will find it for you, as well as determining the detailed flows.

You can start with any flow, including the zero flow, where the flow on every edge is zero. Each edge should be labelled with both the current flow and the excess capacity along that edge. Choose a path from source to sink, and

take the smallest of the flows along any of the edges on that path. On each edge, transfer that amount of flow from the current flow to the excess capacity. Now choose another path and repeat the process, and continue until no path can be found that does not have at least one edge with zero flow. This is often easy to see, as any cut with zero flow guarantees that no such path exists.

The excess capacities on each edge at this stage represent the maximum flow. Note that the maximum flow can often be achieved in a number of different ways, and the choice of a different order of paths from *source to *sink will often lead to a different network solution, but with the same maximum flow.

For example,

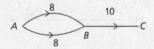

In this very simple network the maximum flow occurs when a total of 10 units flow along the two edges between A and B. However, any x satisfying $2 \le x \le 8$ and $10 - x$ along the two edges will be a maximum flow.

For the network shown here, an initial zero flow is shown in bold, so the excess capacity currently on each edge is the *capacity of the edge.

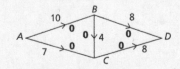

Starting with the path ABD the smallest flow on the route is 8, so transfer 8 to the excess capacity on both AB and BD.

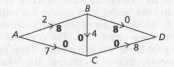

Now choose path ACD which has smallest flow 7, giving

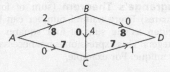

The only route left is $ABCD$ with smallest flow 1, giving

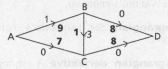

Since all routes into the sink at D have 0 excess capacity, it is obvious immediately that no further increases would be available and the maximum flow is 16, and the bold figures show a way of achieving this maximum flow. If $ABCD$ had been chosen as the second path, before ACD, then 10 would have gone along AB, 2 along BC, and 6 along AC instead of this solution.

lag The time difference between the pairs of *observations used in calculating the *autocorrelation of a sequence. For a monthly *time series the important lags are likely to be 1 (the most recent performance), 3, 6, and 12, reflecting the performance at corresponding points in the business cycle.

Lagrange, Joseph-Louis (1736–1813) Arguably the greatest, alongside *Euler, of 18th-century mathematicians. Although he was born in Turin and spent the early part of his life there, he eventually settled in Paris and is normally deemed to be French. Much of his important work was done in Berlin, where he was Euler's successor at the Academy. His work covers the whole range of mathematics. He is known for results in *number theory and *algebra but is probably best remembered as a leading figure in the development of theoretical *mechanics. His work *Mécanique analytique*, published in 1788, is a comprehensive account of the subject. In particular, he was mainly responsible for the methods of the *calculus of variations and the consequent Lagrangian method in mechanics.

Lagrange interpolation If a function has n known values (x_1, y_1), (x_2, y_2), ..., (x_n, y_n), where the x_i are distinct, then there is s unique polynomial $f(x)$ of degree less than n such that $f(x_i) = y_i$ for each i. It is given by the formula

$$f(x) = \frac{y_1(x - x_2)\ldots(x - x_n)}{(x_1 - x_2)(x_1 - x_3)\ldots(x_1 - x_n)} + \ldots$$
$$+ \frac{y_n(x - x_1)\ldots(x - x_{n-1})}{(x_n - x_1)(x_n - x_2)\ldots(x_n - x_{n-1})}$$

Lagrange multiplier A method of evaluating *maxima and *minima of a function, f, where one or more constraints $g_i = 0$ have to be satisfied. A new function is constructed as

$$L = f + \lambda_1 g_1 + \lambda_2 g_2 + \ldots + \lambda_n g_n.$$

Then the *partial derivatives of L with respect to the original variables and each λ_i are taken, and the *stationary points found by solving the set of simultaneous equations obtained by setting each derivative to zero.

For example, to maximize $f(x,y) = 3x + 4y$ subject to $g(x,y) = x^2 + y^2 - 1 = 0$, we define

$$L(x, y, \lambda) = 3x + 4y + \lambda(x^2 + y^2 - 1),$$

and the equations

$$0 = \frac{\partial L}{\partial x} = 3 + 2\lambda x,$$

$$0 = \frac{\partial L}{\partial y} = 4 + 2\lambda y,$$

$$0 = \frac{\partial L}{\partial \lambda} = x^2 + y^2 - 1$$

have solutions $(x,y) = (3/5, 4/5)$, a maximum, and $(x,y) = (-3/5, -4/5)$, a minimum.

Lagrange point In *mechanics, a Lagrange point is a position near two large orbiting bodies where the gravitational *forces cancel out, so that a small object at a Lagrange point would be at *equilibrium relative to the bodies' *centre of mass. There are five such points, three of which are collinear with the centres of the bodies and unstable, the other two coplanar with the motion of the bodies and *stable.

Lagrange's equations For a mechanical system with n *degrees of freedom in which *conservative forces act and all constraints are *holonomic, the Lagrangian is $L = T - V$, where T denotes *kinetic energy and V denotes *potential energy. For a light pendulum of length l with a mass m at the end and making an angle θ with the downward vertical,

$$L = \frac{1}{2} ml^2 \dot{\theta}^2 + mgl \cos\theta.$$

For such a system, defined by *generalized coordinates $q_1, q_2, \ldots, q_n$, Lagrange's equations state that

$$\frac{d}{dt}\left(\frac{\partial L}{\partial \dot{q_i}}\right) - \frac{\partial L}{\partial q_i} = 0 \quad \text{for } 1 \leq i \leq n.$$

In the example of the pendulum, this gives

$$0 = \frac{d}{dt}(ml^2 \dot{\theta}) + mgl \sin\theta$$

$$= ml^2 \ddot{\theta} + mgl \sin\theta.$$

See also HAMILTONIAN MECHANICS.

Lagrange's Theorem (group theory) For a *group G with *subgroup H, the (left or right) *cosets of H *partition G and are *equinumerous. Thus, if G is finite, the *order of H divides the order of G, and the order of any group element divides the order of G. The converse is true for finite *abelian groups, but is not generally true—for example, the *alternating group A_4 has order 12 and no subgroup of order 6.

Lagrange's Theorem (sum of four squares) Every natural number can be written as the sum of four squares of integers. This expression is not necessarily unique. For example,

$$1 = 1^2 + 0^2 + 0^2 + 0^2,$$
$$10 = 3^2 + 1^2 + 0^2 + 0^2 \text{ or } 2^2 + 2^2 + 1^2 + 1^2.$$

See WARING'S PROBLEM.

Lagrangian See LAGRANGE'S EQUATIONS.

Lagrangian derivative A synonym for CONVECTIVE DERIVATIVE.

Lakatos, Imre (1922–74) Hungarian philosopher of mathematics, best known for his book *Proofs and Refutations* in which he discusses the nature of progress in mathematical understanding. There Lakatos, as a challenge to *formalism, posits that theories evolve in light of so-called 'local counterexamples', by which he means that formal definitions and statements need to be tweaked to include or exclude such a refutation.

Lambert, Johann Heinrich (1728–77) Swiss–German mathematician, scientist, philosopher, and writer on many subjects, who in 1761 proved that π is an *irrational number. He introduced the standard notation for *hyperbolic functions, investigated hyperbolic triangles, and came close to discovering *non-Euclidean geometry.

lamina An object considered as having a 2-dimensional shape and density but having no thickness. It is used in a *mathematical model to represent an object such as a thin plate or sheet.

Lami's Theorem Suppose that three *forces act on a particle and that the *resultant of the three forces is zero. Then the *magnitude of each force is proportional to the *sine of the angle between the other two forces.

Landau's problems At the 1912 *ICM, Edmund Landau listed four problems about *prime numbers which he considered unapproachable using the mathematics of the time. These were *Goldbach's conjecture, the *twin prime conjecture, *Legendre's conjecture, and the conjecture that there are infinitely many primes of the form $n^2 + 1$. While there has been progress on these conjectures, they are all still *open problems.

Langlands, Robert (1936–) Canadian mathematician who was awarded the *Abel Prize in 2018 'for his visionary program connecting *representation theory to *number theory'.

Langlands programme The Langlands programme is a collection of far-sighted, hugely influential conjectures connecting *number theory and *geometry proposed by Robert *Langlands in 1967. The profound *modularity theorem is subsumed within the programme.

Laplace, Pierre-Simon, Marquis de (1749–1827) French mathematician, often referred to as France's *Newton, best known for his work on planetary motion, enshrined in his five-volume *Mécanique céleste*, and for his fundamental contributions to the theory of *probability. It was Laplace who extended Newton's gravitational theory to the study of the whole solar system. He developed the strongly deterministic view that, once the starting conditions of a closed *dynamical system such as the universe are known, its future development is then totally determined.

Laplace expansion *See* DETERMINANT.

Laplace's equation The partial differential equation

$$\nabla^2 V = \frac{\partial^2 V}{\partial x^2} + \frac{\partial^2 V}{\partial y^2} + \frac{\partial^2 V}{\partial z^2} = 0 \quad \text{where}$$

$\nabla^2 = \nabla \cdot \nabla =$ div grad is the Laplacian *differential operator. This equation is important in *potential theory. The *inhomogeneous version of Laplace's equation, $\nabla^2 V = f$, is called Poisson's equation.

Laplace transform An *integral transform $\mathcal{L}$ of a function $f(x)$ into another function $\mathcal{L}f$ or $\bar{f}$ of a different variable p. The Laplace transform $\bar{f}(p)$ of $f(x)$ is defined as

$$\bar{f}(p) = \int_0^\infty f(x)\mathrm{e}^{-px} \, \mathrm{d}x.$$

The transform is useful in solving differential equations, as it handles derivatives well; for example, the transform of $f'(x)$ equals $p\bar{f}(p) - f(0)$. At its simplest, a differential equation in $f(x)$ is transformed into an algebraic equation involving $\bar{f}(p)$ which is solved to find $\bar{f}(p)$; by recognizing $\bar{f}(p)$ as the transform of a function, or by applying an inverse transform theorem, the solution $f(x)$ can then be found. For a list of Laplace transforms, *see* APPENDIX 10. *See* also FOURIER TRANSFORM.

Laplacian *See* LAPLACE'S EQUATION.

latent root A synonym for EIGENVALUE.

latent vector A synonym for EIGENVECTOR.

LaTeX *See* TeX.

Latin square A square array of symbols arranged in rows and columns in such a way that each symbol occurs exactly once in each row and once in each column. Two examples are given in the figure. A *Cayley table is a Latin square, though the converse is not generally true.

A	B	C	D	A	B	C	D
B	A	D	C	D	C	B	A
C	D	A	B	C	D	A	B
D	C	B	A	B	A	D	C

latitude Suppose that the *meridian through a point P on the Earth's surface meets the equator at P'. Let O be the centre of the Earth. The latitude of P is then the angle $P'OP$, measured in degrees north or south. Latitude and *longitude uniquely fix the position of a point on the Earth's surface. The term also describes the circle of points of a given latitude; and more generally, on a *surface of revolution, a latitude is any circle produced by the rotation of a point of the generating curve of the surface.

lattice (set theory) A *partially ordered set X in which every pair of elements x,y has a *least upper bound $x \vee y$ and a *greatest lower bound $x \wedge y$. The lattice is complete if every subset of X has a least upper bound and greatest lower bound. A *power set with $\vee = \cup$ and $\wedge = \cap$ is a complete lattice. $\mathbb{N}$ with $\leq \, = \mid$ (*divides), $\vee = $ *lcm, and $\wedge = $ *hcf is an incomplete lattice as the elements of $\mathbb{N}$ have no upper bound.

lattice (geometry) *Linearly independent vectors $\mathbf{v}_1, \ldots, \mathbf{v}_k$ in $\mathbb{R}^n$ generate a lattice of vectors $a_1\mathbf{v}_1 + \cdots + a_k\mathbf{v}_k$ where $a_1, \ldots, a_k$ are integers. $\mathbb{Z}^n$ in $\mathbb{R}^n$ is often referred to as the integer lattice or cubic lattice; it is the lattice generated by the *canonical basis.

latus rectum The *chord through the focus of a *parabola and perpendicular to the axis. For the parabola $y^2 = 4ax$, the latus rectum has length $4a$.

Laurent expansion For a function $f(z)$ which is *holomorphic within the *annulus centred at a in the complex plane, defined by $r_1 < |z - a| < r_2$ the Laurent expansion is

$$f(z) = \sum_{k=-\infty}^{\infty} c_k (z - a)^k$$

where

$$c_k = \int_C \frac{f(w)}{(w - a)^{k+1}} \, dw$$

with the *path integral being around C, the *positively oriented circle $|z - a| = r$ where $r_1 < r < r_2$. *Compare* TAYLOR'S THEOREM; *see* ISOLATED SINGULARITY.

law A general theorem or principle. For example, *Newton's laws of motion and the *weak law of large numbers.

law of averages False conception that an event becomes more likely if it has been under-represented so far in a sequence of *observations. For example, if a fair coin has been tossed six times, and has come up with five heads and only one instance of tails, the 'law of averages' incorrectly suggests that tails is more likely on the next toss. But as each toss is *independent, the probability of tails remains unchanged.

laws of large numbers (in statistics) The generic name given to a group of theorems relating to the behaviour of $\frac{S_n}{n} = \frac{1}{n}(X_1 + X_2 \ldots + X_n)$ for large n, where $\{X_i\}$ are identically distributed, *independent *random variables, usually identically distributed. *See* CENTRAL LIMIT THEOREM, STRONG LAW OF LARGE NUMBERS, WEAK LAW OF LARGE NUMBERS.

lcm An abbreviation for LEAST COMMON MULTIPLE.

leading coefficient The *coefficient of the term of highest exponent in a *polynomial so if $f(x) = 7x^2 - 3x + 2$, the leading coefficient of $f(x)$ is 7.

leading diagonal A synonym for MAIN DIAGONAL.

least action, principle of Consider a *dynamical system with *generalized coordinates $q_1, q_2, \ldots, q_n$, solely acted on by *conservative forces, with *Lagrangian L. The action as the system moves from position at time t_1 to a position at time t_2 equals

$$S = \int_{t_1}^{t_2} L(\mathbf{q}, \dot{\mathbf{q}}, t) \; dt.$$

The principle of least action states that a trajectory of the dynamical system is an actual motion of the system if and only if the value of S is a *stationary value. *See also* CALCULUS OF VARIATIONS.

least common denominator The *least common multiple of the *denominators of a set of numerical or algebraic *fractions that are to be added. For example, 12 and 8 have an lcm of 24 and hence $\frac{5}{12} + \frac{3}{8} = \frac{10}{24} + \frac{9}{24} = \frac{19}{24}$.

least common multiple For two non-zero *integers a and b, an integer that is a multiple of both is a *common multiple. Of all the positive common multiples, the least is the least common multiple (lcm), denoted by $[a,b]$ or $\text{lcm}(a,b)$. The lcm of a and b has the property of dividing any other common multiple of a and b. If the *prime decompositions of a and b are known, the lcm is easily obtained: for example, if $a = 168 = 2^3 \times 3 \times 7$ and $b = 180 = 2^2 \times 3^2 \times 5$, then the lcm is $2^3 \times 3^2 \times 5 \times 7 = 2520$. For positive integers a and b, the lcm is equal to $ab/(a, b)$, where (a,b) is the *greatest

common divisor (gcd). The gcd can be efficiently found using the *Euclidean Algorithm.

Similarly, any finite set of non-zero integers, $a_1, a_2, \ldots, a_n$ has an lcm denoted by $[a_1, a_2, \ldots, a_n]$.

least squares The method of least squares is used to *estimate *parameters in statistical models such as those that occur in *regression. Estimates for the parameters are obtained by minimizing the sum of the squares of the differences between the observed values and the predicted values under the model. For example, suppose that, in a case of linear regression, where $Y = \alpha + \beta X + \varepsilon$, there are n paired observations (x_1, y_1), $(x_2, y_2), \ldots, (x_n, y_n)$. Then the method of least squares gives a and b as *estimators for α and β, where a and b are chosen so as to minimize $e(a,b) = \sum (y_k - a - bx_k)^2$. Solving the *simultaneous equations $\partial e/\partial a = 0 = \partial e/\partial b$ gives

$$b = \frac{\sum_1^n (x_k - \bar{x})(y_k - \bar{y})}{\sum_1^n (x_k - \bar{x})^2},$$

and $a = \bar{y} - b\bar{x}$, where $\bar{x}$ and $\bar{y}$ denote the *means of the x_k and y_k respectively.

least squares theorem A synonym for GAUSS–MARKOV THEOREM.

least upper bound A synonym for SUPREMUM.

least value *See* GLOBAL MINIMUM.

leaving variable *See* SIMPLEX METHOD.

Lebesgue, Henri (Léon) (1875–1941) French mathematician who revolutionized the theory of *integration. Building on the earlier work of *Borel and *Jordan, he developed the *Lebesgue integral, one of the outstanding concepts in modern *analysis. His two major texts on the subject were published in the first few years of the 20th century.

Lebesgue integration (Lebesgue measure) A theory of *integration and *measure which greatly extended the *Riemann integral. Riemann integrable functions are *bounded and defined on bounded *intervals, whereas Lebesgue's theory was much more general and also included powerful convergence theorems (*see* DOMINATED CONVERGENCE THEOREM, MONOTONE CONVERGENCE THEOREM). The existence of a set or function which is not Lebesgue measurable requires axioms beyond the *Zermelo-Fraenkel axioms, such as the *axiom of choice (*see* VITALI SET).

left and right derivative For the *real function f, if

$$\lim_{h \to 0-} \frac{f(a+h) - f(a)}{h}$$

exists, this limit is the left derivative of f at a. Similarly, if

$$\lim_{h \to 0+} \frac{f(a+h) - f(a)}{h}$$

exists, this limit is the right derivative of f at a. (*See* LIMIT FROM THE LEFT AND RIGHT.) The *derivative $f'(a)$ exists if and only if the left derivative and the right derivative of f at a exist and are equal. An example where the left and right derivatives both exist but are not equal is provided by the function f, where $f(x) = |x|$ for all x. At 0, the left derivative equals -1 and the right derivative equals $+1$.

left-handed system *See* RIGHT-HANDED SYSTEM.

left inverse A left inverse $g{:}Y{\to}X$ of a map $f{:}X{\to}Y$ satisfies $g(f(x)) = x$ for all x in X. If X is non-empty, the existence of a left inverse is equivalent to f being *one-to-one. *See* INVERSE FUNCTION, RIGHT INVERSE.

Legendre, Adrien-Marie (1752–1833) French mathematician who, with

*Lagrange and *Laplace, formed a trio associated with the period of the French Revolution. He was well known in the 19th century for his highly successful textbook on the *geometry of *Euclid. But his real work was concerned with *calculus. He was responsible for the classification of *elliptic integrals into their standard forms. The so-called *Legendre polynomials, solutions of *Legendre's differential equation, are among the most important of the *special functions. In an entirely different area, he, along with *Euler, conjectured and partially proved an important result in *number theory known as the law of *quadratic reciprocity.

Legendre polynomials A set of *polynomial solutions $\{P_n(x)\}$ to *Legendre's differential equation are generated by the expansion of

$$\frac{1}{\sqrt{(1 - 2xt + t^2)}}$$

as the coefficients of t^n.

Alternatively they can be found by *Rodrigues' formula

$$P_n(x) = \frac{1}{2^n n!} \frac{d^n}{dx^n} (x^2 - 1)^n.$$

Legendre's conjecture Currently an *open problem, the conjecture states that there is a *prime number between every pair of consecutive square numbers. *See* LANDAU'S PROBLEMS.

Legendre's differential equation The differential equation

$$(1 - x^2)y'' - 2xy' + n(n+1)y = 0$$

where n is a *natural number. The *polynomial solutions are the *Legendre polynomials.

Legendre symbol For an odd *prime p and integer a, the Legendre symbol $(a|p)$ is defined to be -1 if $x^2 \equiv a \bmod p$

has no solution, 0 if p divides a, 1 if $x^2 \equiv a$ mod p has a solution and p does not divide a. Thus, the number of solutions mod p to $x^2 \equiv a$ is $1 + (a|p)$. *See* QUADRATIC RECIPROCITY.

Leibniz, Gottfried Wilhelm (1646–1716) German mathematician, philosopher, scientist, and writer on a wide range of subjects, who was, with *Newton, the founder of the *calculus. Newton's discovery of differential calculus was perhaps ten years earlier than Leibniz's, but Leibniz was the first to publish his account, written independently of Newton, in 1684. Soon after, he published an exposition of integral calculus that included the *Fundamental Theorem of Calculus. He also wrote on other branches of mathematics, making significant contributions to the development of *symbolic logic.

Leibniz's integral rule The following rule for differentiating integrals:

$$\frac{d}{dt} \int_{a(t)}^{b(t)} f(x,t) \ dx = f\Big(b(t),t\Big)b'(t)$$
$$-f\Big(a(t),t\Big)a'(t) + \int_{a(t)}^{b(t)} \frac{\partial f}{\partial t}(x,t) \ dx.$$

If $a(t)$ and $b(t)$ are constant, then this amounts to *differentiation under the integral sign. *Compare* REYNOLD'S TRANSPORT THEOREM.

Leibniz's Theorem If $h(x) = f(x)g(x)$ for all x, the nth derivative of h is given by

$$h^{(n)}(x) = \sum_{r=0}^{n} \binom{n}{r} f^{(r)}(x)g^{(n-r)}(x),$$

where the coefficients $\binom{n}{r}$ are *binomial coefficients.

For example, to find $h^{(8)}(x)$, when $h(x) = x^2 \sin x$, let $f(x) = x^2$ and $g(x) = \sin x$. Then $f'(x) = 2x$ and $f''(x) = 2$, with higher derivatives being zero; and

$g^{(8)}(x) = \sin x$, $\quad g^{(7)}(x) = -\cos x$ and $g^{(6)}(x) = -\sin x$. So

$$h^{(8)}(x) = x^2 \sin x + \binom{8}{1}2x(-\cos x)$$
$$+ \binom{8}{2}2(-\sin x)$$
$$= x^2 \sin x - 16x\cos x - 56\sin x.$$

lemma A mathematical statement proved primarily for use in the proof of a subsequent, more significant *theorem. Exceptionally, there are some named results that are nonetheless referred to as lemmas: *Gauss' Lemma, *Steinitz Exchange Lemma, *Urysohn's Lemma, and *Zorn's Lemma.

length (of a line segment) The length $|AB|$ of the *line segment AB, and the length $|\overrightarrow{AB}|$ of the *directed line segment, is equal to the *distance between A and B. It is equal to zero when A and B coincide and is otherwise always positive.

length (of a vector) *See* VECTOR.

length of an arc *See* ARC LENGTH.

Leonardo da Vinci (1452–1519) Italian polymath best known for his magnificent paintings such as the *Mona Lisa* and the *Last Supper*, Leonardo was also prolific as an architect, inventor, scientist, and mathematician where he was particularly interested in *geometry, *mechanics, *aerodynamics, and *hydrodynamics.

SEE WEB LINKS
• An online exhibition about da Vinci.

Leonardo of Pisa *See* FIBONACCI.

leptokurtic *See* KURTOSIS.

level set Given a real-valued *function f on a set S, a level set is a subset of S of the form

$$\{s \ \varepsilon \ S | f(s) = c\},$$

for a given real constant c. For example, with $f(x,y,z) = x^2 + y^2 + z^2$, the level sets (which may be termed level surfaces in this case) are *spheres when $c > 0$.

lever A rigid bar or rod free to rotate about a point or axis, called the fulcrum, normally used to transmit a *force at one point to a force at another. In diagrams the fulcrum is usually denoted by a small

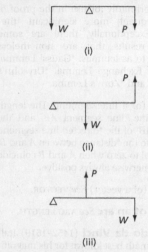

(i)

(ii)

(iii)

triangle. In the three examples shown in the figure, the lever is being used as a machine in which the effort is the force **P** and the load is **W**. In each case, the mechanical advantage can be found by using the *principle of moments. Every-day examples corresponding to the three kinds of lever shown are (i) an oar as used in rowing, (ii) a stationary wheelbarrow, and (iii) the jib of a certain kind of crane.

L'Hôpital, Guillaume François Antoine, Marquis de (1661–1704) French mathematician who in 1696 produced the first textbook on *differential calculus. This and a subsequent book on *analytic geometry were standard texts for much of the 18th century. The first contains *L'Hôpital's rule, known to be

due to Jean *Bernouilli, who is thought to have agreed to keep the Marquis de L'Hôpital informed of his discoveries in return for financial support.

L'Hôpital's rule A rule for evaluating *indeterminate forms.

Theorem Suppose that $f(x) \to 0$ and $g(x) \to 0$ as $x \to a$. Then

$$\lim_{x \to a} \frac{f(x)}{g(x)} = \lim_{x \to a} \frac{f'(x)}{g'(x)}$$

if the limit on the right-hand side exists.

For example,

$$\lim_{x \to a} \frac{\sqrt{1+x} - 1}{x} = \lim_{x \to a} \frac{\frac{1}{2}(1+x)^{-1/2}}{1} = \frac{1}{2}.$$

The result also holds if $f(x) \to \infty$ and $g(x) \to \infty$ as $x \to a$. Moreover, the rule applies if '$x \to a$' is replaced by '$x \to +\infty$' or '$x \to -\infty$'.

LHS Abbreviation for left-hand side, especially in reference to an *equation, *identity, *congruence, or *inequality. As opposed to *RHS, the right-hand side.

liar paradox The paradox that arises from considering the truth of the statement 'This statement is false'.

Lie, (Marius) Sophus (1842–99) (pronounced 'lee') Norwegian mathematician responsible for advances in *differential equations and *differential geometry, and primarily remembered for a three-volume treatise on transformation groups. Like *Klein, he brought *group theory to bear on geometry. *See* LIE CORRESPONDENCE.

Lie algebra A *vector space V together with a Lie bracket, a map $V \times V \to V$ denoted by $[x,y]$ where $x,y \in V$, which is *bilinear, alternating so that $[x,x] = 0$ for all x, and which satisfies *Jacobi's identity. One example is $V = \mathbb{R}^3$ with the Lie bracket being the *vector product. For any *Lie group, the *tangent space at the *identity element is a Lie algebra. The Lie algebra of a Lie group G is often denoted in gothic as $\mathfrak{g}$.

Lie correspondence Sophus *Lie proved three important theorems about *Lie groups and *Lie algebras. The third theorem details the correspondence between the two and states that every finite-dimensional real Lie algebra is the Lie algebra of some *simply connected Lie group. There is also a correspondence between *homomorphisms of algebras and groups and between subalgebras and *subgroups.

For example, let G denote the group of $n \times n$ matrices with positive *determinant. This has Lie algebra $M_{n \times n}(\mathbb{R})$, the space of $n \times n$ matrices with Lie bracket $[\mathbf{A},\mathbf{B}] = \mathbf{AB}\text{-}\mathbf{BA}$. The exponential map $\exp\colon M_{n \times n}(\mathbb{R}) \to G$, taking a matrix to its *exponential, has an important role in Lie's theory. Now the determinant map $\det\colon G \to (0,\infty)$ is a homomorphism of Lie groups, which must correspond to a homomorphism of the Lie algebras $M_{n \times n}(\mathbb{R}) \to \mathbb{R}$; in this case that homomorphism is *trace. Further, as

$$
\begin{array}{ccc}
M_{n \times n}(\mathbb{R}) & \overset{\exp}{\to} & G \\
\text{tr} \downarrow & & \downarrow \det \\
\mathbb{R} & \overset{\exp}{\to} & (0,\infty)
\end{array}
$$

is a *commutative diagram, this means that

$$\det(\exp\mathbf{A}) = \exp(\text{trace}\mathbf{A}).$$

Lie group A Lie group G is a *group G that also has the structure of a smooth *manifold. Further, the group operations of multiplication $G \times G \to G$ and inversion $G \to G$ must be smooth. An example is S^1 considered as unit *modulus *complex numbers under multiplication. Real and complex *matrix groups are further examples. The *Lie correspondence connects Lie groups and *Lie algebras, and there is a rich *representation theory of *compact Lie groups largely due to *Weyl. *See also* TOPOLOGICAL GROUP.

life tables Tables that show life expectancies from current age for different groups in the population. While these tables are based on historical observations, they are adjusted according to expert opinion to take account of new conditions such as improving health care, increasing incidence of cancer, etc. They form part of the basis on which actuaries carry out their analysis for insurance companies when setting premiums and annuity rates.

⊕ SEE WEB LINKS
• Gives access to the latest life tables.

lift *See* AERODYNAMIC DRAG.

light In *mechanics, an object such as a string or a rod is said to be light if its weight may be considered negligible compared with that of other objects involved.

light framework A set of light rods that make a rigid framework. As the masses of the rods are negligible the forces acting in the rods can be determined by considering the forces acting at the joints.

Lighthill, Sir Michael James (1924–98) British mathematician, and Lucasian Professor of Mathematics at Cambridge for ten years between Paul *Dirac and Stephen *Hawking. His main interests lay in *fluid mechanics, and his work was important in projects as diverse as the Thames Barrier and Concorde and in the reduction of noise from jet engines as well as biofluiddynamics.

light year The distance travelled by light in 1 year. A light year is approximately 9.46×10^{12} km. Because of the vast distances in space, the light year is the basic unit of distance in astronomy.

likelihood function For a *sample from a *population with an unknown *parameter, the likelihood function is the *probability that the sample could have occurred at random; it is a function of the unknown parameter. The method of the *maximum likelihood estimator selects the value of the parameter that maximizes the likelihood function. The concept applies equally well when there are two or more unknown parameters.

likelihood ratio test A sophisticated *hypothesis test which compares the likelihood of the observed value happening for different possible values of the parameter under investigation. In its simplest form, the comparison is between two possible values, but more generally the alternatives are ranges of the *parameter value and the comparison is for maximum likelihoods in the interval.

like terms Terms in an algebraic expression which are the same in relation to all variables and their powers and which are separated by addition or subtraction signs. Like terms can then be combined into a single term. So in $3x^2y + 6xy - 5xy^2 - 2xy$ there are two like terms, the terms involving xy, but the other two terms are not 'like terms'. This expression could be simplified to $3x^2y + 4xy - 5xy^2$ but no further.

lim inf Abbreviation for *limit inferior.

limit (of a function) Informally, the limit, if it exists, of a real *function $f(x)$ as x tends to a is a number l with the property that, as x gets closer to a, $f(x)$ gets closer to l. This is written

$$\lim_{x \to a} f(x) = l.$$

It is important to realize that this limit may not equal $f(a)$; indeed, $f(a)$ may not necessarily be defined.

More precisely, $f(x)$ tends to l as x tends to a, written $f(x) \to l$, as $x \to a$, if, given any $\varepsilon > 0$ (however small), there exists $\delta > 0$ (which may depend on ε) such that, for all x, except possibly a itself, lying between $a - \delta$ and $a + \delta$, $f(x)$ lies between $l - \varepsilon$ and $l + \varepsilon$.

Notice that a itself may not be in the *domain of f. For example, let f be the function defined by

$$f(x) = \frac{\sin x}{x} \quad (x \neq 0).$$

Then 0 is not in the domain of f, but it can be shown that

$$\lim_{x \to 0} \frac{\sin x}{x} = 1.$$

If a is in the domain of f, the f is continuous (see CONTINUOUS FUNCTION) at a if the limit of f at a is $f(a)$.

In the above, l is a real number. We write $f(x) \to \infty$ as $x \to a$ if, given any K (however large), there is a positive number δ (which may depend on K) such that, for all x, except possibly a itself, lying between $a - \delta$ and $a + \delta$, $f(x)$ is greater than K. For example, $1/x^2 \to \infty$ as $x \to 0$. There is a similar definition for $f(x) \to -\infty$ as $x \to a$.

If $f : M \to N$ is a function between *metric spaces, we write that f has limit l if given any $\varepsilon > 0$, there exists $\delta > 0$ such that whenever $0 < d_M(x, a) < \delta$, then $d_N(f(x), l) < \varepsilon$.

See ALGEBRA OF LIMITS.

limit (of a sequence) Informally, the limit, if it exists, of an infinite real sequence $a_1, a_2, a_3, \ldots$ is a number l with the property that a_n gets closer to l as n gets indefinitely large.

More precisely, the sequence $a_1, a_2, a_3, \ldots$ has the limit l if, given any $\varepsilon > 0$ (however small), there is a number N (which may depend on ε) such that, for all $n > N$, a_n lies between $l - \varepsilon$ and $l + \varepsilon$. This is written $a_n \to l$. A sequence's limit, if it exists, is unique.

For example, the sequence $0, \frac{1}{2}, \frac{3}{4}, \frac{7}{8}, \frac{15}{16}, \ldots$, has limit 1, and the sequence $-1, \frac{1}{2}, -\frac{1}{3}, \frac{1}{4}, -\frac{1}{5}, \ldots$ has the limit 0; since this is the sequence whose nth term is $(-1)^n/n$; this fact can be stated as $(-1)^n/n \to 0$.

There are, of course, real sequences that do not have a limit. These can be classified into different kinds.

(i) a_n tends to ∞, written $a_n \to \infty$ if, given any K (however large), there is an integer N (which may depend on K) such that, for all $n > N$, $a_n > K$. For example, $a_n \to \infty$ for the sequence $a_n = n^2$.

(ii) There is a similar definition for $a_n \to -\infty$, and an example is the sequence $-4, -5, -6, \ldots$, in which $a_n = -n - 3$.

(iii) The sequence does not have a limit but is *bounded, such as the sequence $-\frac{1}{2}, \frac{2}{3}, -\frac{3}{4}, \frac{4}{5}, \ldots$, in which $a_n = (-1)^n n/(n+1)$.

(iv) The sequence is not bounded, but it is not the case that $a_n \to \infty$ or $a_n \to -\infty$. The sequence $1, 2, 1, 4, 1, 8, 1, \ldots$ is an example.

If a sequence a_n converges to l, then all *subsequences of a_n also converge to l.

More generally, a sequence a_n in a *metric space M converges to a limit l if $d(a_n, l) \to 0$ as $n \to \infty$. Thus a complex sequence (*see* COMPLEX NUMBER) a_n converges to the complex number l if $|a_n - l| \to 0$ as $n \to \infty$. This is equivalent to $\mathrm{Re}(a_n) \to \mathrm{Re}(l)$ and $\mathrm{Im}(a_n) \to \mathrm{Im}(l)$.

Sequential convergence determines the *topology of a metric space, in the sense that a point x is in the *closure of a set A if there exists a sequence a_n in A which converges to x. This is not true more generally in *topological spaces.

See ALGEBRA OF LIMITS, BOLZANO-WEIERSTRASS THEOREM.

limit from the left and right

The statement that $f(x)$ tends to l as x tends to a from the left can be written: $f(x) \to l$ as $x \to a^-$. Another way of writing this is

$$\lim_{x \to a^-} f(x) = l.$$

The formal definition says that this is so if, given $\varepsilon > 0$, there is a number $\delta > 0$

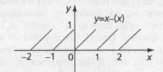

The graph of $x - [x]$

such that, for all x strictly between $a - \delta$ and a, $f(x)$ lies between $l - \varepsilon$ and $l + \varepsilon$. In place of $x \to a^-$, some authors use $x \nearrow a$. In the same way, the statement that $f(x)$ tends to l as x tends to a from the right can be written: $f(x) \to l$ as $x \to a^+$. Another way of writing this is

$$\lim_{x \to a^+} f(x) = l.$$

The formal definition says that this is so if, given $\varepsilon > 0$, there is a number $\delta > 0$ such that, for all x strictly between a and $a + \delta$, $f(x)$ lies between $l - \varepsilon$ and $l + \varepsilon$. In place of $x \to a^+$, some authors use $x \searrow a$. For example, if $f(x) = x - [x]$, then

$$\lim_{x \to 1^-} f(x) = 1, \quad \lim_{x \to 1^+} f(x) = 0.$$

limit inferior *See* LIMIT SUPERIOR.

limit of integration

In the definite *integral

$$\int_a^b f(x) \, dx,$$

the lower limit (of integration) is a and the upper limit (of integration) is b.

limit point *See* ACCUMULATION POINT.

limit superior

Given a real *sequence (a_n) which is *bounded above, the sequence $b_n = \sup\{a_n, a_{n+1}, a_{n+2}, \ldots\}$ is *decreasing and converges to a limit, the limit superior, denoted as $\limsup a_n$. Similarly if (a_n) is *bounded below, the sequence $c_n = \inf\{a_n, a_{n+1}, a_{n+2}, \ldots\}$ is *increasing and converges to a limit, the limit inferior, denoted as $\liminf a_n$.

If b_n or c_n tends to $\pm\infty$, then we define limsup a_n or liminf a_n accordingly. Equivalently, limsup a_n (liminf a_n) is the greatest (least) *limit point of the sequence (a_n). Note limsup a_n = liminf a_n if and only if the sequence a_n converges (including to $\pm\infty$).

lim sup Abbreviation for *limit superior.

Lindemann, (Carl Louis) Ferdinand von (1852–1939) German mathematician, professor at Königsberg and then at Munich, who proved in 1882 that π is a *transcendental number.

line Defined in *Euclid as a length without breadth. The term 'straight line' is used where appropriate. However, where no ambiguity results, the word straight is omitted. Similarly, a line is usually taken as extending indefinitely in both directions, and the term 'line segment' is used to specify the part of a line joining two points.

line (in two dimensions) A line can be determined by two points on the line or by a single point with the direction of the line. In two dimensions the direction can be specified by the *gradient. There are a number of related standard formats for lines in a plane.

Slope-intercept form. $y = mx + c$. The line has a gradient m and intercept at $(0,c)$. For a vertical line the gradient is infinite and the equation is $x = k$ (in this case there is no intercept on the y-axis).

Two-point form. The condition that the general point $P(x,y)$ lies on the line through $A(x_1,y_1)$ and $B(x_2,y_2)$ is equivalent to requiring the gradient of AP = the gradient of AB or $\dfrac{y - y_1}{x - x_1} = \dfrac{y_2 - y_1}{x_2 - x_1}$.

Point-slope form. A line with gradient m passing through the point $A(x_1,y_1)$ has equation $y - y_1 = m(x - x_1)$.

Two-intercept form. When the equation is written as $\dfrac{x}{a} + \dfrac{y}{b} = 1$ the line cuts the axes at $(a,0)$ and $(0,b)$, making it very easy to draw the line.

line (in three dimensions) As in two dimensions there are a number of standard formats which can be used.

Two-point form. The condition that the general point $P(x,y,z)$ lies on the line through $A(x_1,y_1,z_1)$ and $B(x_2,y_2,z_2)$ is equivalent to requiring the *direction ratios of AP and AB being equal. That is:
$$\frac{x - x_1}{x_2 - x_1} = \frac{y - y_1}{y_2 - y_1} = \frac{z - z_1}{z_2 - z_1}.$$

Parametric vector form. If the vector (l,m,n) is parallel to the line and the point $A(x_1,y_1,z_1)$ lies on it, the position vector of a general point P on the line has the form
$$\mathbf{r} = \overrightarrow{OP} = \begin{pmatrix} x_1 \\ y_1 \\ z_1 \end{pmatrix} + \lambda \begin{pmatrix} l \\ m \\ n \end{pmatrix}.$$

If $\mathbf{a}$ and $\mathbf{b}$ are the position vectors of points A and B on the line, the position vector of a general point is
$$\mathbf{r} = \mathbf{a} + \lambda(\mathbf{b} - \mathbf{a}).$$

Vector form. The equation $\mathbf{r} = \mathbf{a} + \lambda(\mathbf{b} - \mathbf{a})$ can be rewritten as $(\mathbf{r} - \mathbf{a}) \times (\mathbf{b} - \mathbf{a}) = \mathbf{0}$, where $\times$ denotes the *vector product. More generally the equation
$$\mathbf{r} \times \mathbf{a} = \mathbf{b},$$
defines a line when $\mathbf{a} \neq \mathbf{0}$ and $\mathbf{a}\cdot\mathbf{b} = 0$. This line is parallel to $\mathbf{a}$ and passes through the point $(\mathbf{a} \times \mathbf{b})/|\mathbf{a}|^2$.

linear action An *action of a *group G on a *vector space V is linear if
$$g.(c_1 v_1 + c_2 v_2) = c_1 g.v_1 + c_2 g.v_2$$
for any group elements g, vectors v_1, v_2 and scalars c_1, c_2. Equivalently, this

means that the action's *representation $G \to GL(V)$ is a *homomorphism.

linear algebra The topics of *linear equations, *matrices, vectors, and the algebraic structure known as a *vector space are intimately linked, and this area of mathematics is known as linear algebra. *Compare* ABSTRACT ALGEBRA.

linear code A linear code is a subspace C of F^n, where F is a *finite field; in the case of a *binary linear code $F = \{0,1\}$. The elements of C are the *codewords. Two advantages of linear codes are that C can be described by a *basis and that the least distance between any codewords is the smallest length (in the sense of *Hamming distance) of any non-zero codeword.

linear combination A linear combination of vectors $v_1, \ldots, v_k$ in a *vector space is a sum of the form $a_1 v_1 + \cdots + a_k v_k$, where $a_1, \ldots, a_k$ are *scalars in the *base field. The *span of a finite set of vectors is the set of linear combinations.

linear complexity *See* ALGORITHMIC COMPLEXITY.

linear congruence equation *See* CONGRUENCE EQUATION.

linear convergence *See* RATE OF CONVERGENCE.

linear differential equation A *differential equation of the form

$$a_n y^{(n)} + a_{n-1} y^{(n-1)} + \ldots a_1 y' + a_0 y = f$$

where $a_0, a_1, \ldots, a_n$ and f are given functions of x, and $y', \ldots, y^{(n)}$ are *derivatives of $y(x)$. *See also* LINEAR DIFFERENTIAL EQUATION WITH CONSTANT COEFFICIENTS, LINEAR FIRST-ORDER DIFFERENTIAL EQUATION.

linear differential equation with constant coefficients For simplicity, consider such an equation of second-order,

$$a\frac{d^2 y}{dx^2} + b\frac{dy}{dx} + cy = f(x), \qquad \mathbf{1}$$

where a, b, and c are given constants and f is a given function. (Higher-order equations can be treated similarly.) Suppose that f is not the zero function. Then the equation

$$a\frac{d^2 y}{dx^2} + b\frac{dy}{dx} + cy = 0 \qquad \mathbf{2}$$

is the *homogeneous equation that corresponds to the inhomogeneous equation **1**. The two are connected by the following result:

Theorem If $y = G(x)$ is the general solution of **2** and $y = y_1(x)$ is a particular solution of **1**, then $y = G(x) + y_1(x)$ is the general solution of **1**.

Thus the problem of solving **1** is reduced to the problem of finding the complementary function (C.F.) $G(x)$, which is the general solution of **2**, and a particular solution $y_1(x)$ of **1**, usually known in this context as a particular integral (P.I.).

The complementary function is found by looking for solutions of **2** of the form $y = e^{mx}$ and obtaining the auxiliary equation $am^2 + bm + c = 0$. If this equation has distinct real roots m_1 and m_2, the C.F. is $y = Ae^{m_1 x} + Be^{m_2 x}$; if it has one (repeated) root m, the C.F. is $y = (A + Bx)e^{mx}$; if it has non-real roots $\alpha \pm \beta i$, the C.F. is $y = e^{\alpha x}(A\cos\beta x + B\sin\beta x)$.

The most elementary way of obtaining a particular integral is to try something similar in form to $f(x)$. Thus, if $f(x) = e^{kx}$, try as the P.I. $y_1(x) = pe^{kx}$. If $f(x)$ is a polynomial in x, try a polynomial of the same degree. If $f(x) = \cos kx$ or $\sin kx$, try $y_1(x) = p\cos kx + q\sin kx$. In each case, the values of the unknown coefficients are found by substituting the possible P.I. into the equation **1**. If $f(x)$ is the sum of two terms, a P.I. corresponding to each may be found and the two added together.

For example, the general solution of $y'' - 3y' + 2y = 4x + e^{3x}$, is found to be $y = Ae^x + Be^{2x} + 2x + 3 + \frac{1}{2}e^{3x}$.

linear equation Consider, in turn, linear equations in one, two, and three variables. A linear equation in one variable x is an equation of the form $ax + b = 0$. If $a \neq 0$, this has the solution $x = -b/a$. A linear equation in two variables x and y is an equation of the form $ax + by + c = 0$. If a and b are not both zero, and x and y are taken as *Cartesian coordinates in the plane, the equation is that of a *straight line. A linear equation in three variables x, y, and z has the form $ax + by + cz + d = 0$. If a, b, and c are not all zero, and x, y, and z are taken as Cartesian coordinates in 3-dimensional space, the equation is that of a *plane. *See also* SIMULTANEOUS LINEAR EQUATIONS.

linear first-order differential equation A *differential equation of the form

$$\frac{dy}{dx} + P(x)y = Q(x),$$

where P and Q are given functions. One method of solution is to multiply both sides of the equation by an integrating factor $\mu(x)$, given by

$$\mu(x) = \exp\left(\int P(x) \ dx\right).$$

This choice is made because then the left-hand side of the equation becomes the derivative of $\mu(x)y$, and the solution can be found by integration.

In the case where $Q(x) = 0$, this can also be treated as a *separable-variable first-order differential equation.

linear function A *polynomial function of *degree one.

linear group A *group G which is isomorphic to a *matrix group of invertible matrices over a *field K and so admits a

*faithful *representation $G \rightarrow GL(n, K)$ for some n.

linear interpolation *See* INTERPOLATION.

linearization For a small quantity x, higher order terms x^2, x^3, ... are yet smaller by comparison. Linearization considers such higher-order terms negligible, so as to simplify an equation with the aim of finding an approximate solution. For example, a *simple pendulum that makes the angle θ with the downward vertical satisfies

$$\ddot{\theta} = -\frac{g}{l}\sin\theta,$$

where g denotes acceleration due to *gravity and l is the *length of the pendulum. The solutions involve *elliptic integrals. However, for small oscillations θ about the vertical, a linear approximation of $\sin\theta$ is θ, which gives the equation for *simple harmonic motion, and so the exact solution can instead be approximated with *trigonometric functions. *See* LINEAR THEORY OF EQUILIBRIA.

linearly dependent and independent A set of *vectors $\mathbf{u}_1, \mathbf{u}_2, \ldots, \mathbf{u}_r$ is linearly independent if $x_1\mathbf{u}_1 + x_2\mathbf{u}_2 + \ldots + x_r\mathbf{u}_r = \mathbf{0}$ implies $x_1 = 0, x_2 = 0, \ldots, x_r = 0$. Otherwise, the set is linearly dependent. A set of two vectors is linearly independent if and only if the two are not *parallel or, in other words, if and only if neither is a *scalar multiple of the other. In 3-dimensional space, any set of four or more vectors must be linearly dependent, and a set of three vectors is linearly independent if and only if the three are not *coplanar. *See* BASIS.

linear map A linear map T between two *vector spaces V and W over the same *field F satisfies

$$T(c_1v_1 + c_2v_2) = c_1T(v_1) + c_2T(v_2)$$

for all vectors v_1 and v_2 in V and all scalars c_1 and c_2 in F. The invertible (*see* INVERTIBLE FUNCTION) linear maps from V to V form the *general linear group GL(V) under composition.

If V and W are finite-dimensional with given *bases, then vectors in V and W are represented by *column vectors of *coordinates and T is represented by *premultiplication by a *matrix. The group GL(V) is *isomorphic to GL(n,F), where n is the *dimension of V.

linear momentum The linear momentum of a particle is the product of its *mass and its *velocity. It is a vector quantity, usually denoted by **p**, and **p** = m**v**. *See also* CONSERVATION OF LINEAR MOMENTUM.

Linear momentum has the dimensions MLT^{-1}. *See also* IMPULSE.

linear programming The branch of mathematics concerned with maximizing or minimizing a *linear function subject to a number of linear constraints. It has applications in economics, industry, and commerce, for example. In its simplest form, with two variables, the constraints determine a feasible region, which is the interior of a *polygon in the plane. The objective function to be maximized or minimized attains its maximum or minimum value at a vertex of the feasible region. For example, consider the problem of maximizing $4x_1 - 3x_2$ subject to

$x_1 - 2x_2 \geq -4$, $2x_1 + 3x_2 \leq 13$,

$x_1 - x_2 \leq 4, x_1 = 0$, $x_2 \geq 0$.

The feasible region is the interior of the polygon $OABCD$ shown in the figure, and the objective function $4x_1 - 3x_2$ attains its maximum value of 17 at the point B with coordinates (5, 1).

Often integer values are required, and in such cases the vertex is not always admissible and it will be necessary to test all points with integer values which lie close to the vertex. It is also possible that the objective function may be parallel to a constraining condition. In this case, all points on the boundary representing that constraint will be optimal.

linear regression *See* REGRESSION.

linear scale A scale in which the same difference in value is always represented by the same size of interval. *Compare* LOGARITHMIC SCALE.

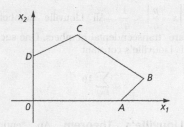

linear space A synonym for VECTOR SPACE.

linear system *See* SIMULTANEOUS LINEAR EQUATIONS.

linear theory of equilibria For small *perturbations about an *equilibrium, terms of higher order than linear may be considered negligible (*see* LINEARIZATION). Small angles θ, that a *simple pendulum makes about the vertical, satisfy $\ddot{\theta} = -(g/l)\theta$; this is *SHM, which has small *bounded *trigonometric solutions, and so $\theta = 0$ is a stable equilibrium. For $\theta = \pi + \varepsilon$, that is, small perturbations about the upward vertical, we have

$$\ddot{\varepsilon} = \ddot{\theta} = -\frac{g}{l}\sin\theta = -\frac{g}{l}\sin(\pi + \varepsilon)$$

$$= \frac{g}{l}\sin\varepsilon \approx \frac{g}{l}\varepsilon,$$

x_2 axis, C, D, B, O, A, x_1 axis

The feasible region

which has unbounded exponential solutions, and so $\theta = \pi$ is an unstable equilibrium.

Linearizing about $(0,0,0)$, the three equations governing the *Lorenz attractor give

$$\frac{dx}{dt} = \alpha(y - x), \quad \frac{dy}{dt} = \beta x - y, \quad \frac{dz}{dt} = -\gamma z,$$

where $\alpha, \beta, \gamma > 0$. So $\mathbf{r}(t) = (x(t), y(t), z(t))^T$ satisfies

$$\frac{d\mathbf{r}}{dt} = \begin{pmatrix} -\alpha & \alpha & 0 \\ \beta & -1 & 0 \\ 0 & 0 & -\gamma \end{pmatrix} \mathbf{r}.$$

If the matrix has distinct real *eigenvalues $\lambda_1, \lambda_2, \lambda_3$ with *eigenvectors $\mathbf{v}_1, \mathbf{v}_2, \mathbf{v}_3$, then the solution is

$$\mathbf{r}(t) = A_1 e^{\lambda_1 t} \mathbf{v}_1 + A_2 e^{\lambda_2 t} \mathbf{v}_2 + A_3 e^{\lambda_3 t} \mathbf{v}_3.$$

This is a small bounded solution if $\lambda_1, \lambda_2, \lambda_3$ are all negative; so the origin is stable when $\beta < 1$.

linear transformation See LINEAR MAP.

line integral An *integral along a *curve. Say a curve C is parametrized by $\mathbf{r}(t)$ where $a \leq t \leq b$. For a scalar function f, defined on C, we define the line integral

$$\int_C f \, ds = \int_a^b f(\mathbf{r}(t)) \left| \frac{d\mathbf{r}}{dt} \right| dt,$$

where s denotes *arc length. It is independent of the parametrization. If f represents the *density of a wire in the shape of C, then the line integral would equal the *mass of the wire.

For a vector function $\mathbf{F}$, defined on C, we define the line integral

$$\int_C \mathbf{F} \cdot d\mathbf{r} = \int_a^b \mathbf{F}(\mathbf{r}(t)) \cdot \frac{d\mathbf{r}}{dt} dt,$$

which equals the *work done by $\mathbf{F}$ along C. It is independent of the parametrization up to sign but changes sign if the orientation of the parametrization is reversed.

Line integrals are also called path integrals, particularly in *complex analysis.

line of action The line of action of a force is a straight line through the *point of application of the force and parallel to the direction of the force.

line of symmetry See SYMMETRICAL ABOUT A LINE.

line segment If A and B are two points on a line, the part of the line between and including A and B is a line segment. This may be denoted by AB or BA. The length of the line segment is the *distance between A and B. Compare DIRECTED LINE SEGMENT.

Liouville, Joseph (1809–82) Prolific and influential French mathematician, who was founder and editor of a notable French mathematical journal and proved important results in the fields of *number theory, *differential equations, *differential geometry, and *complex analysis. In 1842 he began reading *Galois' unpublished work and did much to make his work known. In 1844, he proved the existence of *transcendental numbers (see LIOUVILLE NUMBERS).

Liouville numbers (Liouville's constant) Irrational number x with the property that for every integer n, there is a rational number $\frac{p}{q}$ such that $\left| x - \frac{p}{q} \right| < \frac{1}{q^n}$. All Liouville numbers are *transcendental numbers. One such is Liouville's constant

$$\sum_{k=1}^{\infty} 10^{-k!}.$$

Liouville's Theorem An *entire *bounded function is constant. This is a result of *complex analysis that is in stark contrast to real *analysis, where sine is analytic, bounded, and non-constant.

Lipschitz condition A function f between two *metric spaces M and N satisfies the Lipschitz condition if there exists a constant k for which $d(f(x), f(y)) \leq k\, d(x, y)$ for all points x, y in M, i.e. that the distance between the function values is bounded by a constant multiple of the distance between the values. Lipschitz functions are *uniformly continuous and *absolutely continuous. *Compare* CONTRACTION; *see* RADEMACHER'S THEOREM.

Lissajous curve (Lissajous figure) Named after Jules Antoine Lissajous (1822–80), a Lissajous curve is a planar curve that can be parameterized as

$$x(t) = A_1\sin(\omega_1 t + \epsilon),\ y(t) = A_2\sin(\omega_2 t).$$

The curve is closed if and only if ω_1/ω_2 is *rational.

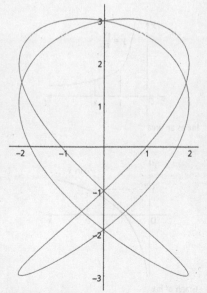

A Lissajous curve

In the figure $A_1 = \omega_2 = 2$, $A_2 = \omega_1 = 3$, $\epsilon = \pi/6$.

litre A measure of capacity equalling 1000 cm³, abbreviated to 'l'. Previously, a litre was defined as the volume of 1 *kilogram of water at 4°C (the *temperature at which water's *density is greatest) at standard atmospheric pressure.

Littlewood, John Edensor (1885–1977) British mathematician, based at Cambridge University. Most famous for his collaboration with Godfrey *Hardy on summability theory, *Fourier series, *analytic number theory, and the *zeta function.

little o notation Notation for comparing the relative order of functions. We write $f(x) = o(g(x))$ if $f(x)/g(x)$ approaches 0 as $x\to\infty$, so, for example, $(x+1)(x+2) = o(x^3)$. The notation is also used in the neighbourhood of finite values; for example, $\sin x = x + o(x^2)$ denotes the fact that $(\sin x - x)/x^2$ approaches 0 as $x\to 0$. *Compare* BIG O NOTATION.

LMS Abbreviation for the London Mathematical Society, one of the UK's learned societies for mathematics alongside the *IMA and *RSS. It was founded in 1865, its first president being *De Morgan.

ln Abbreviation for *natural logarithm, the logarithm to base e.

load *See* MACHINE.

Lobachevsky, Nikolai Ivanovich (1792–1856) Russian mathematician who in 1829 published his discovery, independently of Bolyai, of *hyperbolic geometry. He continued to publicize his ideas, but his work received wide recognition only after his death.

locally compact A *topological space is locally compact if every point has a *neighbourhood which is *compact. *Euclidean space is locally compact, but an infinite-dimensional *Hilbert space is not.

local maximum (local minimum) For a function f, a local maximum (resp. minimum) is a point c that has a *neighbourhood at every point of which $f(x) \leq f(c)$ (resp. $f(x) \geq f(c)$). If f is *differentiable at an *interior local maximum or minimum c, then $f'(c) = 0$; that is, an interior local maximum or minimum is a *stationary point. This is *Fermat's Theorem.

located vector A *vector only requires a *magnitude and *direction to specify it. A located vector further has a specified starting point.

location In *statistics, a measure of location is a single figure which gives a typical or, in some sense, central value for a *distribution or *sample. The most common measures of location are the *mean, the *median, and the *mode. For several reasons, the mean is usually the preferred measure of location, but when a distribution is *skew the median may be more appropriate.

locus (loci) A locus is the set of all points that satisfy some given condition or property. For example, in the *plane, the locus of all points that are a given *distance from a fixed point is a circle, and the locus of all points equidistant from two given points, A and B, is the perpendicular *bisector of AB.

log Abbreviation for *logarithm.

logarithm Let a, x be positive numbers with $a \neq 1$. Then $\log_a x$, the logarithm of x to base a, is defined by

$$\log_a x = \frac{\ln x}{\ln a},$$

where ln denotes the *logarithmic function. This is equivalent to saying that $y = \log_a x$ if and only if $x = a^y$ or that the function $\log_a x$ is the *inverse function of a^x. The following properties hold, where x, y, and r are real, with $x,y > 0$:

(i) $\log_a(xy) = \log_a x + \log_a y$.

(ii) $\log_a(1/x) = -\log_a x$.

(iii) $\log_a(x^r) = r \log_a x$.

(iv) Logarithms to different bases are related by the formula

$$\log_b x = \frac{\log_a x}{\log_a b}.$$

(v) $\dfrac{d}{dx} \log_a x = \dfrac{1}{x \ln a}$.

Logarithms to base 10 are called common logarithms. Logarithms to base e are called natural logarithms. *See* COMPLEX LOGARITHM.

logarithmic function (ln) The logarithmic function $\ln x$ may be defined for positive x as follows:

$$\ln x = \int_1^x \frac{1}{t}\,dt.$$

It can equally be defined as the *inverse function of the *exponential function. $\ln x$ is *differentiable and *increasing with derivative $1/x$.

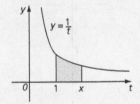

ln$x$ as an area

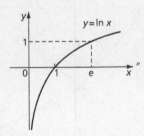

Graph of ln$x$

Further, $\ln x = \log_e x$ satisfies properties (i)-(v) listed under *logarithm.

logarithmic plotting Two varieties of logarithmic graph paper are commonly available. One, known as semi-log or single log paper, has a standard scale on the x-axis and a *logarithmic scale on the y-axis. If given *data satisfy an equation $y = ba^x$ (where a and b are constants), exhibiting *exponential growth or *exponential decay, then on this special graph paper the corresponding points lie on a line. The other variety is known as log–log or double log paper and has a logarithmic scale on both axes. In this case, data satisfying $y = bx^m$ (where b and m are constants) produce points that lie on a line.

logarithmic scale A method of representing positive numbers on a line as follows. Taking one direction along the line as positive, and a point O as origin, the number x is represented by the point P in such a way that OP is proportional to $\log x$, where logarithms are to base 10. Thus the number 1 is represented by O; and, if the point A represents 10, the point B that represents 100 is such that $OB = 2OA$.

The measurement of sound using decibels and of the size of earthquakes using the Richter scale are two examples of logarithmic scales.

See also SLIDE RULE.

logarithmic series The *power series

$$x - \frac{x^2}{2} + \frac{x^3}{3} - \cdots + (-1)^{n+1}\frac{x^n}{n} + \cdots .$$

For $-1 < x \leq 1$ the series converges to $\ln(1 + x)$. For x a *complex number, with $|x| \leq 1$ and $x \neq -1$, the series converges to a *branch of $\log(1 + x)$, where log denotes the *complex logarithm.

logarithmic spiral A synonym for EQUIANGULAR SPIRAL.

logic The study of deductive reasoning, by which conclusions are derived from sets of premises. Informally, the term is also used to refer to the essential reasoning process in a mathematical proof. *See also* FIRST ORDER LOGIC, GÖDEL'S INCOMPLETENESS THEOREMS, PREDICATE, MODUS PONENS, MODEL THEORY, PROOF THEORY.

logically equivalent Two *compound statements involving the same components are logically equivalent if they have the same *truth tables. For example, the truth table for the statement $(\neg p) \vee q$ is:

p	q	$\neg p$	$(\neg p) \vee q$
T	T	F	T
T	F	F	F
F	T	T	T
F	F	T	T

By comparing the last column here with the truth table for $p \Rightarrow q$ (*see* IMPLICATION), it can be seen that $(\neg p) \vee q$ and $p \Rightarrow q$ are logically equivalent. This equivalence is the basis of *proof by contradiction.

logic gate An idealized representation of an electronic device that executes a logical operation such as *and, *or, or *not. The standard symbols for those gates are drawn below. Logic gates may then be combined in *circuits to produce

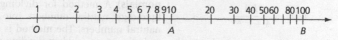

A logarithmic scale

more complicated n-ary logical operations. *See* COMBINATORIAL LOGIC.

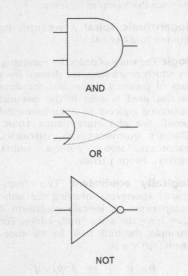

AND

OR

NOT

logicism A *philosophy of mathematics that mathematics is an extension of *logic and that mathematics is reducible to logic. *Frege, *Dedekind, and *Russell were proponents of this philosophy.

logistic map (continuous case) In 1838 the Belgian mathematician Verhulst suggested the *differential equation

$$\frac{\mathrm{d}N}{\mathrm{d}t} = rN\left(1 - \frac{N}{K}\right)$$

as a *model for population growth. $N(t)$ is the population at time t. For small N, we have $\mathrm{d}N/\mathrm{d}t \approx rN(t)$ and so approximately *exponential growth with growth rate r. However, as N becomes comparable to a carrying capacity K, the effective growth rate reduces. The general solution is

$$N(t) = \frac{KA}{A + e^{-Krt}}$$

where A is a positive constant. The graph of $N(t)$ is an increasing S-shaped curve

with $N(t)$ tending to 0 as $t \to -\infty$ and $N(t)$ tending to K as $t \to \infty$.

logistic map (discrete case) A discrete version of the logistic map is given by the *recurrence relation

$$x_{n+1} = rx_n(1 - x_n)$$

where $0 < r < 4$ and $0 < x_0 < 1$.

(Rewrite the continuous logistic equation in terms of N/K to see how this discrete version is arrived at.) By restricting r as above, the sequence remains positive.

This simple iteration leads to surprisingly complex behaviour. If $r < 1$, then the sequence x_n *tends to 0, signifying extinction. If $1 < r < 2$, then the sequence *monotonically *converges to $1 - 1/r$. If $2 < r < 3$, then the sequence converges to $1 - 1/r$ in an oscillatory fashion. But at $r = 3$ this *equilibrium becomes unstable and *bifurcates; for $3 < r < 1 + \sqrt{6}$ the stable behaviour is an oscillation between two values of r. Then at $r = 1 + \sqrt{6}$ we get a further bifurcation, and the stable behaviour is oscillations between four values of r. Further period-doubling bifurcations occur until the behaviour becomes *chaotic at around $r = 3.57$, but then oscillations still occur at some greater values of r.

By *cobwebbing with the graphs of $y = rx(1 - x)$ and $y = x$ for different values of r, a qualitive appreciation of how these behaviours arise is possible.

lognormal distribution A positive valued *random variable X is said to have a lognormal distribution if the random variable $Y = \ln X$ has a *normal distribution.

log paper *See* LOGARITHMIC PLOTTING.

long division (long division of polynomials) A method for dividing multiple-digit natural numbers into other natural numbers. The method is ultimately the same as when dividing single-digit natural numbers but presented

```
              2  4  4
        2  3 │ 5  6  1  5
              4  6
             ─────
              1  0  1
                 9  2
              ─────
                 9  5
                 9  2
                 ─────
                    3
```

long division of integers

```
                        x   −1
        x²+2x+2 │ x³ +x² −x −2
                  x³ +2x²+2x
                  ──────────
                     −x² −3x −2
                     −x² −2x −2
                     ──────────
                          −x
```

long division of polynomials

differently as the remainder at each stage may be multiple-digit. For example, the division of 23 into 5615 is presented above.

This shows that $5615 = 244 \times 23 + 3$. As with normal division of single-digit numbers, 23 does not divide into 5, but does twice into 56 with remainder 10, which is written beneath. The process continues until a remainder is found.

This notation can similarly be used for long division of polynomials. The second long division shows that

$$x^3 + x^2 - x - 2 = (x-1)(x^2 + 2x + 2) - x.$$

See DIVISION ALGORITHM.

longitude The longitude of a point P on the Earth's surface is the *angle, measured in degrees east or west, between the *meridian through P and the *prime meridian through Greenwich. If the meridian through P and the meridian through Greenwich meet the equator at P' and G' respectively, the longitude is the angle $P'OG'$, where O is the centre of the Earth. Longitude and *latitude uniquely fix the position of a point on the Earth's surface.

longitudinal study A study of the development of characteristics of the same individuals over a period of time. If different groups are studied simultaneously at different ages, it is difficult to attribute observed differences to any factor, so longitudinal studies are designed to remove the individual differences in an analogous manner to *paired-sample tests.

longitudinal wave A form of wave motion in which energy is transmitted along the direction of the wave motion as the wave moves through a medium; a sound wave is an example of a longitudinal wave.

loop In a *graph, an *edge that begins and ends in the same *vertex.

Lorentz, Hendrik Antoon (1853–1928) Dutch mathematician and theoretical physicist whose work on the mathematical theory of the electron was awarded the Nobel Prize for Physics in 1902, jointly with his student Pieter Zeeman who had verified the mathematical results experimentally. He is also famous for the *Lorentz–Fitzgerald contraction and *Lorentz transformations.

Lorentz–Fitzgerald contraction To an observer a body moving at high speed appears shorter (in the direction of motion) than it really is. Since the factor is $\sqrt{1 - \dfrac{v^2}{c^2}} : 1$ where c is the *speed of light, the speed has to be very high or the measurement very precise for this to be observable. It was observed 20 years before *Einstein's special theory of relativity provided the explanation for it. The contraction is a consequence of the *observer measuring between events that are not *simultaneous in the *rest frame of the moving object.

Lorentz force law The *force experienced by a charged particle in an electromagnetic field. The law states that a particle with charge q, moving with a velocity **v** through an *electric field **E** and a *magnetic field **B**, will experience a force **F** given by $\mathbf{F} = q(\mathbf{E} + \mathbf{v} \times \mathbf{B})$.

Lorentz group See LORENTZ TRANSFORMATION.

Lorentz transformation Transformations between two *frames of reference in *special relativity. If the frames share an origin and x is taken in the direction of the *relative velocity of the two frames and v is the relative velocity of the frames, then their coordinates are related by

$$\begin{pmatrix} ct' \\ x' \\ y' \\ z' \end{pmatrix} = \begin{pmatrix} \gamma & -\gamma v/c & 0 & 0 \\ -\gamma v/c & \gamma & 0 & 0 \\ 0 & 0 & 1 & 0 \\ 0 & 0 & 0 & 1 \end{pmatrix} \begin{pmatrix} ct \\ x \\ y \\ z \end{pmatrix}$$

where $\gamma = (1 - v^2/c^2)^{-1/2}$. The Lorentz transformations form a *group under *composition called the Lorentz group. Generally, two frames of reference are connected by a Lorentz transformation and a translation, such elements forming the Poincaré group. See also GALILEAN RELATIVITY, SIMULTANEITY, TIME DILATION.

Lorenz attractor Edward Lorenz first studied the equations below in 1963 while *modelling weather.

$$\frac{dx}{dt} = \alpha(y - x)$$

$$\frac{dy}{dt} = x(\beta - z) - y$$

$$\frac{dz}{dt} = xy - \gamma z$$

Solutions to the equations exhibit *chaos. The Lorenz attractor is an example of a strange attractor, an infinite invariant set towards which nearby motions converge.

Lotka-Volterra equations See PREDATOR-PREY EQUATIONS.

lower bound See BOUND.

lower limit See LIMIT OF INTEGRATION.

lower triangular matrix See TRIANGULAR MATRIX.

lowest common denominator A synonym for LEAST COMMON DENOMINATOR.

lowest terms See IRREDUCIBLE FRACTION.

lub A synonym for LEAST UPPER BOUND.

Lucas numbers The sequence 2, 1, 3, 4, 7, 11, 18, 29 which is derived from the same relation as the *Fibonacci sequence, i.e. that each number is the sum of the previous two numbers in the sequence, but it has different starting values.

Lucas theorem Given a *polynomial $P(z)$ with complex coefficients (see COMPLEX NUMBER), the *roots of the *derivative $P'(z)$ lie within the *convex hull of the roots of $P(z)$. Also known as the Gauss-Lucas theorem.

LU decomposition A *square matrix **A** has an LU decomposition if $\mathbf{A} = \mathbf{LU}$, where **L** is a lower *triangular matrix with all *diagonal entries equalling 1 and **U** is an upper triangular matrix. The system of equations $\mathbf{Ax} = \mathbf{b}$ can then be solved separately as $\mathbf{Ly} = \mathbf{b}$ and then $\mathbf{Ux} = \mathbf{y}$, which is computationally more efficient. The system $\mathbf{Ly} = \mathbf{b}$ efficiently yields $y_1, y_2, \ldots, y_n$ in order and is known as forward substitution. The system $\mathbf{Ux} = \mathbf{y}$ likewise yields $x_n, \ldots, x_2, x_1$ in that order and is known as backward substitution. Not all matrices **A** have LU decompositions, but there is always a *permutation matrix **P** such that **PA** has an LU decomposition; **P** effectively just reorders the equations in the linear system.

lune A region of a *sphere bounded by two *great circles.

lurking variable A potential *confounding variable that has not been measured and not discussed in the interpretation of an *experiment or *observational study.

Lyapunov, Aleksandr (1857–1918) Russian mathematician and physicist, who developed the theory of stability in *dynamical systems and also proved the *Central Limit Theorem by applying *characteristic functions.

Lyapunov stability An *equilibrium of a *dynamical system is Lyapunov stable if any suitably small *perturbation from the equilibrium results in the system staying near the equilibrium forever. More precisely, given a *metric d on the *phase space, an equilibrium e is Lyapunov stable if for all $\varepsilon > 0$ there exists $\delta > 0$ so that if $d(x(0),e) < \delta$, then $d(x(t),e) < \varepsilon$ for all t, where $x(t)$ denotes the evolution of the system following the perturbation. Further, the equilibrium is said to be asymptotically stable if $x(t) \to e$ as $t \to \infty$ if $d(x(0),e) < \delta$.

m Abbreviation and symbol for *metre. Also abbreviation for *milli-.

M Roman *numeral for 1000.

M Abbreviation for *mega-.

μ (mu) Abbreviation for *micro-.

machine A device that enables *energy from one source to be modified and transmitted as energy in a different form or for a different purpose. Simple examples, in *mechanics, are *pulleys and *levers, which may be used as components in a more complex machine. The *force applied to the machine is the effort, and the force exerted by the machine is equal and opposite to the load.

If the effort has magnitude P and the load has magnitude W, the mechanical advantage of the machine is the ratio W/P. The velocity ratio is the ratio of the distance travelled by the effort to the distance travelled by the load. It is a constant for a particular machine, calculable from first principles. For example, in the case of a pulley system, the velocity ratio equals the number of ropes supporting the load or, equivalently, the number of pulleys in the system.

The efficiency is the ratio of the *work obtained from the machine to the work put into the machine, often stated as a *percentage. In the real world, the efficiency of a machine is strictly less than 1 because of factors such as *friction. In an ideal machine for which the efficiency is 1, the mechanical advantage equals the velocity ratio.

See also TURING MACHINE.

machine learning An interdisciplinary field centred on artificial intelligence with the aim of constructing programs or *algorithms that learn from experience, that is, from introduction to further *data. *See* NEURAL NETS.

Maclaurin, Colin (1698–1746) Scottish mathematician who was the outstanding British mathematician of the generation following *Newton's. He developed and extended the subject of calculus. His textbook on the subject contains important original results, but the Maclaurin series, which appears in it, is just a special case of the *Taylor series known considerably earlier. He also obtained notable results in *geometry and wrote a popular textbook on *algebra.

Maclaurin series (expansion) *See* TAYLOR'S THEOREM.

magic square A square array of numbers arranged in rows and columns in such a way that all the *rows, all the *columns, and both *diagonals add up to the same total. Often the numbers used are $1, 2, \ldots, n^2$, where n is the number of rows and columns. The example on the left in the figure is said to be of ancient Chinese origin; that on the right is featured in an engraving by Albrecht Dürer.

4	9	2
3	5	7
8	1	6

16	3	2	13
5	10	11	8
9	6	7	12
4	15	14	1

(🌐) SEE WEB LINKS

• An interactive 3 × 3 magic square.

magma A set with a *binary operation. *Compare* MONOID, SEMIGROUP.

magnetic field A *vector field which exerts a *force on moving electric charges. *See* MAXWELL'S EQUATIONS.

magnitude (of a vector) A *vector $\mathbf{v} = (v_1, v_2, \ldots, v_n)$ in $\mathbb{R}^n$ has magnitude

$$|\mathbf{v}| = \sqrt{(v_1)^2 + (v_2)^2 + \ldots + (v_n)^2}.$$

This is also referred to as the *length of $\mathbf{v}$. A vector, represented by a *directed line segment $\overrightarrow{PQ}$ has magnitude equal to the *distance between P and Q.

magnitude (of a statistically significant result) A statistically significant result is a function of both the *sample size and of the magnitude of the difference being measured. With advances in technology, sample sizes now can be extremely large, and in these cases almost any difference will turn out to be significant. It becomes important in such cases to consider whether the difference is large enough to be of practical importance.

main diagonal In the $n \times n$ matrix $[a_{ij}]$, the entries $a_{11}, a_{22}, \ldots, a_{nn}$ form the main diagonal. *Compare* OFF DIAGONAL.

major arc *See* ARC.

major axis *See* ELLIPSE.

Mandelbrot, Benoît (1924–2010) Polish mathematician who showed the range of applications of *fractals in mathematics and in nature and whose work with computer graphics popularized fractal images.

Mandelbrot set The set of points c in the *complex plane for which the repeated application of the function $f(z) = z^2 + c$ to the point $z = 0$ produces a *bounded *sequence. The set is an extremely complicated object whose boundary is a *fractal. It can also be defined as the set of points c for which the *Julia set of $f(z)$ is a *connected set. The different regions of the Mandelbrot set relate to the different characters of the Julia sets; for example, for c in the main *cardioid of the Mandelbrot set $f(z)$ has a single *attracting fixed point.

Manhattan norm A synonym for TAXICAB NORM.

manifold *Curves are 1-dimensional manifolds and *surfaces are 2-dimensional manifolds. Formally an n-dimensional manifold is a *Hausdorff *topological space X in which every point has a *neighbourhood which is *homeomorphic to n-dimensional *Euclidean space. Note that the double cone $x^2 + y^2 = z^2$ is not a manifold as no Euclidean neighbourhood of the origin exists.

The above defines what a *topological* manifold is, but manifolds can have further smooth or metric structure. *See* CLASSIFICATION THEOREM FOR SURFACES, ORIENTABILITY, TRANSITION MAP, RIEMANN SURFACE, RIEMANNIAN MANIFOLD.

Mann–Whitney U test *See* WILCOXON RANK-SUM TEST.

MANOVA *See* MULTIVARIATE ANALYSIS OF VARIANCE.

mantissa *See* CHARACTERISTIC.

many-to-one A function for which more than one element of the domain maps onto an element in the range. So $f(x) = x^2$ is many-to-one if x is defined on the real numbers since $f(a) = f(-a)$. A many-to-one function is not invertible, since there is no way to define a function to return more than one value. However, a many-to-one function can be restricted to a domain within which it is a one-to-one function, and for which the inverse function exists. If $f(x) = x^2$ is restricted to $x \geq 0$ then $f^{-1}(x) = \sqrt{x}$ exists. *See* PRINCIPAL VALUES.

many-valued function *See* MULTI-FUNCTION.

map (mapping) A synonym for FUNCTION.

map projection A representation of the *Earth's surface, or a part thereof, in the *plane. As a consequence of the *Theorema Egregium, such a projection cannot be isometric. (The plane has zero *Gaussian curvature and a *sphere has positive curvature.) In different contexts, different projections might be preferred; Mercator's projection was preferred for navigation as it is *conformal with latitudes remaining horizontal, but area is severely distorted at the poles; equiareal projections (such as Mollweide's projection) arguably give a truer geopolitical representation of the world; alternatively, a projection may trade off between the possible extremes.

marginal distribution Suppose that X and Y are discrete *random variables with *joint probability mass function p. Then the marginal distribution of X is the distribution with *probability mass function p_1, where

$$p_X(x) = \sum_y p(x, y).$$

A similar formula gives the marginal distribution of Y.

Suppose that X and Y are continuous random variables with *joint probability density function f. Then the marginal distribution of X is the distribution with *probability density function f_X, where

$$f_X(x) = \int_{-\infty}^{\infty} f(x, y) \, dy.$$

Markov, Andrei Andreevich (1856–1922) Russian mathematician and student of *Chebyshev who proved important results in *probability, developing the notion of a *Markov chain and initiating the study of *stochastic processes.

Markov chain Consider a *stochastic process $X_1, X_2, X_3, \ldots$ in which the state space is discrete. This is a Markov chain if the probability that X_{n+1} takes a particular value depends only on the value of X_n and not on the values of $X_1, X_2, \ldots, X_{n-1}$. (This definition can be adapted so as to apply to a stochastic process with a continuous state space, or to a more general stochastic process $\{X(t), t \in T\}$, to give what is called a Markov process.) Examples include *random walks and problems in *queuing theory.

Most Markov chains are homogeneous, so the probability that $X_{n+1} = j$ given that $X_n = i$, denoted by p_{ij}, does not depend on n. In that case, if there are N states, these values p_{ij} are called the transition probabilities and form the transition matrix $[p_{ij}]$, an $N \times N$ row *stochastic matrix. *See* COMMUNICATING CLASS, RECURRENT, STATIONARY DISTRIBUTION.

martingale A sequence of *random variables $\{X_i\}$ such that the expected value of the $(n+1)$th random variable, once the first n outcomes are known, is just the current value X_n. The simple *random walk, which moves left or right one step with probability 0.5, is an example of a martingale.

mass With any *body there are associated two parameters: the *gravitational mass, which occurs in the *inverse square law of gravitation, and the *inertial mass, which occurs in *Newton's second law of motion. While they are distinct in definition, no experiment has shown a difference between gravitational and inertial mass.

The SI unit of mass is the *kilogram. *See also* REST MASS. *Compare* WEIGHT.

mass–energy equation $E = mc^2$ where E denotes *energy, m denotes *mass, and c is the *speed of light in a vacuum. This is the equation proposed by Albert *Einstein as part of the special theory of *relativity showing the relationship between mass and energy. The implication of this equation is that mass can be considered as potential energy, though it is extremely difficult to make the conversion and release the stored energy. Two such methods are nuclear fusion and nuclear fission, both of which involve creating a new molecular structure from existing molecules where the new structure contains less mass, with the difference being converted into energy.

matched-pairs design An *experimental design structure which matches subjects as closely as possible and then assigns one of each pair to each of the treatment group and control group. This might involve cutting pieces of cloth in two when testing detergents, or pairing patients in a clinical trial by age, gender, weight, etc. as closely as possible. The allocation of treatments to the paired subjects should be randomized. Then the differences in outcome are measured. The strength of this design lies in reducing the amount of variation between subjects so that any actual differences due to the experimental conditions are more easily identified.

matching In a *graph, a matching is a set of *edges with no *vertices in common. *See also* HALL'S THEOREM.

material derivative A synonym for CONVECTIVE DERIVATIVE.

mathematical biology The area of mathematics modelling biological systems, including such *models as *SIR epidemiology models, population growth as with the *logistic map, and *predator-prey models such as the Lotka-Volterra

equations. The increase in large data sets and computing power has meant much more complex biological systems can now be modelled.

mathematical induction The method of proof 'by mathematical induction' is based on the following principle:

Principle of mathematical induction Let there be associated, with each positive *integer n, a proposition $P(n)$, which is either true or false. If

(i) $P(1)$ is true,
(ii) for all k, $P(k)$ implies $P(k + 1)$,

then $P(n)$ is true for all positive integers n.

The following are typical of results that can be proved by induction:

(a) For all positive integers n,
$$\sum_{r=1}^{n} r^2 = \tfrac{1}{6} n(n + 1)(2n + 1).$$

(b) For all positive integers n, the nth derivative of $\frac{1}{x}$ is $(-1)^n \frac{n!}{x^{n+1}}$.

(c) For all positive integers n, $(\cos\theta + i\ \sin\theta)^n = \cos n\theta + i\ \sin n\theta$. *See* DE MOIVRE'S THEOREM.

In each case, it is clear what the proposition $P(n)$ should be and that (i), the base case, can be verified. The method by which the so-called inductive step (ii), where the inductive hypothesis $P(k)$ is assumed, is proved depends upon the particular result to be established.

There is a so-called 'strong form' of the principle of induction which is equivalent. It states:

If

(i') $P(1)$ is true,
(ii') for all k, the truth of $P(1), P(2), \ldots,$ $P(k\text{-}1), P(k)$ implies $P(k + 1)$,

then $P(n)$ is true for all positive integers n.

This is a useful alternative when the inductive step proving $P(k + 1)$ relies on the truth of some previous proposition $P(i)$ which is not necessarily $P(k)$.

mathematical model A mathematical model is a mathematical description of a real world problem, which may be *deterministic or *stochastic. Typically the model makes simplifying assumptions about the problem so that the mathematical equations describing the model may be solved or more easily analysed. For example, the two-body problem of the Earth going around the Sun treats both as *particles and ignores other celestial bodies. But these are reasonable assumptions to make in light of the characteristics and strength of the gravitational field (*see* GRAVITY). It may be reasonable to consider viscosity negligible in a *fluid but not within a *boundary layer of an object; so, in such cases different simplifications are made in different parts of the model.

mathematical notation *See* NOTATION.

mathematics The branch of human enquiry involving the study of numbers, quantities, data, shape, and space and their relationships, especially their *generalizations and *abstractions and their application to situations in the real world. As a broad generalization *pure mathematics is the study of the relationships between abstract quantities according to a well-defined set of rules, and *applied mathematics is the application and use of *mathematical models in the real-world contexts. Pure mathematics includes *algebra, *analysis, *calculus, *geometry, *number theory, *topology, *logic, *set theory, etc. Applied mathematics includes *numerical analysis, *mechanics, *mathematical biology, *probability and statistics, *networks, *quantum theory, *relativity, etc.

mathematics education The pedagogy or education theory of mathematics—the study of the teaching and learning of mathematics from basic numeracy skills to more abstract concepts. Increasingly, there is interest in the use of technology in education. The differently perceived purposes of mathematics education lead to different preferences of teaching methods.

matrix (matrices) A rectangular array of entries displayed in rows and columns and enclosed in brackets. The entries are elements of some suitable set, either specified or understood. They are often numbers, perhaps integers, real numbers, or complex numbers, but they may be, say, polynomials or other expressions. An $m \times n$ matrix has m rows and n columns and can be written

$$\begin{bmatrix} a_{11} & a_{12} & \cdots & a_{1n} \\ a_{21} & a_{22} & \cdots & a_{2n} \\ \vdots & \vdots & \ddots & \vdots \\ a_{m1} & a_{m2} & \cdots & a_{mn} \end{bmatrix}.$$

Round brackets may be used instead of square brackets. The subscripts are read as though separated by commas: for example, a_{23} is read as 'a, two, three'. The matrix above may be written in abbreviated form as $[a_{ij}]$, where the number of rows and columns is understood and a_{ij} denotes the entry in the ith row and jth column. *See* ADDITION (of matrices), BLOCK MATRIX, DETERMINANT, DIAGONAL MATRIX, IDENTITY MATRIX, INVERSE MATRIX, MATRIX GROUPS, MATRIX OF A LINEAR MAP, MULTIPLICATION (of matrices), SQUARE MATRIX, TRACE, TRANSPOSE, TRIANGULAR MATRIX.

matrix game A *game in which a *matrix contains numerical data giving information about what happens according to the strategies chosen by two opponents or players. By convention, the matrix $[a_{ij}]$ gives the *payoff to one of the players, R: when R chooses the ith row and the other player, C, chooses the jth column, then C pays to R an amount of a_{ij} units. (If a_{ij} is negative, then in fact R pays C a certain amount.) This is an example of a zero-sum game because the total amount received by the two players is zero: R receives a_{ij} units and C receives $-a_{ij}$ units.

If the game is a *strictly determined game, and R and C use *conservative strategies, C will always pay R a certain amount, which is the value of the game. If the game is not strictly determined, then by the *Fundamental Theorem of Game Theory, there is again a value for the game, being the expectation when R and C use *mixed strategies that are optimal.

matrix groups A term loosely applying to groups of *matrices under matrix multiplication, particularly important in *geometry. Examples include:

GL(n,F)	The general linear group. *Invertible $n \times n$ matrices with entries in the *field F.
SL(n,F)	The special linear group. *Determinant 1 $n \times n$ matrices with entries in F.
O(n)	*orthogonal $n \times n$ matrices.
SO(n)	orthogonal $n \times n$ matrices with determinant 1.
U(n)	*unitary $n \times n$ matrices.
SU(n)	unitary $n \times n$ matrices with determinant 1.
AGL(n,F)	*affine transformations on F^n.
PGL(n,F)	*projective transformations on $n-1$ dimensional *projective space over F.
Lorentz group	the linear isometries (see LINEAR MAP) of *Minkowski space.
Sp($2n,F$)	*symplectic group of those $2n \times 2n$ matrices M over F satisfying $M^T \Omega M = \Omega$, where $$\Omega = \begin{pmatrix} 0 & I_n \\ -I_n & 0 \end{pmatrix}.$$

See also LIE GROUPS, LINEAR GROUPS.

matrix norm A *norm on the space of real or complex *square matrices of a given *order, with the further property that

$$\|AB\| \leq \|A\| \times \|B\|.$$

matrix of a linear map A linear map $T: V \rightarrow W$ between finite-dimensional *vector spaces can be represented by a *matrix if bases are chosen for V and W. Given bases $v_1, \ldots, v_m$ and $w_1, \ldots, w_n$, the matrix for T, with respect to these bases, is the $n \times m$ matrix that sends the coordinate vector for v (*wrt the v_i) to the coordinate vector for Tv (wrt the w_i). Precisely, this means the matrix has entries a_{ij}, where

$$T(v_i) = \sum_{j=1}^{n} a_{ji} w_j.$$

Any two matrices representing T will be *equivalent. *See* CHANGE OF BASIS, DIAGONALIZATION.

matrix of coefficients For a given set of m *linear equations in n unknowns $x_1, x_2, \ldots, x_n$:

$$a_{11}x_1 + a_{12}x_2 + \ldots + a_{1n}x_n = b_1,$$
$$a_{21}x_1 + a_{22}x_2 + \ldots + a_{2n}x_n = b_2,$$
$$\vdots$$
$$a_{m1}x_1 + a_{m2}x_2 + \ldots + a_{mn}x_n = b_m,$$

the matrix of coefficients is the $m \times n$ matrix $[a_{ij}]$. *See* AUGMENTED MATRIX.

matrix of cofactors For **A**, a *square matrix $[a_{ij}]$, the matrix of cofactors is the matrix, of the same order, that is obtained by replacing each entry a_{ij} by its *cofactor A_{ij}. It is used to find the *adjoint of **A** and hence the *inverse $\mathbf{A}^{-1}$.

Maupertuis, Pierre-Louis Moreau de (1698–1759) French scientist and mathematician known as the first to formulate a proposition called the *principle of least action. He was an active supporter in France of *Newton's gravitational theory. He led an expedition to Lapland to measure the length of a degree along a meridian, which confirmed that the Earth is an oblate *spheroid.

max-flow/min-cut The value of the total flow in any *network is less than or equal to the *capacity of any *cut, and the maximum flow will equal the minimum cut for any network.

maximal element (maximum) In a *partially ordered set $(S, \leq)$, a maximum is an element m such that $s \leq m$ for all $s \in S$. A maximal element is an element M such that there is no element s with $M < s$. So a maximum is a maximal element, but a maximal element need not be a maximum.

maximal ideal A proper (see PROPER SUBSET) *ideal I of a *ring R which is maximal with respect to containment, that is, if J is an ideal such that $I \subseteq J \subseteq R$, then $I = J$ or $I = R$. An ideal I is maximal if and only if the *quotient ring R/I is a *field.

maximal matching See MATCHING.

maximal torus See TORUS.

maximin strategy See CONSERVATIVE STRATEGY.

maximum See GLOBAL MAXIMUM, LOCAL MAXIMUM.

maximum likelihood estimator The *estimator for an unknown parameter given by the method of maximum likelihood. See LIKELIHOOD.

maximum modulus theorem A *holomorphic function $f:D \to \mathbb{C}$, defined on an open (see OPEN SET) *connected domain D such that $|f(z)|$ has a *local maximum in D, is constant. A similar result, known as the *maximum principle*, applies to *harmonic functions.

Maxwell, James Clerk (1831–79) Influential Scottish mathematical physicist. Of considerable importance in *applied mathematics are the differential equations known as *Maxwell's equations, which are fundamental to the field theory of *electromagnetism. He was also one of the first to be interested in the classification of *knots.

Maxwell's equations A set of *partial differential equations summarizing the relationships between electricity and magnetism:

$\text{div}\mathbf{E} = \frac{\rho}{\varepsilon_0}$	(*Gauss' flux theorem)
$\text{div}\mathbf{B} = 0$	(Gauss' law for magnetism)
$\text{curl}\mathbf{B} - \frac{1}{c^2}\frac{\partial \mathbf{E}}{\partial t} = \mu_0 \mathbf{J}$	(Maxwell-Ampère circuital law)
$\text{curl}\mathbf{E} + \frac{\partial \mathbf{B}}{\partial t} = 0$	(Faraday's Law of Induction)

Here $\mathbf{B}$ is the magnetic field, $\mathbf{E}$ is the electric field, $\mathbf{J}$ is the current density, ρ is the charge density, ε_0 denotes vacuum permittivity, μ_0 denotes vaccuum permeability, c is the *speed of light, and t is time. Maxwell's equations unified theories of magnetism, electricity, light, and radiation. Also, importantly, they are not invariant under *Galilean relativity but are invariant under *Lorentz transformations.

mean The mean of the numbers $a_1, a_2, \ldots, a_n$ is equal to

$$\frac{a_1 + a_2 + \ldots + a_n}{n}.$$

This is the number most commonly used as the average. It may be called the arithmetic mean to distinguish it from other means such as those described below. When each number a_i is to have weight $w_i > 0$, the weighted mean is equal to

$$\frac{w_1 a_1 + w_2 a_2 + \ldots + w_n a_n}{w_1 + w_2 + \ldots + w_n}.$$

The geometric mean of the positive numbers $a_1, a_2, \ldots, a_n$ is

$$\sqrt[n]{a_1 a_2 \ldots a_n}.$$

It is a theorem of elementary algebra that, for any positive numbers $a_1, a_2, \ldots, a_n$, the arithmetic mean is greater than or equal to the geometric mean; that is,

$$\frac{1}{n}(a_1 + a_2 + \ldots + a_n) \geq \sqrt[n]{a_1 a_2 \ldots a_n}.$$

The harmonic mean of positive numbers $a_1, a_2, \ldots, a_n$ is the number h such that $1/h$ is the arithmetic mean of $1/a_1, 1/a_2, \ldots, 1/a_n$. Thus

$$h = \frac{n}{(1/a_1) + \ldots + (1/a_n)}.$$

For any set of positive numbers, $a_1, a_2, \ldots, a_n$ the harmonic mean is less than or equal to the geometric mean, and therefore also the arithmetic mean, and equality takes place for all three measures if and only if all the terms in the set are equal.

In *statistics, the mean of a *sample of *observations $x_1, x_2, \ldots, x_n$ is called the sample mean and is denoted by $\bar{x}$ so that

$$\bar{x} = \frac{x_1 + x_2 + \ldots + x_n}{n}.$$

The mean of a *random variable X is equal to the *expected value $E(X)$. This may be called the *population mean and denoted by μ. A *sample mean $\bar{x}$ may be used to estimate μ.

   See also ARITHMETIC-GEOMETRIC MEAN ITERATION.

mean absolute deviation For the sample $x_1, x_2, \ldots, x_n$, with mean $\bar{x}$, the mean absolute deviation is equal to

$$\frac{\sum |x_i - \bar{x}|}{n}.$$

It is a measure of *dispersion but is not often used as *variance has preferable algebraic properties.

mean curvature The mean curvature of a *surface is

$$H = \frac{LG - 2MF + NE}{2(EG - F^2)}$$

where $E, 2F, G$ and $L, 2M, N$ are respectively the coefficients of the *first and *second fundamental forms. *See* MINIMAL SURFACE.

mean deviation A synonym for MEAN ABSOLUTE DEVIATION.

mean squared deviation A synonym for VARIANCE.

mean squared error For an *estimator X of a parameter θ, the mean squared error is equal to the *expected value of $(X - \theta)^2$. For an unbiased estimator this equals the *variance of X.

mean value (of a function) Let f be a function that is *integrable on the closed interval $[a, b]$. The mean value of f over the interval $[a, b]$ is equal to

$$\frac{1}{b - a} \int_a^b f(x) \mathrm{d}x.$$

The mean value $\bar{y}$ has the property that the area under the curve $y = f(x)$ between $x = a$ and $x = b$ is equal to the area under the horizontal line $y = \bar{y}$ between $x = a$ and $x = b$, which is the area of a rectangle with width $b - a$ and height $\bar{y}$.

mean value theorem Let f be a function that is *continuous on $[a, b]$ and *differentiable in (a, b). The mean value theorem then states that there is a number c with $a < c < b$ such that

$$f'(c) = \frac{f(b) - f(a)}{b - a}.$$

This can be equivalently expressed as: there is a point C on the graph of f

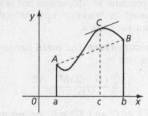

Mean gradient achieved at *C*

where the *tangent is parallel to the line segment joining $A(a,f(a))$ to $B(b,f(b))$. If A, with coordinates $(a, f(a))$, and B, with coordinates $(b, f(b))$, are the points on the graph corresponding to the end-points of the interval, there must be a point C on the graph between A and B at which the tangent is parallel to the chord AB.

*Rolle's Theorem is a special case of the mean value theorem. *Taylor's Theorem is an extension of the mean value theorem. The mean value theorem has two immediate corollaries:

(i) if $f'(x) = 0$ for all x, then f is a *constant function,
(ii) if $f'(x) > 0$ for all x, then f is *strictly increasing.

mean value theorem (for integrals)
Let f and g be two *real functions on the interval $[a,b]$ and assume that $g(x) > 0$ for all x.
If $m \leq f(x) \leq M$ for all x, then

$$m \int_a^b g(x) \ dx \leq \int_a^b f(x)g(x)dx$$
$$\leq M \int_a^b g(x) \ dx.$$

If f is a *continuous function on $[a,b]$, then there exists c in (a,b) such that

$$\int_a^b f(x)g(x) \ dx = f(c)\int_a^b g(x) \ dx.$$

measurable function A *function $f{:}X \rightarrow Y$ between two *measure spaces X and Y is measurable if whenever $S \subseteq Y$ is a *measurable set, the *pre-image $f^{-1}(S) \subseteq X$ is a measurable set.

measurable set If μ is a *measure defined on a *sigma algebra, Σ, of a set S, then the elements of Σ are called μ-measurable sets or measurable sets. *See* VITALI SET.

measure Measure is a generalization of the notions of *length, *area, and *volume. Given a set S with a *sigma algebra Σ of subsets of S, a measure is a map $\mu{:} \Sigma \rightarrow [0,\infty]$ such that $\mu(\varnothing)$, and if $\{A_k\}$ is a sequence of *disjoint subsets in Σ, then

$$\mu\left(\bigcup_{k=1}^{\infty} A_k\right) = \sum_{k=1}^{\infty} \mu(A_k).$$

This last requirement is called 'countable additivity'. A Banach measure need only be finitely additive. *Compare* HAAR MEASURE, HAUSDORFF DIMENSION, LEBESGUE MEASURE, OUTER MEASURE.

measurement The role of measurement from the point of view of *quantum theory, compared with *classical mechanics, is strikingly different. In classical mechanics, an experiment may be designed to measure a physical system with negligible effect on the system. In quantum theory, measurement can fundamentally change the state of a system. A particle's state is described by a wave function $\psi(\mathbf{x},t)$ which satisfies *Schrödinger's equation. An *observable $\mathbf{A}$ is represented by a *Hermitian operator which has *orthonormal *eigenstates $\psi_n(\mathbf{x},t)$. According to the *Copenhagen interpretation, the particle is in a superposition of these eigenstates so that

$$\psi = \sum_{n=0}^{\infty} c_n \psi_n$$

with $\sum_0^{\infty} |c_n|^2 = 1$. Measurement of the system then randomly selects from these eigenstates and measures the observable as the eigenvalue of ψ_n with probability $|c_n|^2$ and collapses the wave function to $c_n\psi_n$. *See* HEISENBERG UNCERTAINTY PRINCIPLE.

measure space A set S, with a *sigma algebra, Σ of subsets of S, and a *measure μ defined on Σ. The measure space is said to be finite if $\mu(S)$ is finite. *See also* PROBABILITY SPACE.

measure theory The area of *mathematics relating to the study of *measurable sets, *measurable functions, and *measure spaces.

measure zero set A synonym for NULL SET.

mechanical (mechanistic) A procedure which does not require interpretation or judgemental decisions to be made; for example, the solutions of *quadratic equations by the *quadratic formula. Such procedures should be relatively easy to *program into a *computer.

mechanical advantage *See* MACHINE.

mechanics The study of the behaviour of systems of *particles or *rigid bodies acted on by *forces, whether in *equilibrium or in motion. *Classical mechanics is restricted to systems where *Newton's laws of motion are adequate—broadly speaking, any system where great *gravity or speeds are not encountered, where the general and special theories of *relativity apply. *See also* QUANTUM THEORY.

median (in probability and statistics) Suppose that the *observations in a set of numerical data are ranked in ascending order. Then the (sample) median is the middle observation if there are an odd number of observations; it is the average of the two middlemost observations if there are an even number.

The median of a continuous *distribution with *probability density function f is any number m such that

$$\int_{-\infty}^{m} f(x)\,\mathrm{d}x = 0.5.$$

In this case the median need not be unique.

median (of a triangle) A *line through a *vertex of a *triangle and the *midpoint of the opposite side. The three medians are *concurrent at the *centroid.

median–median regression line The *least squares line of *regression is relatively easy to compute and has a sound theoretical foundation to justify its use, but *outliers can have a large effect on the line. The median–median regression line is a more *resistant alternative. To obtain the line, divide the *data points into three equal groups by size of x (if there is one extra, make the middle group one larger, and if there are two extra make the low and high groups larger). For the low and high groups, obtain the medians of the x and y values separately, giving one summary point for each group. The median–median regression line is the line joining these two points.

median triangle The triangle found by joining the *midpoints of the three sides of a triangle. It is always *similar to the original triangle with each side half the length of the original.

Compare PEDAL TRIANGLE.

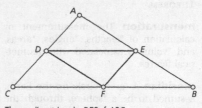

The median triangle *DEF* of *ABC*

mega- Prefix used with *SI units to denote multiplication by 10^6. Abbreviated as M. For binary mega, *see* KILO (BINARY).

member A synonym for ELEMENT.

memoryless source In the fields of *coding theory and *information theory, a memoryless source generates identically distributed, *independent source words. The (Shannon) *entropy of the source is then a lower bound to how quickly the source can be encoded. *See* SHANNON'S THEOREM.

Menelaus of Alexandria (about AD 100) Greek mathematician whose only surviving work is an important treatise on spherical geometry (*see* SPHERICAL TRIGONOMETRY) with applications to astronomy. In this, he was the first to use *great circles extensively and to consider the properties of *spherical triangles. He extended to spherical triangles the theorem about planar triangles that is named after him.

Menelaus' Theorem Let L, M, and N be points on the sides BC, CA, and AB of a triangle (possibly extended). Then Menelaus' Theorem states that L, M, and N are collinear if and only if

$$\frac{BL}{LC} \cdot \frac{CM}{MA} \cdot \frac{AN}{NB} = -1.$$

(Note that BC, for example, is considered to be directed from B to C so that LC, for example, is positive if LC is in the same direction as BC and negative if it is in the opposite direction.) *Compare* CEVA'S THEOREM.

mensuration The measurement or calculation of *lengths, *angles, *areas, and *volumes associated with geometrical figures.

meridian A *great circle on the Earth, assumed to be a *sphere, through the north and south poles or such a semi-circle between poles. More generally, a meridian on a *surface of revolution is a copy of the generating curve that has been rotated through some fixed angle. The meridians are then *normal to the *latitudes.

meromorphic function A complex-valued function on an *open subset S of the complex plane that is *holomorphic on all of S except for a set of *isolated points which are *poles of the function. Equivalently a meromorphic function can be considered as a holomorphic function from S to the *Riemann sphere.

Mersenne, Marin (1588–1648) French monk, philosopher, and mathematician who provided a valuable channel of communication between such contemporaries as *Descartes, *Fermat, *Galileo, and *Pascal: 'To inform Mersenne of a discovery is to publish it throughout the whole of Europe.' In an attempt to find a formula for *prime numbers, he considered the numbers $2^p - 1$, where p is prime. Not all such numbers are prime: $2^{11} - 1$ is not. Those that are prime are called *Mersenne primes.

Mersenne prime A *prime of the form $2^p - 1$, where p is a prime. The number of known primes of this form is over 50 and keeps increasing as they are discovered by using computers. In 2020, the largest known was $2^{82589933} - 1$. Each Mersenne prime gives rise to an even *perfect number. It is an *open problem whether there are infinitely many Mersenne primes.

⊕ SEE WEB LINKS
• Home page of the Great Internet Mersenne Prime Search.

mesh *See* PARTITION (OF AN INTERVAL).

mesokurtic *See* KURTOSIS.

metalogic (metamathematics) *See* PROOF THEORY.

method of differences *See* TELESCOPING SERIES.

method of exhaustion *See* ARCHIMEDES' METHOD.

method of images A technique for *partial differential equations with *boundary conditions. For example, in the figure
there is a *source in the upper half-plane with a wall or barrier along the *x*-axis. The boundary condition at the wall is

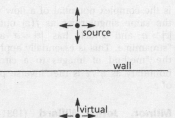

A bounding wall

that there is no *flux through the wall. The solution in the upper half-plane might also be arrived at without the wall if there were a virtual source of

wall

virtual vortex   virtual vortex

wall

virtual vortex   virtual vortex

Two perpendicular bounding walls

equal strength at the source's mirror image. This latter equivalent problem is then easier to solve.

Likewise, if a *vortex was placed in the first *quadrant, with walls along the positive *x*- and *y*-axes, then three virtual vortices—of equal strength, but two with opposite *circulation—can be positioned appropriately in the other three quadrants so to meet the zero flux boundary conditions at the walls.

method of least squares *See* LEAST SQUARES.

metre In *SI units, the base unit used for measuring length, abbreviated to 'm'. It was once defined as one ten-millionth of the distance from one of the poles to the equator, and later as the length of a platinum–iridium bar kept under specified conditions in Paris. Now it is defined in terms of the wavelength of a certain light wave produced by the krypton-86 atom.

metric (distance function) The nonnegative function d used to define a *distance between two points in constructing a *metric space, with the following properties:

$$d(x, y) = 0 \text{ if and only if } x = y,$$
$$d(x, y) = d(y, x),$$
$$d(x, y) + d(y, z) \geq d(x, z)$$

for any points x, y, z.
See also NORM, PSEUDOMETRIC, QUASI-METRIC, SEMI-METRIC, ULTRAMETRIC.

metric space A set of points together with a *metric (distance function) defined on it. *See* EUCLIDEAN SPACE. *Compare* TOPOLOGICAL SPACE.

metrizable A *topological space is metrizable if the *open sets arise from a metric on the set. A *separable *Hausdorff space is metrizable if and only if it is *regular and has a *countable *basis.

micro- Prefix used with *SI units to denote multiplication by 10^{-6}. Abbreviated as μ.

midpoint Points A and B, with *position vectors **a** and **b**, have midpoint with position vector $\frac{1}{2}(\mathbf{a}+\mathbf{b})$. *See* INTERNAL DIVISION, EXTERNAL DIVISION.

midpoint rule Possibly the simplest version of numerical *integration. If a definite integral $\int_a^b f(x)dx$ is to be calculated using n strips of width $h = (b-a)/n$ then the area under the curve is replaced by a series of *rectangles of width h with height equal to the value of the function at its midvalue. Labelling the midpoints $m_1, m_2, \ldots, m_n$ gives the approximation

$$\int_a^b f(x)dx = h \times \{f(m_1) + f(m_2) + \ldots + f(m_n)\}.$$

midpoint theorem The line joining the midpoints of two sides of a triangle is parallel to the third side and half its length.

In the figure for the *median triangle DE, EF, and DF are parallel to and half the length of BC, AC, and AB respectively.

Millennium Prize problems A set of seven classic unsolved problems, which carry a prize of 1 million US dollars each. They were published by the Clay Mathematics Institute on 24 May 2000. They are: P versus NP; the Hodge conjecture; the *Riemann hypothesis; Yang–Mills existence and mass gap; *Navier–Stokes existence and smoothness; the Birch and Swinnerton-Dyer conjecture. The Poincaré conjecture was solved in 2006 by Grigori Perelman. For further details, *see* APPENDIX 23.

⊕ SEE WEB LINKS
• The Millennium Prize problems home page.

milli- Prefix used with *SI units to denote multiplication by 10^{-3}. Abbreviated as m.

Milne-Thomson circle theorem This theorem describes planar fluid flow around a circle. If $w = f(z)$ is the *complex potential of a flow, with all its *singularities satisfying $|z| > a$, then

$$w = f(z) + \overline{f\left(\frac{a^2}{\bar{z}}\right)}$$

is the complex potential of a flow with the same singularities as $f(z)$ outside $|z| > a$ and which has $|z| = a$ as a *streamline. This is essentially applying the *method of images to a circular boundary as $a^2/\bar{z}$ is the *inverse point of z.

Milnor, John Willard (1931–) American mathematician, awarded the *Fields Medal in 1962 for work in *differential topology: he had shown that there are different smooth structures on the 7-dimensional sphere and had also disproved that two *triangulations of a space can be made isomorphic by subdivision. He was also the winner of the *Wolf Prize in 1989. He won the 2011 *Abel Prize 'for pioneering discoveries in *topology, *geometry, and *algebra'. More recently his interests have been in *dynamical systems.

Minding's theorem Any two smooth *surfaces with equal, constant *Gaussian curvature are locally *isometric.

minimal element (minimum) In a *partially ordered set $(S, \leq)$, a minimum is an element m such that $m \leq s$ for all $s \in S$. A minimal element is an element M such that there is no element s with $s < M$. So a minimum is a minimal element, but a minimal element need not be a minimum.

minimal polynomial (of an algebraic number) The least degree *monic polynomial which has the *algebraic number as a *root.

minimal polynomial (of a square matrix) The least degree *monic polynomial which the *square matrix satisfies. The minimal polynomial divides all other polynomials that the matrix satisfies and so, by the *Cayley-Hamilton theorem, has degree at most the *order of the matrix. A matrix is *diagonalizable if and only if the minimal polynomial is a product of distinct linear factors.

minimal surface A surface with zero *mean curvature, such as the *catenoid or *helicoid. Such surfaces locally minimize their area and so can be physically investigated using soap films.

minimax strategy See CONSERVATIVE STRATEGY.

Minimax Theorem A synonym for FUNDAMENTAL THEOREM OF GAME THEORY.

minimum See GLOBAL MINIMUM, LOCAL MINIMUM.

minimum capacity (in a network) The minimum flow along an edge. Only of interest when it is more than zero.

minimum connector problem The problem in *graph theory to find a *minimum cost spanning tree of a *connected *weighted graph. In practical terms, this is the situation faced by service companies laying cables or pipes, which is essentially different from constructing any sort of transport links between the points. *Kruskal's algorithm and *Prim's algorithm are two standard approaches to solving this problem.

minimum cost spanning tree In a connected *weighted graph, a *spanning tree such that the sum of the weights of its edges is minimal.

Minkowski, Hermann (1864–1909) Mathematician, born in Lithuania of German parents, a close friend of *Hilbert, who made significant contributions to the development of *relativity and obtained important results in *number theory.

Minkowski's inequality Minkowski's *inequality on $\mathbb{R}^n$ states that

$$\|\mathbf{x} + \mathbf{y}\|_p \leq \|\mathbf{x}\|_p + \|\mathbf{y}\|_p$$

where $p \geq 1$. This is the *triangle inequality for the *p-norm.

Minkowski space The four-dimensional space-time of special *relativity. It is $\mathbb{R}^4$ with the Minkowski 'metric'

$$g(\mathbf{u}, \mathbf{v}) = u_0 v_0 - u_1 v_1 - u_2 v_2 - u_3 v_3.$$

The linear isometries (see LINEAR MAP) of Minkowski space are the *Lorentz transformations.

minor arc See ARC.

minor axis See ELLIPSE.

minus The term may refer to *subtraction so that 3–5 may be read as '3 minus 5'. The term may also refer to *negative quantities with –5 read as 'minus five'.

minus or plus See PLUS OR MINUS.

minus sign The *symbol − used to denote *subtraction of numbers. It can also be used to denote a *negative quantity.

minute (angular measure) See DEGREE (angular measure).

mirror-image See REFLECTION (of the plane).

Mirzakhani, Maryam (1977–2017) Iranian mathematician whose work was in the *geometry and *dynamics of *Riemann surfaces. In 2014 she became the first female winner of a *Fields Medal. Her birthday, 12 May, is now

International Women in Mathematics Day, which coincides with Florence *Nightingale's birthday.

mixed boundary condition *See* ROBIN BOUNDARY CONDITION.

mixed derivative (mixed partial derivative)
A higher *partial derivative involving more than one *independent variable; for example $\frac{\partial^3 f}{\partial^2 x \partial y}$. If the mixed partial derivatives $\frac{\partial^2 f}{\partial x \partial y}$ and $\frac{\partial^2 f}{\partial y \partial x}$ both exist and are *continuous functions, then they are equal; that is, the order of partial differentiation does not matter.

mixed numbers *See* FRACTION.

mixed strategy
In a *matrix game, if a player chooses the different rows (or columns) with certain *probabilities, the player is using a mixed strategy. If the game is given by an $m \times n$ matrix, a mixed strategy for one player, R, is given by an m-tuple $\mathbf{x} = (x_1, x_2, \ldots, x_m)$ which is a *probability vector. Here x_i is the probability that R chooses the ith row. If $\mathbf{x} = \left(0, \frac{1}{4}, \frac{3}{4}\right)$, then R never chooses the first row and chooses the third row three times as often as the second row. A mixed strategy for the other player, C, is defined similarly.

Möbius, August Ferdinand (1790–1868)
German astronomer and mathematician, whose work in mathematics was mainly in *geometry and *topology. One of his contributions was his conception of the *Möbius band.

Möbius band (Möbius strip)
A continuous band with one twist in it. Between any two points on it, a continuous curve can be drawn on the *surface demonstrating that the band is one-sided and likewise has only one edge. Formally, the Möbius band is *non-orientable. The Möbius band was independently discovered by *Möbius and Johann Listing in 1858.

(((@))) SEE WEB LINKS
• The Möbius band in art.

Möbius function
An *arithmetic function important in *number theory. For a

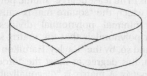
Möbius band

*natural number n, then $\mu(n)$ equals the sum of the *primitive *nth roots of unity. So $\mu(n) = 0$ if n has a repeated *prime factor, and otherwise $\mu(n) = 1$ or -1, depending on whether n has an even or odd number of prime factors.

Möbius transformation
A *bijective *conformal map of the *extended complex plane or *Riemann sphere. A Möbius transformation has the form $f(z) = (az + b)/(cz + d)$, where a,b,c,d are complex numbers such that $ad \neq bc$. Note that $f(z)$ is understood to be ∞ if $cz + d = 0$ and $f(\infty) = a/c$ if $c \neq 0$ and $f(\infty) = \infty$ if $c = 0$. The Möbius transformations form a group under composition; if the extended complex plane is identified with the projective complex line (*see* PROJECTIVE SPACE), then this group is $PGL(2, \mathbb{C})$.

mode
The most frequently occurring *observation in a *sample or, for grouped data, the group with the highest frequency. For a continuous *random variable, a value at which the *probability density function has a *global maximum is called a mode.

model *See* MATHEMATICAL MODEL.

model theory
The branch of *logic concerned with mathematical structures—interpretations or 'models' of formal systems, such as *groups, *rings, or

*graphs, with close links to *abstract algebra.

modular arithmetic *See* MODULO *N* ARITHMETIC.

modular form A modular form of weight k is a *holomorphic function on the upper half plane Im$z > 0$, such that $f((az + b)/(cz + d)) = (cz + d)^k f(z)$ and $f(z)$ can be expressed as a *Fourier series

$$f(z) = \sum_{n=0}^{\infty} c(n)e^{2\pi inz},$$

where the Fourier coefficients $c(n)$ are integers. Note that $f(z + 1) = f(z)$ and so modular forms have *period 1. The only function with odd weight is the zero function.

modular group The group of *Möbius transformations $z \mapsto (az + b)/(cz + d)$ under *composition, where a,b,c,d are integers and $ad - bc = 1$. The group is generated by the maps $z \mapsto z + 1$ and $z \mapsto -1/z$ and *acts on the upper half plane Im$z > 0$. The modular group can identified with the group SL(2,$\mathbb{Z}$) of 2×2 *matrices with integer entries and *determinant 1, provided a matrix is considered equivalent to its negative. *See* MODULAR FORM.

modularity theorem A profound theorem, stating that *elliptic curves over the *rational numbers are connected with *modular forms. The result was first conjectured by Yutaka Taniyama and Goro Shimura in 1957. Andrew *Wiles proved the theorem for an important subset of elliptic curves (the so-called semistable curves) and deduced from this *Fermat's Last Theorem (as *Frey's curve is not modular). The theorem was completely proved by 2001, as a result of papers from Wiles's students, including

Richard *Taylor. *See also* LANGLANDS PROGRAMME.

module An equivalent idea to a *vector space but where *scalars are from a *ring which need not be a *field. Given a ring R with a multiplicative identity 1, an R-module M is an *abelian group together with a notion of scalar multiplication $(r,m) \mapsto r \cdot m$, which further satisfies

$$(r_1 + r_2) \cdot m = r_1 \cdot m + r_2 \cdot m,$$
$$(r_1 r_2) \cdot m = r_1 \cdot (r_2 \cdot m),$$
$$r \cdot (m_1 + m_2) = r \cdot m + r \cdot m_2,$$
$$1 \cdot m = m,$$

for all r_1, r_2, r in R and m_1, m_2, m in M. Typically R is further assumed to be a *principal ideal domain or a *Euclidean domain. $\mathbb{Z}$-modules are abelian groups and vice versa. An $F[x]$-module, where F is a field, is a vector space V over F together with a *linear map $x{:}V \to V$ defining scalar multiplication by x. Modules are intimately connected with *representation theory. *See* FREE MODULE, JORDAN NORMAL FORM, RATIONAL CANONICAL FORM, SMITH NORMAL FORM, STRUCTURE THEOREM, TORSION ELEMENT.

modulo *n* arithmetic The word 'modulo' means 'to the modulus'. For any $n \geq 2$, let $\mathbb{Z}_n$ be the *complete set of residues $\{0, 1, 2, \ldots, n-1\}$. Then addition modulo n on $\mathbb{Z}_n$ is defined as follows. For a and b in $\mathbb{Z}_n$, take the usual sum of a and b as integers, and let r be the element of $\mathbb{Z}_n$ to which the result is *congruent (modulo n); the sum $a + b$ (mod n) is equal to r. Similarly, multiplication modulo n is defined by taking ab (mod n) to be equal to s, where s is the element of $\mathbb{Z}_n$ to which the usual product of a and b is congruent (modulo n). For example, addition and multiplication modulo 5 are given by the following tables:

+	0	1	2	3	4
0	0	1	2	3	4
1	1	2	3	4	0
2	2	3	4	0	1
3	3	4	0	1	2
4	4	0	1	2	3

Addition modulo 5

×	0	1	2	3	4
0	0	0	0	0	0
1	0	1	2	3	4
2	0	2	4	1	3
3	0	3	1	4	2
4	0	4	3	2	1

Multiplication modulo 5

When n is prime, as in these tables, then $\mathbb{Z}_n$ is a *field. Generally $\mathbb{Z}_n$ is *isomorphic to the *quotient ring $\mathbb{Z}/n\mathbb{Z}$.

modulus (moduli) If z is a *complex number and $z = x + yi$, the modulus of z, denoted by $|z|$ (read as 'mod z'), is equal to $\sqrt{x^2 + y^2}$

Modulus is multiplicative, which means $|zw| = |z||w|$ for complex numbers z, w, and also satisfies the *triangle inequality $|z + w| \leq |z| + |w|$. If z is represented by the point P in the *complex plane, the modulus of z equals the distance $|OP|$. Thus $|z| = r$, where (r, θ) are the polar coordinates of P. If z is real, the modulus of z equals the *absolute value of the real number, so the two uses of the same notation are consistent, and the term 'modulus' may be used for 'absolute value'.

The real modulus function $|x|$ is an example of a function which is not *differentiable. Its left derivative at 0 is –1, but its right derivative is 1.

modulus of a congruence See CONGRUENCE, MODULO N ARITHMETIC.

modulus of elasticity A parameter, usually denoted by λ associated with the material of which an *elastic string is made. It occurs in the constant of proportionality in *Hooke's law.

modus ponens The rule of inference which states that if a *conditional statement and the *antecedent contained in it are both true, then the *consequent statement in it must also be true. Also known as the rule of detachment. In logical notation (see LOGIC), if p and $p \Rightarrow q$ are both true, then q is true.

modus tollens The rule of inference which states that if a *conditional statement and the *negation of the *consequent contained in it are both true, then the *antecedent statement must be false. In logical notation (see LOGIC), if $p \Rightarrow q$ is true, and q is false, then p is false.

moment (in mechanics) A means of describing the turning effect of a *force about a point. For a system of *coplanar forces, the moment of one of the forces **F** about any point A in the plane can be defined as the product of the *magnitude of **F** and the distance from A to the *line of action of **F** and is considered to be acting either clockwise or anticlockwise. For example, suppose that forces with magnitudes F_1 and F_2 act at B and C, as shown in the figure. The moment of the first force about A is $F_1 d_1$ clockwise, and the moment of the second force about A is $F_2 d_2$ anticlockwise. The *principle of moments considers when a system of coplanar forces produces a state of *equilibrium.

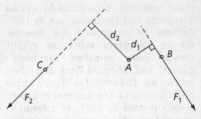

Moments about A

However, a better approach is to define the moment of the force **F**, acting at a point B, about the point A as the vector $(\mathbf{r}_B - \mathbf{r}_A) \times \mathbf{F}$, where $\times$ denotes the *vector product and $\mathbf{r}_A$, $\mathbf{r}_B$ are the position vectors of A and B. The use of vectors not only eliminates the need to distinguish between clockwise and anticlockwise directions but facilitates the measuring of the turning effects of non-coplanar forces acting on a 3-dimensional body.

Similarly, for a particle P with position vector $\mathbf{r}$ and *linear momentum $\mathbf{p}$, the moment of the linear momentum of P about the point A is the vector $(\mathbf{r} - \mathbf{r}_A) \times \mathbf{p}$. This is the *angular momentum of the particle P about the point A.

Suppose that a *couple consists of a force $\mathbf{F}$ acting at B and a force $-\mathbf{F}$ acting at C. The moment of the couple about A is equal to

$$(\mathbf{r}_B - \mathbf{r}_A) \times \mathbf{F} + (\mathbf{r}_C - \mathbf{r}_A) \times (-\mathbf{F})$$
$$= (\mathbf{r}_B - \mathbf{r}_C) \times \mathbf{F},$$

which is independent of the position of A.

moment (in statistics) For a set of *observations $x_1, x_2, \ldots, x_n$, the jth (sample) moment about p is equal to

$$\frac{\sum (x_1 - p)^j}{n}.$$

For a *random variable X, the jth (population) moment about p is equal to $E((X-p)^j)$ (see EXPECTED VALUE). The first moment about 0 is the *mean. The second moment about the mean is the *variance. The jth moment of X, without further specification, refers to the jth moment about 0.

Suppose that a sample is taken from a population with k unknown parameters. In the method of moments, the first k population moments are equated to the first k sample moments to find k equations in k unknowns. These can be

solved to find the moment estimates of the parameters.

moment estimator The *estimator found by the method of *moments.

moment generating function $M(t) = E(e^{tX})$ is the moment generating function (mgf) for a *random variable X. When $M(t)$ exists then, since

$$e^{tX} = \sum_{r=0}^{\infty} \frac{t^r X^r}{r!}$$

and E is an *additive function, it follows that the coefficient of t^r is $r! E(X^r)$, allowing the *moments of X to be recovered from knowledge of the mgf.

moment of inertia A quantity relating to a *rigid body and a given axis, derived from the way in which the *mass of the rigid body is distributed relative to the axis. It arises in the calculation of the *kinetic energy and the *angular momentum of the rigid body in general motion. It replaces the role of mass in formulae for linear motion.

For a planar rigid body, rotating with *angular speed ω about an axis perpendicular to the plane, the moment of inertia is given by

$$I = \sum_i m_i r_i^2 \quad \text{or} \quad I = \int_R \rho(\mathbf{r}) |\mathbf{r}|^2 \, dA,$$

where the first expression is for a discrete collection of particles of mass m_i at distance r_i from the axis, and the second expression is for a continuous distribution of matter within a region R of density $\rho(\mathbf{r})$, where $\mathbf{r}$ is the position vector of a point from the axis. The angular momentum of the body equals $I\omega$, and its (rotational) kinetic energy equals $\frac{1}{2}I\omega^2$.

Note that the same rigid body will have different moments of inertia for different axes. For example, the moment of inertia of a uniform disc will be greater for an axis through a point of the circumference than through the centre, as the mass is distributed farther from the axis. Moments of inertia about other axes may be calculated by using the *parallel axis theorem and the *perpendicular axis theorem. For a three-dimensional rigid body, moments of inertia I_{xx}, I_{yy}, and I_{zz} can be defined for the x-axis, the y-axis, and the z-axis. Here r_i and **r** denote the distance and displacement from the axis. *See* INERTIA MATRIX, PRODUCT OF INERTIA.

For a list of various moments of inertia, *see* APPENDIX 3.

moment of momentum A synonym for ANGULAR MOMENTUM.

momentum *See* ANGULAR MOMENTUM, LINEAR MOMENTUM. Often 'momentum' is used to mean linear momentum.

Monge, Gaspard (1746–1818) French mathematician who was a high-ranking minister in Napoleon's government and influential in the founding of the École Polytechnique in Paris. The courses in *geometry that he taught there were concerned to a great extent with the subjects of his research: he was the originator of what is called *descriptive geometry and the founder of modern analytical solid geometry.

Monge point For a *tetrahedron, there are six planes which perpendicularly cut one edge and pass through the *midpoint of the opposite edge; these six planes are *concurrent at the Monge point.

monic polynomial A *polynomial in one variable in which the coefficient of the highest power is 1.

monoid A set with an *associative *binary operation which has an *identity element, for example, the *natural numbers and *addition. *Compare* GROUP, MAGMA, SEMIGROUP.

monomorphism A *morphism $f:X \to Y$ with the property that for all morphisms g_i between Z and X, $f \circ g_1 = f \circ g_2 \Rightarrow g_1 = g_2$. In *category theory, monomorphisms are the analogue of *injections. *Compare* EPIMORPHISM.

monotone convergence theorem Let $f_n:\mathbb{R} \to \mathbb{R}$, $n \geq 1$, be a sequence of Lebesgue integrable functions (*see* LEBESGUE INTEGRATION) which is increasing—that is, $f_{n+1}(x) \geq f_n(x)$ for all n,x—and such that the integrals $\int_{-\infty}^{\infty} f_n(x)\ dx$ are *bounded above. Then there exists an integrable function f such that $f_n(x)$ converges to $f(x)$ *almost everywhere and

$$\int_{-\infty}^{\infty} f(x)\ dx = \lim_{n\to\infty} \int_{-\infty}^{\infty} f_n(x)\ dx.$$

Compare DOMINATED CONVERGENCE THEOREM.

monotonic function (monotone function) A *real function f is monotonic on or a domain I if it is either increasing on I, so $f(x_1) \leq f(x_2)$ whenever $x_1 \leq x_2$, or decreasing on I, so $f(x_1) \geq f(x_2)$ whenever $x_1 < x_2$. Also, f is strictly monotonic if it is either *strictly increasing or strictly decreasing.

monotonic sequence (monotone sequence) A real sequence a_1, a_2, $a_3, \ldots$ is *monotonic if it is either increasing, so $a_i \leq a_{i+1}$ for all i, or decreasing, so $a_i \geq a_{i+1}$ for all i and strictly monotonic if it is either strictly increasing or strictly decreasing. A *bounded monotonic sequence *converges.

Monte Carlo methods The approximate solution of a problem using repeated *sampling *experiments and observing the *proportion of times some property is satisfied. Estimates of *probabilities may be computed by running a *simulation repeatedly and using the *relative frequency.

Monty Hall problem (three door problem) A counterintuitive problem in probability; the problem is loosely based on a game show once hosted by Monty Hall. A contestant is presented with three doors, behind one of which is a prize. The contestant chooses a door, and then the host opens a different door to reveal no prize. The contestant now has the chance to change their initial choice for the one remaining door: should they change?

Intuitively it may seem that there is no benefit in changing, as the choice is now just one of two doors. However, the chance of the initial choice being right is still the same as it was initially, namely 1 in 3. So there is a 2 in 3 chance that the prize is behind the other door, and so the contestant improves their chance of winning by changing.

Moore-Penrose inverse A synonym for PSEUDOINVERSE.

Morera's Theorem A *continuous function $f:U \to \mathbb{C}$ on an *open set U of $\mathbb{C}$, such that $\int_C f(z)\ dz = 0$ for all closed curves C in U, is *holomorphic. This is a partial converse of *Cauchy's integral theorem.

Mordell Conjecture The conjecture, made in 1922, stated that a curve over the *rational numbers of *genus more than 1 has only finitely many rational points. This was proven by Gerd Faltings in 1983 and so is often termed Faltings' Theorem now.

Morley's Theorem In any triangle, the points of intersection of adjacent *trisectors of the angles form the vertices of an *equilateral triangle.

It is remarkable that this elegant fact of *Euclidean geometry was not discovered until 1899 by Frank Morley (1860–1937).

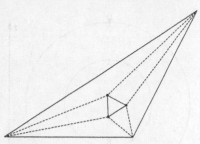
Morley's theorem

morphism An abstraction of a *function between two spaces. *See* CATEGORY.

Morse Theory Theory relating the *topology of a smooth *surface (or more generally a *manifold) with properties of smooth real functions on the surface, named after Marston Morse (1892–1977), who first published on the subject in 1925.

A Morse function on a smooth surface is a smooth real function with only nondegenerate *stationary points (*see* HESSIAN), and so these are local maxima, local minima, or saddle points. For any smooth function f on a closed, smooth manifold X, it is the case that

$$\chi(X) = \#\text{maxima} - \#\text{saddles} + \#\text{minima},$$

where $\chi(X)$ denotes the *Euler characteristic of X and #maxima denotes the number of maxima, etc. In the figure is a cross-section of a *torus and $f(x,y,z) = z$ is a Morse function on the torus. f has a maximum at D, a minimum at A, and saddle points at B and C. This correctly gives $\chi = 1 - 2 + 1 = 0$.

Morse theory includes results for the subsets of the surface of the form $f \leq k$. In the figure, such a subset is that part of the torus below the dotted lines. For any z-coordinate k between B and C Morse theory states the subsets are *homotopy equivalent. Further, the theory states how the homotopy type of the subset $f \leq k$ changes as k takes a stationary value.

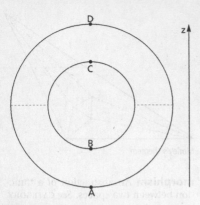

The height function on the torus

Moser's circle problem Named after Leo Moser, this result is commonly cited to highlight the need to prove that perceived patterns do indeed continue forever. The maximum number of regions that the *chords between n points on a circle divide the interior into is 1, 2, 4, 8, 16 for $n = 1, 2, 3, 4, 5$. (In the figure $n = 4$.)

It would seem reasonable to suppose the pattern continues as 2^{n-1}, but in fact

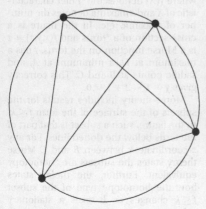

4 points and 8 regions

the next term is 31, and the nth term of the sequence is

$$1 + \binom{n}{2} + \binom{n}{4}.$$

moving average For a *sequence of *observations $x_1, x_2, x_3, \ldots$, the moving average of order n is the sequence of *arithmetic means

$$\frac{x_1 + \ldots + x_n}{n}, \frac{x_2 + \ldots + x_{n+1}}{n},$$
$$\frac{x_3 + \ldots + x_{n+2}}{n}, \ldots$$

A weighted moving average such as

$$\frac{x_1 + 2x_2 + x_3}{4}, \frac{x_2 + 2x_3 + x_4}{4},$$
$$\frac{x_3 + 2x_4 + x_5}{4}, \ldots$$

may also be used. Its main purpose is to offer a view of the underlying behaviour of volatile sequences, especially those which have periodic variation.

((()) SEE WEB LINKS

• Illustrations of the moving average from financial markets.

Müller, Johann *See* REGIOMONTANUS.

multi- Prefix denoting many, for example in *multicollinearity, *multinomial, *multivariate.

multicollinearity In *statistics, when two or more of the *explanatory variables in a multiple *regression analysis are very strongly *correlated.

multifunction (multivalued function) A multifunction is clearly not a *function in the usual sense, as functions may only take a single value at a given argument. However, such multifunctions arise naturally, particularly in *complex analysis, and various approaches can be taken to describe them.

Common examples are *square roots, *inverse trigonometric functions, and the *complex logarithm. For a positive real x, there are two choices of a square root $\pm\sqrt{x}$, and the function $\sqrt{x}$ is *well defined, as a choice is made to take the positive square root as the *principal value. Likewise, $\cos^{-1}x$ is well defined on $[-1,1]$ by taking principal values in $[0,\pi]$.

For the complex square root, aside from the positive real axis, there are no obvious principal values. Further, it is impossible to continuously (see CONTINUOUS FUNCTION) extend the notion of $\sqrt{x}$ for the positive reals to the entire complex plane. One approach is to limit the domain to a *cut plane and define a *branch (a continuous set of principal values) on the cut plane, which may lead to *discontinuities across the cut.

Alternatively, all the possible values of a multifunction may be considered simultaneously on a *Riemann surface. So the Riemann surface of $\sqrt{z}$ is the surface $w^2 = z$ in $\mathbb{C}^2$. On such a surface will be present both the points $(z, \sqrt{z})$ and $(z, -\sqrt{z})$.

multigraph Some authors use the term *graph to mean a *simple graph, so that loops and multiple edges are not permitted. By contrast, 'graphs' with loops and/or multiple edges are then referred to as multigraphs.

multilinear Let $V_1, V_2, \ldots, V_k, W$ be *vector spaces over the same *field. Generalizing the notion of *bilinear, a multilinear map $f{:}V_1 \times V_2 \times \ldots \times V_k \rightarrow W$ is a *function $f(v_1,\ldots,v_k)$ which is linear (see LINEAR MAP) in each variable v_i. As an example, the *determinant is multilinear in its *rows (and likewise in its *columns).

multinomial The general description for an algebraic expression that is a sum of two or more terms. However, where powers of variables, such as in $3x^4 + x^2 - x + 1$, are involved, 'polynomial' is more common, and for two or three terms 'binomial' and 'trinomial' are more typically used.

multinomial coefficient The number

$$\binom{n}{n_1 \quad n_2 \ \ldots \ n_k} = \frac{n!}{n_1!n_2!\ldots n_k!}$$

when $n_1 + n_2 + \ldots + n_k = n$.

This is the coefficient of $x_1^{n_1}x_2^{n_2}\ldots x_k^{n_k}$ in the expansion of $(x_1 + x_2 + \ldots x_k)^n$. It also gives the number of ways of choosing n_i of each of k types when choosing a total of n objects, without regard to the order in which they are chosen.

multinomial distribution The *distribution of the multinomial *random variable which is the generalized form of the *binomial distribution to allow more than two outcomes in each of the individual trials. So if $\{Y_i\}$, $i = 1, 2, \ldots, n$, is a series of *independent, identically distributed random variables with $\Pr\{Y_i = x_j\} = p_j$ for $j = 1, 2, \ldots, k$ with $\sum p_j = 1$ then the probabilities of the possible outcomes are given by the *coefficients in the expansion of $(p_1x_1 + p_2x_2 + \ldots + p_kx_k)^n$.

multinomial theorem A generalization of the binomial theorem, giving the expansions of positive integral powers of a *multinomial expression

$$(x_1 + x_2 + \ldots + x_k)^n =$$

$\sum \left(\frac{n!}{n_1!n_2!\ldots n_k!} x_1^{n_1} x_2^{n_2} \ldots x_k^{n_k} \right)$ where the sum is over all k-tuples $n_1, n_2, \ldots, n_k$ of non-negative integers which total n.

multiple See DIVIDES.

multiple integral (repeated integral) An integral involving two or more integrations with respect to different variables, where the integrations are carried out sequentially and variables other than the one being integrated are treated as constants. A double integral is the special case where two integrations are involved. Multiple integrals can be definite or indefinite. For example,

$$\int_0^3 \int_0^2 (x + 3y + 5)\mathrm{d}x\,\mathrm{d}y$$

$$= \int_0^3 \left[\frac{x^2}{2} + 3xy + 5x\right]_0^2 \mathrm{d}y$$

$$= \int_0^3 (6y + 7)\mathrm{d}y = [3y^2 + 7y]_0^3 = 48.$$

This double integral represents the *volume under the plane $z = x + 3y + 5$ that lies above the rectangle $0 \le x \le 2$, $0 \le y \le 3$.

multiple precision See PRECISION.

multiple regression See REGRESSION.

multiple root See ROOT (of an equation).

multiplication Generally, a *binary operation in which the two entities, which can both be *numbers, *matrices, real *functions, etc. See POINTWISE MULTIPLICATION, SCALAR MULTIPLICATION.

multiplication (of complex numbers) For complex numbers z_1 and z_2, given by $z_1 = a + bi$ and $z_2 = c + di$, the product is defined by

$$z_1 z_2 = (ac - bd) + (ad + bc)i.$$

In some circumstances it may be more convenient to write z_1 and z_2 in *polar form, thus:

$$z_1 = r_1 e^{i\theta_1} = r_1(\cos\theta_1 + i\sin\theta_1),$$
$$z_2 = r_2 e^{i\theta_2} = r_2(\cos\theta_2 + i\sin\theta_2).$$

Then

$$z_1 z_2 = r_1 r_2 e^{i(\theta_1 + \theta_2)}$$
$$= r_1 r_2 \Big(\cos(\theta_1 + \theta_2) + i\sin(\theta_1 + \theta_2)\Big).$$

Thus 'you multiply the *moduli and add the *arguments'.

multiplication (of fractions) Fractions are multiplied by multiplying the *numerators and *denominators separately, that is,

$$\frac{a}{b} \times \frac{c}{d} = \frac{ac}{bd}.$$

If *mixed numbers are involved, they can be converted into *improper fractions before multiplication.

multiplication (of real numbers) The multiplication of *natural numbers is defined by repeated addition, so $3 \times 2 = 2 + 2 + 2 = 6$. Multiplication of (potentially negative) *integers can then be determined as with $-3 \times 2 = -(3 \times 2) = -6$. And this may be extended to *multiplication of fractions, and every *rational number may be expressed as a fraction. Finally, *real numbers can be considered as *limits of rational *sequences and their product taken to be the limit of the products of those sequences. See ALGEBRA OF LIMITS.

multiplication (of matrices) Suppose that matrices $\mathbf{A}$ and $\mathbf{B}$ are *conformable for multiplication; say that $\mathbf{A}$ has order $m \times n$ and $\mathbf{B}$ has order $n \times p$. Let $\mathbf{A} = [a_{ij}]$ and $\mathbf{B} = [b_{ij}]$. Then (matrix) multiplication is defined by taking the product $\mathbf{AB}$ to be the $m \times p$ matrix $\mathbf{C}$, where $\mathbf{C} = [c_{ij}]$ and

$$c_{ij} = a_{i1}b_{1j} + a_{i2}b_{2j} + \dots + a_{in}b_{nj} = \sum_{r=1}^n a_{ir}b_{rj}.$$

The product $\mathbf{AB}$ is not defined if $\mathbf{A}$ and $\mathbf{B}$ are not conformable for multiplication. Matrix multiplication is not *commutative; for example, if

$$\mathbf{A} = \begin{bmatrix} 0 & 1 \\ 0 & 0 \end{bmatrix} \text{ and } \mathbf{B} = \begin{bmatrix} 1 & 0 \\ 0 & 0 \end{bmatrix},$$

then $\mathbf{AB} \ne \mathbf{BA}$. Moreover, it is not true that $\mathbf{AB} = \mathbf{0}$ implies that either $\mathbf{A} = \mathbf{0}$ or $\mathbf{B} = \mathbf{0}$, as the above example shows. However, matrix multiplication is *associative: $\mathbf{A}(\mathbf{BC}) = (\mathbf{AB})\mathbf{C}$, and the *distributive laws hold: $\mathbf{A}(\mathbf{B} + \mathbf{C}) = \mathbf{AB} + \mathbf{AC}$ and $(\mathbf{A} + \mathbf{B})\mathbf{C} = \mathbf{AC} + \mathbf{BC}$.

multiplication (of polynomials) Polynomials are multiplied out by using the *distributive law of multiplication over addition, so each term in the first polynomial is multiplied by each term in the second. For example,

$$(3x^3 + 5x^2 - 3x + 2)(x^2 - 2x + 3)$$
$$= 3x^3(x^2 - 2x + 3) + 5x^2(x^2 - 2x + 3)$$
$$- 3x(x^2 - 2x + 3) + 2(x^2 - 2x + 3)$$

and then each of these brackets can be expanded before collecting like terms together.

multiplication modulo *n* *See* MODULO *N* ARITHMETIC.

multiplication sign The symbol $\times$ used to denote multiplication of number or any other binary operation with similar properties. Where characters and/or brackets are involved rather than just numbers, it is common to use a dot, or to use no symbol at all. So 3×5, $5(3 + 6)$, $4x^2y^3$, $x \cdot y$ all contain instructions to multiply quantities together.

multiplicative group A *group in which the operation is called multiplication. The mutliplicative group of a *field F is the group of non-zero elements under multiplication, denoted F^*.

multiplicative identity An *identity element under *multiplication, so for real and complex numbers 1 is the multiplicative identity and for 3×3 *matrices it is $\begin{pmatrix} 1 & 0 & 0 \\ 0 & 1 & 0 \\ 0 & 0 & 1 \end{pmatrix}$.

multiplicative inverse *See* INVERSE ELEMENT.

multiplicity *See* ROOT (of an equation).

multiplying factor (integrating factor) (in differential equations) *See* LINEAR FIRST-ORDER DIFFERENTIAL EQUATION.

multiply out To expand an expression which is in a contracted form. For example, $(3x + 1)(2x - 3) = 6x^2 - 7x - 3$.

multivariable calculus The branch of *calculus that deals with the *differentiation and *integration of functions of more than one variable. *See* DIFFERENTIAL (MULTIVARIATE), DIRECTIONAL DERVATIVE, DIVERGENCE THEOREM, FUBINI'S THEOREM, JACOBIAN, PARTIAL DIFFERENTIATION, STOKES' THEOREM, TONELLI'S THEOREM.

multivariate Relating to more than two random variables. *See* JOINT CUMULATIVE DISTRIBUTION FUNCTION, JOINT DISTRIBUTION, JOINT PROBABILITY DENSITY FUNCTION, JOINT PROBABILITY MASS FUNCTION.

multivariate analysis of variance (MANOVA) In *statistics, a generalized *analysis of variance (ANOVA) but where there are two or more *dependent variables.

mutually disjoint A synonym for PAIRWISE DISJOINT.

mutually exclusive A synonym for DISJOINT.

mutually exclusive events Events A and B are mutually exclusive if A and B cannot both occur; that is, if $A \cap B = \emptyset$. For mutually exclusive events, the probability that either one event or the other occurs is given by the addition law, $\Pr(A \cap B) = \Pr(A) + \Pr(B)$.

For example, when a die is thrown, the probability of obtaining a 'one' is $\frac{1}{6}$, and the probability of obtaining a 'two' is $\frac{1}{6}$. These events are mutually exclusive. So the probability of obtaining a 'one' or a 'two' is equal to $\frac{1}{6} + \frac{1}{6} = \frac{1}{3}$.

mutually prime A synonym for COPRIME.

mystic hexagram *See* PASCAL'S THEOREM.

n Abbreviation for *nano-.

N Symbol for *newton.

n- Prefix when denoting a finite but unspecified number in a description. For example, an n-gon is a polygon with n sides. n-ary relations generalize *binary relations. *See also* N-TUPLE.

nabla A synonym for DEL.

naïve set theory As opposed to *axiomatic set theory, a treatment of *set theory described in natural language such as the early work of *Cantor, *Frege, and *Russell. However, such treatment leads to paradoxes such as *Russell's paradox.

nand A synonym for NOT AND.

nano- Prefix used with *SI units to denote multiplication by 10^{-9}. Abbreviated as n.

Napier, John (1550–1617) Scottish mathematician responsible for the publication in 1614 of the first tables of *logarithms. His mathematical interests, pursued in spare time from church and state affairs, were in *spherical trigonometry and, notably, in computation. His logarithms were defined geometrically rather than in terms of a base. He defined the logarithm of x to be y, with

$$x = 10^7 \left(1 - \frac{1}{10^7}\right)^y.$$

As $(1 - 10^{-7})^{10^7} \approx 1/e$, Napier was effectively using base $1/e$ for his logarithms.

Naplerian logarithm *See* LOGARITHM.

Napier's bones An invention by Napier for carrying out multiplication. It consisted of a number of rods made of bone or ivory, which were marked with numbers from the multiplication tables. Multiplication using the rods involved some addition of digits and was very similar to the familiar method of long multiplication. *Logarithms were not involved at all.

Napoleon's theorem For an arbitrary triangle ABC, three *equilateral triangles ADB, BEC, CFA are placed on the triangle's edges; the *centroids of these

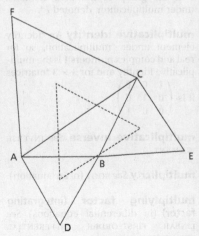

The centroids form the dotted equilateral triangle

three new triangles define a fourth equilateral triangle.

Nash, John Forbes (1928–2015)
American mathematician who won the
Nobel Prize for Economic Science in
1994 for his work in non-cooperative
*game theory which was published in
1950. Despite the importance of this
work in economics and in the diplomatic
and military strategy to which Nash
applied it for the RAND Corporation
during the Cold War, Nash viewed himself as a pure mathematician. He made
important contributions in real *algebraic geometry, the theory of *Riemannian manifolds and particularly on
parabolic and elliptic *partial differential
equations. He was thought to be in con-
tention for the *Fields Medal in 1958 but
was not successful. By the time the 1962
awards were being made, Nash's long
battle with schizophrenia was dominat-
ing his life. The book and film *A Beauti-
ful Mind* catalogued the story of his
brilliance and his struggle with mental
illness. He was awarded the *Abel Prize
in 2015 along with Louis Nirenberg.

Nash equilibrium A situation in a
non-cooperative game, as with the *prisoner's dilemma, where neither player
can unilaterally improve on their chosen
*mixed strategy, even though this
impasse may be worse than what is possible with cooperation. Nash showed in
1950 that there is a Nash equilibrium for
every finite game.

**National Centre for Excellence in
the Teaching of Mathematics
(NCETM)** Set up following the publication of Adrian *Smith's inquiry into the
state of post-14 mathematics education
in the UK to provide effective strategic
leadership for mathematics-specific professional development for teachers.

(⊕) **SEE WEB LINKS**
• The NCETM home page.

natural frequency *See* NORMAL
MODE.

natural logarithm *See* LOGARITHM.

natural number One of the numbers
1, 2, 3, ... Some authors also include 0.
The set of natural numbers is often
denoted by $\mathbb{N}$.

Navier-Stokes equations The Navier-Stokes equations for the flow of a
viscous *incompressible fluid with flow
*velocity **u** state

$$\text{div}\,\mathbf{u} = 0, \qquad \frac{D\mathbf{u}}{Dt} = -\frac{1}{\rho}\nabla p + \nu\nabla^2\mathbf{u} + \mathbf{g}.$$

Here ρ denotes the (constant) *density, p
denotes *pressure, ν denotes the constant viscosity, **g** denotes acceleration
due to *gravity, and D/Dt is the *convective derivative. Note that if $\nu = 0$, then
the Navier-Stokes equations become
*Euler's equations.

n-cube A *simple graph, denoted Q_n,
whose vertices and edges correspond to
the vertices and edges of an n-dimen-
sional *hypercube.
   Thus, there are 2^n vertices that can be
labelled with the *binary words of length
n, and there is an *edge between two
vertices if they are labelled with words
that differ in exactly one digit. The
graphs Q_2 and Q_3 are shown in the
figure.

n-dimensional space *See* EUCLIDEAN
SPACE.

**necessary and sufficient condi-
tion** *See* CONDITION, NECESSARY AND
SUFFICIENT.

necessary condition A condition
which is required to be true for a statement to be true. For example, for a number to be *divisible by 10 it is necessary
that the number be *even. Note that if A
is a necessary condition for B, B is a
*sufficient condition for A, and vice

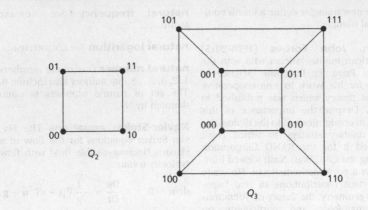

Q_2

Q_3

versa, so in the above example knowing that a number is divisible by 10 is sufficient to know that the number is even.

needle problem *See* BUFFON'S NEEDLE.

negation If p is a statement, then the statement 'not p', denoted by $\neg p$, is the negation of p. It states, in some suitable wording, the opposite of p. For example, if p is 'It is raining', then $\neg p$ is 'It is not raining'; if p is '$r > 2$', then $\neg p$ is '$r \leq 2$'. If p is true then $\neg p$ is false, and vice versa. So the *truth table of $\neg p$ is:

p	$\neg p$
T	F
F	T

See DE MORGAN'S LAWS.

negative *See* INVERSE ELEMENT.

negative binomial distribution The generalization of the *geometric distribution to the waiting time for the kth success in a sequence of *independent experiments, all with the same probabilities of success. The *probability mass function is given by

$$\Pr\{X = r\} = \binom{r-1}{k-1} p^k (1-p)^{r-k}.$$

This is because the kth success will be on the rth trial if the rth trial is a success and there are $k-1$ successful trials in the previous $r-1$ trials. *See* APPENDIX 15.

negative correlation *See* CORRELATION.

negative direction *See* DIRECTED LINE.

negative number A *real number which is less than zero.

neighbourhood (graph theory) The neighbourhood $N_G(v)$ of a vertex v of a *graph G consists of those *vertices which are *adjacent to v. For a subset W of the vertex set, $N_G(W)$ consists of those vertices which are adjacent to some vertex in W.

neighbourhood (topology) On the *real line, a neighbourhood of the real number a is a subset which contains an open interval $(a-\delta, a+\delta)$ for some $\delta > 0$.

In a *Euclidean space or *metric space, a neighbourhood of a point is a subset which contains an *open ball centred at the point. More generally, in a *topological space, a neighbourhood of a point is a subset which contains an *open set which includes the point.

nested multiplication *See* HORNER'S RULE.

nested sets A *sequence of sets A_1, A_2, A_3, ... is said to be nested and increasing if

$$A_1 \subseteq A_2 \subseteq A_3 \subseteq ...$$

and is said to be nested and decreasing if

$$A_1 \supseteq A_2 \supseteq A_3 \supseteq ....$$

net Remaining after all deductions have been made. For example, the net income from the sale of a house will be the sale price less fees paid to lawyers, estate agents, etc. involved in the sale. *Compare with* GROSS.

net A generalization of a *sequence. Whilst the topology of a *metric space can be characterized by *sequences, this is not generally true in *topological spaces. A net (x_λ) in a topological space X consists of points indexed by a set $(\Lambda, \leq)$ with the properties (i) $\lambda \leq \lambda$ for all λ (ii) if $\lambda_1 \leq \lambda_2$ and $\lambda_2 \leq \lambda_3$, then $\lambda_1 \leq \lambda_3$ (ii) for any λ_1, λ_2 there exists λ_3 such that $\lambda_1 \leq \lambda_3$ and $\lambda_2 \leq \lambda_3$. A net (x_λ) then converges to x if for any *open set U containing x, there exists μ such that whenever $\mu \leq \lambda$, then x_λ is in U. Then, for example, x is in the *closure of a set A if and only there is a net in A which converges to x.

net (of a solid) A plane figure which can be folded to construct a *polyhedron. There is typically more than one net for a given polyhedron.

A net of a square-based pyramid:

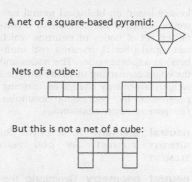

Nets of a cube:

But this is not a net of a cube:

network A *digraph in which every arc is assigned a *weight (normally some non-negative number). In some applications, something may be thought of as flowing or being transported between the vertices of a network, with the weight of each arc giving its *capacity. In some other cases, the vertices of a network may represent steps in a process and the weight of the arc joining u and v may give the time that must elapse between step u and step v.

network flow A set of non-negative values assigned to each arc of a *network which does not exceed the *capacity of that arc and for which the total amount entering and leaving each vertex is the same. Many *optimization problems can be characterized by networks in which it is required to maximize the flow.

Neumann, John Von *See* VON NEUMANN.

Neumann condition A *boundary condition for a *partial differential equation which specifies the outward *directional derivative $\partial u / \partial n = \nabla u \cdot \mathbf{n}$ at the *boundary. Here $\mathbf{n}$ denotes the outward-pointing *unit *normal. *Compare* DIRICHLET PROBLEM, ROBIN BOUNDARY CONDITION.

neural nets (neural networks) A neural net is a computational system loosely based on biological neural networks. A neural net consists of a connected set of nodes or neurons which can signal (that is, transmit real numbers) to adjacent nodes. The nodes and the edges connecting them are assigned weights which may change as 'learning' progresses by processing examples. *Compare* MACHINE LEARNING.

neutral element A synonym for IDENTITY ELEMENT. *See also* ZERO ELEMENT.

neutral geometry Geometric theorems which can be proved without reference to the *parallel postulate. Hence results of neutral geometry are true of both *Euclidean geometry and *non-Euclidean geometry. Also known as absolute geometry.

newton The SI unit of *force, abbreviated to 'N'. One newton is the force required to give a *mass of 1 *kilogram an acceleration of 1 metre per second per second.

Newton, Isaac (1642–1727) English physicist and mathematician who dominated and revolutionized mathematics and physics in the 17th century. He was responsible for the essentials of the *calculus, the theory of *mechanics, the law of *gravity, the theory of planetary motion, the *binomial series, *Newton's method in numerical analysis, and many important results in the theory of equations. A morbid dislike of criticism held him back from publishing much of his work. For example, in 1684, Edmond *Halley suggested to Newton that he investigate the law of attraction that would yield *Kepler's laws. Newton replied immediately that it was the *inverse square law—he had worked that out years earlier. Rather startled by this, Halley set to work to persuade Newton to publish his results. This Newton eventually did in the form of his *Principia*, published in 1687.

Newton quotient The quotient $(f(a + h) - f(a))/h$, which is used to determine whether f is *differentiable at a and, if so, to find the *derivative.

Newton–Raphson method A synonym for NEWTON'S METHOD.

Newton's identities Given a *polynomial $a_n x^n + a_{n-1} x^{n-1} + \cdots + a_0 = 0$, let s_k denote the sum of the kth powers of the polynomial's roots. Then Newton's identities state for $k \geq 1$ that

$$a_n s_k + a_{n-1} s_{k-1} + \ldots + a_{n-k+1} s_1 + k a_{n-k} = 0.$$

These identities recursively determine s_k in terms of $s_{k-1}, \ldots, s_1$. In these identities a_r is understood to be 0 if $r < 0$. *See also* ELEMENTARY SYMMETRIC POLYNOMIALS.

Newton's interpolating polynomial The *polynomial of degree n, which interpolates $n + 1$ evenly spaced *data points (x_0, y_0), $(x_1, y_1), \ldots, (x_n, y_n)$, where $x_k = x_0 + kh$, is given by

$$y_0 + \frac{(x - x_0)}{h} \Delta y_0 + \frac{(x - x_0)(x - x_1)}{2! h^2} \Delta^2 y_0$$
$$+ \ldots + \frac{(x - x_0) \ldots (x - x_n)}{n! h^n} \Delta^n y_0$$

where $\Delta y_0 = y_1 - y_0$, $\Delta^2 y_0 = \Delta y_1 - \Delta y_0$, etc. which are obtained by constructing a *forward difference table.

Newton's interpolation formula *See* GREGORY–NEWTON FORWARD DIFFERENCE.

Newton's law of cooling States that the *heat loss between two bodies is *proportional to the difference in the bodies' *temperatures and the *surface area of the region of contact between the bodies. *See also* FOURIER'S LAW.

Newton's law of gravitation *See* INVERSE SQUARE LAW OF GRAVITATION.

Newton's law of restitution *See* COEFFICIENT OF RESTITUTION.

Newton's laws of motion Three laws of motion, applicable to *particles of constant *mass, formulated by *Newton in his *Principia*. They can be stated as follows:

(i) A particle continues to be at rest or moving with constant *velocity if the total *force acting on it is zero.

(ii) The *rate of change of *linear momentum of a particle is proportional to the total force acting on the particle. The linear momentum $\mathbf{p}$ of a particle is $\mathbf{p}=m\mathbf{v}$, where m is the mass and $\mathbf{v}$ is the velocity. So $d\mathbf{p}/dt = m(d\mathbf{v}/dt) = m\mathbf{a}$, where $\mathbf{a}$ is the acceleration. Thus, the second law is usually stated as $\mathbf{F} = m\mathbf{a}$, where $\mathbf{F}$ is the total *force acting on the particle.

(iii) Whenever a particle exerts a force on a second particle, the second particle exerts an equal and opposite force on the first. This is often phrased as 'To every action there is an equal and opposite reaction.'

Newton's method The following method of finding successive approximations to a *root of an equation $f(x) = 0$. Suppose that x_0 is a first approximation, known to be quite close to a root. If the root is in fact $x_0 + h$, where h is 'small', taking the first two terms of the *Taylor series gives $f(x_0 + h) \approx f(x_0) + hf'(x_0)$.

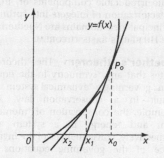

Iterations converge to root

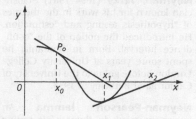

Iterations diverge

Since $f(x_0 + h) = 0$, it follows that $h \approx -f(x_0)/f'(x_0)$. Thus x_1, given by

$$x_1 = x_0 - \frac{f(x_0)}{f'(x_0)},$$

is likely to be a better approximation to the root. To see the geometrical significance of the method, suppose that P_0 is the point $(x_0, f(x_0))$ on the curve $y = f(x)$, as shown in the first figure. The value x_1 is given by the point at which the *tangent to the curve at P_0 meets the x-axis. It may be possible to repeat the process to obtain successive approximations $x_0, x_1, x_2, \ldots$, where

$$x_{n+1} = x_n - \frac{f(x_n)}{f'(x_n)}.$$

These may be successively better approximations to the root as required, but in the second figure is shown the graph of a function, with a value x_0 close to a root, for which x_1 and x_2 do not give successively better approximations. However, if $f''(x)$ is a *continuous function and $f'(x)$ is non-zero, then Newton's method *quadratically converges to the root in some *neighbourhood of the root.

For a system of k equations in k variables, $\mathbf{f}(\mathbf{x}) = \mathbf{0}$, the iteration becomes

$$\mathbf{x}_{n+1} = \mathbf{x}_n - \mathbf{J}^{-1}(\mathbf{x}_n)\mathbf{f}(\mathbf{x}_n)$$

where $\mathbf{J}$ denotes the *Jacobian matrix of $\mathbf{f}$.

Neyman, Jerzy (1894–1981) Statistician known for his work in the theories of *hypothesis testing and *estimation. He introduced the notion of the *confidence interval. Born in Romania, he spent some years at University College London before going to the University of California at Berkeley in 1938.

Neyman–Pearson lemma An important result in *hypothesis testing that gives a *sufficient condition for choosing a *critical region for a given *significance level that maximizes the *power of the test.

Nightingale, Florence (1820–1910) Known to many as the 'Lady with the Lamp', amongst statisticians she is remembered for her pioneering work on data analysis, the representation of *data, and the standardizing of data recording; the implementation of her findings in hospitals; and being the founder of modern nursing. International Nurses Day is celebrated on 12 May, her birthday.

nilpotent A square matrix **A** is nilpotent if $A^n = 0$ for some n. Strictly *triangular matrices are nilpotent. More generally, an element x in a *ring is nilpotent if $x^n = 0$ for some n.

nilradical In a *commutative ring, the *nilpotent elements form a *radical *ideal, the nilradical of the ring.

nine-point circle For a triangle, let M_1, M_2, M_3 denote the *midpoints of the three sides. These three points define a circle known as Feuerbach's circle or the nine-point circle. It gets its name because it passes through six further points of note. It passes through A_1, A_2, A_3, the three feet of the *altitudes. The altitudes meet at the *orthocentre O, and Feuerbach's circle also passes through the midpoints of the line segments from the orthocentre to each vertex. The centre of the circle lies on the *Euler

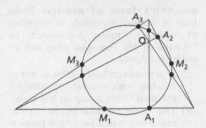

The nine points and circle

line, halfway between the orthocentre and *circumcentre. Further, Feuerbach's circle *touches the *incircle and the three *excircles of the triangle.

node *See* DOUBLE POINT.

node *See* GRAPH, TREE.

Noether, Amalie ('Emmy') (1882–1935) German mathematician, known for her highly creative work in the theory of *rings, non-commutative algebras, and other areas of *abstract algebra. She was responsible for the growth of an active group of algebraists at Göttingen in the period up to 1933.

Noetherian ring A *ring is Noetherian if any ascending chain of ideals $I_1 \subseteq I_2 \subseteq \ldots \subseteq I_k \subseteq \ldots$ eventually becomes constant. The condition guarantees finite decomposition in different senses, for example decomposing *varieties into finite irreducible components or, in $\mathbb{Z}$, to factorization of integers into *primes. *Principal ideal domains are Noetherian. *See* HILBERT'S BASIS THEOREM.

Noether's theorem The theorem states that any *symmetry in the equations governing a *dynamical system will result in a conservation law. For example, the conservation of *momentum and *energy in a system are, respectively, consequences of the invariance of the governing equations in *space and *time.

noise A descriptive name for *random error or variation in observations which is not explained by the *model.

nominal *See* DATA.

nominal scale A set of categories of data which cannot be ordered but by which data can be classified, e.g. by gender or ethnicity. *Compare with* ORDINAL.

nonagon A nine-sided polygon.

non-basic variables *See* BASIC SOLUTION.

non-constructive Not *constructive.

non-denumerable A synonym for UNCOUNTABLE.

non-empty Not *empty.

non-Euclidean geometry After unsuccessful attempts had been made at proving that the *parallel postulate could be deduced from the other postulates of Euclid's, the matter was settled by the discovery of non-Euclidean geometries by *Lobachevsky and *Bolyai. In these, all *Euclid's postulates hold except the parallel postulate. In hyperbolic geometry, given a point not on a given line, there are at least two lines through the point parallel to the line (i.e. these parallel lines do not intersect the given line). In elliptic geometry, given a point not on a given line, there are no parallels through the point. *See* ELLIPTIC PLANE, HYPERBOLIC PLANE, NEUTRAL GEOMETRY.

non-homogeneous linear differential equation *See* LINEAR DIFFERENTIAL EQUATION WITH CONSTANT COEFFICIENTS.

non-homogeneous set of linear equations A set of m linear equations in n unknowns $x_1, x_2, \ldots, x_n$ that has the form

$$a_{11}x_1 + a_{12}x_2 + \ldots + a_{1n}x_n = b_1,$$
$$a_{21}x_1 + a_{22}x_2 + \ldots + a_{2n}x_n = b_2,$$
$$\vdots$$
$$a_{m1}x_1 + a_{m2}x_2 + \ldots + a_{mn}x_n = b_m,$$

where $b_1, b_2, \ldots, b_m$ are not all zero. (*Compare with* HOMOGENEOUS SET OF LINEAR EQUATIONS.) In matrix notation, this set of equations may be written as $\mathbf{Ax} = \mathbf{b}$, where $\mathbf{A}$ is the $m \times n$ matrix $[a_{ij}]$, and $\mathbf{b} \neq \mathbf{0}$ and $\mathbf{x}$ are column vectors:

$$\mathbf{b} = \begin{bmatrix} b_1 \\ b_2 \\ \vdots \\ b_m \end{bmatrix}, \quad \mathbf{x} = \begin{bmatrix} x_1 \\ x_2 \\ \vdots \\ x_m \end{bmatrix}.$$

Such a set of equations may be *inconsistent, have a unique solution, or have infinitely many solutions (*see* SIMULTANEOUS LINEAR EQUATIONS). If $m = n$, $\mathbf{A}$ is *square and the set of equations has a unique solution, namely $\mathbf{x} = \mathbf{A}^{-1}\mathbf{b}$, if and only if $\mathbf{A}$ is *invertible.

non-linear Describing an equation or expression that is not linear or a *dynamical system governed by non-linear equations. Many areas of mathematics focus on linear aspects, in part because there are many important linear *differential equations and a rich theory, as in *linear algebra, can be developed. Also, even non-linear systems can be approximated to some extent by linear ones (*see* LINEARIZATION, LINEAR THEORY OF EQUILIBRIA). However, there are many important non-linear equations and systems, and their general properties and behaviour are completely different, often exhibiting *chaos (*see* LORENZ ATTRACTOR) with little hope of analytical solutions, though it is possible to qualitatively analyse them.

non-negative Not *negative, so it can be either *zero or *positive.

non-orientable *See* ORIENTABLE.

n

non-parametric methods Methods of *inference that make no assumptions about the underlying population *distribution. Non-parametric tests are often concerned with hypotheses about the *median of a population and use the ranks of the *observations. Examples of non-parametric tests are the *Wilcoxon rank-sum test, the *Kolmogorov–Smirnov test, and the *sign test.

non-response bias Unless a survey carries some element of compulsion in it, even if a *sample is well constructed, there is a likelihood that some people will choose not to respond to it or that some people may not be able to be contacted. Strong opinions are likely to be over-represented in any survey with a voluntary response, and the response rate is important. The lower the response rate, the less reasonable it is to try to *extrapolate from the survey results to say something about the *population.

non-significant result interpretation If a statistical test does not show a *significant result, it is important to interpret this not as positive support for the *null hypothesi, but as an absence of strong evidence against it. Thus, we 'fail to reject' the null hypothesis, rather than 'accept' it.

non-singular (curve) A *curve (or geometric object more generally) is non-singular if it has no *singular points. So a *parabola is non-singular, but the curve with equation $y^2 = x^3$ is not, having a singular point (a *cusp) at the origin.

non-singular (matrix) A square matrix A is non-singular if it is not *singular; that is, if $\det A \neq 0$, where $\det A$ is the *determinant of A. *See also* INVERSE MATRIX.

non-standard analysis *See* HYPERREALS.

non-symmetric (of a relation) Not *symmetric, or *asymmetric. The relation has to hold for some pairs in both orders and hold for only one order for some other pairs, i.e. there exist elements a, b, c, d for which $a \sim b$, $b \sim a$, whereas $c \sim d$, but $d \sim c$ does not hold.

non-transitive (of a relation) Neither *transitive nor *intransitive. The transitive relationship has to hold for some triples and not for others, i.e. there exist elements a, b, c, d, e, f for which $a \sim b$, $b \sim c$, $a \sim c$, whereas $d \sim e$, $e \sim f$, but $d \sim f$ does not hold.

non-transitive dice A set of differently numbered dice designed to demonstrate non-transitivity. For example, three dice A, B, C are non-transitive if A beats B on average, B beats C, but C beats A. Four such dice are called Efron dice after the American statistician Bradley Efron.

non-trivial *See* TRIVIAL.

non-zero Not equal to *zero.

nor A synonym for NOT OR.

norm A norm on a real or complex *vector space V is a map $\|.\| : V \to \mathbb{R}$ such that (i) $\|v\| \geq 0$ for all v and $\|v\| = 0$ if and only if $v = 0$; (ii) $\|kv\| = |k| \|v\|$ for all v and all scalars k; (iii) $\|v + w\| \leq \|v\| + \|w\|$. Examples include the *absolute value of real numbers, *modulus of a complex number, *p-norms, and *matrix norms. The term norm is confusingly used in other contexts: *see* EUCLIDEAN DOMAIN, PARTITION (of an interval).

normal (to a curve) Suppose that P is a point on a *curve in the plane. Then the normal at P is the *line through P *perpendicular to the *tangent at P.

normal (to a plane) A line *perpendicular to the plane.

normal (to a surface) *See* TANGENT PLANE.

normal coordinates *See* NORMAL MODE.

normal distribution (Gaussian distribution) The continuous probability *distribution with *probability density function f given by

$$f(x) = \frac{1}{\sqrt{2\pi\sigma^2}} \exp\left(-\frac{(x-\mu)^2}{2\sigma^2}\right),$$

denoted by $N(\mu, \sigma^2)$. It has *mean μ and *variance σ^2. The distribution is widely used in *statistics because many experiments produce *data that are approximately normally distributed; the sum of *random variables from non-normal distributions is approximately normally distributed (*see* CENTRAL LIMIT THEOREM), and it is the limiting distribution of distributions such as the *binomial, *Poisson, and *chi-squared distributions. It is called the standard normal distribution when $\mu = 0$ and $\sigma^2 = 1$.

If X has the distribution $N(\mu, \sigma^2)$ and $Z = (X-\mu)/\sigma$, then Z has the distribution $N(0,1)$. The diagram shows the graph of the probability density function of $N(0,1)$.

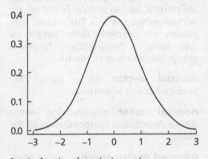

Density function of standard normal

The table gives, for each value z, the *percentage of *observations which exceed z, for the standard normal distribution $N(0,1)$. Thus the values are to be used for *one-tailed tests. *Interpolation may be used for values of z not included.

normal form A synonym for CANONICAL FORM.

normal form of conics The *ellipse, *parabola, and *hyperbola have the following normal forms:

- ellipse (and circle): $\frac{x^2}{a^2} + \frac{y^2}{b^2} = 1$ where $0 < b \le a$.
- parabola: $y^2 = 4ax$ where $0 < a$.
- hyperbola: $\frac{x^2}{a^2} - \frac{y^2}{b^2} = 1$ where $0 < a, b$.

By means of a *rotation and *translation, any non-degenerate *conic can be transformed into one of the above forms, and no two distinct normal forms are *congruent. *See also* QUADRICS.

normalize (normalizable) To normalize a *vector **v** is to scale a non-zero vector **v** so that it has *unit length, that is, to create $v/|v|$. To be normalizable is for a vector to have finite *norm so that it can be normalized; this is a term particularly used in *probability of *probability density functions and *quantum theory of *wave functions.

normalizer Given a *group G with *subgroup H, the normalizer $N_G(H)$ of H in G equals

$$N_G(H) = \{g \in G : gH = Hg\}.$$

Then $N_G(H)$ is a subgroup of G which contains H as a *normal subgroup. Indeed, $N_G(H)$ is the largest subgroup

z	0.0	0.5	1.0	1.28	1.5	1.64	1.96	2.33	2.57	3.0	3.5
%	50	30.9	15.9	10.0	6.7	5.0	2.5	1.0	0.5	0.14	0.02

$P(Z \ge z)$ as percentages

containing H as a normal subgroup. *Compare* CENTRALIZER.

normal mode An oscillating mechanical system with n *degrees of freedom has normal coordinates $q_1, q_2, \ldots, q_n$ about an *equilibrium $q_1 = q_2 = \ldots = q_n = 0$ if

$$T = \frac{1}{2}(\dot{q}_1^2 + \dot{q}_2^2 + \ldots + \dot{q}_n^2),$$

$$V = \frac{1}{2}(\omega_1^2 q_1^2 + \omega_2^2 q_2^2 + \ldots + \omega_n^2 q_n^2),$$

where T denotes *kinetic energy and V denotes *potential energy. Here $\omega_1, \omega_2, \ldots, \omega_n$ are the natural frequencies of the system. The normal modes of the system are given by $q_k(t) = A_k \cos(\omega_k t + \varepsilon_k)$ with all other q_j being 0.

For a simple light *pendulum of length l with a mass m at the end, performing small oscillations about the downward vertical $\theta = 0$, then

$$T = \frac{1}{2}ml^2\dot{\theta}^2, \quad V = mgl(1 - \cos\theta) \approx \frac{1}{2}mgl\theta^2.$$

Thus, the normal coordinate is $\sqrt{ml}\theta$ and the natural frequency is $\sqrt{g/l}$. For a string $y(x,t)$ fixed at $y(0,t) = y(l,t) = 0$ and performing small transverse oscillations, there are infinitely many degrees of freedom and the normal modes are

$$A_k \sin\left(\frac{k\pi x}{l}\right)\cos\left(\frac{k\pi ct}{l} + \varepsilon_k\right) \quad k = 1, 2, 3, \ldots$$

and the corresponding natural frequency is $k\pi c/l$. Here $c^2 = T/\rho$ where T is the *tension in the string and ρ is its *density.

normal number A *real number is normal in a *base if its expansion does not favour any digit or string of digits. So, to be normal in *binary means that, on average, there are as many 0s as 1s in the expansion, the string 101 appears one-eighth of the time, etc. A number is normal if it is normal in all bases. It is

relatively straightforward to show that *almost all numbers are normal but can be difficult to show a specific number is normal; it is not known whether π or $\sqrt{2}$ is normal. *Rational numbers are not normal.

normal operator A *linear map N: $V \to V$ of an *inner product space such that $NN^* = N^*N$ where N^* denotes the *adjoint of N. *Self-adjoint and unitary maps (*see* UNITARY MATRIX) are normal. The *spectral theorem applies to normal operators.

normal reaction *See* CONTACT FORCE.

normal space A *topological space X in which every two disjoint *closed sets of X have disjoint *neighbourhoods. A *compact *Hausdorff space is normal. *See* SEPARATION AXIOMS.

normal subgroup If H is a *subgroup of a *group G, and for any element, x of G, the left and right *cosets of H are equal, then H is a normal subgroup. This is denoted $H \lhd G$. Equivalently, H is closed under *conjugation; that is, for every x in G and h in H, then $x^{-1}hx$ is in H. In an *abelian group, all subgroups are normal, and in general G and $\{e\}$ are normal subgroups of G. The normal subgroups are precisely those subgroups that form a *well-defined *quotient group. *See also* SIMPLE GROUP.

normal vector (to a curve) *See* SERRET-FRENET FORMULAE.

normal vector (to a plane) A *vector whose direction is *perpendicular to the plane.

normed vector space A real or complex *vector space with a *norm.

not *See* NEGATION.

not and If p and q are statements then 'p not and q', denoted by $p \uparrow q$, is true unless both p and q are true. Since 'not and' is cumbersome language, this is

often referred to as 'nand' in logic. The
*truth table is as follows:

p	q	$p \uparrow q$
T	T	F
T	F	T
F	T	T
F	F	T

notation (mathematical notation)
The symbolic encoding of mathematical
objects and numerical quantities, ideally
with the intentions of clarity, easy
manipulation, and argumentation. The
use of Arabic *numerals was a marked
improvement over Roman numerals for
calculations. *Viète's introduction of letters to represent *variables in the 16th
century simplified algebraic arguments
(see ALGEBRA). *Euler introduced or
popularized the use of many now standard symbols in the 18th century. *Set
theoretic and logical notation (see
LOGIC) appeared in the late 19th and
early 20th centuries.

For tables of commonly used mathematical symbols, see APPENDIX 19.

not or If p and q are statements, then 'p
not or q', denoted by $p \downarrow q$, is true only
when both p and q are false. Since 'not
or' is cumbersome language, this is often
referred to as 'nor' in logic. The *truth
table is as follows:

p	q	$p \downarrow q$
T	T	F
T	F	F
F	T	F
F	F	T

nought A synonym for ZERO.

nowhere dense Describing a set in a
*topological space whose *closure has
empty *interior.

nowhere-differentiable function
A *continuous real function that is not
differentiable at a single point. See
BLANCMANGE FUNCTION.

NP problem This is short for 'nonde-
terministic *polynomial time'. For some
problems, there is no known *algorithm
that solves the problem in polynomial
time, but there is a known algorithm
which verifies a potential solution as a
genuine solution in polynomial time.
Such a problem is the *travelling salesman problem. It is an open *Millennium
Prize problem whether NP = P, which
means that NP problems can in fact
also be solved in polynomial time.

NRICH The NRICH website was begun
by the Millennium Mathematics Project
in Cambridge. It contains free mathematical enrichment material intended for
students aged 5–19.

(((●))) SEE WEB LINKS
• The official NRICH website.

nth derivative See HIGHER
DERIVATIVE.

nth-order partial derivative See
HIGHER-ORDER PARTIAL DERIVATIVE.

nth root For any real number a, an nth
root of a is a number x such that $x^n = a$.
When $n = 2$, it is called a *square root,
and when $n = 3$ a cube root.

First consider n even. If $a < 0$, there is
no real number x such that $x^n = a$. If
$a > 0$, there are two such numbers, one
positive and one negative. For $a \geq 0$, the
notation $\sqrt[n]{a}$ is used to denote quite specifically the non-negative nth root of a.
For example, $\sqrt[4]{16} = 2$, and 16 has two
real fourth roots, namely 2 and −2.

Next consider n odd. For all values of
a, there is a unique number x such that
$x^n = a$, and it is denoted by an $\sqrt[n]{a}$. For
example, $\sqrt[3]{-8} = -2$.

For a non-zero *complex number a, there are n complex numbers z such that $z^n = a$, which may also be referred to as *nth roots of a. In the *Argand diagram these nth roots form the vertices of a *regular n-gon.

nth root of unity A complex number z such that $z^n = 1$. The n distinct nth roots of unity are $e^{i2k\pi/n}(k = 0, 1, \ldots, n-1)$, or

$$\cos\frac{2k\pi}{n} + i\sin\frac{2k\pi}{n} \quad (k = 0, 1, \ldots, n-1).$$

They are represented in the *complex plane by points that lie on the unit circle and are vertices of a *regular n-sided polygon. The nth roots of unity for $n = 5$ and $n = 6$ are shown in the figure. The nth roots of unity always include the real number 1 and also include the real number -1 if n is even. The non-real nth roots of unity form pairs of *conjugates.

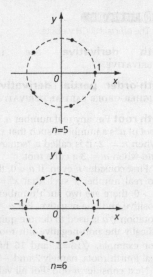

$n=5$

$n=6$

n-tuple An n-tuple consists of n objects normally taken in a particular order, denoted, for example, by $(x_1, x_2, \ldots, x_n)$. The term is a generalization of a triple when $n = 3$ and a quadruple when $n = 4$. When $n = 2$ it is an *ordered pair.

null hypothesis *See* HYPOTHESIS TESTING.

nullity The nullity of a matrix $\mathbf{A}$ is the *dimension of the solution space to the equation $\mathbf{Ax} = \mathbf{0}$; equivalently, it is the minimum number of *parameters required to describe the system's *general solution. The general solution can be found by transforming the matrix into *reduced echelon form and assigning a parameter to the variable of each column not containing a leading 1. *See* RANK-NULLITY THEOREM.

null matrix A synonym for ZERO MATRIX.

null measure A synonym for NULL SET (measure theory).

null sequence A *sequence whose *limit is 0.

null set A synonym for EMPTY SET.

null set (measure theory) A subset of a *measure space having measure zero. A subset X of $\mathbb{R}$ is a null set if, for any $\varepsilon > 0$, there exists intervals $I_1, I_2, I_3, \ldots$, of total length less than ε, such that X is contained in the *union of the I_i. Any *countable set is null, and the *Cantor set is an example of an *uncountable null set.

null space A synonym for KERNEL.

Nullstellensatz The following important result in *commutative algebra due to *Hilbert. Let F be an *algebraically closed *field (such as $\mathbb{C}$) and n be a positive integer. For an *ideal J in the *polynomial ring $F[x_1, \ldots, x_n]$ and for a subset S of F^n we may define

$V(J) = \{\mathbf{x} \in F^n; f(\mathbf{x}) = 0 \text{ for all } f \in J\};$
$I(S) = \{f \in F[x_1, \ldots, x_n];$
   $f(\mathbf{x}) = 0 \text{ for all } \mathbf{x} \in S\}$

Then $I(V(J)) = \sqrt{J}$, the *radical of J. By comparison $V(I(S)) = \bar{S}$, the *closure of S in the *Zariski topology.

null vector A vector that has zero *magnitude. Since vectors normally have a magnitude and a *direction, the null vector is unusual in that it does not have a defined direction.

number *See* CARDINAL NUMBER, COMPLEX NUMBER, INTEGER, NATURAL NUMBER, ORDINAL NUMBER, RATIONAL NUMBER, REAL NUMBER.

number field A number field is a *field F containing the *rational numbers, such that $F: \mathbb{Q}$ has finite degree. Number fields are important in algebraic *number theory.

number line A synonym for REAL LINE.

Numberphile A mathematical YouTube channel begun by Brady Haran in 2011. There have been many guest presenters and contributors including John *Conway, Hannah *Fry, Donald *Knuth, Matt *Parker, and Roger *Penrose.

⊕ SEE WEB LINKS
• The Numberphile YouTube channel.

number systems The early Egyptian number system used different symbols for 1, 10, 100, and so on, with each symbol repeated the required number of times. Later, the Babylonians had symbols for 1 and 10 repeated similarly, but for larger numbers they used a positional notation with base 60, so that groups of symbols were positioned to indicate the number of different powers of 60.

The Greek number system used letters to stand for numbers. For example, $\alpha, \beta, \gamma,$ and δ represented 1, 2, 3, and 4; and $\iota,$ $\kappa, \lambda,$ and μ represented 10, 20, 30, and 40. The Roman number system is still known today and used for some special purposes. Roughly speaking, each Roman numeral is repeated as often as necessary to give the required total, with the larger numerals appearing before the smaller, except that if a smaller precedes a larger its value is subtracted. For example, IX, XXVI, and CXLIV represent 9, 26, and 144. *See* APPENDIX 21.

The Hindu–Arabic number system, in which numbers are generally written today, uses the Arabic numerals and a positional notation with base 10. It originated in India, where records of its use go back to the 6th century. It was introduced to Europe in the 12th century, promoted by *Fibonacci and others.

number theory (higher arithmetic) The area of mathematics concerning the study of the arithmetic properties of *integers and related number systems such as *prime numbers. Representations of numbers as sums of squares (*see* LAGRANGE'S THEOREM) etc. appear very abstract, but number theory has provided the basis for secure encryption (*see* RSA).

numeral A symbol used to denote a number. The Roman numerals I, V, X, L, C, D, and M represent 1, 5, 10, 50, 100, 500, and 1000 in the Roman number system. The Arabic numerals 0, 1, 2, 3, 4, 5, 6, 7, 8, and 9 are used in the Hindu–Arabic number system to give numbers in the form generally familiar today. *See* NUMBER SYSTEMS.

numerator *See* FRACTION.

numerical analysis The solutions to many mathematical problems cannot be described exactly in full generality. Instead, a numerical method may be employed; that is, a solution—or an approximation to one—is obtained instead. The study of relevant theory and methods is called numerical analysis.

The advent of very powerful computers means that such computational methods are increasingly important in *modelling *complex systems. For example, the design of aircraft and high-performance cars can now be extensively tested by simulations prior to physically building and testing models.

numerical differentiation The use of formulae, usually expressed in terms of *finite differences, for estimating derivatives of a function $f(x)$ from values of the function. For example

$$f'(x) \approx \frac{f(x+h) - f(x)}{h}$$

and

$$f''(x) \approx \frac{f(x+h) - 2f(x) + f(x-h)}{h^2}$$

where h is a small positive increment.

numerical integration The methods of numerical integration are used to find approximate values for *definite integrals and are useful when there is no analytical method of finding an *antiderivative of the *integrand. Among the elementary methods of numerical integration are the *midpoint rule, *trapezium rule, *Simpson's rule, and *Gaussian quadrature.

numerical stability See STABLE NUMERICAL ANALYSIS.

numerical value A synonym for ABSOLUTE VALUE.

O See BIG O NOTATION.

o See LITTLE O NOTATION.

𝕆 See OCTONION.

objective function See LINEAR PROGRAMMING.

objective row See SIMPLEX METHOD.

oblate See SPHEROID.

oblique A pair of intersecting *lines are oblique if they are not *perpendicular.

oblong A synonym for RECTANGLE or RECTANGULAR.

observable In *quantum theory, a quantity which may be measured such as position, *energy, or *momentum (*see* MEASUREMENT). Observables are associated with *Hermitian operators, and the possible measurements of the observable are the *eigenvalues of the operator. Two observables are simultaneously measurable if the operators commute, which is not the case with position and momentum (*see* HEISENBERG'S UNCERTAINTY PRINCIPLE).

observation A particular value taken by a *random variable. Normally, n observations of the random variable X are denoted by $x_1, x_2, \ldots, x_n$.

observational study A statistical study in which the researchers observe or question participants but without *designing an experiment in which interventions are used. The lack of *randomization in allocating participants into groups means that the issue of *confounding variables is present. *See also* CASE CONTROL STUDIES.

observer Given a *frame of reference, it is convenient to associate an observer who is thought of as viewing the motion with respect to this frame of reference and having the capability to assign events *space and *time *coordinates. *See* GALILEAN RELATIVITY, RELATIVITY.

obtuse angle An *angle that is greater than a *right angle and less than two right angles. An obtuse-angled triangle is one in which at least one angle is obtuse.

oct- Prefix denoting eight.

octagon An eight-sided *polygon.

octahedron (octahedra) A *polyhedron with 8 faces. The regular octahedron (*see* REGULAR POLYHEDRON) is one of the *Platonic solids and its faces are equilateral triangles. It has 6 vertices and 12 edges.

octal A number system using the base of eight, often used in computing.

octant In a 3-dimensional *Cartesian coordinate system, the *axial planes divide the rest of the space into eight regions called octants. The set of points $\{(x, y, z)| x,y,z > 0\}$ may be called the *positive (or first) octant. *Compare* QUADRANT.

octonion An 8-dimensional *division algebra, denoted 𝕆. While not *associative, it is one of only four normed division algebras, the others being the *real numbers, the *complex numbers, and the *quaternions.

odd function The *real function f is an odd function if $f(-x) = -f(x)$ for all x (in the domain of f). Thus the graph $y = f(x)$ of an odd function has *rotational symmetry of order 2 about the origin because whenever (x,y) lies on the graph then so does $(-x,-y)$. The following are odd functions of x: $2x$, x^3, $x^7 - 8x^3 + 5x$, $1/(x^3 - x)$, $\sin x$, $\tan x$.

odd integer Not exactly *divisible by 2. The general form of an odd number is $2n + 1$ where n is an *integer. *Compare* EVEN INTEGER.

odd part The largest odd factor of a given integer, so the odd part of 14, which can be written as odd(14) is 7; odd(9) = 9 and odd(2^n) = 1 for any positive integer n.

odd permutation A *permutation which can be written as the *composition of an odd number of *transpositions. *See* EVEN PERMUTATION.

odds Betting odds are expressed in the form $r{:}s$ corresponding in theory to a *probability of $\frac{r}{r+s}$ of winning. In reverse, if p is the probability of an event happening, the odds on it will be $p{:}(1-p)$. In practice, odds offered by a bookmaker will only be an approximation to this probability, as the bookmaker wishes to make money and will change the odds to reflect the amount of money being bet on particular events even if they have no reason to change their view of the probabilities of those events.

odds ratio Defined as the ratio of the *odds of an event happening in two distinct groups. If the probabilities of the event in the two groups are p and q then the odds are $p{:}(1-p)$ and $q{:}(1-q)$ and the odds ratio is $\dfrac{p/(1-p)}{q/(1-q)} = \dfrac{p(1-q)}{q(1-p)}$.

ODE Short for 'ordinary differential equation'. *See* DIFFERENTIAL EQUATION.

off diagonal In a square matrix, the *diagonal from the bottom-left to the top-right corner. *Compare* MAIN DIAGONAL.

omega The last Greek letter, 'long' o. Its lower case symbol ω, is often used to represent *angular velocity, *angular frequency, and the smallest infinite *ordinal number, i.e. that of the natural numbers. Its upper case symbol Ω often denotes *sample space or *solid angle.

one The smallest positive *integer, symbol 1. The *multiplicative identity for the real numbers, and more generally the term and symbol used for the multiplicative identity of a *ring.

one-sided surface A surface, such as the *Klein bottle, which is not *orientable. This is equivalent to the surface containing a *Möbius band. *See* CLASSIFICATION THEOREM FOR SURFACES.

one-sided test *See* HYPOTHESIS TESTING.

one-tailed test *See* HYPOTHESIS TESTING.

one-to-many A rule which assigns to a single *element of the *domain more than one element in the *range. Consequently, any such assignment is not a *function. For example, $\tan^{-1}x$ takes not only the *principal value, α, lying between $-90°$ and $+90°$, but also $\alpha \pm 180n$ for any integer n. *See* MULTIFUNCTION.

one-to-one A *function $f{:}S \rightarrow T$ is one-to-one if, whenever s_1 and s_2 are distinct elements of S, their images $f(s_1)$ and $f(s_2)$ are distinct elements of T. Equivalently, f is one-to-one if $f(s_1) = f(s_2)$ implies that $s_1 = s_2$. Synonymous with injective.

one-to-one correspondence A synonym for BIJECTION.

one-way function A one-way function is an *invertible function, which itself involves relatively little computation, but the inverse is computationally much more onerous. For example, multiplying two large *prime numbers is straightforward for a computer, but factorizing the *product is considerably more time-consuming. A second example is the *discrete logarithm. One-way functions are important in *public key cryptography. *See* CRYPTOGRAPHY, RSA.

only if *See* CONDITION, NECESSARY AND SUFFICIENT.

onto A *function $f: S \rightarrow T$ is onto (or surjective) if every element of the *codomain T is the *image under f of at least one element of the domain S. So f is onto if the image (or *range) $f(S)$ equals the codomain. *Compare* ONE-TO-ONE.

open ball In a *metric space, the set of all points whose distance δ from a fixed point a is strictly less than a specified value, the *radius of the ball i.e. the set of points x for which $d(x, a) < \delta$. The open balls form a *basis for the *topology.

open cover An open cover of a *metric space (or *topological space) X is a collection of *open sets whose union contains X as a subset. *See* COMPACT.

open disc *See* DISC.

open interval The open interval (a, b) is the set $\{x \mid x \in \mathbb{R}$ and $a < x < b\}$. *See also* INTERVAL.

open mapping A *function $f: X \rightarrow Y$ between *topological spaces is open if, whenever U is an open subset (*see* OPEN SET) of X, then $f(U)$ is an open subset of Y. In particular, *homeomorphisms are open mappings.

open mapping theorem A *surjective, continuous (*see* CONTINUOUS FUNCTION) *linear map $T: X \rightarrow Y$ between *Banach spaces is an *open mapping.

open problem A problem in mathematics which is not currently known to be true or false, such as *Goldbach's conjecture or the *Riemann hypothesis.

open set A subset A of a *metric space M is open if every point of A is an *interior point of A or, equivalently, around each point of A there exists an *open ball of M which is contained in A. The open sets form the *topology of M. In a *topological space the open sets are, by definition, those sets in the topology.

Note that openness is a relative term and that open sets are open *in* another set. So $[0,1)$ is not open in $\mathbb{R}$ as 0 is not an interior point but $[0,1)$ is open in $[0,\infty)$. *See* CLOPEN SET, CLOSED SET.

operation An operation on a set S is a rule that associates with some number n of elements of S a resulting element in S. If $n = 1, 2, 3$, the operation is respectively called unary, binary, ternary and more generally is referred to as n-ary.

operational research (OR) Often considered an area of *applied mathematics, operational research encompasses the application of various analytical methods such as stochastic modelling, *simulation, *optimization, or *decision analysis to resolve complex decision-making scenarios.

operator A term variously used in mathematics and which is typically synonymous with *function. The terms 'linear operator' and 'differential operator' are common, where the elements of the *domain are themselves functions.

opposite angles *See* VERTICAL ANGLES.

opposite side In a *right-angled *triangle, where one of the other angles has been specified, the opposite side is the side which is not an *arm of the specified angle. *Compare* ADJACENT SIDE.

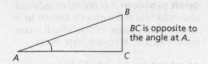

BC is opposite to the angle at A.

optimal A value or solution is optimal if it is, in a specified sense, the best possible. For example, in a *linear programming problem the maximum or minimum value of the objective function, whichever is required, is the optimal value and a point at which it is attained is an optimal solution.

optimality condition See SIMPLEX METHOD.

optimal strategy See FUNDAMENTAL THEOREM OF GAME THEORY.

optimization The process of finding the best possible solution to a problem. In mathematics, this often consists of maximizing or minimizing the value of a certain function, perhaps subject to given constraints. *Compare* CONSTRAINED OPTIMIZATION.

or See DISJUNCTION.

OR An abbreviation for OPERATIONAL RESEARCH.

orbit (mechanics) The path traced out by a *body experiencing a *central force, such as the gravitational force associated with a second body. When the force is given by the *inverse square law of gravitation, the orbit is either elliptic, parabolic, or hyperbolic, with the second body at a focus of the *conic.

orbit (group theory) For a *group action G on a set S, the orbit of s ε S is the subset {g·s | g ε G}. The orbits *partition the set S.

orbit-counting formula Also known as the Cauchy-Frobenius lemma and,

incorrectly, as Burnside's lemma. For a *group action of a finite *group G on a finite *set S, the number of *orbits equals

$$\frac{1}{|G|} \sum_{g \in G} \text{fix}(g),$$

where fix(g) is the number of elements s ∈ S fixed by g, that is g.s = s. The formula is useful in *combinatorics.

orbit-stabilizer theorem For a finite *group G *acting on a finite set S, and s ∈ S, the product of the *cardinality of the *orbit of s and the *order of the *stabilizer of s equals the order of G. For example, the rotational *symmetry group of a cube acts *transitively on its six faces; so taking s to be any face, there is one orbit of size 6 and the stabilizer has order 4 (the rotations through the face's midpoint). Hence the rotational symmetry group has order $6 \times 4 = 24$.

order (of a differential equation) See DIFFERENTIAL EQUATION.

order (of a group) The order of a *group G is the number of elements in G. For a list of groups up to order 15, *see* APPENDIX 17.

order (of a group element) The order of a *group element g is the smallest positive *integer n such $g^n = e$, the group's *identity. If the group has finite order, then group elements have finite order, and that order must divide the group's *order by *Lagrange's theorem; in an infinite group an element may have finite or infinite order.

order (of a matrix) An $m \times n$ matrix is said to have order $m \times n$ (read as 'm by n'). An $n \times n$ matrix may be called a square matrix of order n.

order (of a partial derivative) See HIGHER-ORDER PARTIAL DERIVATIVE.

order (of a root) See ROOT.

order (real numbers) The order $<$ on the *real numbers satisfies the following *axioms; here x, y, z denote real numbers:

(i) (Trichotomy) Precisely one of $x < y$, $y < x$, $x = y$ holds.
(ii) (Transitivity) If $x < y$ and $y < z$, then $x < z$.
(iii) If $x < y$, then $x + z < y + z$.
(iv) If $0 < z$ and $x < y$, then $zx < zy$.

Together with the *field axioms, (i)–(iv) above, state $\mathbb{R}$ is an ordered field. Other examples are the *rational numbers and the *hyperreals. The real numbers also satisfy the *completeness axiom and are the unique complete ordered field.

order (set theory) *See* ORDINAL NUMBER, PARTIAL ORDER, TOTAL ORDER, WELL ORDERED.

ordered field *See* ORDER (real numbers).

ordered pair An ordered pair consists of two objects considered in a particular order. Thus, if $a \neq b$, the ordered pair (a,b) does not equal the ordered pair (b,a). *See* CARTESIAN PRODUCT, PAIR.

order isomorphism A *one-to-one correspondence f between two *partially ordered sets $(S, \leq_S)$ and $(T, \leq_T)$ such that $f(s_1) \leq_T f(s_2)$ if and only if $s_1 \leq_S s_2$. *See also* ORDINAL NUMBER.

order notation *See* BIG O NOTATION *and* LITTLE O NOTATION.

order of contact *See* CONTACT

order of convergence *See* RATE OF CONVERGENCE.

order of magnitude Loosely speaking, the power of 10 that can be used to represent a number—so n is the order of magnitude of $k \times 10^n$. Different conventions restrict k to the ranges $1/\sqrt{10} \leq$ $k < \sqrt{10}$ or $0.5 < k \leq 5$ or $1 \leq k < 10$, the last being *scientific notation.

order statistics Statistics which depend on their value for the place they occupy in an ordering of the data. So *minimum, *maximum, *median, *quartiles, and *percentiles are all order statistics.

ordinal data *Data that can be ordered. For example, measurements on a scale, preferences in some contexts, but where multiple criteria are concerned it is not always possible to construct an ordered set of preferences. *Compare* NOMINAL.

ordinal number Informally, a number denoting the position in a sequence, so 'first', 'second', 'third' is the start of the ordinal numbers, thus designating order as opposed to the *cardinal numbers. More generally, in *set theory, two *well-ordered sets have the same ordinal number if they are *order isomorphic. As with *cardinal arithmetic, ordinal numbers can be added, multiplied, and exponentiated. The ordinal number of the *natural numbers with the usual ordering is denoted ω. Perhaps surprisingly $1 + \omega = \omega$, but $\omega + 1 \neq \omega$, as the natural numbers do not have a largest element.

ordinal scale A scale in which *data are ranked but without units of measurement to indicate the size of differences. So the values used in calculating *Spearman's or another *rank correlation coefficient are on an ordinal scale.

ordinary differential equation *See* DIFFERENTIAL EQUATION.

ordinate The y-coordinate in a Cartesian coordinate system in the plane. *Compare* ABSCISSA.

orientable A *surface is orientable if it is two-sided—such as the *plane or *sphere—and is otherwise non-orientable—such as the *Möbius band or *Klein bottle, which are one-sided surfaces. A *closed surface in $\mathbb{R}^3$ has an inside and outside and so is orientable. The notion of orientability can be extended to higher-dimensional *manifolds. *See also* CLASSIFICATION THEOREM FOR SURFACES.

orientation A continuous choice of the sense of handedness in a space. A line may be assigned one of two *directions. A *simple, *closed plane curve may be *positively oriented or assigned an outward-pointing *normal. Left- and right-handed systems are two different orientations of 3-dimensional space. Two *bases of $\mathbb{R}^n$ give the same orientation if the *change of basis matrix has positive *determinant. Some spaces, such as the *Möbius band and *Klein bottle, are non-orientable and so have no orientation. *See also* STOKES' THEOREM.

origin *See* COORDINATES (IN EUCLIDEAN SPACE).

orthocentre The point at which the *altitudes of a triangle are *concurrent. The orthocentre lies on the *Euler line of the triangle.

orthogonal At *right angles to one another. The term 'perpendicular' is synonymous.

orthogonal complement For a *subspace U of an *inner product space V, the orthogonal complement

$$U^\perp = \{v \in V : \langle v, u \rangle = 0 \text{ for all } u \in U\}$$

is a subspace of V. It is often referred to as 'U perp'. If V is finite-dimensional, then $V = U \oplus U^\perp$, which generalizes to *closed subspaces of *Hilbert spaces.

orthogonal curves (orthogonal surfaces) A *family of curves (or surfaces) which meet at *right angles whenever they intersect, for example as with the surfaces of constant ρ, θ, z in *cylindrical

polar coordinates. *See also* CURVILINEAR COORDINATES.

orthogonal matrix A square matrix $\mathbf{A}$ is orthogonal if $\mathbf{A}^T\mathbf{A} = \mathbf{I}$, where $\mathbf{A}^T$ is the *transpose of $\mathbf{A}$. The following properties hold:

(i) If $\mathbf{A}$ is orthogonal, $\mathbf{A}^{-1} = \mathbf{A}^T$ and so $\mathbf{A}\mathbf{A}^T = \mathbf{I}$.
(ii) If $\mathbf{A}$ is orthogonal, det $\mathbf{A} = \pm 1$.
(iii) If $\mathbf{A}$ and $\mathbf{B}$ are orthogonal matrices of the same order, then $\mathbf{A}\mathbf{B}$ is orthogonal, as is $\mathbf{A}^{-1}$.

The orthogonal $n \times n$ matrices form a *group $O(n)$, and those with *determinant 1 form a *subgroup $SO(n)$. The orthogonal matrices are the linear (*see* LINEAR MAP) *isometries of $\mathbb{R}^n$.

2×2 orthogonal matrices have the form

$$\begin{pmatrix} \cos\theta & -\sin\theta \\ \sin\theta & \cos\theta \end{pmatrix}, \text{ or } \begin{pmatrix} \cos\theta & \sin\theta \\ \sin\theta & -\cos\theta \end{pmatrix}.$$

The first matrix represents *rotation by θ anticlockwise about the origin. The second represents reflection in the line $y = x\tan(\theta/2)$. *See* UNITARY MATRIX.

orthogonal polynomials A set of *polynomials $\{p_k(x)\}$, $k \geq 0$ are said to be orthogonal over an interval $[a,b]$ if for some positive weighting function $w(x)$,

$$\int_a^b w(x)p_i(x)p_j(x)\, dx = 0$$

when $i \neq j$. *See* CHEBYSHEV POLYNOMIALS, HERMITE POLYNOMIAL, LEGENDRE POLYNOMIALS.

orthogonal projection *See* PROJECTION (of a point on a line or plane)

orthogonal set (orthogonal vectors) A set of pairwise orthogonal *vectors, that is, the vectors are mutually perpendicular. Note that such vectors are necessarily *linearly independent.

orthonormal (orthonormal basis) An orthonormal set of vectors is a set of pairwise *orthogonal *unit vectors. Every finite-dimensional *inner product space has an orthonormal basis. The *canonical basis **i**, **j**, and **k** in $\mathbb{R}^3$ is orthonormal. *See* GRAM-SCHMIDT METHOD; *compare* COMPLETE ORTHONORMAL BASIS.

Osborne's rule The rule which summarizes the correspondence between trigonometric and hyperbolic function identities. It states that the identities are the same except where a product of two 'sin's or 'sinh's is involved, where a change of sign is required. For example,

$\sin(A+B) = \sin A \cos B + \sin B \cos A$
gives
$\sinh(A+B) = \sinh A \cosh B + \sinh B \cosh A$,
but
$\cos(A+B) = \cos A \cos B - \sin A \sin B$
gives
$\cosh(A+B) = \cosh A \cosh B + \sinh B \sinh A$.

oscillations A *particle or *rigid body performs oscillations if it travels to and fro in some way about a central position, usually a point of stable *equilibrium. Examples include the swings of a simple or compound *pendulum, the bobbing up and down of a particle suspended by a spring, and the vibrations of a violin string. The oscillations are *damped when there is a resistive force and *forced when there is an applied force. *See* LINEAR THEORY OF EQUILIBRIA, SIMPLE HARMONIC MOTION.

osculate Two curves osculate when they meet at a point where they share a common *tangent. *See* CONTACT.

The parabola $y=x^2$ and the circle $x^2+(y-1)^2=1$ share a common tangent at the origin.

osculation *See* CUSP.

osculinflection *See* CUSP.

outer measure An outer measure on a set X is a map m^* from the *power set of X to $[0,\infty]$ such that

$m^*(\varnothing) = 0$;
m^* is monotonic: if $A \subseteq B$, then $m^*(A) \le m^*(B)$;
m^* is countably *subadditive: $m^*(\bigcup_1^\infty A_k) \le \sum_1^\infty m^*(A_k)$.

Given an outer measure m^*, a set E is called Carathéodory measurable if

$$m^*(A) = m^*(A \cap E) + m^*(A \setminus E)$$

for all subsets A of X. The Carathéodory measurable sets form a *sigma algebra on which m^* is a *measure.

outer product Given an $m \times 1$ column vector **v** and $n \times 1$ column vector **w**, their outer product is $\mathbf{vw}^T$, which is an $m \times n$ matrix. The outer product is written as $\mathbf{v} \otimes \mathbf{w}$, as it is equivalent to the *tensor product.

outlier An *observation that is deemed to be unusual and possibly erroneous because it does not follow the general pattern of the data in the *sample. However, in some contexts it is the outliers which are the most important observations. *See* EXTREME VALUE DISTRIBUTION.

output *See* FUNCTION.

p Abbreviation for *pico-.

P Abbreviation for *peta-.

π See PI.

p-adic numbers The p-adic numbers were introduced by Kurt Hensel in 1897 with the aim of applying *power series methods to *number theory. Here p is a *prime number.

A p-adic integer is a sequence of the form $(x_0, x_1, x_2, \ldots)$ such that $x_n \equiv x_{n-1}$ mod p^n for each n. Such sequences arise naturally when investigating congruences such as $x^2 \equiv c$ mod p^n. A more natural representation of a p-adic integer is as

$$\sum_{i=0}^{\infty} a_i p^i \quad \text{where} \quad x_n = \sum_{i=0}^{n} a_i p^i$$

and $0 \le a_i < p$ for all i. A p-adic number is then a series of the form

$$\sum_{i=k}^{\infty} a_i p^i \quad \text{with} \quad 0 \le a_i < p \text{ for all } i,$$

where k is any integer (possibly negative). The p-adic numbers then naturally form a *field denoted $\mathbb{Q}_p$.

There is an alternative analytic construction of $\mathbb{Q}_p$. Any non-zero *rational number x can be uniquely written as $p^n(a/b)$, where neither a nor b is divisible by p. The order $|x|_p$ of x is defined to be p^{-n}, with the order of 0 being 0, and a *metric can be defined on $\mathbb{Q}$ by $d(x,y) = |x-y|_p$. The p-adic numbers can then be constructed as the *completion of $\mathbb{Q}$ using this metric.

pair A set with two elements, often with an order explicit in it, such as the coordinate pair (x, y), which would be referred to as an *ordered pair. So, as ordered pairs $(1,2) \ne (2,1)$ but as sets $\{1,2\} = \{2,1\}$. See CARTESIAN PRODUCT.

paired-sample tests (in statistics) A situation in which there is a specific link between the two *observations in each pair, for example if pieces of cloth are divided in two and the two halves of each piece make a pair. Paired-sample tests are generally much more powerful than the analogous two-sample test because they remove a major source of variation. By looking only at the difference in outcome between the two members of each pair, the variability between subjects has been removed, making it easier to identify any genuine difference in the experimental treatments.

pairwise Applying to all possible pairs which can be constructed from distinct elements of a set, as with *pairwise disjoint.

pairwise disjoint Sets $A_1, A_2, \ldots, A_n$ are said to be pairwise disjoint if $A_i \cap A_j = \varnothing$ for all $i \ne j$. The term can also be applied to an infinite collection of sets.

Pappus of Alexandria (about AD 320) Considered to be the last great ancient Greek geometer. He wrote commentaries on *Euclid and *Ptolemy, but most valuable is his *Synagoge* ('Collection'), which contains detailed accounts of much Greek mathematics, some of which would otherwise be unknown.

Pappus' Centroid Theorems First Theorem Suppose that an arc of a plane *curve is rotated through one *revolution about a line in the plane that does not cut the arc. Then the area of the surface of revolution obtained is equal to the length of the arc times the distance travelled by the *centroid of the curve.

Second Theorem Suppose that a plane region is rotated through one revolution about a line in the plane that does not cut the region. Then the volume of the solid of revolution obtained is equal to the area of the region times the distance travelled by the centroid of the region.

Pappus' Hexagon Theorem (projective geometry) Given three *collinear points A,B,C and three further collinear points A′,B′,C′, then the points

$$AB′ \cap BA′, \quad AC′ \cap CA′, \quad BC′ \cap CB′$$

are collinear. This is a degenerate case of *Pascal's theorem.

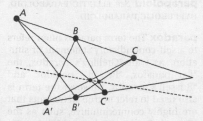

Pappus' Hexagon Theorem

parabola A *conic with eccentricity equal to 1. Thus a parabola is the *locus of all points P such that the distance from P to a fixed point F (the focus) is equal to the distance from P to a fixed line l (the directrix). It is obtained as a plane section of a cone in the case when the plane is *parallel to a generator of the cone (see CONIC). A line through the focus perpendicular to the directrix is the axis of the parabola, and the point where the axis cuts the parabola is the vertex. It is possible to take the vertex as

origin, the axis of the parabola as the x-axis and the focus as the point $(a,0)$. In this coordinate system, the directrix has equation $x = -a$ and the parabola has equation $y^2 = 4ax$ (see NORMAL FORM OF CONICS).

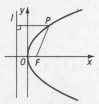

A parabola with focus and directrix

Different values of a give parabolas of different sizes, but all parabolas are the same shape (i.e. are *similar to one another). The equations $x = at^2$, $y = 2at$ are *parametric equations for the parabola $y^2 = 4ax$.

For a point P on the parabola, let α be the angle between the tangent at P and a line through P parallel to the axis of the parabola, and let β be the angle between the tangent at P and a line through P and the focus, as shown in the figure below; then $\alpha = \beta$. This is the basis of the parabolic reflector: if a source of light is placed at the focus of a parabolic reflector, each ray of light is reflected parallel to the axis, so producing a parallel beam of light.

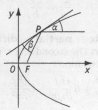

A parabolic reflector

• Properties of the parabola with links to animated illustrations.

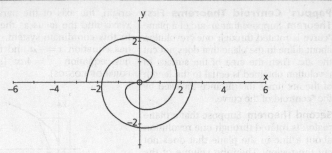

The parabolic spiral $r^2 = \theta$

parabolic coordinates Orthogonal *curvilinear coordinates (s,t) in the *half-plane $y > 0$ defined by

$$x = (s^2 - t^2)/2, \quad y = st \quad (s, t > 0).$$

The curves where s or t are constant are parabolas. *Laplace's equation is invariant under a change to parabolic coordinates. Note that these coordinates arise from the complex, *conformal map $z \mapsto z^2/2$ from the first *quadrant to the upper half-plane.

parabolic cylinder A *cylinder in which the fixed curve is a *parabola and the fixed line to which the generators are parallel is perpendicular to the plane of the parabola. It is a *quadric and in a suitable coordinate system has equation

$$\frac{x^2}{a^2} = \frac{2y}{b}.$$

parabolic partial differential equation The second order *partial differential equation

$$A\frac{\partial^2 f}{\partial x^2} + B\frac{\partial^2 f}{\partial x \partial y} + C\frac{\partial^2 f}{\partial y^2} + D\frac{\partial f}{\partial x}$$

$$+ E\frac{\partial f}{\partial y} + Ff + G = 0,$$

where A, B, C, D, E, F, G are functions of x and y is parabolic, elliptic, or hyperbolic when $B^2 - 4AC$ is zero, negative, or positive, respectively. So the *heat equation

is parabolic, *Laplace's equation is elliptic, and the *wave equation is hyperbolic.

parabolic spiral A spiral defined in *polar coordinates by the equation $r^2 = k\theta$. In the figure $k = 1$.

(((●))) SEE WEB LINKS

• An animation of a parabolic spiral.

paraboloid See ELLIPTIC PARABOLOID, HYPERBOLIC PARABOLOID.

paradox The term paradox often refers to a self-contradictory statement or situation, as with *Grelling's paradox, the *liar paradox, *Perron's paradox, and *Russell's paradox. However the term is also used to refer to true statements that are highly counterintuitive, such as the *Banach-Tarski paradox and *Simpson's paradox. See also ZENO OF ELEA.

parallel In *Euclidean geometry, two or more lines or planes that are always equidistant from one another, however far they are extended. Consequently they will never meet, though this is not a sufficient condition for lines in three dimensions to be parallel, as *skew lines also never meet. Two parallel lines have to lie in a plane. Sometimes curves or surfaces are described as parallel if they satisfy the equidistant condition, where the distance between the curves or surfaces is defined as the shortest distance from a point on one curve or surface to any point lying on the other.

In the *hyperbolic plane, two lines are said to be parallel if they do not meet. No two lines are parallel in the *elliptic plane.

parallel axis theorem Let I_G be the *moment of inertia of a *rigid body about an axis through G, the *centre of mass of the rigid body. Then the moment of inertia of the body about some other axis *parallel to the first axis equals $I_G + md^2$, where m is the mass of the rigid body and d is the distance between the two parallel axes.

Rephrasing the theorem in terms of the *inertia matrix, for a point P at (column) position vector $\mathbf{r}$ from G, then

$$\mathbf{I}_P = \mathbf{I}_G + m(|\mathbf{r}|^2\mathbf{I}_3 - \mathbf{rr}^\mathrm{T})$$

where $\mathbf{I}_3$ denotes the 3×3 *identity matrix.

parallel computation The simultaneous execution of parts of the same task on multiple processors to obtain faster results. *Compare* SERIAL COMPUTATION.

parallelepiped A *polyhedron with six faces, each of which is a *parallelogram. *See* SCALAR TRIPLE PRODUCT.

parallelogram A quadrilateral in which (i) both pairs of opposite sides are parallel and (ii) the lengths of opposite sides are equal. Either property, (i) or (ii), in fact implies the other. The area of a parallelogram equals 'base times height'. That is to say, if one pair of parallel sides, of length b, are a distance h apart, the area equals bh. Alternatively, if the other pair of sides have length a

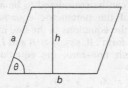

A parallelogram

and θ is the angle between adjacent sides, the area equals $ab\sin\theta$, so the area also equals $|\mathbf{a} \times \mathbf{b}|$, the magnitude of the *vector product of the sides as vectors.

parallelogram law This law states for a *parallelogram that the sum of the squares of the diagonals' lengths equals the sum of the squares of the sides' lengths. Parallelograms are the only *quadrilaterals for which the law holds. For a parallelogram OABC, where OA = $\mathbf{v}$ and OC = $\mathbf{w}$, the law states

$$|\mathbf{v} + \mathbf{w}|^2 + |\mathbf{v} - \mathbf{w}|^2 = 2|\mathbf{v}|^2 + 2|\mathbf{w}|^2.$$

This identity holds in *inner product spaces; a *Banach space is a *Hilbert space if and only if the parallelogram law holds.

parallel postulate The axiom of *Euclidean geometry which says that, if two straight lines are cut by a *transversal and the interior angles on one side add up to less than two right angles, then the two lines meet on that side. It is equivalent to *Playfair's axiom, which says that, given a point not on a given line, there is precisely one line through the point *parallel to the line. The parallel postulate was shown to be independent of the other axioms of Euclidean geometry in the 19th century, when *non-Euclidean geometries were discovered in which the other axioms hold but the parallel postulate does not.

parallel transport A smooth *vector field $\mathbf{v}$ along a parameterized curve $\gamma(t)$ in a *surface X is parallel if the *covariant derivative $D\mathbf{v}/dt$ is zero. For a plane X, this means that $\mathbf{v}$ is constant. $\gamma(t)$ is a *geodesic if and only if the tangent vectors $\gamma'(t)$ are parallel.

If $\mathbf{v}$ is parallel and $t_1 < t_2$, then $\gamma(t_2)$ is the parallel transport of $\gamma(t_1)$ along γ.

parameter (in pure mathematics) A variable taking different values, thereby giving different values to certain other variables. For example, a parameter t could be used to write the *solutions of the equation $5x_1 + 4x_2 = 7$ as

$$x_1 = 3 - 4t, \quad x_2 = -2 + 5t (t \in \mathbb{R}).$$

More generally parameters can be systematically used to describe the general solution of a system of *simultaneous linear equations. See also PARAMETERIZATION, PARAMETRIC EQUATIONS (of a line in space).

parameter (in statistics) A parameter for a *population is some quantity that relates to the population, such as its *mean or *median. A parameter for a population may be estimated from a *sample by using an appropriate *statistic as an *estimator. For a *distribution, a constant that appears in the *probability mass function or *probability density function of the distribution is called a parameter. In this sense, the *normal distribution has two parameters and the *Poisson distribution has one parameter, for example.

parameterization (of a curve) A method of associating, with every value of a *parameter t in some interval I (or some other subset of $\mathbb{R}$), a point $P(t)$ on the curve such that every point of the curve corresponds uniquely to some value of t. Often this is done by giving the x- and y-coordinates of P as functions of t, so that the coordinates of P may be written $(x(t), y(t))$. The equations that give x and y as functions of t are parametric equations for the curve. For example, $x = at^2, y = 2at (t \in \mathbb{R})$ are parametric equations for the parabola $y^2 = 4ax$; and $x = a\cos\theta, y = b\sin\theta (\theta \in [0, 2\pi))$ are parametric equations for the ellipse

$$\frac{x^2}{a^2} + \frac{y^2}{b^2} = 1.$$

The gradient dy/dx of the curve at any point can be found, if $x'(t) \neq 0$, from $dy/dx = y'(t)/x'(t)$.

parameterized surface A surface defined by a single *chart (or coordinate map). Specifically, a parameterized surface $X \subseteq \mathbb{R}^3$ is a function $\mathbf{r}: U \rightarrow \mathbb{R}^3$ defined on an *open set $U \subseteq \mathbb{R}^2$ such that

- $\mathbf{r}$ is a continuous (see CONTINUOUS FUNCTION) *bijection from U onto X and $\mathbf{r}^{-1}$ is continuous; that is, $\mathbf{r}$ is a homeomorphism between U and X;
- $\partial\mathbf{r}/\partial x, \partial\mathbf{r}/\partial y$ exist and are continuous;
- (smoothness condition) $\partial\mathbf{r}/\partial x$ and $\partial\mathbf{r}/\partial y$ are linearly independent.

The last condition means that the *tangent vectors $\partial\mathbf{r}/\partial x$ and $\partial\mathbf{r}/\partial y$ span a 2-dimensional vector space called the *tangent space. The smooth condition is equivalent to $EG - F^2 \neq 0$, where $E, 2F, G$ are the coefficients of the *first fundamental form. See also ATLAS, TRANSITION MAP.

parametric equations (of a curve) See PARAMETERIZATION.

parametric equations (of a line in space) Given a line in 3-dimensional space, let (x_1, y_1, z_1) be coordinates of a point on the line, and l, m, n be *direction ratios of a direction along the line. Then the line consists of all points P whose coordinates (x, y, z) are given by

$$x = x_1 + tl, \quad y = y_1 + tm, \quad z = z_1 + tn,$$

for some value of the *parameter t. These are parametric equations for the line. They are most easily established by using the *vector equation of the line and taking components. If none of l, m, n is zero, the equations can be written

$$\frac{x - x_1}{l} = \frac{y - y_1}{m} = \frac{z - z_1}{n} \ (= t),$$

which can be considered to be another form of the parametric equations, or called the equations of the line in 'symmetric form'. If, say, $n = 0$ and l and m are both non-zero, the equations are written

$$\frac{x - x_1}{l} = \frac{y - y_1}{m}, \quad z = z_1.$$

if, say, $m = n = 0$, they become $y = y_1$, $z = z_1$.

More generally in n-dimensional space, if **p** and **a** are in $\mathbb{R}^n$ with $\mathbf{a} \neq \mathbf{0}$, then

$$\mathbf{r}(t) = \mathbf{p} + t\mathbf{a} \quad (t \in \mathbb{R})$$

is a *parameterization of the line passing through **p** which is parallel to **a**.

parametric statistics The branch of *statistics dealing with *inference about *parameters of a *population on the basis of *observations and measurements taken from a *sample.

parentheses *Brackets, usually (), which can be used in algebraic expressions to make explicit the order in which arithmetic operations are to be carried out or to associate an *argument with a *function such as in $f(x)$.

Pareto chart A *bar chart in which the categories are arranged in the order of their frequencies, starting with the most frequent. This reveals what are the most important factors in any given situation, and enables a realistic cost–benefit analysis of what measures might be undertaken to improve performance.

parity The attribute of an integer of being *even or *odd. Thus, it can be said that 6 and 14 have the same parity (both are even), whereas 7 and 12 have opposite parity. The term also relates to *even permutations and *odd permutations.

parity check *See* CHECK DIGIT.

Parker, Matt (1980–) An Australian popularizer of mathematics and comedian whose first book was *Things to Make and Do in the Fourth Dimension*. He began MathsJam in London in 2008 as an informal gathering of maths enthusiasts, and it has since spread to over 50 cities.

(⊕) SEE WEB LINKS

• Matt Parker's official website.

Parseval's identity The identity states that

$$\|x\|^2 = \sum_{i=1}^{n} |\langle x, x_i \rangle|^2$$

for a *orthonormal basis $x_1, x_2, \ldots, x_n$ of a *finite-dimensional *inner product space X and $x \in X$. This is a generalization of *Pythagoras' Theorem. The identity holds for an orthonormal sequence in an infinite-dimensional *Hilbert space if and only if the sequence is *complete, or equivalently, a *Schauder basis. *Compare* BESSEL'S INEQUALITY.

partial derivative Suppose that $f(x_1, x_2, \ldots, x_n)$ is a real *function of n variables. If

$$\frac{f(x_1 + h, x_2, \ldots, x_n) - f(x_1, x_2, \ldots, x_n)}{h}$$

tends to a limit as $h \to 0$, this limit is the partial derivative of f, at the point $(x_1, x_2, \ldots, x_n)$, with respect to x_1; it is denoted by $f_1(x_1, x_2, \ldots, x_n)$ or $\partial f/\partial x_1$ (read as 'partial d f by d x_1'). This partial derivative may be found using the normal rules of *differentiation, by differentiating as though the function were a function of x_1 only and treating $x_2, \ldots, x_n$ as constants. The other partial derivatives,

$$\frac{\partial f}{\partial x_2}, \ldots, \frac{\partial f}{\partial x_n},$$

are defined similarly. The partial derivatives may also be denoted by $f_{x_1}, f_{x_2}, \ldots, f_{x_n}$. For example, if $f(x,y) = xy^3$, then the partial derivatives are $f_x = y^3$ and $f_y = 3xy^2$. *See also* CHAIN RULE (multivariable), DIFFERENTIAL (multivariate case), DIRECTIONAL DERIVATIVE, HIGHER-ORDER PARTIAL DERIVATIVE, JACOBIAN MATRIX.

partial differential equation Let z be a *function of variables x and y let z_x, z_y, z_{xx}, z_{xy}, z_{yy}, z_{xxx} ... denote the *partial derivatives and *higher-order partial derivatives of z. A partial differential equation (PDE for short) is an equation involving x, y, z and its partial derivatives (*compare* ORDINARY DIFFERENTIAL EQUATION, STOCHASTIC DIFFERENTIAL EQUATION). The order of the PDE is the highest total order of any partial derivative that appears. These notions readily generalize to functions of more than two variables.

As the differential operators $\partial/\partial x$ and $\partial/\partial y$ respectively send functions of y only and x only to 0, a PDE of order n might be expected to have n arbitrary functions, rather than n arbitrary constants, in its general solution. As with ODEs, this is at best a rule of thumb and does not always hold. For example, the first order PDE $z_x + z_y = 0$ has general solution $z = f(x-y)$, and the second order PDE $z_{xy} = 0$ has general solution $z = f(x) + g(y)$.

See HEAT EQUATION, LAPLACE'S EQUATION, PARABOLIC DIFFERENTIAL EQUATION, WAVE EQUATION.

partial differentiation The process of obtaining one of the *partial derivatives of a function of more than one variable. The partial derivative $\partial f/\partial x_i$ is said to be obtained from f by 'differentiating partially with respect to x_i'.

partial fractions Suppose that $f(x)/g(x)$ defines a *rational function, so that $f(x)$ and $g(x)$ are *polynomials, and suppose that the degree of $f(x)$ is less than the degree of $g(x)$. For real polynomials, $g(x)$ can be factorized into a product of linear factors and *irreducible quadratic factors. Then the original expression $f(x)/g(x)$ can be written as a sum of terms: corresponding to each $(x-\alpha)^n$ in $g(x)$, there are terms

$$\frac{A_1}{x-\alpha} + \frac{A_2}{(x-\alpha)^2} + \dots + \frac{A_n}{(x-\alpha)^n},$$

and corresponding to each irreducible factor $ax^2 + bx + c$ in $g(x)$, there are terms

$$\frac{B_1 x + C_1}{ax^2 + bx + c} + \frac{B_2 x + C_2}{(ax^2 + bx + c)^2} + \dots$$
$$+ \frac{B_n x + C_n}{(ax^2 + bx + c)^n},$$

where the scalars A_k, B_k, C_k are uniquely determined. The expression $f(x)/g(x)$ is then said to have been written in partial fractions. If the degree of $f(x)$ is greater than that of $g(x)$, then *long division of $g(x)$ into $f(x)$ can be done first to reduce the numerator to less than that of the denominator.

As examples:

$$\frac{3}{(x-1)(x+2)} = \frac{A}{x-1} + \frac{B}{x+2},$$

$$\frac{3x^2 + 2x + 1}{(x-1)^3} = \frac{A}{x-1} + \frac{B}{(x-1)^2} + \frac{C}{(x-1)^3},$$

$$\frac{3x+2}{(x-1)(x^2+x+1)^2} = \frac{A}{x-1} + \frac{Bx+C}{x^2+x+1}$$
$$+ \frac{Dx+E}{(x^2+x+1)^2}, \frac{3x+2}{(x-1)^2(x^2+x+1)}$$
$$= \frac{A}{x-1} + \frac{B}{(x-1)^2} + \frac{Cx+D}{x^2+x+1}.$$

The values for the numbers A, B, C, ... can be found by multiplying both sides of the equation by the denominator $g(x)$. In the last example, this gives

$$3x + 2 = A(x - 1)(x^2 + x + 1) +$$
$$B(x^2 + x + 1) + (Cx + D)(x - 1)^2.$$

This has to hold for all values of x, so the coefficients of corresponding powers of x on the two sides can be equated, and this determines the unknowns. In some cases, setting x equal to particular values (in this example, $x = 1$) may determine some of the unknowns more quickly. The method of partial fractions is useful in the *integration of rational functions or to express rational functions as *power series.

partial order A partial order on a set X, usually denoted $\leqslant$, is a *binary relation which is *reflexive, *anti-symmetric, and *transitive. Examples include $\subseteq$ on the *power set of a set and *divides on the set of positive integers. *See also* TOTAL ORDER.

partial product When an infinite product

$$\prod_{r=1}^{\infty} a_r$$

is formed from a sequence $a_1, a_2, a_3, \ldots$, the product $a_1 a_2 \ldots a_n$ of the first n terms is called the nth partial product.

partial sum The nth partial sum s_n of a *series $a_1 + a_2 + \ldots$ is the sum of the first n terms; thus

$$s_n = a_1 + a_2 + \ldots + a_n.$$

particle An object considered as having no size but having *mass, *displacement, *velocity, *acceleration, and so on. It is used in *mathematical models to represent an object in the real world of negligible size. *See also* POINT MASS.

particular integral *See* LINEAR DIFFERENTIAL EQUATION WITH CONSTANT COEFFICIENTS.

particular solution *See* DIFFERENTIAL EQUATION.

partition (of an interval) Let $[a,b]$ be a closed interval. A set of $n+1$ points $x_0, x_1, \ldots, x_n$ such that

$$a = x_0 < x_1 < x_2 < \ldots < x_{n-1} < x_n = b$$

is a partition of the interval $[a,b]$. A partition divides the interval into n subintervals $[x_i, x_{i+1}]$. The *norm (or mesh) of the partition P is equal to the length of the largest subinterval and is denoted by $\|P\|$. Such partitions are used in defining *Riemann sums (*see* INTEGRAL).

partition (of a number) A partition of the positive integer n is obtained by writing

$$n = n_1 + n_2 + \ldots + n_k,$$

where $n_1, n_2, \ldots, n_k$ are positive integers, and the order in which $n_1, n_2, \ldots, n_k$ appear is unimportant. The number of partitions of n is denoted by $p(n)$. For example, the partitions of 5 are

$$5, \quad 4+1, \quad 3+2, \quad 3+1+1, \quad 2+2+1,$$
$$2+1+1+1, \quad 1+1+1+1+1,$$

and hence $p(5) = 7$. The values of $p(n)$ for small values of n are as follows:

n	1	2	3	4	5	6	7	8	9	10
$p(n)$	1	2	3	5	7	11	15	22	30	42

Table of values of $p(n)$

The *asymptotic approximation (*see* ASYMPTOTICALLY EQUAL)

$$p(n) \sim \frac{1}{4n\sqrt{3}} \exp\left(\pi\sqrt{\frac{2n}{3}}\right),$$

for large n, was famously obtained by *Hardy and *Ramanujan in 1918.

Compare COMPOSITION (of a number).

partition (of a set) A partition of a set S is a collection of *non-empty *disjoint subsets of S whose union is S. Equivalently, every element of S belongs to exactly one of the subsets in the collection. Given a partition of a set S, an *equivalence relation $\sim$ on S can be obtained by defining $a \sim b$ if a and b belong to the same subset in the partition. Conversely, given any equivalence relation on S, the equivalence classes partition S.

pascal The SI unit of *pressure, abbreviated to 'Pa'. One pascal is equal to one *newton per square metre.

Pascal, Blaise (1623–62) French mathematician and religious philosopher who in mathematics is noted for his work in *geometry, *hydrostatics, and *probability. *Pascal's triangle, the arrangement of the binomial coefficients, was not invented by him, but he did use it in his studies in probability, on which he corresponded with *Fermat. He also joined in the work on finding the areas of curved figures, work which was soon to lead to the calculus. Here, his main contribution was to find the area of an arch in the shape of a *cycloid.

Pascal's theorem (Pascal's mystic hexagram) A theorem of *projective geometry: given two triples of points A, B,C and A',B',C' on a non-degenerate *conic, the points of intersection

$$AB'' \cap BA', \quad BC'' \cap CB', \quad AC'' \cap CA',$$

are *collinear. *Pappus' Hexagon Theorem can be seen as a version of Pascal's theorem for *degenerate conics.

The dual of Pascal's theorem is Brianchon's theorem: if a *hexagon with vertices P_1, P_2, P_3, P_4, P_5, P_6 circumscribes a non-degenerate conic, then the hexagon's diagonals P_1P_4, P_2P_5, P_3P_6 are concurrent. *See* DUALITY.

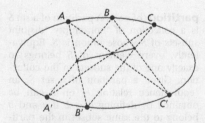

Pascal's 'mystic hexagram'

Pascal's triangle The figure shows the first seven rows of the arrangement of numbers known as Pascal's triangle. In general, the nth row consists of the *binomial coefficients $\begin{pmatrix} n \\ r \end{pmatrix}$, or ^nC_r, with $0 \leq r \leq n$. With the numbers set out in this fashion, it can be seen how the number $\begin{pmatrix} n+1 \\ r \end{pmatrix}$ is equal to the sum of the two numbers $\begin{pmatrix} n \\ r-1 \end{pmatrix}$ and $\begin{pmatrix} n \\ r \end{pmatrix}$, which are situated above it to the left and right. For example, $\begin{pmatrix} 7 \\ 3 \end{pmatrix}$ equals 35, and is the sum of $\begin{pmatrix} 6 \\ 2 \end{pmatrix}$, which equals 15, and $\begin{pmatrix} 6 \\ 3 \end{pmatrix}$, which equals 20.

$$
\begin{array}{ccccccccccccc}
&&&&&& 1 &&&&&& \\
&&&&& 1 && 1 &&&&& \\
&&&& 1 && 2 && 1 &&&& \\
&&& 1 && 3 && 3 && 1 &&& \\
&& 1 && 4 && 6 && 4 && 1 && \\
& 1 && 5 && 10 && 10 && 5 && 1 & \\
1 && 6 && 15 && 20 && 15 && 6 && 1 \\
\end{array}
$$
1 7 21 35 35 21 7 1

See also SINGMASTER'S CONJECTURE.

path (in a graph) Let u and v be vertices of a *graph. A path is a *walk where no vertex is visited more than once (and consequently no edge traced more than once).

path component An *equivalence relation can be defined on a *topological space (or *metric space) X by $x \sim y$ if there is a path between x and y; that is, there is a *continuous function f:$[0,1] \rightarrow X$ such that $f(0) = x$ and $f(1) = y$. The equivalence classes of $\sim$ are the path components. A path-connected space is a space with a single path component.

path-connected Describing a *topological space (or *metric space) X in which there is a path between any two points; that is for any x,y in X there is a *continuous function f:$[0,1] \rightarrow X$ such that $f(0) = x$ and $f(1) = y$. Path-connected spaces are necessarily *connected but the *converse is not true.

path integral *See* LINE INTEGRAL.

Pauli, Wolfgang Ernst (1900–58) Austrian theoretical physicist. He introduced the notion of spin (relating to *angular momentum) into *quantum theory and demonstrated the Pauli exclusion principle, which states that two electrons (or more generally particles with half-integer spin) cannot occupy the same quantum state, for which he was awarded the Nobel Prize for Physics in 1945.

payoff In *game theory, the positive or negative amount each player is credited with once all players have chosen their *strategy.

PDE Short for *partial differential equation.

pdf An abbreviation for PROBABILITY DENSITY FUNCTION.

Peano, Giuseppe (1858–1932) Italian mathematician whose work towards the axiomatization of mathematics was very influential. He was responsible for the development of *symbolic logic, for which he introduced important notation. His *axioms for the integers were an important development in the formal analysis of *arithmetic. He was also the first person to produce examples of so-called space-filling curves (*see* PEANO CURVE).

Peano axioms In 1889 Giuseppe *Peano published a set of five axioms which attempted to provide a rigorous basis for the *natural numbers. In 1888 Richard *Dedekind had published a set of axioms on the same topic, but Peano's were more precisely stated. However, they are sometimes referred to as the Peano–Dedekind axioms in recognition of the foundation laid by Dedekind.

The axioms are:
  (i) Zero is a natural number.
 (ii) Every natural number has a successor in the natural numbers.
(iii) Zero is not the successor of any natural number.

 (iv) If the successors of two natural numbers n, m are the same number, then $n = m$.
  (v) If a set S contains zero and the successor of every number in S is also in S, then the set S contains the natural numbers.

The last of these is known as the principle of *mathematical induction.

Peano curve Take the diagonal of a square, as shown in the first diagram. Replace this by the nine diagonals of smaller squares, as in the second diagram (drawn here in a way that indicates the order in which the diagonals are to be traced out). Now replace each straight section of this by the nine diagonals of even smaller squares, to obtain the result shown in the third diagram. The Peano curve is the curve obtained when this process is continued indefinitely. It has the remarkable property that it passes through every point of the square, and it is therefore described as space-filling.

First iteration

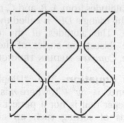

Second iteration

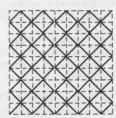

Third iteration

Pearson, Karl (1857–1936) British statistician influential in the development of statistics for application to the biological and social sciences. Stimulated by problems in evolution and heredity, he developed fundamental concepts such as the *standard deviation, the *coefficient of variation, and, in 1900, the *chi-squared test. This work was contained in a series of important papers written while he was professor of applied mathematics, and then eugenics, at University College, London.

Pearson's product moment correlation coefficient *See* CORRELATION.

pedal curve The *curve generated by taking the foot of the perpendiculars from a fixed point to all the *tangents to a given curve.

(⊕) SEE WEB LINKS
• An animation of a pedal curve.

pedal triangle In the triangle with vertices *A*, *B*, and *C*, let *D*, *E*, and *F* be the feet of the *altitudes from *A*, *B*, and *C*, respectively. The triangle *DEF* is called the pedal triangle. The altitudes of the original triangle *bisect the angles of the pedal triangle. *Compare* MEDIAN TRIANGLE.

Pell's equation The *Diophantine equation $x^2 = ny^2 + 1$, where n is a positive integer that is not a perfect square. Methods of solving such an equation have been sought from as long ago as the time of *Archimedes. *Bhāskara solved particular cases. *Fermat apparently understood that infinitely many solutions always exist, a fact that was proved by *Lagrange. The equation was mistakenly attributed to the English mathematician John Pell by *Euler.

pendulum *See* COMPOUND PENDULUM, CONICAL PENDULUM, FOUCAULT PENDULUM, SIMPLE PENDULUM.

Penrose, Sir Roger (1931–) British mathematician and theoretical physicist who was Rouse Ball Professor of Mathematics at Oxford University for 25 years. He was awarded the *Wolf Prize for Physics in 1988 jointly with Stephen *Hawking and shared the Noble Prize in 2020 for physics. He also produced important papers on cosmology, *topology, *manifolds, and twistor theory, which uses *geometry and *algebra in an attempt to unite *quantum theory and *relativity. He is also well known as an author of popular science books.

Penrose tiling In the 1970s, Roger *Penrose discovered various aperiodic *tilings of the plane. The most famous of these uses two different *rhombi for tiles. A tiling is aperiodic if there are no nontrivial *translations that are symmetries of the tiling and if there are not arbitrarily large periodic sections of the tiling.

penta- Prefix denoting five, as does *quin-.

pentagon A five-sided polygon.

pentagonal number A number which can be arranged in the shape of nested pentagons as shown in the figure. The sequence is 1, 5, 12, 22, 35, . . . which can be generated by $\frac{n}{2}(3n - 1)$ for $n \geq 1$.

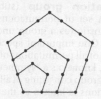

First four pentagonal numbers

pentagram The plane figure shown below, formed by joining alternate vertices of a regular *pentagon.

A pentagram

per cent (%) Literally from the Latin 'in every hundred', so it is a fraction with the denominator of 100 replaced by this phrase or by the symbol %. So 20 per cent = 20% = 20/100 which is 1/5 or 0.2.

percentage A *proportion, *interest rate, or *ratio expressed with a *denominator of 100.

percentage error The *relative error expressed as a percentage. When 1.9 is used as an *approximation for 1.875, the relative error equals 0.025/1.875 (or 0.025/1.9) = 0.013, to 2 *significant figures. So the percentage error is 1.3%, to 2 significant figures.

percentile *See* QUANTILE.

perfectly normal A *topological space X in which, given any two *disjoint, non-empty, *closed subsets A and B, there exists a *continuous function $f:X \rightarrow [0,1]$ such that A is the *pre-image of 0 and B is the pre-image of 1. *See* SEPARATION AXIOMS.

perfect number An integer that is equal to the sum of its positive *divisors (not including itself). Thus, 6 is a perfect number, since its positive divisors (not including itself) are 1, 2, and 3, and $1 + 2 + 3 = 6$; so too are 28 and 496, for example. At present there are over 50 known perfect numbers, all even. If $2^p - 1$ is prime (so that it is a *Mersenne prime), then $2^{p-1}(2^p - 1)$ is perfect; moreover, these are the only even perfect numbers. It is an *open problem whether an odd perfect number exists.

((⊕)) SEE WEB LINKS

• A brief history of perfect numbers.

perfect square An *integer of the form n^2, where n is an integer. The perfect squares are 0, 1, 4, 9, 16, 25, ...

perigee *See* APSE.

perigon A synonym for FULL ANGLE.

perihelion *See* APSE.

perimeter The perimeter of a plane figure is the length of its boundary. Thus, the perimeter of a *rectangle of length L and width W is $2L + 2W$. The perimeter of a circle is the *arc length of its *circumference. *See* ISOPERIMETRIC INEQUALITY.

period, periodic If, for some value p, $f(x + p) = f(x)$ for all x, the real function f is periodic and has period p. For example, $\cos x$ is periodic with period 2π *radians, since $\cos(x + 2\pi) = \cos x$ for all x; or, using *degrees, $\cos x°$ is periodic with period 360. Some authors restrict the use of the term 'period' to the smallest positive value of p with this property.

In mechanics, any phenomenon that repeats regularly may be called periodic, and the time taken before the phenomenon repeats itself is then called the period. Suppose that $x = A\sin(\omega t + \alpha)$, where $A > 0$, ω and α are constants. This may, for example, give the *displacement x of a particle, moving in a line, at time t. The particle is thus oscillating about the origin. The period is the time taken for one complete *oscillation and is equal to $2\pi/\omega$.

periodic point Given a function $f:X \rightarrow X$ of a set X, then x in X is a periodic point if the sequence

$$x, f(x), f^2(x), f^3(x), \ldots$$

eventually returns to x. (Here $f^k(x)$ denotes k applications of f to x.) The least positive n such that $f^n(x) = x$ is called the period of x and

$$\{x, f(x), f^2(x), \ldots, f^{n-1}(x)\}$$

is called the orbit of x.

peripheral vertex (in a graph) Any vertex in a *graph for which the *distance between it and a *central vertex is equal to the *radius of the graph. In other words, any point which is as far away from the centre as it is possible to get.

permutation In *combinatorics, a permutation of n objects is an arrangement n objects. The number of permutations of n objects is equal to $n!$. The number of 'permutations of r objects from n objects' is denoted by ^nP_r, and equals

$$n(n-1)\ldots(n-r+1) = \frac{n!}{(n-r)!}.$$

For example, there are 12 permutations of A, B, C, D taken two at a time: AB, AC, AD, BA, BC, BD, CA, CB, CD, DA, DB, DC.

Suppose that n objects are of k different kinds, with r_1 alike of one kind, r_2 of a second kind, and so on. Then the number of distinct permutations of the n objects equals the *multinomial coefficient

$$\frac{n!}{r_1! r_2! \ldots r_k!},$$

where $r_1 + r_2 + \ldots + r_k = n$. For example, the number of different anagrams of the word '$CHEESES$' is $7!/3! \, 2!$, which equals 420. *Compare* COMBINATION.

permutation A permutation of a set S is a *bijection from S to S. The permutations of S form a group, the *symmetric group Sym(S) of S, under *composition.

permutation group (substitution group) A set of permutations of a set which constitutes a group under *composition. One important special case is the set S_n of all permutations of any set of n objects, known as a *symmetric group. The *subgroup consisting of all the *even permutations is known as the *alternating group A_n.

permutation matrix An $n \times n$ *matrix with a single 1 in each row and each column and 0s everywhere else which can be used to represent a *permutation of the n objects in a set. For example, to map 1, 2, 3 to 1, 3, 2 the

matrix $\begin{pmatrix} 1 & 0 & 0 \\ 0 & 0 & 1 \\ 0 & 1 & 0 \end{pmatrix}$ would be used

because

$$\begin{pmatrix} 1 & 0 & 0 \\ 0 & 0 & 1 \\ 0 & 1 & 0 \end{pmatrix} \begin{pmatrix} 1 \\ 2 \\ 3 \end{pmatrix} = \begin{pmatrix} 1 \\ 3 \\ 2 \end{pmatrix}.$$

perpendicular At *right angles to one another. The term 'orthogonal' is synonymous.

perpendicular axis theorem Suppose that a *rigid body is in the form of a *lamina lying in the plane $z = 0$. Let I_x, I_y, and I_z be the moments of inertia of the body about the three coordinate axes. The perpendicular axis theorem then states that $I_z = I_x + I_y$.

perpendicular bisector The perpendicular bisector of a *line segment AB is the straight line *perpendicular to AB through the *midpoint of AB. The perpendicular bisectors of a triangle's sides are *concurrent at the *circumcentre.

perpendicular distance The distance from a point to a *line or *plane measured along the line *perpendicular to the line or plane which passes through the given point. It is the shortest distance between the point and any given point on the line or plane.

• A

AP is the perpendicular
distance from A to BC.

B   P   C

perpendicular lines In *coordinate geometry of the plane, a useful *necessary and sufficient condition that two lines, with *gradients m_1 and m_2, are perpendicular is that $m_1 m_2 = -1$. (This is taken to include the cases when $m_1 = 0$ and m_2 is infinite, and vice versa.)

perpendicular planes Two *planes in 3-dimensional space are *perpendicular if *normals to the two planes are perpendicular. If $\mathbf{n}_1$ and $\mathbf{n}_2$ are *vectors normal to the planes, this requires that $\mathbf{n}_1 \cdot \mathbf{n}_2 = 0$.

Perron's paradox The following paradox: let n be the largest *integer; if $n > 1$, then $n^2 > n$, which contradicts n being the largest integer and so $n = 1$ is the largest integer. The ridiculous conclusion shows the danger of simply assuming a solution to an *optimization problem exists.

perspective The art and mathematics of realistically representing three dimensions in a 2-dimensional diagram. Straight lines in space appear as straight in the image. Lines which are parallel to the image plane appear as *parallel in the image. Other parallel lines in space meet at a vanishing point, and each new direction will require a separate vanishing

point. See diagram for an example with two vanishing points. *See also* POINTS AT INFINITY, PROJECTIVE GEOMETRY.

perspectivity Given two lines L and L' in a projective plane (*see* PROJECTIVE SPACE) and a point O on neither line, each point P on L can be be associated with a point P' on L' such that O, P, P' are *collinear (as in the figure). The *bijection from L to L' sending P to P' is known as a perspectivity, or the perspectivity from O. By assigning *homogeneous coordinates to L and L', the lines can be identified, and so we can consider perspectivities from a line to itself. The perspectivities of a line generate the *projective transformations of the line. *See* DESARGUES' THEOREM.

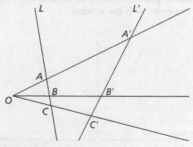

A perspectivity between two lines

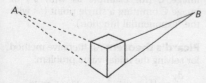

Parallel edges meeting at vanishing points

perturbation A small change in the parameter values in an *equation, *dynamical system, or *optimization problem, usually done to investigate the *stability of a solution. *See* LINEAR THEORY OF EQUILIBRIA.

peta- Prefix used with *SI units to denote multiplication by 10^{15}. Abbreviated as P.

pgf An abbreviation for PROBABILITY GENERATING FUNCTION.

p-group Where p is a *prime number, a *group G in which every element has *order which is a power of p. If G is finite, then this is equivalent to the order of G being a power of p. See SYLOW's THEOREMS.

phase Suppose that $x = A\sin(\omega t + \alpha)$, where $A > 0$, ω and α are constants. This may, for example, give the *displacement x of a particle, moving in a line, at time t. The particle is thus oscillating about the origin. The constant α is the phase. Two particles oscillating like this with the same *amplitude and *period but with different phases are executing the same motion apart from a shift in time.

phase plane (phase space) Given a *dynamical system describing the evolution of a vector $\mathbf{x}(t)$, the possible states of the system make up the phase space (or phase plane if it is two-dimensional). For example, with the system

$$\frac{dx}{dt} = x + y, \qquad \frac{dy}{dt} = 4x + y,$$

the phase plane is the xy-plane and the possible evolutions or trajectories of the system can be sketched as in the figure. This system has an unstable *equilibrium at the origin.

philosophy of mathematics The branch of philosophy that deals with the nature of mathematics, its notions, and objects. It commonly relates to the *foundations of mathematics and also to the question of whether mathematics is discovered or invented. See FINITISM, FORMALISM, INTUITIONISM, LOGICISM, PLATONISM, REALISM, STRUCTURALISM.

pi For any circle, the length of the circumference divided by the length of the diameter is the same, and this number is the value of π. It is equal to 3.14159265 to 8 decimal places. Sometimes π is approximated as $\frac{22}{7}$. In 1761 Lambert showed, using *continued fractions, that π is *irrational. In 1882, Lindemann proved further that π is *transcendental. π appears in some contexts that seem to have no connection with the definition relating to a circle, as in the *Basel problem

$$\frac{\pi^2}{6} = 1 + \frac{1}{2^2} + \frac{1}{3^2} + \frac{1}{4^2} + \dots$$

and in *Wallis' product

$$\frac{\pi}{2} = \frac{2}{1} \times \frac{2}{3} \times \frac{4}{3} \times \frac{4}{5} \times \frac{6}{5} \times \frac{6}{7} \times \dots.$$

See also TAU.

Picard's little theorem A non-constant *holomorphic function on $\mathbb{C}$ has image $\mathbb{C}$ (for example as with z^2) or image $\mathbb{C}$ omitting a single point (as with the *exponential function).

Picard's theorem An *iterative method for solving the initial-value problem:

$$\frac{dy}{dx} = f\big(x, y(x)\big), \qquad y(x_0) = y_0.$$

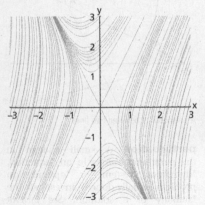

Trajectories in a phase plane

Assume that f is a *continuous function in the first variable and *Lipschitz in the second variable. Then there exists an open interval about x_0 on which there is a unique solution $y(x)$. The solution can be constructively found as the limit of functions $y_n(x)$, where

$$y_{n+1}(x) = y_0 + \int_{x_0}^{x} f\left(t, y_n(t)\right) \, dt,$$
$$y_0(x) = y_0.$$

As an example, suppose $dy/dx = y$ and $y(0) = 1$. Then

$$y_1(x) = 1 + \int_0^x 1 \, dt = 1 + x,$$
$$y_2(x) = 1 + \int_0^x (1+t) \, dt = 1 + x + \frac{x^2}{2}.$$

So $y_n(x)$ is the first $n+1$ terms of the *exponential series and converges to the solution $y(x) = e^x$.

Pick's theorem The *area of a *polygon with *vertices on a planar integer *lattice, as drawn below, equals $i + \frac{b}{2} - 1$, where i denotes the number of lattice points interior to the polygon and b denotes the number of lattice points on the polygon's *boundary. In the diagram, $i = 2$ and $b = 7$, so that the area is 4½.

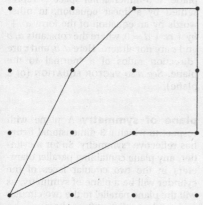

2 interior and 7 boundary points

pico- Prefix used with *SI units to denote multiplication by 10^{-12}. Abbreviated as p.

PID An abbreviation for PRINCIPAL IDEAL DOMAIN.

piecewise continuous A *function is piecewise continuous on a *bounded *interval if the interval can be subdivided into a finite number of subintervals on which the function is continuous (see CONTINUOUS FUNCTION) with finite *limits at each subinterval's endpoints.

pie chart Suppose that some finite set is partitioned into subsets. A pie chart is a diagram consisting of a disc divided into *sectors whose areas are in the same proportion as the sizes of the subsets. Florence *Nightingale helped popularize the pie chart, though it was first introduced by William Playfair in 1801. The figure shows the kinds of vehicles recorded in a small traffic survey.

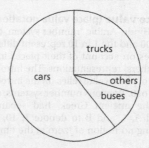

A pie chart

pigeonhole principle The observation that, if m objects are distributed into n boxes, with $m > n$, then at least one box receives at least two objects. This can be applied in some obvious ways; for example, if you take any 13 people, then at least two of them have birthdays that fall in the same month. It also has less trivial applications.

pilot survey A small-scale study conducted before a complete survey to test the effectiveness of the questionnaire.

pivot In the *simplex method or in *Gaussian elimination and *Gauss–Jordan elimination, a pivot is an entry of a matrix selected first at each stage of the *algorithm. In Gauss-Jordan elimination the pivots are the entries that become the leading 1s.

pivot table An interactive table in an electronic spreadsheet that computes summary statistics for a table of *data. It is of particular use in removing the tedium of repeated calculations of the same statistic in a 2-way table, or n-way table.

placebo A placebo appears the same as the treatment to a participant in a study, often a clinical trial of a drug or other treatment, but has no active ingredients. They are used in *control groups because of the positive effects patients can experience simply from taking something they believe will make them better.

place value (place value notation) In the Hindu-Arabic *number system, the 1 in 100 and the 1 in 10 represent different values on account of their places in the numbers' representations. The introduction of place value relies on the introduction of 0. Previous number systems, such as the ancient Greek, had separately used A, I, and P to denote 1, 10, 100, having no notion of *zero at the time.

plaintext *See* CRYPTOGRAPHY.

planar graph A *graph that can be drawn in a plane in such a way that no

Two planar representations of K_4

two edges cross. The *complete graph K_4, for example, is planar as either drawing of it in the figure shows. Neither the complete graph K_5 nor the complete *bipartite graph $K_{3,3}$ is planar. *See* DUAL GRAPH, EULER'S THEOREM, KURATOWSKI'S THEOREM.

Planck, Max Karl Ernst Ludwig (1858–1947) German mathematician and theoretical physicist whose work in thermodynamics concerning the relationship between energy and wavelength won him the Nobel Prize for Physics in 1918 and provided some of the foundations on which *quantum theory was later developed.

Planck's constant The energy E of a photon equals hv, where h is Planck's constant and v denotes the light's frequency. This alternatively is written $E = \hbar\omega$, where $\hbar = h/(2\pi)$ is the reduced Planck constant and ω denotes angular frequency. Its approximate value is 6.6×10^{-34} Js. The constant appears in *Schrödinger's equation and *Heisenberg's Uncertainty Principle.

plane (in Cartesian coordinates) A plane in 3-dimensional space is represented by a linear equation, in other words by an equation of the form $ax + by + cz + d = 0$, where the constants a, b and c are not all zero. Here a, b, and c are *direction ratios of a *normal to the plane. *See also* VECTOR EQUATION (of a plane).

plane of symmetry A plane with respect to which a 3-dimensional figure has reflective *symmetry. So for a cylinder, any plane containing parallel diameters in the two circular faces of the cylinder will be a plane of symmetry, as will the plane parallel to the two circular ends and equidistant from them.

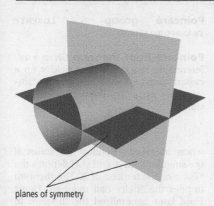

planes of symmetry

Platonic solid A convex *polyhedron is regular if all its faces are alike and all its vertices are alike. More precisely, this means that

(i) all the faces are *regular polygons having the same number p of edges, and
(ii) the same number q of edges meet at each vertex.

Notice that the polyhedron shown here, with 6 triangular faces, satisfies (i) but is not regular because it does not satisfy (ii).

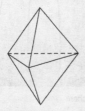

This solid is face-transitive but not regular

There are five regular convex polyhedra, known as the Platonic solids:

(i) the regular *tetrahedron, with 4 triangular faces ($p = 3$, $q = 3$),
(ii) the *cube, with 6 square faces ($p = 4$, $q = 3$),
(iii) the regular *octahedron, with 8 triangular faces ($p = 3$, $q = 4$),

(iv) the regular *dodecahedron, with 12 pentagonal faces ($p = 5$, $q = 3$),
(v) the regular *icosahedron, with 20 triangular faces ($p = 3$, $q = 5$).

See also EDGE-TRANSITIVE, FACE-TRANSITIVE, LUDWIG, SCHLÄFLI, SCHLÄFLI SYMBOL, VERTEX-TRANSITIVE.

Platonism Mathematical Platonism is a *philosophy of mathematics which maintains that mathematical objects and notions are abstract and eternal but have no spatial or temporal existence. An issue with such a philosophy is the (epistemological) question of how mathematicians interact with such abstractions.

platykurtic *See* KURTOSIS.

Playfair's axiom The axiom which says that, given a point not on a given line, there is precisely one line through the point parallel to the line. It is equivalent to *Euclid's parallel postulate. The name arises from its occurrence in a book by John Playfair (1748–1819), a Scottish geologist and mathematician.

plot To mark points on a coordinate graph. Also used in relating to constructing the *graph of a function—by plotting a number of points and inferring the behaviour of the graph between the plotted points.

plus The term may refer to *addition so that $3 + 5$ may be read as '3 plus 5'. The term may also refer to positive quantities and excess quantities as with 'the temperature was 30 plus', which means that the temperature was at least 30.

***Plus* magazine** *Plus* is an online popular mathematics magazine, run by the Millennium Mathematics Project.

(((•))) SEE WEB LINKS
• The official website of *Plus*.

plus or minus The symbol $\pm$ is read as 'plus or minus' and denotes the two possibilities of $+$ and $-$ as in the the *quadratic formula, when either the positive or negative square root of the *discriminant is valid. It can be efficiently used alongside the $\mp$ symbol, read 'minus or plus', as in the *identity

$$\cos(x \pm y) = \cos x \cos y \mp \sin x \sin y,$$

to signify that the $+$ (respectively $-$) in $\pm$ corresponds to the $-$ (respectively $+$) in $\mp$.

plus sign The *symbol $+$ used to denote addition of numbers or any other *binary operation with similar properties, for example in *rings and *abelian groups. It can also be used to denote a positive quantity, though the absence of $+$ or $-$, as in $\sqrt{2}$ is taken to imply the positive.

pmf An abbreviation for PROBABILITY MASS FUNCTION.

p-norm In $\mathbb{R}^n$, the norm defined by

$$||\mathbf{x}||_p = \left(\sum_{i=1}^{n} |x_i|^p \right)^{\frac{1}{p}}.$$

for $1 \leq p \leq \infty$. Note that when $p = 2$, this is just the standard *Euclidean norm, or the length of the vector, and for $p = 1$ it is the Manhattan or *taxicab norm. See l^p.

Poincaré, (Jules) Henri (1854–1912) French mathematician, who can be regarded as the last mathematician to be active over the entire field of mathematics. Within pure mathematics, he is seen as one of the main founders of *topology. In applied mathematics, he is remembered for his theoretical work on the qualitative aspects of celestial mechanics, which was probably the most important work in that area since *Laplace and *Lagrange. Connecting and motivating both was his qualitative theory of *differential equations. He was also known for a stream of popular books for the aspiring young mathematician.

Poincaré group *See* LORENTZ TRANSFORMATION.

Poincaré-Hopf theorem Given a differentiable tangent *vector field **v** on a closed *surface X with finitely many *singularities $p_1, \ldots, p_n$, Poincaré's theorem states

$$\sum_i \text{index}(\mathbf{v}; p_i) = \chi(X)$$

where index$(\mathbf{v}; p_i)$ denotes the *index of the singularity at p_i and $\chi(X)$ denotes the *Euler characteristic of X. The theorem implies the *hairy ball theorem. Heinz Hopf later generalized the theorem to higher-dimensional *manifolds.

In the figure is a sphere with *sink (index 1) at north pole and *source (index 1) at south pole giving. $\chi = 2$.

A flow on a sphere

point A geometrical construct which has position but no size. Its position is often specified by *coordinates.

point estimate *See* ESTIMATE.

point mass (in mechanics) The simplifying assumption that all the *mass of a body is concentrated at a single point.

This can be a reasonable assumption, even when applied to a planet or star, as the *gravitational potential outside a *spherically symmetric massive body is indistinguishable from a point mass with the same total mass placed at the centre of the body.

point of application In the real world, a *force acting on a body is usually applied over a region, such as a part of the surface of the body. When someone is pushing a trolley by hand, there is a region of the handle of the trolley where the hand pushes. When a tugboat is towing a ship, the rope that is being used is attached to a bollard on the deck of the ship. Some forces, such as *gravitational forces, act throughout a body. In a *mathematical model, it may be assumed that a force acts at a particular point, the point of application of the force. In other mathematical models, each point of the surface of a body may experience a force (when the surface is experiencing *pressure), or every point throughout the volume occupied by a body may experience a force (when the body is experiencing a *field of force).

point of inflexion (flex) A point on a graph $y = f(x)$ at which the *concavity (or sense of rotation of the graph) changes. Thus, there is a point of inflexion at a if, for some $\delta > 0$, $f''(x)$ exists and is positive in the open interval $(a - \delta, a)$ and $f''(x)$ exists and is negative in the open interval $(a, a + \delta)$, or vice versa.

If the change (as x increases) is from concave up (anticlockwise rotation) to concave down (clockwise rotation), the graph and its tangent at the point of inflexion look like one of the first three graphs in the figure. If the change is from concave down to concave up, the graph and its tangent look like one of the last three graphs. The second and fourth graph show points of inflexion which are also *stationary points.

If f'' is *continuous at a, then for $y = f(x)$ to have a point of inflexion at a it is necessary that $f''(a) = 0$, and so this is the usual method of finding possible

Inflections that are concave up to concave down

Inflections that are concave down to concave up

points of inflexion. However, the condition $f''(a) = 0$ is not sufficient to ensure that there is a point of inflexion at a: it must be shown that $f''(x)$ is positive to one side of a and negative to the other side. Thus, if $f(x) = x^4$ then $f''(0) = 0$; but $y = x^4$ does not have a point of inflexion at 0 since $f''(x)$ is positive to both sides of 0. Finally, there may be a point of inflexion at a point where $f''(x)$ does not exist, for example at the origin on the curve $y = x^{1/3}$, as shown in the figure.

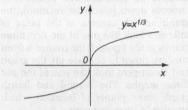

A flex with infinite gradient

If $f(x,y,z)$ is a *homogeneous *polynomial defining a curve C in a projective plane (see PROJECTIVE SPACE), then a point P on C is a point of inflexion if the *Hessian

$$\begin{vmatrix} f_{xx} & f_{xy} & f_{xz} \\ f_{yx} & f_{yy} & f_{yz} \\ f_{zx} & f_{zy} & f_{zz} \end{vmatrix}$$

is zero at P.

points at infinity The point added to the set of *complex numbers to obtain the *extended complex plane. More generally points at infinity may be added to n-dimensional space to obtain n-dimensional *projective space, or to an affine *curve to obtain the projective curve, etc.

point-set topology A synonym for GENERAL TOPOLOGY.

pointwise addition (pointwise multiplication) Given two real-valued functions f,g on a set S, their *sum $f + g$

and *product fg can be defined pointwise on S by

$$(f + g)(s) = f(s) + g(s), \quad (fg)(s) = f(s)g(s),$$

where $s \in S$. Addition and *scalar multiplication of functionals in a *dual space are likewise defined pointwise. *Compare* COMPONENTWISE.

pointwise convergence A *sequence of functions $f_n : S \to \mathbb{R}$ on a given set S *converges pointwise if the real sequence $f_n(x)$ converges for all $x \in S$. This means for all x in S and for any $\varepsilon > 0$, there is an integer N such that for all $m,n \geq N$ $|f_m(x) - f_n(x)| < \varepsilon$. Note that N may depend on how both x and ε are defined. This is in contrast to *uniform convergence, where, for any given ε, a single N has to exist for all values of x.

Poisson, Siméon-Denis (1781–1840) French mathematician who was a pupil and friend of *Laplace and *Lagrange. He extended their work in celestial mechanics and made outstanding contributions in the fields of electricity and magnetism. In 1837, in an important paper on *probability, he introduced the *distribution that is named after him and formulated a *law of large numbers.

Poisson distribution The discrete probability *distribution with *probability mass function given by

$$\Pr(X = r) = \frac{e^{-\lambda}\lambda^r}{r!}, \quad (r \geq 0)$$

where λ is a positive parameter. The *mean and *variance both equal λ. The Poisson distribution gives the number of occurrences in a certain time period of an event which occurs randomly but at a constant average rate. It can be used as an approximation to the *binomial distribution, where n is large and p is small, by taking $\lambda = np$.

Poisson process A *random process where the number of *events occurring in a time interval of length t has a

*Poisson distribution with parameter λt for some positive constant λ. Poisson processes naturally arise when *modelling traffic flow and queueing.

Poisson's equation *See* LAPLACE'S EQUATION.

Poisson's integral formula Poisson's solution to the *Dirichlet problem in the upper half-plane, $\nabla^2 V = 0$ for $y > 0$ and $V(x,0) = f(x)$, is given by the integral formula

$$V(x,y) = \frac{y}{\pi} \int_{-\infty}^{\infty} \frac{f(t) \, dt}{(t-x)^2 + y^2}.$$

polar Given a point P and a non-degenerate *conic C, a line L can be associated with P as follows. Two *tangent lines to C pass through P, and the points of tangency define the line L. This line is known as the polar of P, and P is known as the pole of L.

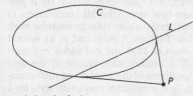

A pole P and polar L

In a projective plane (*see* PROJECTIVE SPACE), this process gives a *one-to-one correspondence between points and lines. If the conic has equation $\mathbf{x}^T B \mathbf{x} = 0$, where B is a 3×3 *symmetric *invertible matrix, and P has coordinate vector $\mathbf{p}$, then the equation of the polar is $\mathbf{x}^T B \mathbf{p} = 0$.

polar coordinates Suppose that a point O in the plane is chosen as *origin, and let Ox be a *directed line through O, with a given unit of length. For any point P in the plane, let $r = |OP|$ and, if P is not O, let θ be the angle (in *radians) that OP

makes with Ox, the angle being given a positive *sense anticlockwise from Ox. The angle θ satisfies $0 \le \theta < 2\pi$. Then (r,θ) are the polar coordinates of P.

Suppose that *Cartesian coordinates are taken with the same origin and the same unit of length, with positive x-axis along the directed line Ox. Then the Cartesian coordinates (x,y) of a point P can be found from (r,θ), by $x = r\cos\theta$, $y = r\sin\theta$. Conversely, the polar coordinates can be found from the Cartesian coordinates by $r = \sqrt{x^2 + y^2}$, and θ is such that

$$\cos\theta = \frac{x}{\sqrt{x^2+y^2}}, \quad \sin\theta = \frac{y}{\sqrt{x^2+y^2}}.$$

(Note that $\theta = \tan^{-1}(y/x)$ but this is not, of itself, sufficient to uniquely determine θ.) In certain circumstances, authors may allow r to be negative, in which case the polar coordinates (r,θ) give the same point as $(-r, \theta + \pi)$.

In mechanics, it is useful, when a point P has polar coordinates (r,θ), to define unit vectors $\mathbf{e}_r$ and $\mathbf{e}_\theta$ by

$$\mathbf{e}_r = \mathbf{i}\cos\theta + \mathbf{j}\sin\theta \text{ and}$$
$$\mathbf{e}_\theta = -\mathbf{i}\sin\theta + \mathbf{j}\cos\theta,$$

where $\mathbf{i}$ and $\mathbf{j}$ are unit vectors in the directions of the positive x- and y-axes. Then $\mathbf{e}_r$ is a unit vector along OP in the direction of increasing r, and $\mathbf{e}_\theta$ is a unit vector perpendicular to this in the direction of increasing θ. These vectors satisfy $\mathbf{e}_r \cdot \mathbf{e}_\theta = 0$ and $\mathbf{e}_r \times \mathbf{e}_\theta = \mathbf{k}$, where $\mathbf{k} = \mathbf{i} \times \mathbf{j}$. *See* RADIAL AND TRANSVERSE COMPONENTS.

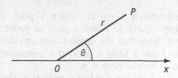

The polar coordinates of P

Curve		Polar equation	
Circle	$x^2 + y^2 = 1$	$r = 1$	
Half-line	$y = x \, (x > 0)$	$\theta = \pi / 4$	
Line	$x = 1$	$r = \sec\theta$	$(-\tfrac{1}{2}\pi < \theta < \tfrac{1}{2}\pi)$
Line	$y = 1$	$r = \operatorname{cosec}\theta$	$(0 < \theta < \pi)$
Circle	$x^2 + y^2 - 2ax = 0 \ (a > 0)$	$r = 2a\cos\theta$	$(-\tfrac{1}{2}\pi < \theta \leq \tfrac{1}{2}\pi)$
Circle	$x^2 + y^2 - 2by = 0 \ (b > 0)$	$r = 2b\sin\theta$	$(0 \leq \theta < \pi)$
Cardioid	*See* CARDIOID	$r = 2a(1 + \cos\theta)$	$(-\pi < \theta \leq \pi)$
Conic	*See* CONIC	$l/r = 1 + e\cos\theta$	$(\cos\theta \neq -1/e)$

Polar equations of standard curves

polar decomposition A square complex matrix **A** can be written **A = UH**, where **U** is *unitary and **H** is *Hermitian and *positive semidefinite.

polar equation An equation of a curve in *polar coordinates is usually written in the form $r = f(\theta)$. Polar equations of some common curves are given in the above table.

polar form of a complex number For a nonzero *complex number z, let $r = |z|$ and $\theta = \arg z$. Then $z = r(\cos\theta + i\sin\theta)$, and this is the polar form of z. It may also be written $z = re^{i\theta}$ using *Euler's formula.

pole (in complex analysis) An isolated *singularity in a complex-valued function at which the function's *Laurent series has a finite *principal part. For example,

$$\frac{z^2 + 1}{z}, \qquad \frac{\sin z}{(z - \pi)(z - 2)^4}$$

respectively have an order 1 pole, or simple pole, at 0 and a pole of order 4 at 2. Note that the second function does not have a pole at π, as the numerator has a simple zero at π.

If a function $f(z)$ has a pole at a, then $f(z) \rightarrow \infty$ as $z \rightarrow a$. *See* MEROMORPHIC FUNCTION.

pole (of a line) *See* POLAR.

Polya, George (1887–1985) Hungarian mathematician whose work included collaboration with *Hardy and *Littlewood, with papers on mathematical physics, *geometry, *complex analysis, *combinatorics, and *probability theory. He is perhaps best known for his contribution to *mathematics education through his book *How To Solve It* published in its second edition in 1957 and still widely regarded half a century later as one of the best expositions of the art of doing mathematics. In a remarkably easy prose style for a book about mathematics, Polya argues that problem solving requires the study of *heuristics and summarizes the problem-solving cycle in four stages: understanding the problem, devising a plan, carrying out the plan, and looking back—or 'see, plan, do, check'.

polygon A plane figure bounded by some number of straight sides. This definition would include polygons like the one shown, but often figures of this sort are intended to be excluded and it is implicitly assumed that a polygon is to be *convex. Thus a convex polygon is a plane figure consisting of a finite region bounded by some finite number of lines, in the sense that the region lies entirely

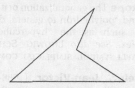

A concave (non-convex) polygon

on one side of each line. The *exterior angles of the polygon shown below are those marked, and the sum of the exterior angles is always equal to 360°. The sum of the *interior angles of an n-sided convex polygon equals $2n-4$ right angles. In a *regular convex polygon, all the sides have equal length and the interior angles have equal size; and the vertices are *concyclic.

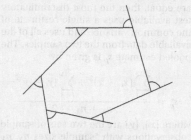

Exterior angles of a convex polygon

5	pentagon	9	enneagon (or nonagon)
6	hexagon	10	decagon
7	heptagon	11	hendecagon
8	octagon	12	dodecagon

Table of polygons' names

polygon of forces *See* TRIANGLE OF FORCES.

polyhedron (polyhedra) A solid figure bounded by some number of plane *polygonal faces. It is often assumed that a polyhedron is *convex, unlike the one in the figure. Thus, a convex polyhedron is a finite region bounded by some number of planes, in the sense that the polyhedron lies entirely on one side of each plane. Each *edge of the polyhedron joins two *vertices, and each edge is the common edge of two *faces.

The numbers of vertices, edges, and faces of a convex polyhedron are related by *Euler's Theorem (for polyhedra).

Certain polyhedra are called regular, and these are the five *Platonic solids; the thirteen *Archimedean solids are semi-regular. *See* POLYTOPE.

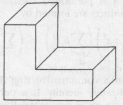

A concave (non-convex) polyhedron

polynomial Let $a_0, a_1, \ldots, a_n$ be real numbers. then

$$a_n x^n + a_{n-1} x^{n-1} + \ldots + a_1 x + a_0$$

is a polynomial in x with coefficients $a_0, a_1, \ldots, a_n$. When $a_0, a_1, \ldots, a_n$ are not all zero, it can be assumed that $a_n \neq 0$ and the polynomial has DEGREE n. For example, $x^2 - \sqrt{2}x + 5$ and $3x^4 + \frac{7}{2}x^2 - x$ are polynomials of degrees 2 and 4 respectively. A polynomial can be denoted by $f(x)$ (so that f is a *function), and then $f(-1)$, for example, denotes the value of the polynomial when x is replaced by -1. In the same way, it is possible to consider polynomials in x with coefficients from the *complex numbers, such as $z^2 + 2(1-i)z + (15 + 6i)$, or more generally from other *fields and *rings. See FUNDAMENTAL THEOREM OF ALGEBRA, POLYNOMIAL RING.

polynomial equation An equation $f(x) = 0$, where $f(x)$ is a *polynomial.

polynomial ring $\mathbb{R}[x]$ denotes the *ring of polynomials with real coefficients in a variable x. The ring is *commutative and has a multiplicative *identity, the constant polynomial 1. Addition is performed coefficient-wise and products are formed by

$$\left(\sum_{i=0}^{m} a_i x^i \right) \left(\sum_{j=0}^{n} b_j x^j \right) = \left(\sum_{k=0}^{m+n} c_k x^k \right)$$

where $c_k = \sum_{i=0}^{k} a_i b_{k-i}$.

Given a commutative ring R with a multiplicative identity 1, a polynomial ring $R[x]$ may be similarly defined. *Factorization may be non-standard, though; note

$$x^2 - 1 = (x - 1)(x + 1)$$
$$= (x - 3)(x - 5) \text{ in } \mathbb{Z}_8[x]$$

and that $x^2 - 1 = 0$ has four roots in $\mathbb{Z}_8$. If R is an *integral domain, then the formula

$$\deg(fg) = \deg(f) + \deg(g)$$

holds. If R is a *UFD, then $R[x]$ is a UFD. If R is a *field, then $R[x]$ is a *Euclidean domain. Polynomials rings such as $R[x_1, x_2, \ldots, x_n]$ in several (or infinitely many) variables may similarly be defined.

polynomial time See ALGORITHMIC COMPLEXITY.

polytope The generalization of a *polygon and *polyhedron to general dimensions, such as the *hypercube and *simplex. See also LUDWIG, SCHLÄFLI, SCHLÄFLI SYMBOL, SIMPLICIAL COMPLEX.

Poncelet, Jean-Victor (1788–1867) French military engineer and mathematician, recognized in mathematics, alongside *Desargues, as a pioneer of *projective geometry.

pons asinorum The Latin name, meaning 'bridge of asses', familiarly given to the fifth proposition in the first book of *Euclid's *Elements*: that the base angles of an *isosceles triangle are equal. It is said to be the first proof found difficult by some readers.

pooled estimate of variance (standard deviation) If it is reasonable to assume that the *variances of two *populations that you wish to compare are equal, then the most discriminatory test available uses a single *estimate of the common variance and uses all of the available data from the two samples. The pooled estimate s_p is given by

$$s_p^2 = \frac{\sum_{i=1}^{n_X} (x_i - \bar{x})^2 + \sum_{j=1}^{n_Y} (y_j - \bar{y})^2}{n_X + n_Y - 2}$$

where $\{x_i\}$, $\{y_j\}$ are the two sets of sample *observations with *sample sizes n_X, n_Y respectively and sample *means $\bar{x}, \bar{y}$.

population A set about which *inferences are to be drawn based on a sample taken from the set. The *sample may be used to reject a *hypothesis about the population or not.

population mean See MEAN.

poset Short for 'partially ordered set', i.e. a set together with a *partial order.

position ratio If A and B are two given points, and A, P, and B are collinear in that order, then the ratio $AP : PB$ may be called the position ratio of P. *See* EXTERNAL DIVISION, INTERNAL DIVISION.

position vector Suppose that a point O is chosen as the *origin (in the plane, or in 3-dimensional space). Given any point P, the position vector **p** of the point P is the *vector represented by the *directed line segment OP.

In *mechanics, the position vector of a particle is often denoted by **r**, which is a *function of the time t, giving the position of the particle as it moves. If Cartesian coordinates are used, then $\mathbf{r} = x\mathbf{i} + y\mathbf{j} + z\mathbf{k}$, where x, y, and z are functions of t.

positive A value greater than zero.

positive angle An *angle measured in an *anticlockwise direction.

positive correlation *See* CORRELATION.

positive definite (positive semidefinite) A *symmetric matrix **A** is positive definite if $\mathbf{x}^T\mathbf{A}\mathbf{x} > 0$ for all non-zero *column vectors **x** and is positive semidefinite if $\mathbf{x}^T\mathbf{A}\mathbf{x} \geq 0$ for all **x**. A symmetric matrix is positive definite (resp. semidefinite) if and only if its *eigenvalues are positive (resp. non-negative). *See* CHOLESKY'S DECOMPOSITION.

positive direction *See* DIRECTED LINE.

positively oriented A *simple, *closed curve in the *plane is positively oriented if it is anticlockwise oriented.

positive semidefinite *See* POSITIVE DEFINITE.

possible Capable of being true, not containing any inherent *contradiction. For example, it is possible to solve $x^2 + 1 = 0$ within the set of *complex numbers but not within the set of reals. It is not possible for x to be both *even and *odd.

posterior distribution *See* PRIOR DISTRIBUTION.

posterior probability *See* PRIOR PROBABILITY.

postmultiplication When the product **AB** of *matrices **A** and **B** is formed (*see* MULTIPLICATION (of matrices)), **A** is said to be postmultiplied by **B**. *Compare* PREMULTIPLICATION.

post-optimal analysis When a computer program is used to solve a *linear programming problem, the solution can provide further information, e.g. about any *slack there is in the variables, or the value attached to relaxing a *constraint. Since these pieces of information are available only after the *algorithm has produced the optimal *solution, the general term 'post-optimal analysis' is used to describe procedures to obtain them.

postulate A synonym for AXIOM.

potential For a *conservative force **F**, a *scalar field ϕ such that $\mathbf{F} = \nabla\phi$. If the domain of **F** is *path-connected, then ϕ is unique up to a constant. If, as is conventional, gravitational potential ϕ is defined as gravitational *potential energy per unit mass, then $\mathbf{F} = -\nabla\phi$, where **F** is the gravitational field.

potential energy With every *conservative force there is associated a potential energy, often denoted as V, defined up to an additive constant. The change in potential energy during a time interval is defined to be the negative of the *work done by the force during that interval. In detail, the change in V during the time interval from $t_1 \leq t \leq t_2$, or equally from position $\mathbf{r} = \mathbf{p}_1$ to $\mathbf{r} = \mathbf{p}_2$ is equal to

$$-\int_{t_1}^{t_2} \mathbf{F} \cdot \mathbf{v} \ dt = -\int_{\mathbf{p}_1}^{\mathbf{p}_2} \mathbf{F} \cdot d\mathbf{r}$$

where $\mathbf{v} = d\mathbf{r}/dt$ is the velocity of the point of application of the conservative force **F**. For example, if $\mathbf{F} = -k(x-l)\mathbf{i}$, as in *Hooke's law, then $V = \frac{1}{2}k(x-l)^2$, taking the potential energy to be zero when $x = l$. In problems involving *gravity near the Earth's surface it may be assumed that the gravitational force is given by $\mathbf{F} = -mg\mathbf{k}$, and then, taking the potential energy to be zero when $z = 0$, $V = mgz$. *See also* GRAVITATIONAL POTENTIAL ENERGY.

Potential energy has the dimensions ML^2T^{-2}, and the *SI unit of measurement is the *joule.

potential infinity In the *philosophy of mathematics, a distinction is made between ideas of actual and potential infinities. For example, the *natural numbers begin 1,2,3, ... Even in a *finitist philosophy, it is recognized that this list does not end, but such a philosophy would exclude discussion of the actual infinity of how many natural numbers there are, as a mathematical object, that might be calculated with as *Cantor did.

power (of a matrix) If $\mathbf{A}$ is a *square matrix, then the products $\mathbf{AA}$, $\mathbf{AAA}$, $\mathbf{AAAA}$, ... are defined, and these powers of $\mathbf{A}$ are denoted by $\mathbf{A}^2, \mathbf{A}^3, \mathbf{A}^4, \ldots$. For all positive integers p and q,

   (i) $\mathbf{A}^p\mathbf{A}^q = \mathbf{A}^{p+q} = \mathbf{A}^q\mathbf{A}^p$, and

   (ii) $(\mathbf{A}^p)^q = \mathbf{A}^{pq}$.

By definition, $\mathbf{A}^0 = \mathbf{I}$. Moreover, if $\mathbf{A}$ is *invertible, then $\mathbf{A}^2, \mathbf{A}^3, \ldots$ are invertible, and it can be shown that $(\mathbf{A}^{-1})^p$ is the inverse of $\mathbf{A}^p$, so that $(\mathbf{A}^{-1})^p = (\mathbf{A}^p)^{-1}$. So either of these is denoted by $\mathbf{A}^{-p}$. Thus, when $\mathbf{A}$ is invertible, $\mathbf{A}^p$ has been defined for all *integers and properties (i) and (ii) above hold for *all* integers p and q.

power (in mechanics) The power associated with a *force is the rate at which the force does *work. The power P is given by $P = \mathbf{F} \cdot \mathbf{v}$, where $\mathbf{v}$ is the *velocity of the point of application of the force $\mathbf{F}$. The work done by the force during a certain time interval equals the *integral of the power, with respect to t over the interval.

Power has the dimensions ML^2T^{-3}, and the *SI unit of measurement is the *watt.

power When a *real number a is raised to the *index p to give a^p, the result is a power of a, and p is often referred to as the power. The same index notation is used in other contexts, for example to give a power $\mathbf{A}^p$ of a square matrix (*see* POWER (of a matrix)) or a power g^p of an element g in a *multiplicative group. For details of how general powers of positive numbers are rigorously defined, *see* INDEX.

power (of a test) In *hypothesis testing, the *probability that a test rejects the null hypothesis when it is indeed false. It is equal to $1 - \beta$, where β is the probability of a *Type II error.

power series A series

$$a_0 + a_1x + a_2x^2 + a_3x^3 + \ldots + a_nx^n + \ldots,$$

in ascending powers of x, with coefficients $a_0, a_1, a_2, \ldots$, is a power series in x. For example, the *geometric series $1 + x + x^2 + \ldots + x^n + \ldots$ is a power series; this has a sum to infinity (*see* SERIES) only if $-1 < x < 1$ when considered as a real power series or for $|x| < 1$ when considered as a complex power series.

Real and complex power series *converge (that is, have a sum to infinity) within a *circle of convergence.

The coefficients a_i may more generally be taken from a *field or commutative *ring R. The formal power series in an *indeterminate x with coefficients in R themselves form a ring denoted $R[[x]]$. These power series may be naturally added and multiplied without reference to convergence.

For further examples of power series, *see* APPENDIX 11. *See also* TAYLOR SERIES.

power set The set of all *subsets of a set S is the power set of S, denoted by $\wp(S)$. Suppose that S has n elements a_1,

$a_2, \ldots, a_n$, and let A be a subset of S. For each element a_i of S, there are two possibilities: either $a_i \, \varepsilon \, A$ or not. Considering all n elements leads to 2^n possibilities in all. Hence, S has 2^n subsets; that is, $\wp(S)$ has 2^n elements. If $S = \{a, b, c\}$, the 8 ($= 2^3$) elements of $\wp(S)$ are

$\emptyset, \{a\}, \{b\}, \{c\}, \{a, b\}, \{a, c\}, \{b, c\}$, and $\{a, b, c\}$.

*Cantor's Diagonal Theorem states that the power set of a set always has greater *cardinality than the cardinality of the set.

precision (in statistics) A measurement of the quality of a *statistic as an *estimator of a *parameter. The precision is measured by the *standard error of the statistic, so in general the precision may be improved by increasing the *sample size. The precision is used in comparing the *efficiency of two or more estimators of a parameter.

precision (numerical analysis) The accuracy to which a given calculation is performed. Single precision is usually up to 16 digits, double precision uses two connected single-precision blocks, and more than two connected blocks is variously described as high, multiple, or extended precision.

predator–prey equations To model the behaviour in a predator–prey *situation, the normal dependencies on birth and death rates for each population are modified by terms reflecting the species' interdependency: since the rate of encounter between predator and prey is approximately proportional to the product of the population sizes, this would give rise to linked differential equations of the form

$$\frac{dX}{dt} = (b_X - d_X)X - kXY,$$

$$\frac{dX}{dt} = (b_Y - d_Y)Y + lXY,$$

These equations are also known as the Lotka–Volterra equations after Alfred Lotka and Vito Volterra who independently introduced them. There is an unstable *equilibrium at $X = Y = 0$ and a single stable equilibrium at non-zero X and Y.

(()) SEE WEB LINKS

• Simulations of fox and rabbit populations.

predicate (in logic) An expression which ascribes a property to one or more subjects. 'Louise is a girl' is a predicate with a single subject. A predicate with more than one subject will be a *relation. So, 'Louise, Joanne, and Laura are sisters' is a predicate involving a ternary *relation.

predicate logic A synonym for FIRST-ORDER LOGIC.

predicted variable A synonym for DEPENDENT VARIABLE.

predictor variable A synonym for EXPLANATORY VARIABLE.

prefix A *word w_1 is a prefix of another w_2 if there is a word w such that $w_2 = w_1 w$ where the product is *concatenation. For example, $w_1 = 0110$ is a prefix of $w_2 = 0110101$ with $w = 101$.

prefix code A *code in which no *codeword is a *prefix of another codeword. Thus, once received, a codeword can be immediately interpreted without further reference to the remainder of the message. Also called an instantaneous code.

pre-image Given a *function $f : X \rightarrow Y$ and $A \subseteq Y$, the pre-image $f^{-1}(A)$ of A is

$$f^{-1}(A) = \{x \in X | f(x) \in A\} \subseteq X.$$

This notation does not imply that f is invertible (*see* INVERTIBLE FUNCTION). Note $f(f^{-1}(A)) \subseteq A$, but equality does not generally hold.

premultiplication When the product **AB** of *matrices **A** and **B** is formed (*see*

MULTIPLICATION (of matrices)), **B** is said to be premultiplied by **A**. *Compare* POSTMULTIPLICATION.

presentation A description of a *group in terms of *generators and *relations. For example, $\langle x \mid x^n = e \rangle$ is a group with a single generator x of *order n and so is the *cyclic group of order n. The *dihedral group D_{2n} has the presentation

$$\langle r, s \mid r^n = e = s^2, sr = r^{-1}s \rangle.$$

*Words in the generators are equal if and only the relations can be used to show this. Formally, the presentation defines a group as the *quotient group of the *free group on the set of generators by the *normal subgroup generated by the relations.

pressure *Force per unit *area. The simplest situation to consider is when a force is uniformly distributed over a plane surface, the direction of the force being perpendicular to the surface. Suppose that a rectangular box, of length a and width b, has *weight mg. Then the area of the base is ab, and the *gravitational force creates a pressure of $mg/(ab)$ acting at each point of the ground on which the box rests.

Now consider a liquid of *density ρ at rest. The weight of liquid supported by a horizontal surface of area ΔA situated at a depth h is $\rho(\Delta A)hg$. The resulting pressure (in excess of *atmospheric pressure) at each point of this horizontal surface at a depth h is ρgh. More general analysis, involving elemental regions, defines pressure as a *scalar field in the liquid as a function of position and time.

Pressure has dimensions $ML^{-1}T^{-2}$, and the *SI unit of measurement is the *pascal.

primary decomposition theorem Let $p(x)$ and $q(x)$ be *coprime real polynomials and **A** a *square real matrix. Then

$$\ker\big(p(\mathbf{A})q(\mathbf{A})\big) = \ker\big(p(\mathbf{A})\big) \oplus \ker\big(q(\mathbf{A})\big),$$

where ker denotes *kernel and $\oplus$ denotes a *direct sum. The theorem generalizes to other *fields of scalars and to *linear maps on *vector spaces.

prime (prime number) A positive *integer p is a prime if $p \neq 1$ and its only positive *divisors are 1 and itself.

It is known that there are infinitely many primes. Euclid's proof argues by *contradiction as follows. Suppose that there are finitely many primes p_1, $p_2, \ldots, p_n$. Consider the number $p_1 p_2 \times \ldots \times p_n + 1$. This is not divisible by any of $p_1, p_2, \ldots, p_n$—division by each leaves a remainder of 1—so it is either another prime itself or is divisible by primes not among the p_i. It follows that the number of primes is not finite.

The notion of a prime number generalizes to that of a *prime element in the theory of *rings. At any time, the largest known prime is usually the largest known *Mersenne prime. There are many deep results in *number theory relating to prime numbers, such as *Fermat's Two Square Theorem, *Bertrand's postulate, and the *Green-Tao theorem, and many important *open problems, such as *Goldbach's conjecture, *Legendre's conjecture, *Riemann hypothesis, and *Twin prime conjecture.

See also FUNDAMENTAL THEOREM OF ARITHMETIC, PRIME NUMBER THEOREM.

prime decomposition (prime factorization, prime representation) *See* FUNDAMENTAL THEOREM OF ARITHMETIC.

prime element A non-zero element p in a commutative *ring is said to be prime if whenever p divides the product ab, then p divides a or p divides b (or both). Thus, prime elements are, in a sense, atomic with regard to multiplication. 6 is not prime as it divides 4×9 without dividing 4 or 9. With the existence of *units in rings, this definition of being prime generalizes better to rings than the notion of being divisible by itself and 1 only. Equivalently, the ideal (p) is a *prime ideal if and only if p is a prime element. *Compare* IRREDUCIBLE.

prime ideal An *ideal I in a *ring R is said to be prime if whenever $ab \in I$, then $a \in I$ or $b \in I$. The ideal I is prime if and only if the *quotient ring R/I is an integral domain. *Maximal ideals are prime.

prime knot A *knot which cannot be written as the *connected sum of simpler knots. The *trefoil knot is prime, whereas the granny knot (in the figure) is not prime, being the connected sum of two trefoils.

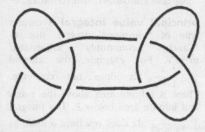

Granny knot

prime meridian The *meridian through Greenwich from which *longitude is measured.

prime number theorem For a positive real number x, let $\pi(x)$ be the number of *primes less than or equal to x. The prime number theorem says that, as $x \to \infty$,

$$\frac{\pi(x)}{x/\ln x} \to 1.$$

This gives, in a sense, an idea of what proportion of integers are prime. It was suspected since the time of *Gauss and *Legendre and first proved in 1896 by Jacques *Hadamard and Charles De La *Vallée-Poussin independently.

prime subfield The smallest subfield of a *field. For *characteristic zero fields, the prime subfield is *isomorphic to $\mathbb{Q}$, and for characteristic p fields the prime subfield is isomorphic to $\mathbb{Z}_p$ the integers modulo p.

primitive A synonym for ANTIDERIVATIVE.

primitive (nth root of unity) An *nth root of unity z is primitive if every nth root of unity is a *power of z. For example, i is a primitive fourth root of unity, but -1 is a fourth root of unity that is not primitive.

primitive element *See* SIMPLE EXTENSION.

primitive polynomial A *polynomial over the integers (or more generally a *unique factorization domain) with the property that the *greatest common divisor of its coefficients is 1. *See* CONTENT, GAUSS' LEMMA.

primitive root *See* DISCRETE LOGARITHM.

Prim's algorithm (to solve the minimum connector problem) This method is more effective than *Kruskal's algorithm when a large number of vertices and/or when the distances are listed in tabular form rather than shown on a *graph. Since all vertices will be on the minimum connected route, it does not matter which vertex you choose as a starting point, so make an arbitrary choice, calling it P_1 say. Now choose a point with the shortest edge connecting it to P_1 and call it P_2. At each stage add the edge and new point P_i which adds the shortest distance to the total and does not create any loops. Once P_n has been reached, the minimum connected path has been identified.

principal axes (in mechanics) *See* INERTIA MATRIX.

principal axes (of a quadric) A set of principal axes of a *quadric is a set of axes of a coordinate system in which the quadric has an equation in *canonical form.

principal ideal An *ideal generated by a single element, that is, the smallest ideal containing that element. In a commutative ring R, the principal ideal generated by a is $\{ar : r \in R\}$ and is usually denoted (a) or $\langle a \rangle$. In $\mathbb{Z}$ every ideal is principal, but in $\mathbb{Q}[x,y]$ the ideal generated by x and y is not principal.

principal ideal domain An *integral domain in which every *ideal is *principal. *Euclidean domains are principal ideal domains which in turn are *unique factorization domains.

principal moments of inertia *See* INERTIA MATRIX.

principle of conservation of energy *See* CONSERVATION OF ENERGY.

principle of conservation of linear momentum *See* CONSERVATION OF LINEAR MOMENTUM.

principle of mathematical induction *See* MATHEMATICAL INDUCTION.

principle of moments The principle that, when a system of coplanar forces acts on a body and produces a state of equilibrium, the sum of the *moments of the forces about any point in the plane is zero.

principle of the excluded middle The principle of *propositional logic that every statement is either true or false. Equivalently, this means for any statement P then $P \vee (\neg P)$ is true. Some *philosophies of mathematics, such as *intuitionism, reject this principle and *proof by contradiction, on the grounds that $\neg P$ being provably false is not the same as a proof of P.

principal part The principal part of a *Laurent expansion $\sum_{-\infty}^{\infty} c_n (z - a)^n$ equals $\sum_{-\infty}^{-1} c_n (z - a)^n$, that is, the series consisting of the negative powers.

principal value A unique member of a set of values a variable can take, especially when seeking to define the *inverse of *trigonometric functions or *square roots. Both 3 and –3 are square roots of 9, but $\sqrt{9}$ is chosen to be 3, the principal value.

Similarly $\sin 30° = \sin 150° = \sin(-210°) = 0.5$, but $\sin^{-1} 0.5 = 30°$. Sine is not an *invertible function, but by choosing principal values a function $\sin^{-1}: [-1,1] \rightarrow [-90°,90°]$ can be defined such that $\sin(\sin^{-1} x) = x$ for $-1 \leq x \leq 1$.

See also ARGUMENT, MULTIFUNCTION.

principal value integral A certain type of *improper integral, due to *Cauchy, commonly abbreviated to PV. For example, the integral $\int_{-\infty}^{\infty} \dfrac{x}{1 + x^2}\, dx$ does not converge. There is infinite area above the x-axis and infinite area below it. The integral $\int_{-R}^{S} \dfrac{x}{1 + x^2}\, dx$ does not have a limit, as R, S tend to infinity. However, the limit $\lim_{R \rightarrow \infty} \int_{-R}^{R} \dfrac{x}{1 + x^2}\, dx$ does exist, equalling 0, and this would be its PV. Similarly, an integral $\int_{0}^{2} \dfrac{x^2 dx}{x^2 - 1}$ can be assigned a PV of 2, by adding the integrals on $(0,1-\varepsilon)$ and $(1 + \varepsilon, 2)$ and letting ε tend to 0.

prior distribution The *distribution attached to a *parameter before certain *data are obtained. The posterior distribution is the distribution attached to it after the data are obtained. The posterior distribution may then be considered as a new prior distribution before further data are obtained.

prior probability The *probability attached to an *event before certain *data are obtained. The posterior probability is the probability attached to it after the data are obtained. The posterior probability may be calculated by using *Bayes' Theorem. It may then be considered as a new prior probability before further data are obtained.

prism A *convex *polyhedron with two 'end' *faces that are *congruent convex *polygons lying in *parallel planes in such a way that, with edges joining corresponding vertices, the remaining faces are *parallelograms. A right-regular prism is one in which the two end faces are regular polygons and the remaining faces are *rectangular. A right-regular prism in which the rectangular faces are square is semi-regular (*see* ARCHIMEDEAN SOLID).

Prism (regular pentagonal ends)

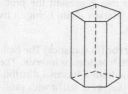

Right-regular (rectangular lateral faces)

prisoner's dilemma In *game theory, any generalized situation which is essentially similar to the following: two prisoners are interviewed separately about a crime. If both deny the charge, there is enough evidence to convict them of a minor charge. If one confesses and implicates the other, who denies it, then the first will be released and the other will be heavily punished. If both

confess, then both receive a medium punishment. The 'dilemma' for either prisoner is that denying the charge can result in either a minor conviction or the worst possible outcome depending on the action of the other prisoner. *See* NASH EQUILIBRIUM.

(⊕) SEE WEB LINKS

• Play the prisoner's dilemma against the computer.

private key crytography *See* KEY.

probability The probability of an *event A, denoted by $\Pr(A)$, is a measure of the possibility of the event occurring as the result of an experiment. For any event A, $0 \leq \Pr(A) \leq 1$. If A never occurs, then $\Pr(A) = 0$; if A always occurs, then $\Pr(A) = 1$. If an experiment is repeated n times and the event A occurs m times, then the *limit of m/n as $n \to \infty$ is equal to $\Pr(A)$.

If the *sample space S is finite and the possible outcomes are all equally likely, then the probability of the event A equals $|A|/|S|$ (where $|\ |$ denotes *cardinality). The probability that a randomly selected element from a finite population belongs to a certain category is equal to the *proportion of the population belonging to that category.

The probability that a discrete *random variable X takes the value x_i is denoted by $\Pr(X = x_i)$ (*see* PROBABILITY MASS FUNCTION). The probability that a continuous random variable X takes a value less than or equal to x is denoted by $\Pr(X \leq x)$ (*see* CUMULATIVE DISTRIBUTION FUNCTION).

The term 'probability' also refers generally to the theory of probability and *stochastic processes and is intimately linked with *statistics.

See also CONDITIONAL PROBABILITY, LAWS OF LARGE NUMBERS, PRIOR PROBABILITY, PROBABILITY DENSITY FUNCTION, PROBABILITY SPACE.

probability density function For a continuous *random variable X, a probability density function (or pdf) of X is the function f_X such that

$$\Pr(a \leq X \leq b) = \int_a^b f_X(x) \; dx.$$

A probability density function is necessarily non-negative with *integral 1. The *Cantor distribution is an example of a continuous random variable which does not have a probability density function.

Suppose that, for the random variable X, a sample is taken and a corresponding *histogram is drawn with class intervals of a certain width. Then, loosely speaking, as the number of *observations increases and the width of the class intervals decreases, the histogram assumes more closely the shape of the *graph of the probability density function.

probability generating function If X is a random variable where X takes only non-negative integer values, then $G_X(s) = E(s^X)$ is the probability generating function (pgf). Then

$$G_X(s) = \sum_{k=0}^{\infty} \Pr(X = k)s^k$$

The *mean and *variance of X can also be computed using the pgf from

$$E(X) = G_X'(1) \quad \text{and}$$
$$\mathrm{Var}(X) = G_X''(1) + G_X'(1) - (G_X'(1))^2.$$

The pgf of the sum $X + Y$ of two independent random variables is given by

$$G_{X+Y}(s) = G_X(s)G_Y(s),$$

and for random sampling, where $Z = X_1 + \ldots + X_N$, with N itself a random variable and the X_i are iid, then $G_Z(s) = G_N\big(G_X(s)\big)$.

probability mass function For a discrete *random variable X, the probability mass function (or pmf) of X is the function p_X such that $p_X(x_i) = \Pr(X = x_i)$, for all i.

probability measure See PROBABILITY SPACE.

probability paper Graph paper scaled so that the *cumulative distribution function for a specified probability distribution makes a line. This then provides an informal test as to whether *data might reasonably be supposed to come from that family of distributions, by considering how closely the observed data fit some line. If it is a satisfactory fit, it can also provide estimates of parameter values. The most commonly tested distribution is the *normal, so this is the most common probability paper, often generated now within a statistical computer package while testing assumptions for sophisticated procedures.

probability space A finite *measure space $(\Omega, \mathcal{F}, P)$ with total measure 1 – that is $P(\Omega) = 1$. The set Ω is the *sample space, elements of the *sigma algebra $\mathcal{F}$ are called *events, and P is the probability measure.

probability vector (stochastic vector) A *vector with non-negative coordinates whose *sum is 1. In particular, such a vector may represent the probabilities of a *Markov chain being in its various states.

probable error (in statistics) The half-width of a 50% *confidence interval. The observed value from a *normal distribution satisfies $|X - \mu| < 0.67\sigma$ with probability 0.5, so the probable error for a normal distribution is 0.67σ.

product The result of multiplying two or more *numbers, *functions (see POINTWISE MULTIPLICATION), *matrices (see MULTIPLICATION (of matrices)), or elements in a *multiplicative group or *ring. See also INFINITE PRODUCT, PRODUCT NOTATION.

product group Given two groups $(G, \cdot)$ and $(H, \circ)$, the product group is the

Cartesian set $G \times H$ together with the group operation $*$ given by

$$(g_1, h_1) * (g_2, h_2) = (g_1 \cdot g_2, h_1 \circ h_2).$$

The product $G \times H$ is *abelian if and only if G and H are both abelian. $G \times H$ is also called the direct product. *See also* SEMI-DIRECT PRODUCT.

product moment correlation coefficient *See* CORRELATION.

product notation For a finite sequence $a_1, a_2, \ldots, a_n$, the product $a_1 a_2 \ldots a_n$ may be denoted, using the capital Greek letter pi, by

$$\prod_{r=1}^{n} a_r.$$

where r is a *dummy variable. For example,

$$\prod_{r=1}^{9}\left(1 - \frac{1}{r+1}\right)$$
$$= \left(1 - \frac{1}{2}\right)\left(1 - \frac{1}{3}\right)\ldots\left(1 - \frac{1}{10}\right).$$

Similarly, from an infinite sequence $a_1, a_2, a_3, \ldots$, an *infinite product $a_1 a_2 a_3 \ldots$ can be formed, and it is denoted by

$$\prod_{r=1}^{\infty} a_r.$$

This notation is also used for the *limit of the infinite product, if it exists. For example

$$\prod_{r=2}^{\infty}\left(1 - \frac{1}{r^2}\right) = \frac{1}{2}.$$

and also as in *Wallis' product.

product of inertia A quantity similar to a *moment of inertia and appearing in the *inertia matrix. Suppose a rigid body consists of n particles $P_1, P_2, \ldots, P_n$, where P_i has mass m_i and position vector $\mathbf{r}_i = (x_i, y_i, z_i)$. Then

$$I_{yz} = \sum_{i=1}^{n} m_i y_i z_i, \quad I_{zx} = \sum_{i=1}^{n} m_i z_i x_i,$$
$$I_{xy} = \sum_{i=1}^{n} m_i x_i y_i.$$

For a continuous distribution of mass, the products of inertia are defined by corresponding *integrals.

When the coordinate planes are planes of symmetry of the rigid body, the products of inertia are zero. In fact, by the *spectral theorem, through any fixed point there is a set of three perpendicular axes, called the principle axes, such that the corresponding products of inertia are all zero.

product rule *See* DIFFERENTIATION.

product set A synonym for CARTESIAN PRODUCT.

product space Given *topological spaces X and Y, the product space topology on the *Cartesian product $X \times Y$ has the sets $U \times V$, where U is open in X and V is open in Y, as a *basis. The product topology on $\mathbb{R} \times \mathbb{R}$ is the same as the topology on $\mathbb{R}^2$, though the basic *open sets for the former might be considered as open rectangles and for the latter open discs.

If (X, d_X) and (Y, d_Y) are *metric spaces, then

$$d\big((x_1, y_1), (x_2, y_2)\big) = d_X(x_1, x_2) + d_Y(y_1, y_2)$$

is a metric on $X \times Y$ which *induces the product topology.

program In computer science, a program is a finite system of commands and statements to be executed by a computer to achieve a desired result, such as the implementation of an *algorithm. *See* COMPUTABILITY THEORY.

progression A *sequence in which each term is obtained from the previous one by some rule. The most common progressions are *arithmetic sequences, *geometric sequences, and *harmonic sequences.

projectile When a small *body is travelling near the Earth's surface, subject to no forces except the *uniform gravitational force and possibly *air resistance, it may be called a projectile. The standard *mathematical model uses a *particle to represent the body and a horizontal plane to represent the Earth's surface.

When there is no air resistance, the trajectory, the path traced out by the projectile, is a *parabola whose vertex corresponds to the point at which the projectile attains its maximum height. Take the origin at the point of projection, the x-axis horizontal and the y-axis vertical with the positive direction upwards. Then the equation of motion is $m\ddot{\mathbf{r}} = -mg\mathbf{j}$, subject to $\mathbf{r} = \mathbf{0}$ and $\dot{\mathbf{r}} = (v\cos\theta)\mathbf{i} + (v\sin\theta)\mathbf{j}$ at $t = 0$, where v is the *speed of projection and θ is the *angle of projection. This gives

$$x = (v\cos\theta)t \text{ and } y = (v\sin\theta)t - \frac{1}{2}gt^2,$$

for $t \geq 0$. It follows that $y = 0$ when $t = (2v/g)\sin\theta$, the time of flight, and $x = (v^2/g)\sin\theta$, the range of the projectile.

The maximum height of $v^2\sin^2\theta/(2g)$ is attained halfway through the flight. The maximum range, for any given value of v, is obtained when $\theta = \pi/4$. Similar investigations can be carried out for a projectile projected from a point a given height above a horizontal plane or projected up or down an inclined plane. For a body projected vertically upwards with initial speed v, set $\theta = \pi/2$.

projection (of a point on a line or plane) Given a line l (resp. a plane Π) and a point P not on l (resp. Π), the projection of P on l (resp. Π) is the point N on l (resp. Π) such that PN is perpendicular to l (resp. Π). The length $|PN|$ is the minimal distance from P to any point of l (resp. Π). The point N is called the foot of the perpendicular from P to l (resp. Π). This map from points P to l (or Π) is called orthogonal projection. *See also* VECTOR PROJECTION.

projection (linear algebra) A projection of a *vector space V is a *linear map $P:V \rightarrow V$ such that $P^2 = P$. Then V is the *direct sum of the *kernel and *image of P. Conversely, if $V = M \oplus N$, every v can be uniquely written as $v = m + n$ (where $m \in M$, $n \in N$), and the map $v \mapsto m$ is a projection. If V is an *inner product space, then $V = M \oplus M^{\perp}$ and $v \mapsto m$ is orthogonal projection onto M.

projective geometry An area of *geometry that grew from the study of *perspective, with *Desargues and *Poncelet being early developers of the field. From different observers' perspectives, *angles and *lengths may change, but collinearity will remain invariant. In the *real projective plane, *ellipses, *parabolae, and *hyperbolae are equivalent, each being viewed as a *conic that has 0, 1, or 2 *points at infinity. *See* BÉZOUT'S THEOREM, DUALITY, HOMOGENEOUS COORDINATES, PROJECTIVE SPACE, PROJECTIVE TRANSFORMATION.

projective space Real (or complex) n-dimensional projective space can be considered as $\mathbb{R}^n$ (or $\mathbb{C}^n$) together with *points at infinity (specifically, a hyperplane at infinity). In terms of *homogeneous coordinates, points $[1:x_1:\ldots:x_n]$ can be identified with $\mathbb{R}^n$ and the hyperplane has equation $x_0 = 0$. From the point of view of *affine geometry, these points at infinity will remain at infinity; from the projective view, the points at infinity are no different from other points, and the hyperplane at infinity need not be invariant under a *projective transformation. Topologically, the real projective line is the *circle, and the complex projective line is the *Riemann sphere.

projective transformation Real n-dimensional projective space can be defined as $(\mathbb{R}^{n+1}-\{0\})/\mathbb{R}^*$ (*see* HOMOGENEOUS COORDINATES). So an *invertible *linear map $T:\mathbb{R}^{n+1}\rightarrow\mathbb{R}^{n+1}$ induces a projective transformation $[T]([\mathbf{v}]) = [T\mathbf{v}]$, where $[\mathbf{v}]$ denotes the projective point represented by $\mathbf{v}$ ε $\mathbb{R}^{n+1}$. The projective transformations form a *group $\text{PGL}(n+1,\mathbb{R})$ under *composition. The group $\text{PGL}(2,\mathbb{C})$ is the group of *Möbius transformations, acting on the *Riemann sphere.

prolate *See* SPHEROID.

proof A chain of reasoning, starting from *axioms or known results, that leads to a conclusion and satisfies the assumed logical rules of inference.

proof by contradiction A *direct proof of a statement is a logically correct argument establishing the truth of the statement. A proof by contradiction instead assumes that a statement is false and derives a conclusion which is false, thus proving the original statement using the *principle of the excluded middle. Two commonly used proofs by contradiction are showing the $\sqrt{2}$ is *irrational and that there are infinitely many *primes.

If wishing to prove $p\Rightarrow q$ by contradiction, we prove $\neg(p\wedge(\neg q))$, that is, the hypotheses p are still assumed, and the negation $\neg q$ of the conclusion, and then $p\wedge(\neg q)$ is shown to be false. This is equivalent to direct proof by *De Morgan's Laws.

proof theory (metalogic, meta-mathematics) The branch of *logic that studies *proofs and the deducibility and independence of results within formal systems. The roots of proof theory may be seen as beginning with *Hilbert's programme.

proof verification *See* THEOREM PROVING.

proper class *See* CLASS.

proper divisor A synonym for PROPER FACTOR.

proper factor A *factor other than the number itself; so 1,2,3,6 are all factors of 6, but only 1, 2, and 3 are proper factors.

proper fraction *See* FRACTION.

proper map A *function between two *topological spaces such that the *pre-image of every *compact subset is compact.

proper subset Let A be a *subset of B. Then A is a proper subset of B if A is not equal to B itself. Equivalently, there is some element of B not in A. This is written $A \subset B$, though be aware that some authors use $A \subset B$ to mean $A \subseteq B$.

proper value *See* EIGENVALUE.

proper vector *See* EIGENVECTOR.

proportion The portion of a set with a certain characteristic. This might be expressed as a *fraction or percentage of the whole.

proportion (proportional) If two quantities x and y are related by an equation $y = kx$, where k is a constant, then y is said to be (directly) proportional to x, which may be written $y \propto x$. The constant k is the constant of proportionality. It is also said that y varies directly as x. When y is plotted against x, the graph is a straight line through the origin.

If $y = k/x$, then y is inversely proportional to x. This is written $y \propto 1/x$, and it is said that y varies inversely as x.

proposition A mathematical *theorem, usually accompanied by a *proof deduced from *axioms or previously demonstrated results. Pure mathematics texts, in particular, tend to present mathematical results as propositions and theorems (and *lemmas and *corollaries) deduced from axioms. Typically, a proposition is of less consequence than a theorem, though this is, of course, subjective. In *logic, a proposition is equivalent to a statement.

propositional logic *See* FIRST-ORDER LOGIC.

pseudoinverse A generalization of the notion of the *inverse of a *square matrix to other matrices. The pseudoinverse A^+ of a matrix A is characterized by the properties

(i) $AA^+A = A$ and $A^+AA^+ = A^+$
(ii) AA^+ and A^+A are *symmetric.

When A is invertible, $A^+ = A^{-1}$. Further, $x_0 = A^+b$ minimizes $|Ax-b|$ as x varies. *See* SINGULAR VALUE DECOMPOSTION.

pseudometric One of the conditions for a distance measure d to be a *metric is that $d(x,y) = 0$ implies that $x = y$. If this condition is relaxed so that the distance between two different points can be zero, then d is said to be a pseudometric, and the space is said to be a pseudometric space.

pseudoprime (Carmichael number) The positive integer n is a pseudoprime if it is *composite, but $a^n \equiv a \pmod{n}$, for all integers a. Thus, *Fermat's Little Theorem cannot be used to demonstrate n is not prime. The first pseudoprime is 561. It was shown in 1994 that there are infinitely many pseudoprimes.

pseudorandom numbers *See* RANDOM NUMBERS.

Ptolemy (Claudius Ptolemaeus) (2nd century AD) Greek astronomer and mathematician, responsible for the most significant work of *trigonometry of ancient times. Usually known by its Arabic name *Almagest* ('The Greatest'), it contains, amongst other things, tables of chords, equivalent to a modern table of sines, and an account of how they were obtained. Use is made of *Ptolemy's Theorem, from which the familiar addition formulae of trigonometry can be shown to follow.

Ptolemy's Theorem Suppose that a *convex *quadrilateral has vertices A, B, C, and D, in that order. Then the quadrilateral is *cyclic if and only if

$$AB.CD + AD.BC = AC.BD.$$

public key cryptography *See* KEY.

pulley A grooved wheel around which a rope can pass. When supported on an axle in some way, the device can be used to change the direction of a *force: for example, pulling down on a rope may enable a load to be lifted. If the contact between the pulley and the axle is smooth, the *magnitude of the force is not changed. A system of pulleys can be constructed that enables a large load to be raised a small distance by a small effort moving through a large distance. In such a *machine, the velocity ratio is equal to the number of ropes supporting the load or, equivalently, the number of pulleys in the system.

pure imaginary A *complex number is pure imaginary if its *real part is zero.

pure mathematics The area of mathematics concerning the relationships between abstract systems and structures and the rules governing their behaviours, motivated by its intrinsic interest or elegance rather than its application to solving problems in the real world. Much modern applied mathematics is based on or employs what was once viewed as esoteric pure mathematics when it was devised. For example, matrix algebra has applications as diverse as computer graphics, *quantum theory, Internet searches, and meteorology.

pure strategy In a *matrix game, if a player always chooses one particular row (or column) this is a pure strategy. Compare this with *mixed strategy.

PV *See* PRINCIPAL VALUE INTEGRAL.

p-value (statistics) The *probability that a given *test statistic would take the observed value, or one which was less likely, if the *null hypothesis had been true. While most text books construct hypothesis tests by finding critical values or constructing *critical regions for a predetermined level of significance for the test, almost all published statistics report test outcomes in terms of p-values.

pyramid A *convex *polyhedron with one face (the base) a convex polygon and all the vertices of the base joined by edges to one other vertex (the apex); thus the remaining faces are all triangular. A right-regular pyramid is one in which the base is a *regular polygon and the remaining faces are *isosceles triangles.

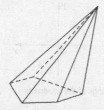

Not right-regular

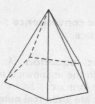

Right-regular pyramid

Pythagoras (died about 500 BC) Greek philosopher, mathematician, and mystic. It is unclear how much mathematics can be reliably attributed to Pythagoras and his school. Certainly *Pythagoras' Theorem and *Pythagorean triples were known to Babylonian and Indian mathematicians centuries earlier.

Pythagoras' Theorem In a *right-angled triangle, the square on the *hypotenuse is equal to the sum of the squares on the other two sides.

Thus, if the hypotenuse, the side opposite the right angle, has length c and the other two sides have lengths a and b, then $a^2 + b^2 = c^2$. One elegant proof is obtained by dividing up a square of side $a + b$ in two different ways as shown in the figure, and equating areas.

Pythagoras's Theorem is a special case of the *cosine rule.

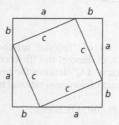

Triangles leave area c^2

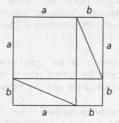

Triangles leave area $a^2 + b^2$

Pythagorean triple A set of three positive integers a, b, and c such that $a^2 + b^2 = c^2$ (*see* PYTHAGORAS' THEOREM). If $\{a,b,c\}$ is a Pythagorean triple, then so is $\{ka, kb, kc\}$ for any positive integer k. Pythagorean triples that have *greatest common divisor equal to 1 include the following: $\{3,4,5\}$, $\{5,12,13\}$, $\{8,15,17\}$, $\{7,24,25\}$ and $\{20,21,29\}$. A general formula for such Pythagorean triples is

$$\{p^2 - q^2, 2pq, p^2 + q^2\}$$

where $p > q$ are *coprime, positive integers which are not both odd.

p

Q The set of *rational numbers. The symbol also denotes the *field of rational numbers, and $\mathbb{Q}^*$ denotes the multiplicative *group of non-zero rational numbers.

$\mathbb{Q}_p$ *See* P-ADIC NUMBERS.

$\mathbf{Q}_8$ *See* QUATERNION GROUP.

QED Abbreviation for 'quod erat demonstrandum', Latin for 'which was to be proved'. Often written at the end of a *proof.

QEF Abbreviation for 'quod erat faciendum', Latin for 'which was to be done'. Often written at the end of a *construction.

QR decomposition Any *square matrix **M** can be decomposed as **M** = **QR**, where **Q** is *orthogonal and **R** is upper *triangular. If **M** is *invertible, then this decomposition is unique if we require the diagonal entries of **R** to be positive, and the decomposition can be achieved by applying the *Gram-Schmidt method to the columns of **M**. *Compare* SINGULAR VALUE DECOMPOSITION.

quad- Prefix, denoting four, as in *quadrilateral. *Compare* TETRA-.

quadrant In a *Cartesian coordinate system in the plane, the axes divide the rest of the plane into four regions called quadrants. They are numbered as in the figure, with the first quadrant being where $x, y > 0$.

The numberings of the quadrants

quadratic complexity *See* ALGORITHMIC COMPLEXITY.

quadratic convergence *See* RATE OF CONVERGENCE.

quadratic equation A quadratic equation in the unknown x is an equation of the form $ax^2 + bx + c = 0$, where a, b and c are given real numbers, with $a \neq 0$. This may be solved by *completing the square or by using the quadratic formula

$$x = \frac{-b \pm \sqrt{b^2 - 4ac}}{2a},$$

which is established by completing the square. The term $b^2 - 4ac$ is known as the *discriminant. The equation has two distinct real *roots, a repeated real root, or two *complex *conjugate roots, depending on whether the discriminant is positive, zero, or negative respectively.

If α and β are the roots of the quadratic equation $ax^2 + bx + c = 0$, then $\alpha + \beta = -b/a$ and $\alpha\beta = c/a$. (*See* VIÈTE'S FORMULAE.) The quadratic formula is still valid over other *fields provided that the discriminant has a *square root in the field.

quadratic field Any field of the form $\mathbb{Q}[\sqrt{d}]$ where d is a square-free integer and $d \neq 0,1$. The elements of $\mathbb{Q}[\sqrt{d}]$ have the form $q_1 + q_2\sqrt{d}$, where q_1, q_2 are *rational. $\mathbb{Q}[\sqrt{d}]$ is a degree 2 *field extension over $\mathbb{Q}$. Its *ring of integers consists of those elements $\zeta \in \mathbb{Q}[\sqrt{d}]$ that satisfy a quadratic equation $\zeta^2 + a\zeta + b = 0$, where a,b are integers. When $d = -19$, the ring of integers is $\mathbb{Z}[(1 + \sqrt{-19})/2]$, which is a *principal ideal domain which is not a *Euclidean domain.

quadratic form In one, two, and three variables, functions in quadratic form are:

$$F(x) = ax^2$$
$$F(x,y) = ax^2 + by^2 + cxy$$
$$F(x,y,z) = ax^2 + by^2 + cz^2 + dxy + exz + fyz$$

Generally, if $\mathbf{x}$ is a column of n variables and $\mathbf{A}$ is an $n \times n$ matrix then $\mathbf{x}^T\mathbf{A}\mathbf{x}$ will be in quadratic form. $\mathbf{A}$ can be chosen to be *symmetric. *See* FORM.

quadratic formula *See* QUADRATIC EQUATION.

quadratic function (quadratic polynomial) A quadratic function is a *real function f such that $f(x) = ax^2 + bx + c$ for all x in $\mathbb{R}$, where a, b and c are real numbers, with $a \neq 0$. The graph $y = f(x)$ of such a function is a *parabola with its axis parallel to the y-axis, and with its vertex downwards if $a > 0$ and upwards if $a < 0$. The term equally applies to *degree two *polynomials over other *fields and *rings.

quadratic reciprocity Given two distinct, odd *primes p and q, then p is a *quadratic residue modulo q if and only if q is a quadratic residue modulo p, unless $p \equiv q \equiv 3 \pmod 4$, in which case precisely one of p,q is a quadratic residue modulo the other. This was proved by *Gauss around 1801 but had been

suspected since *Euler's time. As an application, note that it is not immediately obvious whether $x^2 \equiv 31 \pmod{257}$ has a solution, but now note, in terms of the *Legendre symbol, that $(31|257) = (257|31) = (9|31) = 1$ as 9 is clearly a quadratic residue and so 31 is a quadratic residue modulo 257. The theorem does not provide guidance on what the two square roots are—in this case ± 51.

quadratic residue An integer a is a quadratic residue modulo n if the congruence $x^2 \equiv a \pmod n$ has a solution. If n is an odd prime, then there are $(n+1)/2$ quadratic residues in the range $0 \leq a < n$. *See* LEGENDRE SYMBOL, QUADRATIC RECIPROCITY.

quadratrix A transcendental curve, its main application being that it could be used to *trisect a general angle, an impossible *construction with ruler and compass. In the figure, if the radius OA rotates uniformly to radius OB and at the same rate the line CD moves from the top to the base, then when the radius has reached OP we can define the point P' as the intersection of CD with OP. The quadratrix is the curve traced by P'. The angle OP could then be trisected by

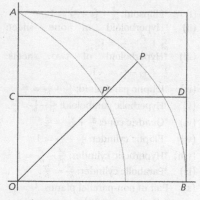

A quadratrix

constructing $C'D'$ at one-third the height of OC, and if $C'D'$ meets the quadratrix at P'', then OP'' trisects OP.

The quadratrix can be parameterized as

$$x(t) = \frac{2at}{\pi \tan t}, \quad y(t) = \frac{2at}{\pi}, \quad 0 < t \leq \frac{\pi}{2},$$

where $a > 0$.

quadrature A method of quadrature is a numerical method that finds an approximation to a definite *integral; in many cases, the integral may then be found by a limiting process. *See* GAUSSIAN QUADRATURE, SIMPSON'S RULE, TRAPEZIUM RULE.

quadric (quadric surface) A *locus in 3-dimensional space that can be represented in a *Cartesian coordinate system by a *polynomial equation in x, y, and z of degree two; that is, an equation of the form

$$ax^2 + by^2 + cz^2 + 2fyz + 2gzx + 2hxy$$
$$+ 2ux + 2vy + 2wz + d = 0,$$

where the constants a, b, c, f, g, and h are not all zero. When the equation represents a non-empty locus, it can be reduced by *translation and *rotation to one of the following *canonical forms:

(i) *Ellipsoid: $\frac{x^2}{a^2} + \frac{y^2}{b^2} + \frac{z^2}{c^2} = 1$.

(ii) *Hyperboloid of one sheet: $\frac{x^2}{a^2} + \frac{y^2}{b^2} - \frac{z^2}{c^2} = 1$.

(iii) *Hyperboloid of two sheets: $\frac{x^2}{a^2} + \frac{y^2}{b^2} - \frac{z^2}{c^2} = -1$.

(iv) *Elliptic paraboloid: $\frac{x^2}{a^2} + \frac{y^2}{b^2} = \frac{2z}{c}$.

(v) *Hyperbolic paraboloid: $\frac{x^2}{a^2} - \frac{y^2}{b^2} = \frac{2z}{c}$.

(vi) *Quadric cone: $\frac{x^2}{a^2} + \frac{y^2}{b^2} = \frac{z^2}{c^2}$.

(vii) *Elliptic cylinder: $\frac{x^2}{a^2} + \frac{y^2}{b^2} = 1$.

(viii) *Hyperbolic cylinder: $\frac{x^2}{a^2} - \frac{y^2}{b^2} = 1$.

(ix) *Parabolic cylinder: $\frac{x^2}{a^2} = \frac{2y}{b}$.

(x) Pair of non-parallel planes: $\frac{x^2}{a^2} = \frac{y^2}{b^2}$ (that is, $y = \pm\frac{b}{a}x$).

(xi) Pair of parallel planes: $\frac{x^2}{a^2} = 1$ (that is, $x = \pm a$).

(xii) Plane: $\frac{x^2}{a^2} = 0$ (that is, $x = 0$).

(xiii) Line: $\frac{x^2}{a^2} + \frac{y^2}{b^2} = 0$ (that is, $x = y = 0$).

(xiv) Point: $\frac{x^2}{a^2} + \frac{y^2}{b^2} + \frac{z^2}{c^2} = 0$ (that is, $x = y = z = 0$).

Forms (i)–(v) are the non-*degenerate quadrics.

quadric cone A *quadric whose equation in a suitable coordinate system is

$$\frac{x^2}{a^2} + \frac{y^2}{b^2} = \frac{z^2}{c^2}.$$

Sections by planes parallel to the xy-plane are *ellipses, and sections by planes parallel to the other axial planes are *hyperbolas.

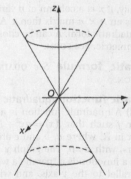

Elliptical sections of a quadric cone

quadrilateral A polygon with four sides. *See also* COMPLETE QUADRILATERAL.

quadrillion A thousand million million (10^{15}). In Britain it used to mean a million to the power of 4 (10^{24}), but this use is largely obsolete.

quadruple *See* N-TUPLE.

quality control The application of statistical methods to the maintenance of quality standards in a production or other process. Methods include *acceptance sampling to make decisions on whether to take delivery of products and *control charts to identify when modification to a process is required.

quantifier The two expressions 'for all...' and 'there exists...' are called quantifiers. A phrase such as 'for all x' or 'there exists x' may stand in front of a sentence involving a symbol x and thereby create a statement that makes sense and is either true or false. There are different ways in English of expressing the same sense as 'for all x', but it is sometimes useful to standardize the language to this particular form. This is known as a universal quantifier and is written in symbols as '$\forall x$'. Similarly, 'there exists x' may be used as the standard form to replace any phrase with this meaning and is an existential quantifier, written in symbols as '$\exists x$'.

For example, the statements 'if x is any number greater than 3 then x is positive' and 'there is a real number satisfying $x^2 = 2$' can be written in more standard form: 'for all x, if x is greater than 3 then x is positive', and 'there exists x such that x is real and $x^2 = 2$'. These can be written, using the symbols of mathematical logic, as: $(\forall x)(x > 3 \Rightarrow x > 0)$, and $(\exists x)(x \in \mathbb{R} \land x^2 = 2)$.

quantile Let X be a continuous *random variable. For $0 < p < 1$, the p-th quantile is the value x_p such that $\Pr(X \leq x_p) = p$. In other words, the fraction of the population less than or equal to x_p is p. For example, $x_{0.5}$ is the (population) median.

Often percentages are used. The n-th percentile is the value $x_{n/100}$ such that n per cent of the population is less than or equal to $x_{n/100}$. For example, 30% of the population is less than or equal to the 30th percentile. The 25th, 50th, and 75th percentiles are the *quartiles.

Alternatively, the population may be divided into tenths. The n-th decile is the value $x_{n/10}$ such that n-tenths of the population is less than or equal to $x_{n/10}$. For example, three-tenths of the population is less than or equal to the 3rd decile.

The terms can be modified, though not always very satisfactorily, to be applicable to a discrete random variable or to a large sample ranked in ascending order.

quantity An entity that has a numerical value or magnitude.

quantum theory The area of physics concerned with the behaviour of particles at small scales where the discrete nature of matter becomes important and more generally with other quantum phenomena such as quantum information and quantum computers. The results of experiments early in the 20th century could only be explained by quantities such as momentum and energy only being able to take certain discrete values, yet the behaviour of particles taken together was that of a wave, giving rise to the wave-particle duality. *See also* HEISENBERG'S UNCERTAINTY PRINCIPLE, MEASUREMENT, OBSERVABLE, SCHRÖDINGER'S EQUATION.

quartic polynomial A polynomial of degree four. A quartic equation is an equation defined by a quartic polynomial.

quartile deviation A synonym for SEMI-INTERQUARTILE RANGE.

quartiles For numerical *data ranked in ascending order, the quartiles are values derived from the data which divide the data into four equal parts. If there are n observations, the first quartile (or lower quartile) Q_1 is the $\frac{1}{4}(n+1)$-th, the second quartile (which is the *median) Q_2 is the $\frac{1}{2}(n+1)$-th, and the third quartile (or upper quartile) Q_3 is the $\frac{3}{4}(n+1)$-th in ascending order. When $\frac{1}{4}(n+1)$ is not a whole number, it is sometimes thought necessary to take the (weighted) average of two observations, as is done for the median. However, unless n is very small, an observation that is nearest will normally suffice. For example, for the sample 15, 37, 43, 47, 54, 55, 57, 64, 76, 98, we may take $Q_1 = 43$, $Q_2 = 54.5$, and $Q_3 = 64$.

For a *random variable, the quartiles are the *quantiles $x_{0.25}$, $x_{0.5}$, and $x_{0.75}$; that is, the 25th, 50th, and 75th percentiles.

quasi-metric One of the conditions for a distance measure d to be a *metric is that $d(x, y) = d(y, x)$. If this condition is relaxed so that the distance between two points can be different depending on the direction of travel, then d is said to be a quasi-metric, and the space is said to be a quasi-metric space. Quasi-metrics are common in real life—the times taken to walk between two points can be quite different if there is a steep gradient between them, for example, and one-way systems in towns mean there is often a different route required to go from B to A than will take you from A to B.

quaternion The complex number system can be obtained by taking a complex number to be an expression $a + bi$, where a and b are real numbers, and defining addition and multiplication in the natural way with the understanding that $i^2 = -1$. In an extension of this idea, *Hamilton introduced the following notion, originally for use in mechanics. Define a quaternion to be an expression $a + bi + cj + dk$, where a, b, c, and d are real numbers, and define addition and multiplication in the natural way, with

$$i^2 = j^2 = k^2 = -1, \quad ij = -ji = k,$$
$$jk = -kj = i, \quad ki = -ik = j.$$

All the normal laws of algebra hold except that multiplication is not *commutative. That is to say, the quaternions form a *division ring. *See also* DIVISION ALGEBRA.

quaternion group The *group Q_8 of quaternions $\{\pm 1, \pm i, \pm j, \pm k\}$ under multiplication. It has a *presentation

$$Q_8 = \{x, y : x^4 = 1, x^2 = y^2, y^{-1}xy = x^{-1}\}.$$

queuing theory A study of the processes relating to customer wait and service time patterns where there is a random element involved in one or other or in most cases both parts. In various contexts, the probability *model for each part can be quite different, and the randomness of each stage makes analysis appropriate for *simulations where the behaviour of the queue can be observed repeatedly for a variety of parameter values in order to investigate the effect of a variety of management strategies.

quick sort algorithm As the name suggests, this method is quicker than a standard *bubble sort algorithm, because it sorts smaller groups only. The method involves choosing a *pivot value for any list or sublist, usually by taking the middle (or one of the two middle items in an even-numbered list) and placing all items smaller than the pivot on the left and the larger items on the right, then the two groups created being treated as lists in the next stage. The process ends when all sublists are of length 1.

Quillen, Daniel Grey (1940–2011) American mathematician whose work on the higher algebraic K-theory, which combined geometric and topological methods in dealing with algebraic problems in *ring theory and *module theory, was awarded the *Fields Medal in 1978.

quin- Prefix, denoting five, as in *quintic. *Compare* PENTA.

Quine, Willard van Orman (1908–2000) American logician and philosopher whose *New Foundations in Mathematical Logic* was influential in *set theory. He was a proponent of *first-order logic. He was also influential in the philosophy of science, where he drew no line between philosophy and empirical science.

quintic polynomial A polynomial of degree five. A quintic equation is an equation defined by a quintic polynomial.

quota sample *See* SAMPLE.

quotient *See* DIVISION ALGORITHM.

quotient group A *group formed by the *cosets of a *normal subgroup. Let G/H denote the set of cosets of a subgroup H of a group G. The group operation on G *induces a *well-defined group operation on G/H given by $(g_1H)(g_2H) = (g_1g_2)H$ if and only if H is a normal subgroup of G, By *Lagrange's theorem, if G is finite then G/H has order $|G|/|H|$.

quotient ring A *ring formed by the *cosets of an *ideal. Let R/I denote the set of cosets of an ideal I of a ring R. The ring operations on R *induce a *well-defined ring operation on R/I given by $(r_1 + I) + (r_2 + I) = (r_1 + r_2) + I$ and $(r_1 + I)(r_2 + I) = r_1r_2 + I$.

quotient rule (for differentiation) *See* DIFFERENTIATION.

quotient space If X is a *topological space and ~ is an *equivalence relation defined on X, then the quotient space $Y = X/\sim$ is defined to be the set of equivalence classes of elements of X. So $Y = \{\{u \in X : u \sim x\} : x \in X\}$, where the open sets are defined to be those sets of equivalence classes whose unions are open sets in X. Quotient spaces are also known as identification spaces and factor spaces.

R The set of *real numbers. The symbol also denotes the *field of real numbers, and $\mathbb{R}^*$ denotes the multiplicative *group of non-zero real numbers.

$\mathbb{R}^n$ *See* EUCLIDEAN SPACE.

Rademacher's Theorem A *Lipschitz function $f{:}\mathbb{R} \to \mathbb{R}$ is *differentiable *almost everywhere.

radial and transverse components When a point P has *polar coordinates (r,θ), the *vectors $\mathbf{e}_r$ and $\mathbf{e}_\theta$ are defined by

$$\mathbf{e}_r = \mathbf{i}\cos\theta + \mathbf{j}\sin\theta,$$
$$\mathbf{e}_\theta = -\mathbf{i}\sin\theta + \mathbf{j}\cos\theta,$$

where $\mathbf{i}$ and $\mathbf{j}$ are *unit vectors in the directions of the positive x- and y-axes. Then $\mathbf{e}_r$ is a unit vector along OP in the direction of increasing r, and $\mathbf{e}_\theta$ is a unit vector perpendicular to this in the direction of increasing θ. Any vector $\mathbf{v}$ can be written uniquely in terms of its components in the directions of $\mathbf{e}_r$ and $\mathbf{e}_\theta$. Thus $\mathbf{v} = v_1\mathbf{e}_r + v_2\mathbf{e}_\theta$, where $v_1 = \mathbf{v}{\cdot}\mathbf{e}_r$ and $v_2 = \mathbf{v}{\cdot}\mathbf{e}_\theta$. The component v_1 is the radial component, and the component v_2 is the transverse component. For a particle with *position vector $\mathbf{r}(t) = r(t)\mathbf{e}_r(t)$, the *velocity $\mathbf{r}'(t)$ has components r' and $r\theta'$ and *acceleration $\mathbf{r}''(t)$ has components $r'' - r(\theta')^2$ and $r^{-1}(r^2\theta')'$.

radial set *See* ABSORBING SET.

radian In elementary work, *angles are measured in *degrees. In more advanced work, particularly involving *calculus, it is essential that angles are measured in radians. For example the *differential of *sine is *cosine only when radians are used.

Suppose that a circle of radius 1 with centre O meets two half-lines from O to A and B. Then the size of $\angle AOB$, measured in radians, equals the length of the *arc AB.

On a unit circle, the arc AB equals $\angle AOB$ in radians

1 radian is approximately $57°$. More accurately, 1 radian $\approx 57.296° \approx 57°$ $17'45''$. Since the circumference of a unit circle is 2π, one revolution or $360°$ measures 2π radians. Consequently, $x° = \pi x/180$ radians.

The radian is the SI unit for measuring angle.

Compare STERADIAN.

radical Given an *ideal I in a *ring R, its radical is

$$\sqrt{I} = \{r \in R | r^n \in I \text{ for some } n\}.$$

This is an ideal of R which contains I. A radical ideal is an ideal which equals its radical. *Prime ideals are radical; indeed, $\sqrt{I}$ equals the intersection of all prime ideals containing I. *See also* NILRADICAL.

radical axis The radical axis of two circles is the line containing all points P such that the lengths of the *tangents from P to the two circles are equal. Each figure shows a point P on the radical axis and tangents touching the two circles at T_1 and T_2, with $PT_1 = PT_2$. If the circles *intersect in two points, as in the second figure, the radical axis is the straight line through the two points of intersection.

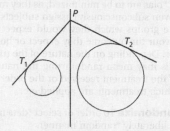

Radical axis of disjoint circles

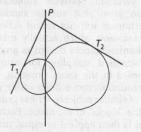

Radical axis of intersecting circles

radical sign The sign $\sqrt{}$ used in connection with *square roots, *cube roots, and, more generally, *nth roots for larger values of n. The *notation $\sqrt{a}$ indicates square root, $\sqrt[3]{a}$ indicates cube root and $\sqrt[n]{a}$ indicates nth root of a. For a detailed explanation of the correct usage of the notation, *see* NTH ROOT, SQUARE ROOT. *See also* SOLVABLE BY RADICALS.

radicand The expression that appears under the *radical sign $\sqrt{}$.

radius (radii) A radius of a *circle is a line segment joining the centre of the circle to a point on the circle. All such line segments have the same length, and this length is also called the radius of the circle. The term also applies in both senses to a sphere. *See also* SPHERE.

radius (of a graph) The minimum *eccentricity of any vertex in a *graph. *See* CENTRAL VERTEX.

radius of convergence *See* CIRCLE OF CONVERGENCE.

radius of curvature *See* CURVATURE.

radius of gyration The square root of the ratio of a *moment of inertia to the *mass of a *rigid body. Thus, if I is the moment of inertia of a rigid body of mass m about a specified axis, and k is the radius of gyration, then $I = mk^2$. It means that, when rotating about this axis, the body has the same moment of inertia as a ring of the same mass and of *radius k.

radius vector A synonym for POSITION VECTOR.

raise (to a power) To *multiply by itself to the number of times specified by the *power. For fuller details, *see* POWER (OF A REAL NUMBER).

Ramanujan, Srinivasa (1887–1920) The outstanding Indian mathematician of modern times. Originally a clerk in Madras, he studied and worked on mathematics totally unaided. Following correspondence with G. H. *Hardy, he accepted an invitation to visit Britain in 1914. He studied and collaborated with Hardy on the subject of *partitions and other topics in *number theory. He was considered a genius for his inexplicable ability in, for example, the handling of *series and *continued fractions. Because of ill health he returned to India a year before he died.

Ramsey, Frank (1903–30) Despite dying tragically young, Ramsey was influential in mathematics, philosophy, and economics. Ramsey's Theorem states that there is a number $R(m,n)$ such that every colouring with two colours of the *edges of a *complete graph with $R(m,n)$ vertices has a collection of m vertices with all edges between them being the first colour or a similar collection with n vertices for the second colour. In practice, these Ramsey numbers are difficult to calculate, even for small values of m and n.

Ramsey Theory An umbrella term relating to results that demonstrate that a certain level of order must be found amongst sufficiently large *random arrangements. For example, *Van der Waerden's Theorem and *Szemerédi's Theorem.

random The informal use of the term is to mean haphazard or irregular, but in *statistics it carries a more precise meaning indicating that the outcome can only be described in probabilistic terms. For example, a *random number generator should generate each of the n numbers in its outcome space uniformly with probability $1/n$, and a random *sample size from the *population means that all possible combinations of that sample size from the population are equally likely.

random error (noise) The difference between predicted values and observed values treated as a *random variable. In effect it is the variation which is unexplained by the *model, and hence the term 'error' is rather unhelpful. In *regression modelling it is also termed the *residual.

random graph Given a set of *vertices, a random graph may be created by independently assigning an *edge between each pair of vertices with a given *probability p. This leads to possible *combinatorics arguments to demonstrate, amongst all such random graphs, the existence of a graph with certain properties.

randomization In *experimental design, it is important that researchers do not have the flexibility to choose which subjects are assigned to different experimental groups if potential sources of *bias are to be minimized, as they may (even subconsciously) assign subjects to the groups which they would expect to favour the outcome they expect or hope for. Depending on the nature of the trial, this may mean randomizing the nature of the treatment received or the order in which treatments are applied etc.

randomize To order or select *data in a deliberately *random manner.

randomized blocks Where groups within a *population are likely to respond differently to an *experiment, the variation between subjects may be so great that simple randomized experiments will not reveal differences in treatments which actually exist. The randomized blocks design is analogous to *stratified sampling for estimation purposes in the same situation, where the randomization is done within groupings of similar subjects—these groupings are the 'blocks' of the name. Each treatment is then applied to equal numbers of subjects in each block, but the choice of which subject gets which treatment is randomized within each block. For example, if two methods of improving memory are to be compared it might be considered likely that age and gender may be factors, in which case the blocks might consist of males or females of different age-groupings.

random numbers Tables of random numbers give lists of the digits $0, 1, 2, \ldots,$ 9 in which each digit is equally likely to occur at any stage. There is no way of predicting the next digit. Such tables can be used to select items at random from a *population. Numbers generated by a deterministic *algorithm that appear to pass statistical tests for randomness are called pseudo-random numbers. Such an algorithm is called a random number generator.

random sample *See* SAMPLE.

random variable A quantity that takes different numerical values according to the result of a particular *experiment. For example, if a die is rolled, then the number rolled is a random variable on the *sample space of possible rolls.

A random variable is *discrete if the set of possible values is finite or *denumerable. For a discrete random variable, the probability of its taking any particular value is given by the *probability mass function.

A random variable is *continuous if possible outcomes are distributed over uncountably many values and the *cumulative distribution function $F_X(x) = P(X \leq x)$ is a *continuous function. If F_X is *piecewise *differentiable, then the derivative f_X is the *probability density function.

Some random variables are mixed in the sense of having both discrete and continuous characteristics. For example, in *modelling the lifetime T of a light bulb, there may be a non-zero probability p of the light bulb blowing immediately (so $T = 0$) and the remaining $1 - p$ is distributed over $T > 0$ continuously.

Formally, a random variable on a *probability space Ω is a *measurable function $X: \Omega \rightarrow \mathbb{R}$.

random vector An *n-tuple of n *random variables, often representing the outcomes of a repeated experiment in order. *Compare* PROBABILITY VECTOR.

random walk With the integers positioned as they occur on the *real line, imagine an object moving from integer to integer a step at a time. The position X_n of the object at the nth step is part of a *Markov chain $X_1, X_2, X_3, \ldots$, with state space being the integers. Such a Markov chain is called a one-dimensional random walk if $X_n = i$, then $X_{n+1} = i-1$, i or $i + 1$, so that the object may stay where it is, or move one to the left or right.

When a gambler, playing a sequence of games, either wins or loses a fixed amount in each game, their winnings give an example of a random walk, with 0 being an *absorbing state.

Random walks in two or more dimensions can be defined similarly.

See BROWNIAN MOTION.

range (of a function or mapping) *See* FUNCTION.

range (in mechanics) The range of a *projectile on a horizontal or inclined plane passing through the point of projection is the greatest distance from the point of projection to the point at which the projectile may land.

range (in statistics) The difference between the maximum and minimum *observations in a set of numerical data. It is a possible measure of *dispersion of a *sample.

rank (of a matrix or linear map) Let **A** be an $m \times n$ *matrix. The row rank and column rank of **A** are respectively the largest number of *linearly independent rows or columns of **A**. Equivalently, they are the dimensions of the *row space and *column space, and the two ranks can be shown to be equal. This common value is the rank of **A** and also equals the *determinantal rank or the number of non-zero rows when **A** is put into *reduced echelon form. An $n \times n$ matrix is *invertible if and only if it has rank n.

Given a *linear map $T: V \rightarrow W$ between finite-dimensional vector spaces, the rank of T is the dimension of the *image of T. If the matrix **A** represents T (*see* MATRIX OF A LINEAR MAP), then the rank of T equals the rank of **A**, as the image of T is now represented by the column space of **A**. *See* RANK-NULLITY THEOREM.

rank (in statistics) The *observations in a *sample are said to be ranked when they are arranged in order according to some criterion. For example, numbers can be ranked in ascending or descending order, people can be ranked according to height or age, and products can be ranked according to their sales. The rank of an *observation is its position in the list when the sample has been ranked. *Non-parametric methods frequently make use of ranking rather than the exact values of the observations in the sample.

rank correlation coefficient A lot of *bivariate *data do not meet the requirements necessary to use the product moment *correlation coefficient (pmcc). This is often because at least one of the variables is measured on a scale which is *ordinal but not *interval. Where the measurement is reported on a non-numerical scale it is transparently not interval, but often gradings, based on subjective evaluations, are reported on a numerical scale.

*Spearman's rank correlation coefficient is the product moment correlation applied to the ranks. In the case of where there are no tied ranks, this is algebraically equivalent to computing the value of

$$1 - \frac{6\sum d_i^2}{n(n^2 - 1)}$$

where $d_i =$ difference in ranks of the ith pair and n is the number of data pairs. When n is reasonably small, this computation is very quick. When there are tied ranks the above formula is not algebraically equivalent to the pmcc applied to the ranks but is often used as a reasonable approximation.

Kendall's rank correlation coefficient: Sir Maurice Kendall proposed a measure of rank correlation based on the number of neighbour-swaps needed to move from one rank ordering to another. If Q is the minimum number of neighbour swaps needed in a set of data with n values then Kendall's coefficient is given by

$$\tau = 1 - \frac{4Q}{n(n-1)}.$$

*Significance tests can then be carried out using critical values from books of statistical tables.

rank-nullity theorem Let T be a *linear map between finite-dimensional *vector spaces V and W. Then the sum of the *rank of T and the *nullity of T equals the dimension of V. If V and W have equal dimensions, this implies T is *one-to-one if and only if T is *onto.

rate of change Suppose that the quantity y is a function of the quantity x, so that $y = f(x)$. If f is *differentiable, the rate of change of y with respect to x is the *derivative dy/dx or $f'(x)$. The rate of change is often with respect to time t. Suppose, now, that x denotes the *displacement of a particle, at time t, on a directed line with origin O. Then the *velocity is dx/dt or $\dot{x}$, the rate of change of x with respect to t, and the acceleration is d^2x/dt^2 or $\ddot{x}$, the rate of change of the velocity with respect to t.

When the motion is in two or three dimensions, say

$$\mathbf{r}(t) = (x(t), y(t), z(t))$$

denoting the *position vector of a particle, the velocity of the particle is the vector

$$\dot{\mathbf{r}}(t) = (\dot{x}(t), \dot{y}(t), \dot{z}(t)),$$

and the acceleration is

$$\ddot{\mathbf{r}}(t) = \left(\ddot{x}(t), \ddot{y}(t), \ddot{z}(t)\right).$$

rate of convergence A *sequence $\{x_n\}$ which *converges to a limit l is said to have rate (or order) of convergence $r \le 1$ if there exists M such that

$$|x_{n+1} - l| \le M|x_n - l|^r$$

for *sufficiently large n. For $r = 1$, it is further required that $M \le 1$. If $r = 2$, then the convergence is said to be quadratic. If $r = 1$ and

$$\lim_{n \to \infty} \frac{|x_{n+1} - l|}{|x_n - l|} = m,$$

then the convergence is said to be superlinear, linear, or sublinear depending on whether $m = 0$, $0 < m < 1$ or $m = 1$.

ratio The *quotient of two numbers or quantities giving their relative size. The ratio of x to y is written as $x : y$ and is unchanged if both quantities are multiplied or divided by the same quantity. So $2 : 3$ is the same as $6 : 9$ and $1 : \frac{3}{2}$. Where a ratio is expressed in the form $1 : a$ it is called a unitary ratio, and this form makes comparison of ratios much easier.

rational canonical form (Frobenius normal form) The rational canonical form, or Frobenius normal form, of a *square matrix is a *block diagonal matrix which is *similar to the matrix and where the blocks are *companion matrices. The blocks are $C(d_1), C(d_2), \ldots, C(d_k)$, where $d_1, d_2, \ldots, d_k$ are *monic polynomials such that d_i divides d_{i+1} and $C(d_i)$ denotes the companion matrix of d_i. The *minimal polynomial equals d_k and the *characteristic polynomial equals the product $d_1 d_2 \ldots d_k$. See JORDAN NORMAL FORM, STRUCTURE THEOREM (for modules).

rational function A rational function is one f of the form $f(x) = g(x)/h(x)$, where $g(x)$ and $h(x)$ are *polynomials, usually with *coefficients from a *field F. The rational functions themselves then make a field denoted $F(x)$. Addition and multiplication are defined as expected by

$$\frac{f_1(x)}{g_1(x)} + \frac{f_2(x)}{g_2(x)} = \frac{g_2(x)f_1(x) + f_2(x)g_1(x)}{g_1(x)g_2(x)},$$

$$\left(\frac{f_1(x)}{g_1(x)}\right)\left(\frac{f_2(x)}{g_2(x)}\right) = \frac{f_1(x)f_2(x)}{g_1(x)g_2(x)}.$$

Compare POLYNOMIAL RING.

rationalize To remove radicals from an expression or part of it without changing the value of the whole expression. For example, in the expression $1/(2 - \sqrt{x})$ the denominator can be rationalized by multiplying the numerator and the denominator by $2 + \sqrt{x}$ to give

$$\frac{2 + \sqrt{x}}{4 - x}.$$

See also CONJUGATE SURDS.

rational number A number that can be written in the form a/b, where a and b are integers, with $b \neq 0$. The set of all rational numbers is usually denoted by $\mathbb{Q}$. A real number is rational if and only if, when expressed as a decimal, it has a finite or recurring expansion (*see* DECIMAL REPRESENTATION). For example,

$$\frac{5}{4} = 1.25, \quad \frac{2}{3} = 0.\dot{6}, \quad \frac{20}{7} = 2.\dot{8}5714\dot{2}.$$

A famous proof, attributed to *Pythagoras, shows that $\sqrt{2}$ is not rational, and e and π are also known to be irrational.

The same rational number can be expressed as a/b in different ways; for example, $\frac{2}{3} = \frac{6}{9} = \frac{-4}{-6}$. In fact, $a/b = c/d$ if and only if $ad = bc$. But a rational number can be expressed uniquely as a/b if it is insisted that a and b are *coprime and that $b > 0$. The field $\mathbb{Q}$ is the *field of fractions of the *ring of *integers $\mathbb{Z}$.

ratio test A test for *convergence or *divergence of an infinite *series $\sum a_n$, due to *D'Alembert. If the ratio $\left|\frac{a_{n+1}}{a_n}\right| \to k$ as $n \to \infty$ then when $k < 1$ the series converges absolutely, and for $k > 1$ the series diverges. If $k = 1$, the ratio test is not sufficient to determine whether it converges. For example, $k=1$ for both the series $\sum 1/n$ and $\sum 1/n^2$, but the first series diverges whereas the second converges.

raw data *Data which have not been analysed. The values of the variables as they were recorded.

ray A synonym for HALF-LINE.

Re Abbreviation and symbol for the *real part of a *complex number.

reachable set A synonym for ATTAINABLE SET.

reaction *See* CONTACT FORCE.

real Relating to the *real numbers. Of a *complex number, having no *imaginary part.

real axis In the *complex plane, the x-axis is called the real axis; its points represent the real numbers.

real function A *function from the set $\mathbb{R}$ of real numbers (or a subset of $\mathbb{R}$) to $\mathbb{R}$. Thus, if f is a real function, then, for every real number x in the domain, a corresponding real number $f(x)$ is defined. If a function $f: S \rightarrow \mathbb{R}$ is defined by giving a formula for $f(x)$, without specifying the *domain S, it is usual to assume that the domain is the largest meaningful subset S of $\mathbb{R}$. For example, if $f(x) = 1/(x-2)$, the domain would be taken to be $\mathbb{R} \setminus \{2\}$; that is, the set of all real numbers not equal to 2. If $f(x) = \sqrt{9-x^2}$, the domain would be the closed interval $[-3, 3]$.

realism In the *philosophy of mathematics, mathematical realism maintains that mathematical objects and concepts exist independent of human knowledge; such a philosophy is mathematical *Platonism. Anti-realist philosophies include *formalism and *structuralism.

real line On a horizontal straight *line, choose a point O as *origin and a point

A, to the right of O, such that $|OA|$ is equal to 1 unit. Each positive real number x can then be uniquely represented by a point on the line to the right of O, whose distance from O equals x units, and each negative number by a point on the line to the left of O. The origin represents zero. The line is called the real line when its points are taken in this way to represent the real numbers. *Compare* ARGAND DIAGRAM.

real number The numbers generally used in mathematics, scientific work, and everyday life are the real numbers. They can be pictured as points of a *real line, with the *integers equally spaced along the line and a real number b to the right of a real number a if $a < b$. The set of real numbers is usually denoted by $\mathbb{R}$. As a set, $\mathbb{R}$ is *uncountable.

Every real number is either *rational or *irrational, and every real number has an expression as a *decimal representation.

The set of real numbers forms an *ordered *field. The *completeness axiom of the real numbers then uniquely characterizes $\mathbb{R}$. *See* ORDER (REAL NUMBERS).

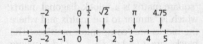

The real number line

real part A *complex number z may be written $x + yi$, where x and y are real, and then x is the real part of z. It is denoted by Rez.

real projective plane An abstract 2-dimensional, *non-orientable (one-sided) surface without boundary. Like the *Klein bottle, there is no way to physically construct (*embed) the real projective plane in three dimensions, so models intersect or pass through themselves. However, 4-dimensional models exist without self-intersection.

One definition of the real projective plane is the space of all lines through the origin in 3-dimensional *Euclidean space. Topologically, this can be defined simply as the direction vectors for the unit sphere. Then, choosing a hemisphere, since each vector in the other hemisphere defines the same line as one in the chosen hemisphere, the identification of opposite points provides an equivalence, whose *quotient space is the real projective plane. *See* PROJECTIVE GEOMETRY, PROJECTIVE SPACE.

real world *See* MATHEMATICAL MODEL.

rearrangement A synonym for PERMUTATION.

reciprocal The *multiplicative inverse of a quantity may, when the operation of multiplication is *commutative, be called its reciprocal. Thus, the reciprocal of 2 is $\frac{1}{2}$, the reciprocal of $3x+4$ is $1/(3x+4)$, and the reciprocal of $1/(3x+4)$ is $3x+4$.

rectangle A *quadrilateral containing four *right angles. Opposite sides will necessarily be *parallel and of equal *length. When all four sides are equal, the figure is a *square.

rectangular (oblong) Shaped like a *rectangle.

rectangular distribution A synonym for UNIFORM DISTRIBUTION.

rectangular hyperbola A *hyperbola whose *asymptotes are *perpendicular. With the origin at the centre and the coordinate x-axis along the *transverse axis, it has equation $x^2-y^2 = a^2$. Instead, the coordinate axes can be taken along the asymptotes in such a way that the two branches of the hyperbola are in the first and third quadrants. The rectangular hyperbola then has equation of the form $xy = c^2$. For example, $y = 1/x$ is a rectangular hyperbola. For $xy = c^2$, it is customary to take $c > 0$ and to use, as parametric equations, $x = ct, y = c/t$ $(t \neq 0)$.

rectangular number Any positive integer which can be expressed nontrivially as a rectangular array of dots, i.e. excluding the trivial case of a single row or column. Thus, rectangular numbers are *composite and vice versa. For many rectangular numbers, the representation is not unique, so 12 could be drawn as a 4×3 or a 6×2 rectangle.

rectilinear motion Motion in a *line, which may be in either *direction.

recurrence relation A synonym for DIFFERENCE EQUATION.

recurrent In a *Markov chain, a state is recurrent if, having been in the state, the chain is certain to return to the state at some future time.

recurring decimal *See* DECIMAL REPRESENTATION.

recursion theory A branch of *logic and computer science concerned with the study of *computable functions.

reduce Modify or simplify a form, expression, or number; for example, $\frac{6}{9}$ can be reduced to the equivalent form $\frac{2}{3}$

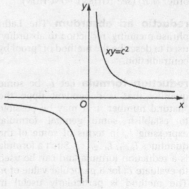

A rectangular hyperbola

and $\dfrac{x^2 - 1}{x - 1}$ can be reduced to $x+1$ provided $x \neq 1$.

reduced echelon form (reduced row echelon form) A *matrix is in reduced echelon form if it is in *echelon form and, further, every column that contains a leading 1 (of some row) otherwise has entries of 0. For example, these two matrices are in reduced echelon form:

$$\begin{bmatrix} 1 & 6 & 0 & 0 & 2 \\ 0 & 0 & 1 & 0 & -3 \\ 0 & 0 & 0 & 1 & 5 \end{bmatrix}, \begin{bmatrix} 0 & 0 & -1 & 4 & 2 \\ 0 & 1 & 2 & -3 & 5 \\ 0 & 0 & 0 & 0 & 0 \end{bmatrix}.$$

Any matrix can be transformed to a unique matrix in reduced echelon form using *elementary row operations, by *Gauss–Jordan elimination. The solutions of a system of linear equations can be immediately obtained from the reduced echelon form to which the *augmented matrix has been transformed.

reduced set of residues For $n \geq 2$ a reduced set of residues modulo n is a set of integers, one *congruent (modulo n) to each of the positive integers less than n which are *coprime with n. Thus, $\{1, 5, 7, 11\}$ is a reduced set of residues modulo 12, as is $\{1, -1, 5, -5\}$. A reduced set of residues forms an *abelian group under multiplication, denoted $\mathbb{Z}_n^*$, which has order $\phi(n)$ (*see* TOTIENT FUNCTION).

reductio ad absurdum The Latin phrase meaning 'reduction to absurdity' used to describe the method of *proof by contradiction.

reduction formula Let I_n be some quantity that is dependent upon the *natural number n. It may be possible to establish some general formula, expressing I_n in terms of some of the quantities $I_{n-1}, I_{n-2}, \ldots$. Such a formula is a reduction formula and can be used to evaluate I_n for a particular value of n. The method is particularly useful in *integration. For example, if

$$I_n = \int_0^{\pi/2} \sin^n x \, dx,$$

it can be shown, by *integration by parts, that

$$I_n = \left(\frac{n-1}{n} \right) I_{n-2} \quad \text{for } n \geq 2.$$

As $I_0 = \pi/2$, the reduction formula can be used, for example, to find that

$$I_6 = \frac{5}{6} I_4 = \frac{5}{6} \times \frac{3}{4} I_2$$
$$= \frac{5}{6} \times \frac{3}{4} \times \frac{1}{2} I_0$$
$$= \frac{5}{6} \times \frac{3}{4} \times \frac{1}{2} \times \frac{\pi}{2} = \frac{5\pi}{32}.$$

redundancy In *coding, the presence of code, such as checksums, that is surplus to the encoded message but provides means to error-detect or *error-correct. The term also relates to unintentional redundancy where a code is simply suboptimal.

redundant If an *equation (or *inequality) does not affect the *solution set of other equations (or inequalities), it is said to be redundant. For example, if

$$2x + y = 7, \quad 3x - y = 3, \quad 5x + y = 13,$$

any one of these equations can be termed redundant because the other two are sufficient to identify $x = 2$, $y = 3$ as the only solution. If $3x + 2y > 4$ and $6x + 4y > 9$ then the first inequality is redundant because if $6x + 4y > 9$ it follows that $3x + 2y > 4.5$ and the first inequality is automatically satisfied. If one or more *simultaneous linear equations are redundant, then in *echelon form there will be a zero row.

re-entrant An *interior angle of a *polygon is re-entrant if it is greater than two *right angles.

reflection Let l be a *line in the *plane. Then the mirror-image of a point P is the point P' such that PP' is perpendicular to l and l cuts PP' at its *midpoint.

Reflection in the line l is the transformation of the plane that maps each point P to its mirror-image P'. Suppose that the line l passes through the origin O and makes an angle α with the x-axis. If P has *polar coordinates (r, θ), its mirror-image P' has polar coordinates $(r, 2\alpha - \theta)$. In terms of Cartesian *coordinates, reflection in the line l maps P with coordinates (x, y) to P' with coordinates (x', y'), where

$$x' = x \cos 2\alpha + y \sin 2\alpha,$$
$$y' = x \sin 2\alpha - y \cos 2\alpha.$$

In 3-dimensional space, an *orthogonal matrix with *determinant -1 and *trace 1 represents a reflection in a plane containing the origin.

In higher dimensions, in $\mathbb{R}^n$, reflection in the *hyperplane with equation $\mathbf{r} \cdot \mathbf{n} = c$ is given by

$$\mathbf{r} \mapsto \mathbf{r} + 2\left(\frac{c - \mathbf{r} \cdot \mathbf{n}}{\mathbf{n} \cdot \mathbf{n}}\right)\mathbf{n}.$$

If $c = 0$, this is represented by an orthogonal matrix. In general, reflections are *isometries.

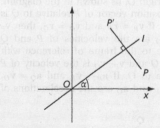

Reflection in l

reflex angle An angle that is greater than 2 *right angles and less than 4 right angles.

reflexive relation A *binary relation $\sim$ on a set S is reflexive if $a \sim a$ for all a in S. *See* EQUIVALENCE RELATION.

reflexive space For a *Banach space V, there is an *one-to-one *linear map i: $V \rightarrow V''$ from V to its second *dual given by $i(v)(\varphi) = \varphi(v)$ for $v \in V$ and $\varphi \in V'$. Then V is said to be reflexive if i is also *onto. Finite-dimensional vector spaces are reflexive (as $\dim V = \dim V''$), as are *Hilbert spaces, but infinite-dimensional Banach spaces need not be.

Regiomontanus (1436–76) A central figure in mathematics in the 15th century. Born Johann Müller, he took as his name the Latin form of Königsberg, his birthplace. His *De triangulis omnimodis* ('On All Classes of Triangles') was the first modern account of trigonometry, and, even though it did not appear in print until 1533, it was influential in the revival of the subject in the West.

region A synonym for DOMAIN (ANALYSIS).

regression A statistical procedure to determine the relationship between a *dependent variable and one or more *explanatory variables. The purpose is normally to enable the value of the dependent variable to be predicted from given values of the explanatory variables. It is multiple regression if there are two or more explanatory variables. Usually the model supposes that Y, where Y is the dependent variable, is given by some formula involving certain unknown parameters. In multiple linear regression, with k explanatory variables $X_1, X_2, \ldots, X_k$, then

$$Y = b_0 + b_1 X_1 + b_2 X_2 + \ldots + b_k X_k + \varepsilon.$$

Here $b_0, b_1, \ldots, b_k$ are the regression coefficients and ε denotes the error term. *See also* LEAST SQUARES.

regression to the mean A synonym for REVERSION TO THE MEAN.

regular function A synonym for HOLOMORPHIC.

regular graph A *graph in which all the vertices have the same *degree. It is *r*-regular or regular of degree *r* if every vertex has degree *r*.

regular polygon *See* POLYGON.

regular polyhedron *See* PLATONIC SOLID.

regular space A *topological space *X* in which every *closed subset *S* of *X* and a point *p* of *X* which is not in *S* have *disjoint *neighbourhoods. *See* SEPARATION AXIOMS.

regular tessellation *See* TESSELLATION.

relation A relation on a set *S* is usually a *binary relation on *S*, though the notion can be extended to involve more than two elements. An example of a ternary relation, involving three elements, is '*a* lies between *b* and *c*', where *a*, *b*, and *c* are real numbers. *See also* EQUIVALENCE RELATION.

relation (group theory) *See* PRESENTATION.

relational understanding In *mathematics education, the term refers to an understanding of a mathematical rule or concept that goes beyond *instrumental understanding, being able to appreciate why the rule works and also to adapt the rule and relate it to other rules.

relative complement A synonym for SET DIFFERENCE.

relative efficiency *See* ESTIMATOR.

relative error Let *x* be an *approximation to a value *X* and let $X = x + e$. The relative error is $|e/X|$. When 1.9 is used as an approximation for 1.875, the relative error equals $0.025/1.875 = 0.013$, to 3 decimal places or 1.3%. The relative error may be a more helpful figure than the absolute *error. An absolute error of 0.025 in a value of 1.9, as above, may be acceptable. But the same absolute error

in a value of 0.2, say, would give a relative error of $0.025/0.2 = 0.125$ or 12.5%, which would probably be considered quite serious.

relative frequency The ratio n/N where *n* is the observed number of a particular *event, and *N* is the number of *trials. As *N* gets large, the *weak law of large numbers says that n/N will tend to the *probability of that event with a probability of 1. In cases where no way of calculating a probability exists, then the value of n/N can be used to estimate this probability. The larger the value of *N*, the better this *statistic is as an *estimator.

relatively prime A synonym for COPRIME.

relative measure of dispersion A measure that indicates the magnitude of a measure of *dispersion relative to the magnitude of the *mean, such as the *coefficient of variation.

relative position (relative velocity, relative acceleration) Let r_P and r_Q be the *position vectors of particles *P* and *Q* with respect to some *frame of reference with origin *O*, as shown in the diagram. The position vector of *P* relative to *Q* is $r_P - r_Q$. If $v_P = \dot{r}_P$ and $v_Q = \dot{r}_Q$, then v_P and v_Q are the velocities of *P* and *Q* relative to the frame of reference with origin *O* and $v_P - v_Q$ is the velocity of *P* relative to *Q*. If $a_P = \dot{v}_P$ and $a_Q = \dot{v}_Q$, then a_P and a_Q are the accelerations of

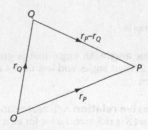

Position of *P* relative to *Q*

P and Q relative to the frame of reference with origin O, and $\mathbf{a}_P - \mathbf{a}_Q$ is the acceleration of P relative to Q. These quantities may be called the relative position, the relative velocity, and the relative acceleration of P with respect to Q.

These notions are important when there are two or more frames of reference, each with an associated *observer.

relative risk The *ratio of the *risks of the same outcome for two different groups or of a different outcome for the same group. For example, one could compute the relative risk of travelling by plane compared to travelling by car.

relativity theory The theory in physics put forward by Albert *Einstein in the early 20th century which fundamentally changed commonly held views of the universe and concepts of space and time which had previously been regarded as independent of one another. The special theory of relativity deals with the behaviour of particles in different *frames of reference which are moving at constant relative speeds (*see* LORENTZ TRANSFORMATION, LORENTZ-FITZGERALD CONTRACTION, MASS-ENERGY EQUATION, MINKOWSKI SPACE, SIMULTANEITY, TIME DILATION). The general theory of relativity describes gravitational forces in terms of the *curvature of space, caused by the presence of mass, and deals with the behaviour of particles in different frames of reference which are accelerating relative to one another. Light then moves along the *geodesics of space-time, which allows for the phenomenon of gravitational lensing, where a distant object may appear as two images to an observer, as light from the object has travelled along two different geodesics around an intervening massive object.

(⊕) SEE WEB LINKS

• An article about relativity with a related video.

reliability The sampling *variance of a *statistic. For example, the reliability of

$$\bar{X}_n = \frac{1}{n}\sum_{k=1}^{n} X_k$$

is σ^2/n where $\sigma^2 = \mathrm{Var}(X)$, so, as the sample size increases, $\bar{X}_n$ becomes a better *estimator in the sense that the variability decreases.

remainder *See* DIVISION ALGORITHM, TAYLOR'S THEOREM.

remainder theorem If a polynomial $f(x)$ is divided by $x-h$, then the remainder equals $f(h)$. That is $f(x) = (x-h)q(x) + f(h)$ for some polynomial $q(x)$. *See also* FACTOR THEOREM.

removable singularity If there is a value which $f(z)$ could be assigned at an isolated *singular point which would make the function *holomorphic at that point, then that point is termed a removable singularity. For example, if $f(z) = 1/z$, then $|f(z)| \to \infty$ as $|z| \to 0$, so this is not a removable singularity (but rather a *pole), but for $f(z) = \frac{\sin z}{z}$, which is undefined at $z = 0$, if we define $f(0) = 1$ the function is holomorphic at 0, and this is an example of a removable singularity. As a holomorphic function can be *analytically continued to a removable singularity, many authors do not consider removable singularities to be genuine singularities.

repeated integral A synonym for MULTIPLE INTEGRAL.

repeated measures designs An *experimental design where the same participants are measured repeatedly under different conditions. By using the same participants for each condition a major source of variability is removed.

repeated root *See* ROOT.

repeating decimal *See* DECIMAL REPRESENTATION.

repelling fixed point *See* FIXED-POINT ITERATION.

repetition codes One of the simplest possible *error-correcting procedures in *data transmission. The principle is that the sender repeats the message several times, with the hope that the noisy channel interferes with only a small proportion of these messages and not in a consistent way. Generally, other error-correcting mechanisms such as *checksums offer better performance.

replicable (replication) Able to be repeated. For example, an *experiment may be repeated several times to reduce potential experimental *errors. If a system is *chaotic, then an experiment may not be replicable, and in *quantum theory experiments are generally not replicable because of the effect of *measurement.

representation (of a vector) When the *directed line segment $\overrightarrow{AB}$ represents the *vector **a**, then $\overrightarrow{AB}$ is a representation of **a**.

representation (representation theory) When a *group G *acts on a set S, there is an associated *homomorphism $\rho : G \rightarrow \text{Sym}(S)$, where $\text{Sym}(S)$ denotes the *symmetry group of S defined by $\rho(g)(s) = g \cdot s$. Here ρ is called a representation, and conversely each such representation gives an action of G on S. If the action is *faithful, then the representation is *one-to-one.

Commonly, a representation is defined as a homomorphism $\rho : G \rightarrow \text{GL}(V)$ or $\rho : G \rightarrow \text{GL}(n,F)$ associating each group element with a *linear map or *matrix. Such a representation can instead be treated as a *module over the *group algebra FG.

Representation theory, as applied to the study of groups and other algebraic structures, is a large part of *abstract algebra. It has many applications, including in physics and chemistry, due to the rich theory of representations of *Lie groups. *See also* CHARACTER, CHARACTER TABLE, LINEAR ACTION.

representative Given an *equivalence relation on a set, any one of the *equivalence classes can be specified by giving one of the elements in it. The particular element a used can be called a representative of the class, and the class can be denoted by $[a]$ or $\bar{a}$.

representative sample A *sample which shares certain characteristics of the *population, usually the *proportions of members of particular groups who might be expected to behave differently in the variable(s) of interest. Quota sampling and *stratified sampling both give representative samples.

residual The difference between an observed value (*see* OBSERVATION) and the value predicted by some *statistical model. The residuals may be checked to assess how well the model fits the *data, perhaps by using a *chi-squared test. A large residual may indicate an *outlier.

residual variation The variation unaccounted for by a *model which has been fitted to a set of *data.

residue *Laurent's theorem guarantees a doubly infinite series expansion

$$\sum_{n=-\infty}^{\infty} c_n (z-a)^n$$

for a complex function $f(z)$ with an isolated *singularity at a in $\mathbb{C}$. The residue of $f(z)$ at a equals c_{-1}. *See* CAUCHY'S RESIDUE THEOREM.

residue class (modulo n) An *equivalence class for the *equivalence relation of *congruence modulo n. So, two integers are in the same class if they have the same remainder upon division by n. If $[a]$ denotes the residue class modulo n containing a, the residue classes modulo

n can be represented as $[0], [1], [2], \ldots,$ $[n-1]$. The sum and product of residue classes are defined by

$$[a] + [b] = [a+b], \quad [a][b] = [ab],$$

where it is necessary to show these definitions are *well defined and independent of the choice of *representatives a and b. With this addition and multiplication, the set, denoted by $\mathbb{Z}_n$, of residue classes modulo n forms a commutative *ring with identity. If n is composite, the ring $\mathbb{Z}_n$ has *zero-divisors, but when p is prime $\mathbb{Z}_p$ is a *field.

resistant statistic A *statistic which is relatively unaffected by unusual *observations and *outliers. The *median and *interquartile range are examples of resistant statistics, while the *mean, *standard deviation, and *range are not.

resistive force A *force that opposes the motion of a body. It acts on the body in a *direction opposite to the direction of the *velocity of the body. Examples include *friction and *aerodynamic drag. When there is a resistive force, the principle of *conservation of energy does not hold, but the *work–energy principle does hold.

resolution (resolve) The process of writing a *vector in terms of its *components in two or three mutually *perpendicular directions.

resonance Suppose that a body capable of performing *oscillations is subject to an applied *force which is itself oscillatory and not subject to any *resistive force. In certain circumstances the body may oscillate with an *amplitude that, in theory, increases indefinitely. This happens when the applied force has the same frequency as a *natural frequency of the body, and then resonance is said to occur. For example, if the equation $m\ddot{x} + kx = F\cos\Omega t$ holds, there is resonance when $\Omega = \sqrt{k/m}$, in

which case the solution of the equation involves the particular integral

$$\frac{Ft}{2\sqrt{km}}\sin\sqrt{\frac{k}{m}}\,t,$$

which gives oscillations of increasing amplitude.

(⊕) SEE WEB LINKS
● Animations and videos of forced oscillation and resonance.

response bias Occurs when participants in a survey respond differently from the way they actually feel. This may be because the questions have been asked in a loaded manner or because the information sought is sensitive and the participant is not prepared to be honest. *See also* ANONYMITY OF SURVEY DATA.

response variable A synonym for DEPENDENT VARIABLE.

rest frame In special *relativity, the rest frame of a particle is the *inertial frame in which the particle is at rest. *See* LORENTZ-FITZGERALD CONTRACTION, REST MASS, SIMULTANEITY.

rest mass In special *relativity, the mass of a particle as measured in its *rest frame. A particle of rest mass m_0 has an *inertial mass (or relativistic mass) of $m_0/\sqrt{1 - v^2/c^2}$ to an observer travelling at speed v relative to the particle. (Here c denotes the *speed of light.)

restriction (of a function) For a *function $f:X \to Y$ and a *subset $A \subseteq X$, the restriction of f to A is the function, denoted $f_{|A}$, with domain A and codomain Y, such that $f_{|A}(a) = f(a)$ for all a in A. Thus, $f_{|A}$ is essentially the same assignment as f but applied to a restricted set of inputs.

resultant The sum of two or more *vectors, particularly if they represent *forces, may be called their resultant.

retardation A synonym for DECELERATION.

retraction A retraction of a *topological space X to a subspace $Y \subseteq X$ is a *continuous function $r:X \rightarrow Y$ such that $r(y) = y$ for all $y \in Y$, and Y is referred to as a retract of X. Every continuous function $f:A \rightarrow Z$ can then be extended to $g: X \rightarrow Z$, where $g = f \circ r$.

reverse mathematics In *logic, reverse mathematics is the study of what *axioms are necessary for a given theorem to hold.

reverse triangle inequality For *vectors (or *real or *complex numbers) $\mathbf{x}, \mathbf{y}$, the inequality $|\mathbf{x} - \mathbf{y}| \leq |\,|\mathbf{x}| - |\mathbf{y}|\,|$. This is an easy consequence of the *triangle inequality. The inequality also holds in *metric spaces, as $|d(b,a) - d(a,c)| \leq d(b,c)$, where d denotes the metric.

reversion to the mean (regression to the mean) In a sequence of *independent *observations of a *random variable, the greater the deviation from its *mean of an observation, the greater the probability that the next observation will be closer to the mean. This appears to be counter-intuitive to the principle of independence, but it is actually a direct consequence of it. So, as you move away from the mean you necessarily increase the proportion of the distribution lying closer to the mean.

revolution One complete turn of a *full angle or a *cycle of a periodic motion (*see* PERIOD).

revolve To *rotate about an axis or point.

Reynolds number For a *viscous flow with speed u, viscosity v passing an object of *length a, the ratio $R = ua/v$ is a dimensionless (*see* DIMENSIONS) quantity called the Reynolds number. For low Reynolds numbers, uniform or laminar flows tend to occur, whereas for high Reynolds numbers turbulent flow is common.

Reynolds transport theorem For a region of *fluid $V(t)$ moving in a flow velocity $\mathbf{u}$ and a *differentiable function $G(\mathbf{x},t)$ defined on $V(t)$,

$$\frac{d}{dt} \int_{V(t)} G \, dV = \int_{V(t)} \left(\frac{DG}{Dt} + G \nabla \cdot \mathbf{u} \right) dV$$

where D/Dt denotes the *convective derivative. This can be viewed as a *generalization of *Leibniz's integral rule.

rhombohedron A *polyhedron with six faces, each of which is a *rhombus. It is thus a *parallelepiped whose edges are all the same length.

rhombus (rhombi) A *quadrilateral all of whose sides have the same length. A rhombus is both a *kite and a *parallelogram.

RHS Short for right-hand side, especially in reference to an *equation, *identity, or *congruence. As opposed to LHS, the left-hand side.

RI Abbreviation for ROYAL INSTITUTION.

Riemann, (Georg Friedrich) Bernhard (1826–66) German mathematician who was a major figure in 19th-century mathematics and arguably the first modern mathematician because of his general, abstract approach to many problems. In *geometry, he started the development of those tools which *Einstein would eventually use to describe the universe and which in the 20th century would be turned into the theory of *Riemannian manifolds. He did much significant work in *analysis, in which his name is preserved in the *Riemann integral, the *Cauchy-Riemann equations, and *Riemann surfaces. He also made connections between prime number theory and analysis: he formulated the *Riemann hypothesis, a conjecture concerning the *zeta function, which, if proved, would give information about the distribution of prime numbers.

Riemann hypothesis (Riemann zeta hypothesis) The hypothesis that the non-trivial zeros of the *zeta function all have real part equal to ½. It was both on *Hilbert's list of 23 problems in 1900 and on the list of *Millennium Prizes in 2000.

Riemannian manifold A *manifold with a *metric structure, that is, a *first fundamental form, is associated with each set of local coordinates. This means *intrinsic properties such as *length, *angle, *area, and *Gaussian curvature may be defined. In two dimensions, the first fundamental form is commonly written as

$$ds^2 = E du^2 + 2F du dv + G dv^2$$

in terms of local coordinates u and v. The length L of a curve $(u(t), v(t))$ and the area A of a region R are then defined as

$$L = \int \sqrt{E\left(\frac{du}{dt}\right)^2 + 2F\frac{du}{dt}\frac{dv}{dt} + G\left(\frac{dv}{dt}\right)^2} \, dt,$$

$$A = \iint_R \sqrt{EG - F^2} \, du dv.$$

If the manifold uses more than one set of coordinates, then *transition maps are necessarily *isometries so that the metric structure is consistent across the manifold.

Riemann integral (Riemann sum) *See* INTEGRAL.

Riemann-Lebesgue lemma Let f: $\mathbb{R} \to \mathbb{R}$ be Lebesgue integrable (*see* LEBESGUE INTEGRATION). Then the integrals

$$\int_{-\infty}^{\infty} f(x)\cos nx\,dx \text{ and } \int_{-\infty}^{\infty} f(x)\sin nx\,dx$$

exist and converge to 0 and $n \to \infty$. *See* FOURIER COEFFICIENTS.

Riemann mapping theorem Every *non-empty, *proper, *simply connected, open subset (*see* OPEN SET) of $\mathbb{C}$ is *conformally equivalent to the open unit disc.

Riemann–Roch Theorem The theorem connects the *complex analysis of a *compact *Riemann surface with the surface's *topology, that is, with its *genus. The theorem is named after *Riemann and his student Gustav Roch (1839–66). There have since been important generalizations of the theorem by *Atiyah and Singer and by *Grothendieck.

As a *holomorphic function on a compact Riemann surface is constant, it is the *meromorphic functions that are of interest. If it is specified at certain points that a meromorphic function should have *poles no worse than some given order, then the theorem gives information about the number of such meromorphic functions in terms of the specifications and the genus of the surface.

Riemann sphere The representation of the *extended complex plane by *stereographic projection. *See also* UNIFORMIZATION THEOREM.

Riemann surface A *connected *surface which is a 1-dimensional complex *manifold, such as the *Riemann sphere or the *complex plane. Riemann surfaces can be thought of as deformations of the complex plane in the sense that the local topology can be that of the complex plane but the global topology may be quite different. Riemann surfaces are necessarily *orientable. They arise naturally in the study of algebraic curves in the complex projective plane (*see* DEGREE-GENUS FORMULA, PROJECTIVE SPACE); and closed, orientable surfaces can be given the structure of a Riemann surface (*see* UNIFORMIZATION THEOREM).

Riemann zeta function A synonym for ZETA FUNCTION.

r

Riesz representation theorem
Given a finite-dimensional *inner product space V and a linear *functional $\varphi \, \varepsilon \, V'$, there exists $w \, \varepsilon \, V$ such that $\varphi(v) = \langle v, w \rangle$ for all $v \, \varepsilon \, V$. Further the *norm $\|\varphi\|$ of φ equals the norm of w. The theorem extends to infinite-dimensional *Hilbert spaces and their (analytic) *dual spaces.

right angle A quarter of a complete *revolution. It is thus equal to 90 *degrees or $\pi/2$ *radians.

right-angled triangle A *triangle containing a right angle. *See* PYTHAGORAS' THEOREM.

right-circular *See* CONE, CYLINDER.

right derivative *See* LEFT AND RIGHT DERIVATIVE.

right-handed system Let Ox, Oy, and Oz be three mutually perpendicular directed lines, intersecting at the point O. In the order Ox, Oy, Oz, they form a right-handed system if a person standing erect in the positive z-direction and facing the positive y-direction would have the positive x-direction to their right. Typically, right-handed systems are used in calculations and diagrams.

The three directed lines Ox, Oy, and Oz (in that order) form a right-handed system, but if taken in the order Oy, Ox, Oz, they form a left-handed system. If the direction of any one of three lines of a right-handed system is reversed, the three directed lines form a left-handed system.

Similarly, an ordered set of three oblique directed lines may be described as forming a right- or left-handed system. Three *vectors, in a given order, form a right- or left-handed system if directed line segments representing them define directed lines that do so. For *independent vectors **u** and **v** the *vector product is defined so that **u**,**v** and **u** × **v** form a right-handed system.

right inverse A right inverse $g : Y \rightarrow X$ of a map $f : X \rightarrow Y$ satisfies $f(g(y)) = y$ for all y in Y. The existence of a right inverse is equivalent to f being *onto, assuming the *axiom of choice. *See* INVERSE FUNCTION, LEFT INVERSE.

right-regular *See* PRISM, PYRAMID.

rigid body An object with the property that it does not change shape whatever *forces are applied to it. It is used in a *mathematical model to represent an object in the real world. It may be a system of *particles held in a rigid formation, or it may be a distribution of mass in the form of a *rod, a *lamina, or some 3-dimensional shape. In general, a rigid body has six *degrees of freedom—three to determine the body's *centre of mass and three more to specify the body's orientation about its centre of mass. The equations of motion for a rigid body are

$$m\frac{\mathrm{d}^2\mathbf{r}_G}{\mathrm{d}t^2} = \sum \mathbf{F}_k, \quad \frac{\mathrm{d}\mathbf{L}_G}{\mathrm{d}t} = \sum \mathbf{r}_k \times \mathbf{F}_k,$$

where m denotes the *mass of the body, $\mathbf{r}_G$ is the position vector of the centre of mass, $\mathbf{L}_G$ is the *angular momentum about the centre of mass, $\mathbf{F}_k$ are the external forces acting on the body, and $\mathbf{r}_k \times \mathbf{F}_k$ is the *moment of the force $\mathbf{F}_k$ (so that $\mathbf{r}_k$ is the position vector of a point on the line of action of the force $\mathbf{F}_k$). *See also* INERTIA TENSOR.

rigid motion A synonym for ISOMETRY.

ring Sets with two *binary operations, often called addition and multiplication, occur commonly in mathematics and sometimes share many of the same properties. One such set of properties is specified in the definition of a ring: a ring is a set R, closed under two operations called addition and multiplication, such that

(i) for all a, b, and c in R,
$a + (b + c) = (a + b) + c$,
(ii) for all a and b in R, $a + b = b + a$,

(iii) there is an element 0 in R such that $a+0 = a$ for all a in R,

(iv) for each a in R, there is an element $-a$ in R such that $a+(-a) = 0$,

(v) for all a, b, and c in R, $a(bc) = (ab)c$,

(vi) for all a, b, and c in R, $a(b+c) = ab + ac$ and $(a+b)c = ac + bc$.

The element 0 guaranteed by (iii) is an additive *identity. It can be shown to be unique and has the extra property that $a0 = 0$ for all a in R; it is called *zero. Also, for each a, the element $-a$ guaranteed by (iv) is unique and is the *negative or *additive inverse of a. The ring is a commutative ring if it is further true that

(vii) for all a and b in R, $ab = ba$,

and it is a commutative ring with identity if also

(viii) there is an element 1 ($\neq 0$) such that $a1 = a$ for all a in R.

The element 1 guaranteed by (viii) is a multiplicative identity. It can be shown to be unique and is referred to as 'one'.

Further properties may be required for other types of ring such as *integral domains and *fields. Examples of rings include the set of 2×2 real matrices and the set of all even integers, each with the appropriate addition and multiplication. Another example is $\mathbb{Z}_n$, the set of integers with addition and multiplication modulo n.

A ring may be denoted by $(R, +, \times)$ when it is necessary to be clear about the ring's operations. But it is sufficient to refer simply to the ring R when the operations intended are clear.

ring of integers *See* ALGEBRAIC INTEGER.

rise The difference between the *ordinates (y values) of a pair of points in a 2-dimensional coordinate system. Used with the *run in calculating the *gradient of the line joining the points.

risk The proportion or rate of incidence of a particular outcome in a group. So the risk of a disease can be expressed as a percentage which equates to the statistical probability that an individual in that category catches the disease.

Robin boundary condition (mixed boundary condition) A combination of the function and the *flux at the boundary. For example, a function φ might satisfy *Laplace's equation inside a region and the Robin condition $c\varphi + \partial\varphi/\partial n = f$ on the boundary where c is a constant and f a specified function; here $\partial\varphi/\partial n$ denotes the *directional derivative of φ in the direction of the outward pointing normal. The existence and uniqueness of any solution depends on the sign of c. *Compare* DIRICHLET PROBLEM, NEUMANN CONDITION.

Robinson, Julia (Hall Bowman) (1919–85) American mathematician who made contributions to *number theory, *computability, and *complexity theory, most notably in work on *Hilbert's tenth problem. In 1983 she became the first female president of the American Mathematical Society.

robust A test or *estimator is robust if it is not sensitive to small discrepancies in assumptions, such as the assumption of the normality of the underlying distribution. A test is said to be robust to *outliers if it is not unduly affected by their presence.

rod An object considered as being 1-dimensional, having length and density but having no width or thickness. It is used in a *mathematical model to represent a thin straight object in the real world. It may be rigid or flexible, *uniform or not, depending on the circumstances.

Rodrigues' formula *See* LEGENDRE POLYNOMIALS.

Rolle, Michel (1652–1719) French mathematician, remembered primarily for the theorem that bears his name, which appears in a book of his published in 1691.

Rolle's Theorem Let f be a *continuous function on $[a,b]$ and *differentiable on (a,b), such that $f(a) = f(b)$. Then there exists c with $a < c < b$ such that $f'(c) = 0$.

The theorem is a special case of the *mean value theorem. A rigorous proof relies on a continuous function on a closed, bounded interval attaining its bounds (*Weierstrass' Theorem).

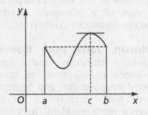

Two stationary points in (a,b)

rolling condition The relationship between the linear and angular *speeds of a *cylinder or *sphere when it is rolling on a plane surface without slipping. If v is the speed of the centre of mass of the cylinder or sphere, with radius r, and ω is the angular speed of the rotation, then $v = r\omega$.

Roman numeral See NUMERAL; see also APPENDIX 21.

root (of an equation) Let $f(x) = 0$ be an equation involving the *variable x. A root of the equation is a value h such that $f(h) = 0$. Such a value is also called a zero of the function f.

For a suitably *differentiable function f, then h is said to be a root of order (or multiplicity) n if

$$0 = f(h) = f'(h) = f''(h) = \dots = f^{(n-1)}(h)$$
$$\text{and} \quad f^{(n)}(h) \neq 0.$$

If $f(x)$ is a polynomial, this is equivalent to $(x - h)^n$ dividing $f(x)$ but no greater power. A root of order greater than 1 is called a multiple root or repeated root.

See also ORDER OF CONTACT, VIÈTE'S FORMULAE.

root (of a tree) *See* TREE.

root mean squared deviation The positive square root of the *mean squared deviation.

root of unity *See* NTH ROOT OF UNITY.

root test (Cauchy's test) For a *series $a_1 + a_2 + a_3 + \dots$ of positive terms:

if limsup $a_n^{1/n} < 1$, then the series *converges.

if limsup $a_n^{1/n} > 1$, then the series *diverges.

if limsup $a_n^{1/n} = 1$, then the series may or may not converge.

See also RADIUS OF CONVERGENCE.

rose A *curve consisting of loops meeting at the origin, resembling the petals of a flower. In *polar coordinates, the equation is $r = a\cos(n\theta)$, where $a > 0$. If n is even, then there will be $2n$ loops, while if n is odd, there are only n loops.

rotation A rotation of the plane about the origin O through an *angle α is the *transformation of the plane in which O is mapped to itself, and a point P with *polar coordinates (r, θ) is mapped to the point P' with polar coordinates $(r, \theta + \alpha)$. In terms of *Cartesian coordinates, P with coordinates (x, y) is mapped to P' with coordinates (x', y'), where

$$x' = x \cos \alpha - y \sin \alpha,$$
$$y' = x \sin \alpha + y \cos \alpha.$$

This change of coordinates can be represented by the *matrix equation

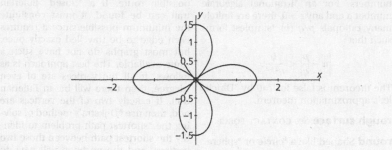

The rose $r = 1.5\cos(2\theta)$

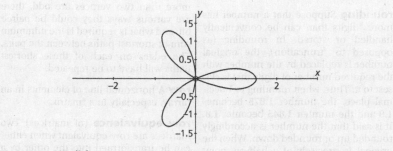

The rose $r = 1.5\cos(3\theta)$

$$\begin{pmatrix} x' \\ y' \end{pmatrix} = \begin{pmatrix} \cos\alpha & -\sin\alpha \\ \sin\alpha & \cos\alpha \end{pmatrix} \begin{pmatrix} x \\ y \end{pmatrix}.$$

Importantly, *distances and *angles are still calculated the same using x' and y' as when using x and y. Note that the matrix is *orthogonal and has *determinant 1. In 3-dimensional space, an orthogonal matrix with determinant 1

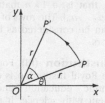

Rotation about O by α

represents a rotation about an axis through the origin.

rotational kinetic energy *See* KINETIC ENERGY.

rotational symmetry A plane figure has rotational symmetry about a point O if *rotation about O through some angle $0° < \theta < 360°$ is a *symmetry of the figure. For example, an *equilateral triangle has rotational symmetry about its centre. The order of rotational symmetry about O is the number of such rotations in the range $0° \leq \theta < 360°$; so, the equilateral triangle has order 3.

rotation of axes *See* ROTATION.

Roth's Theorem A result in *number theory relating to the approximation of *algebraic numbers by *rational numbers. For an *irrational algebraic number α and any ε > 0, there are finitely many rationals p/q (in *simplest form) such that

$$\left| \alpha - \frac{p}{q} \right| < \frac{1}{q^{2+\varepsilon}}.$$

The theorem is false for ε=0 by *Dirichlet's approximation theorem.

rough surface *See* CONTACT FORCE.

round Shaped like a *circle or *sphere.

round angle A synonym for FULL ANGLE.

rounding Suppose that a number has more *digits than can be conveniently handled or stored. In rounding (as opposed to *truncation), the original number is replaced by the number with the required number of digits that is closest to it. Thus, when rounding to 1 decimal place, the number 1.875 becomes 1.9 and the number 1.845 becomes 1.8. It is said that the number is accordingly rounded up or rounded down. When the original is precisely at a halfway point (for example, if 1.85 is to be rounded to 1 decimal place), it is typically rounded up (to 1.9) rather than rounded down (to 1.8). *See also* DECIMAL PLACES, SIGNIFICANT FIGURES.

round-off error (rounding error) When a number X is *rounded to a certain number of digits to obtain an approximation x, the *error is called the round-off error. For some authors this is $X-x$, and for others it is $x-X$. When a number is rounded to k *decimal places, the round-off error lies between $\pm 5 \times 10^{-(k+1)}$. For some authors, the error is $|X-x|$ (*see* ABSOLUTE VALUE) and so, for them, it is always greater than or equal to zero.

route inspection problem (Chinese postman problem) A postman needs to travel along each road, or edge in a *graph, while travelling the shortest

possible route. If a *closed *Eulerian trail can be found, it must constitute the minimum possible since it requires each edge to be travelled exactly once, but most graphs do not have such a route available. The best approach is as follows. If all the vertices are of even *degree, then there will be an Eulerian trail. If exactly two of the vertices are odd, then use *Dijkstra's method of solving the *shortest path problem to identify the shortest path between those two vertices, and the postman will have to walk the edges on that route twice and all other edges once. However, when more than two vertices are odd, there are various ways they could be paired off, and what is required is the minimum sum of shortest paths between the pairs. The edges on each of these shortest paths will have to be repeated.

row A horizontal line of elements in an *array, especially in a *matrix.

row equivalence (of matrices) Two matrices are row-equivalent when either can be transformed into the other by a finite set of *elementary row operations.

row operation *See* ELEMENTARY ROW OPERATION.

row rank *See* RANK.

row space The row space of a *matrix is the *span of its rows. Its *dimension is the *rank of the matrix, and the non-zero rows of the matrix in *reduced echelon form a *basis for it.

row vector A *matrix with exactly one row, that is, a $1 \times n$ matrix of the form $[a_1\, a_2 \ldots a_n]$. The rows of a general matrix can be considered as individual row vectors.

Royal Institution (RI) Founded in 1799, the Royal Institution is dedicated to scientific education and public engagement with science. It is famous for the RI Christmas lectures, which were begun by Michael Faraday in 1825. Each year, the RI runs series of master-classes and summer schools on a range of scientific subjects for schoolchildren.

RRE form An abbreviation for REDUCED ROW ECHELON FORM.

RSA (public-key cryptography) Named after the mathematicians Ron Rivest, Adi Shamir, and Leonard Aldeman, a *cryptographic *algorithm which depends on the difficulty of *factorizing large *semiprimes and a commonly used form of encryption for Internet security.

The *public key consists of natural numbers n and e, where n is a product of two distinct large *primes p and q. As of 2020, most RSA keys have $2^{1024} < n < 2^{4096}$. The number e needs to be *coprime with $k = (p-1)(q-1)$. Then a message M in the range $0 \leq M < n$ is encrypted as $C = M^e$ (mod n). When received, the message is recovered by $M = C^d$ (mod n), where d is the *multiplicative inverse of e mod k. Knowledge of d lies only with the receiver; finding d is equivalent to factorizing n, which should not be practically possible, provided large enough primes are used.

ruled surface A *surface that can be traced out by a moving straight line; in other words, every point of the surface lies on a straight line lying wholly in the surface. Examples are a (double) *cone, a *cylinder, a *hyperboloid of one sheet, a *hyperbolic paraboloid, and a *helicoid.

ruler and compass construction *See* CONSTRUCTION WITH RULER AND COMPASS.

run (in a coordinate system) The difference between the *abscissae (x-values) of a pair of points in a 2-dimensional coordinate system. Used with the *rise in calculating the *gradient of the line joining the points.

run (in statistics) A *sequence of consecutive *observations from a *sample that share a specified property. Often the observations have either one or other of two properties A and B. For example, A may be 'above the *median' and B 'below the median'. If the sequence of observations then gives the sequence $AABAAABBAB$, for example, the number of runs equals 6. This statistic is used in some *non-parametric methods.

Runge–Kutta methods A numerical method of solving *differential equations in the form $\frac{dy}{dx} = f(x,y)$ which uses the midpoint of interval(s) to improve accuracy. So if the value of the function is known at (x_n, y_n), and the estimate y_{n+1} of y is required at $x_{n+1} = x_n + h$, the second-order Runge-Kutta formula is

$$k_1 = h \times f(x_n, y_n)$$
$$k_2 = h \times f\left(x_n + \frac{1}{2}h, y_n + \frac{1}{2}k_1\right)$$
$$y_{n+1} = y_n + k_2$$

and the fourth-order Runge-Kutta formula is

$$k_1 = h \times f(x_n, y_n)$$
$$k_2 = h \times f\left(x_n + \frac{1}{2}h, y_n + \frac{1}{2}k_1\right)$$
$$k_3 = h \times f\left(x_n + \frac{1}{2}h, y_n + \frac{1}{2}k_2\right)$$
$$k_4 = h \times f(x_n + h, y_n + k_3)$$
$$y_{n+1} = y_n + \frac{1}{6}k_1 + \frac{1}{3}k_2 + \frac{1}{3}k_3 + \frac{1}{6}k_4$$

Russell, Bertrand Arthur William (1872-1970) British philosopher, logician, and writer on many subjects. He is remembered in mathematics as the author, with A. N. *Whitehead, of *Principia mathematica*, published between 1910 and 1913 in three volumes, which set out to show that pure mathematics could all be derived from certain fundamental logical axioms. Although the attempt was not completely successful, the work was highly influential. He was also responsible for the discovery of *Russell's paradox.

Russell's paradox By using the notation of set theory, a *set can be defined as the set of all x that satisfy some property. Now it is clearly possible for a set not to belong to itself: any set of numbers, say, does not belong to itself because to belong to itself the set would have to be a number. But it is also possible to have a set that does belong to itself: for example, the set of all sets belongs to itself. In 1901, Bertrand Russell drew attention to the following paradox, by considering the set $R = \{x \mid x \notin x\}$. If $R \in R$, then R fails the condition for being an element of R and so $R \notin R$; and if $R \notin R$, then $R \in R$ by definition. The paradox points out the necessity of defining mathematical objects carefully (*compare* PERRON'S PARADOX). For example, such a set R cannot be defined using the *Zermelo-Frankel axioms, so no paradox arises; likewise the 'set of all sets' does not exist within ZF theory.

Rutherford, Lord (1871–1937) Born in New Zealand to a Scottish father and English mother, Ernest Rutherford went to Cambridge University in 1894 following an undergraduate degree in mathematics and physics from Canterbury College. He then spent 10 years at McGill University in Montreal before returning to England, eventually as Cavendish Professor of Physics at Cambridge. He was awarded the 1908 Nobel Prize in Chemistry 'for his investigations into the disintegration of the elements, and the chemistry of radioactive substances', and he named many of the basic components of atomic physics such as alpha, beta and gamma rays, the proton, the neutron, and half-life. He was the first to realize that almost all of the mass of an atom, and all its positively charged components, were concentrated in a tiny proportion of the atom's size which came to be known as the 'nucleus'.

rv Abbreviation for *random variable.

S_n The *symmetric group of the set $\{1, 2, \ldots, n\}$.

S^n The n-dimensional *sphere, usually identified with those *vectors in $\mathbb{R}^{n+1}$ of length 1. Note S^1, thought of as the unit circle in $\mathbb{C}$, is a subgroup of the *multiplicative group of $\mathbb{C}$.

saddle-point Suppose that a *surface has equation $z = f(x, y)$. A point P on the surface is a saddle-point if the *tangent plane at P is horizontal and if P is a local *minimum on the curve obtained by one vertical cross-section and a local *maximum on the curve obtained by another vertical cross-section. It is so called because the central point on the seat of a horse's saddle has this property. The *hyperbolic paraboloid, for example, has a saddle-point at the origin. *See also* HESSIAN, STATIONARY POINT (IN TWO VARIABLES).

sample A subset of a *population selected in order to make *inferences about the population. (Strictly speaking, it is not a subset because elements may be repeated.) It is a random sample if it is chosen in such a way that every sample of the same size has an equal chance of being selected. If the sample is chosen in such a way that no member of the population can be selected more than once, this is sampling without replacement. In sampling with replacement, an element has a chance of being selected more than once. A quota sample is a sample in which a predetermined number of elements have to be selected from each category in a specified list. *See also* STRATIFIED SAMPLE.

sample space The set of all possible outcomes of an *experiment. For example, suppose that an experiment involves tossing a coin three times. Then the sample space could be expressed as

$$\{HHH, HHT, HTH, HTT, THH,$$
$$THT, TTH, TTT\}$$

where, for example, 'HTT' indicates the tosses came up 'heads, tails, tails'. *See* PROBABILITY SPACE, RANDOM VARIABLE.

sampling distribution Every *statistic is a *random variable. The *distribution of this random variable is a sampling distribution.

sandwich theorem Let a_n, b_n, c_n be real *sequences such that if $a_n \leq b_n \leq c_n$ for all n; the theorem states that if a_n and c_n *converge to some *limit L, then so does b_n. A similar result holds for the limits of *functions. Variously also called the sandwich rule or squeeze theorem.

SAS *See* SOLUTION OF TRIANGLES.

saturated (in networks) An edge where the flow is at its *capacity.

scalar Referring to a quantity that has a magnitude or numerical value, in contrast to a *vector, which also has *direction. Thus, *mass, *speed, and *temperature are scalar quantities, unlike *velocity and *force, which are vector quantities.

scalar (for vectors and matrices) The *coordinates in a *vector or *entries in a *matrix are elements of some set of 'scalars' which are usually a *field (such as the *real numbers) or at least a *ring (such as the *integers). *See* SCALAR MULTIPLICATION.

scalar field A scalar field is a real-valued *function defined on $\mathbb{R}^3$ or some subset of $\mathbb{R}^3$, though the notion can easily be extended to other dimensions. A field will usually be assumed to have continuous derivatives (*see* CONTINUOUS FUNCTION). Scalar fields are common in physical applied mathematics (examples include *temperature, *pressure, *density, *gravitational potential) but may also be studied in their own right in *multivariable calculus. *Compare* VECTOR FIELD.

scalar matrix A *diagonal matrix where all the entries in the main diagonal are equal, to k say, and all other entries are zero. Multiplication by this matrix is equivalent to *scalar multiplication by k.

scalar multiplication One of the two operations in a *vector space V, together with addition. Given any vector v in V and *scalar c (from the *base field), the scalar multiple $cv \, \varepsilon \, V$ may be formed. It may be that $v = \mathbf{v}$ is a *coordinate vector, in which case the coordinates of $c\mathbf{v}$ are those of $\mathbf{v}$ multiplied by c. If $v = \mathbf{A}$ is a *matrix, then the entries of $c\mathbf{A}$ are the entries of $\mathbf{A}$ multiplied by c. Multiplication by scalars has the following properties:

(i) $(h+k)v = hv + kv$.
(ii) $k(v+w) = kv + kw$.
(iii) $h(kv) = (hk)v$.
(iv) $1v = v$.

Geometrically, if v is in $\mathbb{R}^n$, then the scalar multiples cv comprise the line through v and the origin $\mathbf{0}$; when $c > 0$, the scalar mutliples comprise the half-line from $\mathbf{0}$ through v.

scalar product (dot product) The Euclidean *inner product. For *vectors $\mathbf{u}$ and $\mathbf{v}$ in $\mathbb{R}^3$, the scalar product is defined by

$$\mathbf{u} \cdot \mathbf{v} = u_1v_1 + u_2v_2 + u_3v_3,$$

and this generalizes naturally to higher dimensions.

The scalar product has the following properties:

(i) $\mathbf{u} \cdot \mathbf{v} = \mathbf{v} \cdot \mathbf{u}$.
(ii) For non-zero vectors $\mathbf{u}$ and $\mathbf{v}$, $\mathbf{u} \cdot \mathbf{v} = 0$ if and only if $\mathbf{u}$ is perpendicular to $\mathbf{v}$.
(iii) $\mathbf{u} \cdot \mathbf{u} = |\mathbf{u}|^2$; the scalar product $\mathbf{u} \cdot \mathbf{u}$ may be written $\mathbf{u}^2$.
(iv) $\mathbf{u} \cdot (\mathbf{v} + \mathbf{w}) = \mathbf{u} \cdot \mathbf{v} + \mathbf{u} \cdot \mathbf{w}$, the distributive law.
(v) $\mathbf{u} \cdot (k\mathbf{v}) = k(\mathbf{u} \cdot \mathbf{v})$.

The *length $|\mathbf{u}|$ of a vector can be defined in terms of the scalar product from (iii). Further, the *angle θ between $\mathbf{u}$ and $\mathbf{v}$ can be defined by

$$\mathbf{u} \cdot \mathbf{v} = |\mathbf{u}||\mathbf{v}|\cos\theta.$$

This uniquely defines an angle θ in the range $0 \leq \theta \leq \pi$ by the *Cauchy-Schwarz inequality.

scalar quadratic product For vectors $\mathbf{a}$, $\mathbf{b}$, $\mathbf{c}$, and $\mathbf{d}$, the identity

$$(\mathbf{a} \times \mathbf{b}) \cdot (\mathbf{c} \times \mathbf{d}) = (\mathbf{a} \cdot \mathbf{c})(\mathbf{b} \cdot \mathbf{d}) - (\mathbf{a} \cdot \mathbf{d})(\mathbf{b} \cdot \mathbf{c})$$

where $\cdot$ denotes the *scalar product and $\times$ denotes the *vector product.

scalar triple product The scalar triple product of three vectors $\mathbf{a}$, $\mathbf{b}$, and $\mathbf{c}$, is $\mathbf{a} \cdot (\mathbf{b} \times \mathbf{c})$, the *scalar product of $\mathbf{a}$ with the *vector product $\mathbf{b} \times \mathbf{c}$. It is a scalar quantity and is denoted $[\mathbf{a}, \mathbf{b}, \mathbf{c}]$. It has the following properties:

(i) $[\mathbf{a}, \mathbf{b}, \mathbf{c}] = -[\mathbf{a}, \mathbf{c}, \mathbf{b}]$.
(ii) $[\mathbf{a}, \mathbf{b}, \mathbf{c}] = [\mathbf{b}, \mathbf{c}, \mathbf{a}] = [\mathbf{c}, \mathbf{a}, \mathbf{b}]$.
(iii) The vectors $\mathbf{a}$, $\mathbf{b}$, and $\mathbf{c}$ are *linearly dependent if and only if $[\mathbf{a}, \mathbf{b}, \mathbf{c}] = 0$.
(iv) If $\mathbf{a} = (a_1, a_2, a_3)$, etc., then

$$[\mathbf{a}, \mathbf{b}, \mathbf{c}] = a_1(b_2c_3 - b_3c_2)$$
$$+ a_2(b_3c_1 - b_1c_3) + a_3(b_1c_2 - b_2c_1)$$

$$= \begin{bmatrix} a_1 & a_2 & a_3 \\ b_1 & b_2 & b_3 \\ c_1 & c_2 & c_3 \end{bmatrix}.$$

(v) Let $\overrightarrow{OA}$, $\overrightarrow{OB}$, and $\overrightarrow{OC}$ represent $\mathbf{a}$, $\mathbf{b}$ and $\mathbf{c}$. Then the *parallelepiped with OA, OB, and OC as three of its edges has volume equal to the *absolute value of $[\mathbf{a,b,c}]$.

scale A marking of values along a line to use in making measurements or in representing measurements.

scalene triangle A *triangle in which all three sides have different lengths.

scatter diagram (scatter plot) A 2-dimensional diagram showing the points corresponding to n *paired-sample *observations $(x_1,y_1),(x_2,y_2),\ldots,(x_n,y_n)$, where $x_1, x_2, \ldots, x_n$ are the observed values of the *explanatory variable and $y_1, y_2, \ldots, y_n$ are the observed values of the *dependent variable.

Schauder basis An infinite sequence of vectors $\{v_n\}$ in a *Banach space such that every $v \, \varepsilon V$ can be written uniquely as $v = \sum_1^\infty a_n v_n$ for scalars a_n. Every Banach space with a Schauder basis is necessarily *separable, but the converse does not hold. *Compare* BAIRE CATEGORY THEOREM, COMPLETE ORTHONORMAL BASIS, HAMEL BASIS.

scheduling In its simplest form, *critical path analysis assumes each activity requires one worker, but in reality this is often not true. It can be further complicated if certain activities can only be carried out by certain workers, as is the case in building projects involving skilled tradespeople. Scheduling is the task of assigning workers to the activities with the aim of completing the project with as few workers as possible, or if the number

of workers is limited, completing it in the shortest possible time with the available workers.

Schläfli, Ludwig (1814–95) Swiss mathematician mainly remembered now for the classification of regular *polytopes—higher dimensional *Platonic solids, of which there are six in four dimensions and three in higher ones— and generalizing *Euler's theorem, though in his lifetime his work went unrecognized. *See* SCHLÄFLI SYMBOL.

Schläfli symbol A notation Ludwig *Schläfli introduced to help classify the regular *polytopes. The Schläfli symbols for the five *Platonic solids are $\{3,3\}$ tetrahedron, $\{4,3\}$ cube, $\{3,4\}$ octahedron, $\{5,3\}$ dodecahedron, and $\{3,5\}$ icosahedron. The symbol $\{p,q\}$ represents that *faces are p-gons, q of which meet at each *vertex.

In four dimensions, the symbol $\{p,q,r\}$ represents that there are r 3-dimensional faces about each *edge that are polyhedra with symbol $\{p,q\}$. The 4-dimensional regular polytopes are $\{3,3,3\}$ pentatope, $\{3,3,4\}$ orthoplex, $\{4,3,3\}$ tesseract, $\{3,4,3\}$ octaplex, $\{5,3,3\}$ dodecaplex, and $\{3,3,5\}$ tetraplex. Here the tesseract and pentatope are the 4-dimensional *hypercube and *simplex respectively; in n dimensions, the hypercube and simplex have symbols $\{4,3,3,\ldots,3\}$ and $\{3,3,\ldots,3\}$.

Schrödinger, Erwin Rudolf Alexander (1887–1961) Austrian mathematician and theoretical physicist who developed *Heisenberg's initial work in *quantum theory specifically relating to wave mechanics and the general theory of relativity. He shared the Nobel Prize for Physics in 1933 with Paul *Dirac for this work.

Schrödinger's cat A thought experiment devised by Schrödinger to highlight implications of the *Copenhagen interpretation of *quantum theory.

S

According to that interpretation, a *wave function describes the superposition of states a particle is in until the particle is measured, which collapses the wave function. The experiment involves a cat in a box whose life or death is connected to the radioactive decay of a single atom. By the Copenhagen interpretation, the cat is simultaneously alive and dead until the box is opened, which Schrödinger considered absurd.

Schrödinger's equation Schrödinger's equation is a *partial differential equation which describes the state of a particle according to *quantum theory. The particle's state is described by a complex-valued *wave function $\psi(\mathbf{x}, t)$, where $\mathbf{x}$ denotes position and t denotes time. Schrödinger's time-dependent equation states that

$$i\hbar \frac{\partial \psi}{\partial t} = -\frac{\hbar^2}{2m} \nabla^2 \psi + V\psi$$

where $\hbar$ denotes the reduced *Planck's constant, m is the *mass of the particle, V denotes *potential energy, and ∇^2 denotes the *Laplacian. The wave function ψ is a superposition of states $\psi_n(\mathbf{x}, t)$ which satisfy the time-independent Schrödinger equation

$$-\frac{\hbar^2}{2m} \nabla^2 \psi_n + V\psi_n = E_n \psi_n$$

where E_n denotes *energy of the nth state. More concisely, this may be written $H\psi_n = E_n \psi_n$ so that ψ_n is an *eigenstate of the *Hamiltonian H. *See* COPENHAGEN INTERPRETATION, MEASUREMENT, OBSERVABLE.

Schur decomposition Any complex *square matrix $\mathbf{A}$ can be written $\mathbf{A} = \mathbf{Q}^{-1}\mathbf{U}\mathbf{Q}$, where $\mathbf{Q}$ is *unitary and $\mathbf{U}$ is upper *triangular. *See also* TRIANGULARIZABLE.

Schwartz distribution *See* DISTRIBUTION (SCHWARTZ DISTRIBUTION)

scientific notation A number is said to be in scientific notation (or standard form) when it is written as $a \times 10^n$, where $1 \le a < 10$ and n is an integer. Thus 634.8 and 0.00234 are written in scientific notation as 6.348×10^2 and 2.34×10^{-3}. The notation is particularly useful for very large and very small numbers. *See also* ORDER OF MAGNITUDE.

sd Abbreviation for *standard deviation.

SDE Abbreviation for *stochastic differential equation.

se Abbreviation for *standard error.

seasonal variation *See* TIME SERIES.

secant *See* TRIGONOMETRIC FUNCTION.

secant (secant line) (of a curve) A *line that cuts a given *curve, in two or more points.

secant method Given two successive approximations to a function's *root a, the secant method calculates where the secant through these two points meets the x-axis and uses this as the next approximation. So the *iteration is given by

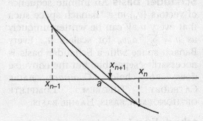

The secant method

$$x_{n+1} = x_n - \frac{(x_n - x_{n-1})}{f(x_n) - f(x_{n-1})} \times f(x_n)$$ for

$$n = 1, 2, 3, \ldots.$$

sech *See* HYPERBOLIC FUNCTION.

second (angular measure) *See* DEGREE (angular measure).

second (time) In *SI units, the base unit used for measuring time, abbreviated to 's'. It was once defined as 1/86 400 of the mean solar day, which is the average time it takes for the Earth to rotate relative to the Sun. Now it is defined in terms of the radiation emitted by the caesium-133 atom.

second derivative See HIGHER DERIVATIVE.

second derivative test See DERIVATIVE TEST.

second fundamental form For a *parametrized surface $\mathbf{r}(u,v)$, smooth at a point p, so that $\mathbf{r}_u = \partial\mathbf{r}/\partial u$ and $\mathbf{r}_v = \partial\mathbf{r}/\partial v$ is a *basis for the *tangent space, the second fundamental form is given by

$$\alpha\mathbf{r}_u + \beta\mathbf{r}_v \mapsto L\alpha^2 + 2M\alpha\beta + N\beta^2$$

where $L = \mathbf{r}_{uu}\cdot\mathbf{n}$, $M = \mathbf{r}_{uv}\cdot\mathbf{n}$, $N = \mathbf{r}_{vv}\cdot\mathbf{n}$, and $\mathbf{n}$ denotes the unit *normal. Unlike the *first fundamental form, which relates to the *intrinsic metric properties of the surface, the second fundamental form relates to the *embedding of the surface in $\mathbb{R}^3$. For example, the *cylinder and *plane are locally isometric but have different second fundamental forms. The two fundamental forms must satisfy certain compatibility equations (one of which is given by the *Theorema Egregium). Two surfaces in $\mathbb{R}^3$ with the same fundamental forms are related by an *isometry of $\mathbb{R}^3$.

second-order logic See FIRST-ORDER LOGIC.

second-order partial derivative See HIGHER-ORDER PARTIAL DERIVATIVE.

section The plane figure obtained when a surface or solid is intersected by a plane. If the figure has an *axis of symmetry and the plane is *perpendicular to the axis, then it is a cross section. The sections of a (double) cone are the *conics.

section formulae See EXTERNAL DIVISION, INTERNAL DIVISION.

sector A sector of a *circle, with centre O, is the region bounded by an *arc AB of the circle and the two *radii OA and OB. The area of a sector is equal to $\frac{1}{2}r^2\theta$, where r is the radius and θ is the angle in radians.

segment A segment of a *circle is the smaller region bounded by an *arc AB of the circle and the *chord AB.

selection The number of selections of n objects taken r at a time (that is, the number of ways of selecting r objects out of n) is denoted by ^nC_r and is equal to

$$\frac{n!}{r!(n-r)!}.$$

(See also BINOMIAL COEFFICIENT.) For example, from four objects A, B, C, and D, there are six ways of selecting two: AB, AC, AD, BC, BD, CD. The property that

$$^{n+1}C_r = {}^nC_{r-1} + {}^nC_r$$

can be directly seen in *Pascal's triangle. Compare PERMUTATION.

selection bias *Bias which occurs because of the method of selection of a *sample. For example, if people are interviewed on a Monday morning in town, various groups such as teachers and rural dwellers are likely to be under-represented.

self-adjoint A *linear map $T:V \to V$ on an *inner product space V is self-adjoint if T equals its *adjoint T^*, that is, $\langle Tv,w \rangle = \langle v,Tw \rangle$ for all $v,\ w \in V$. If V is *finite-dimensional, then the *matrix for T, with respect to an *orthonormal basis, is *symmetric (or *Hermitian over $\mathbb{C}$). See SPECTRAL THEOREM.

self-inverse An element of a *group, *ring, etc. which is its own inverse, i.e. an element a for which $a^2 = e$ where e is the *identity element. So the identity is always self-inverse; any *reflection is self-inverse, as is a *rotation through 180°.

self-reference A *statement which refers to itself. This can give rise to *paradoxes. For example 'this statement is false' makes no sense, because if it is true, then it must be false, and if it is false, then it must be true.

self-selected samples (self-selection) Self-selection involves, probably unintentionally, creating a *sample from a survey, say, where the likelihood or ability to respond means that the sample is unlikely to be random and so *biased. For example, where a newspaper, radio, or television station asks those interested to respond, the results are likely only to reflect the most strongly held opinions, and the results are virtually worthless.

self-similarity See FRACTAL.

semi- Prefix denoting half.

semicircle One half of a *circle cut off by a *diameter.

semi-decidable See DECIDABLE.

semi-direct product A *group G is an internal semi-direct product of a *normal subgroup N and *subgroup H if $G = NH$ and $N \cap H = \{e\}$. This is written $G = N \rtimes H$. In this case every $g \in G$ can be uniquely written $g = nh$, where $n \in N$ and $h \in H$. Multiplication in the group is given by

$$(n_1 h_1)(n_2 h_2) = (n_1 h_1 n_2 h_1^{-1})(h_1 h_2).$$

Such an example is the *dihedral group D_{2n}, where N is the subgroup of *rotations and H is a subgroup generated by a *reflection.

For $h \in H$, the map $\varphi_h: n \mapsto hnh^{-1}$ is an *automorphism of N and $\varphi: h \mapsto \varphi_h$ is a *homomorphism from H to Aut(N). More generally, given two groups N and H and a homomorphism $\varphi: H \rightarrow$ Aut(N), the external semi-direct product $N \rtimes_\varphi H$ can be formed by multiplying elements of the *Cartesian product according to rule

$$(n_1, h_1)(n_2, h_2) = (n_1 \varphi(h_1)(n_2), h_1 h_2).$$

The semi-direct product is a means of creating a larger group from two groups other than the *direct product. If the map φ is the constant map to the *identity map of N, then the semi-direct product agrees with the direct product.

semigroup A set S with an *associative *binary operation. Thus, all *groups are semigroups, but semigroups need not have an *identity or *inverses. See MAGMA, MONOID.

semi-interquartile range A measure of *dispersion equal to half the difference between the first and third *quartiles in a set of numerical *data.

semi-metric One of the conditions for a distance measure d to be a *metric is the *triangle inequality. If this condition is relaxed, then d is said to be a semimetric.

semi-norm This is a generalization of the *norm on a *vector space which removes the restriction that $\|x\| = 0$ only if $x = 0$.

semiprime The product of two primes, so $6 = 2 \times 3$ is a semiprime but 12 is not. Very large semiprimes form the basis of much cryptography.

semi-regular polyhedron See ARCHIMEDEAN SOLID.

semi-regular tessellation See TESSELLATION.

semi-vertical angle See CONE.

sense (sense-preserving) One of the two possible *orientations of an *angle, i.e. clockwise or anticlockwise. A map is sense-preserving if it maintains the sense of angles. So, a *rotation is sense-preserving and a *reflection is sense-reversing. A *square matrix **A** represents a sense-preserving *linear map if det**A** > 0 and a sense-reversing map if det**A** < 0. *See also* POSITIVELY ORIENTED, RIGHT-HANDED SYSTEM.

sensitivity analysis The varying of *parameters in a *simulation to discover which parameters have the greatest influence on the features of interest.

separable (of a function) A function which can be written so that the variables are separated additively or multiplicatively, for example $2x^2y + 4y = 2y(x^2 + 2)$ or $x^2 + y^2 - 4x + 6y + 13 = (x-2)^2 + (y+3)^2$.

separable (of a space) A *topological space (or *metric space) is separable if it has a *countable dense subset (*see* DENSE SET). So $\mathbb{R}$ is separable, as $\mathbb{Q}$ is countable and its *closure is $\mathbb{R}$.

separable first-order differential equation A *first-order differential equation $dy/dx = f(x,y)$ in which the function f can be expressed as the product of a function of x and a function of y. The differential equation then has the form $dy/dx = g(x)h(y)$, and its solution is given by the equation

$$\int \frac{1}{h(y)} dy = \int g(x) dx + c,$$

where c is an arbitrary constant.

separable solution Consider the following *boundary value problem for the *wave equation:

$$c^2 \frac{\partial^2 y}{\partial x^2} = \frac{\partial^2 y}{\partial t^2}, \quad y(0) = y(l) = 0.$$

Solutions may be found by separating variables and seeking a separable solution of the form $y(x,t) = X(x)T(t)$. This leads to the equation and conditions

$$\frac{X''(x)}{X(x)} = \frac{T''(t)}{c^2 T(t)}, \quad X(0) = X(l) = 0.$$

As X''/X is a function of x alone and T''/T is a function of t alone, both must be a

constant k. To meet the boundary conditions $X(0) = X(l) = 0$, it follows that $k = -n^2\pi^2/l^2$ for some positive integer n. The separable solutions are then the *normal modes

$$y(x,t) = \sin\left(\frac{n\pi x}{l}\right) \times \left[A\sin\left(\frac{n\pi ct}{l}\right) + B\cos\left(\frac{n\pi ct}{l}\right) \right].$$

The above argument applies more generally to other *PDEs, and the separable solutions then play an important part in finding the *general solution (*see* SUPERPOSITION PRINICIPLE).

separated sets In a *topological space X, two sets where each set is *disjoint from the *closure of the other. Two separated sets are then necessarily disjoint, but the sets $[0, 1)$ and $(1, 2]$ are disjoint, but not separated, as 1 belongs to both closures. *See* SEPARATION AXIOMS.

separation axioms The axioms of a *topological space do not themselves guarantee sufficient *open sets to separate out the space; there are various separation axioms which address this problem, most notably the T_n axioms. A space is: T_0 if, given distinct points, there is an open set which contains one and not the other; T_1 if, given distinct points, each point is contained in an open set and not the other. This is equivalent to points being *closed; T_2 if it is a *Hausdorff space; T_3 if it is a *regular space and T_1; T_4 if it is a *normal space and T_1; T_5 if it is a *completely normal space and T_1; T_6 if it is a *perfectly normal space and T_1.

A *Tychonoff space is T_3 but need not be T_4 and so is often referred to as $T_{3.5}$. All these separation axioms are strictly nested, that is, if $m > n$, then a T_m space is T_n and there exists a T_n space which is not T_m. *Urysohn's lemma shows that T_4 spaces are $T_{3.5}$. *Metric spaces are T_6.

separation of variables *See* SEPARABLE SOLUTION.

sept- Prefix denoting seven.

sequence A finite sequence consists of n terms $a_1, a_2, \ldots, a_n$, in a given order, where n is the length of the sequence. An infinite sequence consists of terms $a_1, a_2, a_3, \ldots$, one corresponding to each positive integer. Sometimes, it is more convenient to denote the terms of a sequence by $a_0, a_1, a_2, \ldots$. *Addition and *scalar multiplication of sequences are defined *componentwise, and so the set of sequences forms a *vector space (*see c, l^p*). *See also* LIMIT (of a sequence), SERIES.

sequence of functions An ordered list of functions, $\{f_n\}$ each having the same *domain and *codomain. For example, such a sequence might arise as the *partial sums of a *Fourier series. *Convergence of a sequence of functions might be considered *pointwise or *uniform.

sequence space *See l^p, c*.

sequential compactness A *topological space (or *metric space) X is sequentially compact if every *sequence in X has a *subsequence which *converges in X. A metric space is *compact if and only if it is sequentially compact, but this is not generally true for topological spaces. *See* BOLZANO-WEIERSTRASS THEOREM.

sequential sampling A statistical process to decide which of two *hypotheses to reject as false. *Observations are made one at a time, and a test is carried out to decide whether one of the hypotheses is to be rejected or whether more observations should be made. The sampling stops when it has been decided which hypothesis to reject.

serial (of a relation) A *binary relation R on a set S such that for every x in S there exists y in S satisfying xRy.

serial computation The sequential execution of all parts of the same task on a single processor. *Compare* PARALLEL COMPUTATION.

serial correlation A synonym for AUTOCORRELATION.

series A finite series of length n is written as $a_1 + a_2 + \ldots + a_n$, where $a_1, a_2, \ldots, a_n$ are n numbers called the terms in the series. The sum of the series is simply the sum of the n terms.

An infinite series is written as $a_1 + a_2 + a_3 + \ldots$, with terms $a_1, a_2, a_3, \ldots$, one corresponding to each positive integer. The nth *partial sum is s_n, the sum of the first n terms. If the *sequence $s_1, s_2, s_3, \ldots$ has a *limit s, then the value s is called the sum (or sum to infinity) of the infinite series. Otherwise, the infinite series is said to diverge. For a list of important series and convergence tests, *see* APPENDIX 11 and APPENDIX 12. *See also* ALGEBRA OF LIMITS, ARITHMETIC SERIES, GEOMETRIC SERIES, PARTIAL SUM, TAYLOR SERIES.

Serre, Jean-Pierre (1926–) Influential French mathematician who made important contributions in *topology, *complex analysis, *number theory, and *algebraic geometry. Winner of the *Fields Medal in 1954, the *Wolf Prize in 2000, and the first *Abel Prize in 2003.

Serret-Frenet formulae Three *differential equations describing and governing how a *curve in three dimensions evolves. For a curve $\mathbf{r}(s)$, parametrized by *arc length s, the tangent vector $\mathbf{t} = d\mathbf{r}/ds$ is a *unit vector. Necessarily, $d\mathbf{t}/ds$ is perpendicular to $\mathbf{t}$, so $d\mathbf{t}/ds = \kappa\mathbf{n}$, where $\mathbf{n}$ is a unit vector called the normal vector, and $\kappa > 0$ is the curvature. The unit vector $\mathbf{b} = \mathbf{t} \times \mathbf{n}$ is the binormal vector. Then $\{\mathbf{t}, \mathbf{n}, \mathbf{b}\}$ is an *orthonormal basis for each s. The Serret-Frenet formulae state:

$$\frac{d\mathbf{t}}{ds} = \kappa\mathbf{n}, \quad \frac{d\mathbf{n}}{ds} = -\kappa\mathbf{t} + \tau\mathbf{b}, \quad \frac{d\mathbf{b}}{ds} = -\tau\mathbf{n},$$

where τ denotes torsion. If the torsion of a curve is 0, then the curve is *planar. $\kappa(s)$ and $\tau(s)$ determine a curve up to *isometry.

sesquilinear An *inner product on a complex *vector space is not *bilinear, but rather linear in one variable and conjugate linear in the second, and so is referred to as sesquilinear, the prefix 'sesqui-' meaning 'one and a half'.

set A *well-defined collection of objects. It may be possible to define a set by listing the elements: $\{a, e, i, o, u\}$ is the set consisting of the vowels of the alphabet, $\{1, 2, \ldots, 100\}$ is the set of the first 100 positive *integers. The meaning of $\{1, 2, 3, \ldots\}$ is also clear: it is the set of all positive integers. It may be possible to define a set as consisting of all elements, from some *universal set, that satisfy some property. Thus the set of all real numbers that are greater than 1 can be written as either $\{x \mid x \in \mathbb{R} \text{ and } x > 1\}$ or $\{x : x \in \mathbb{R} \text{ and } x > 1\}$, both of which are read as 'the set of x such that x belongs to $\mathbb{R}$ and x is greater than 1'. The same set is sometimes written $\{x \in \mathbb{R} \mid x > 1\}$. *See* ALGEBRA OF SETS, RUSSELL'S PARADOX, ZERMELO-FRAENKEL AXIOMS.

set difference (difference of two sets) The set difference $A\backslash B$ of sets A and B, also known as the relative complement, is the set consisting of all elements of A that are not elements of B. The

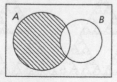

The set difference $A\backslash B$

notation $A - B$ is also used. The set is represented by the shaded region of the *Venn diagram shown in the figure.

set theory The study of the properties of sets and their relations, originally developed by Georg *Cantor. *see* AXIOM OF CHOICE, AXIOMATIC SET THEORY, RUSSELL'S PARADOX, ZERMELO-FRAENKEL AXIOMS.

sex- Prefix denoting six, as does hexa-.

sexagesimal Based on the number 60, so measurement of *time in hours, minutes, and seconds and of *angles in *degrees, minutes, and seconds are sexagesimal measurements.

sf Abbreviation for *significant figures.

sgn *See* SIGNUM FUNCTION.

Shannon, Claude (1916–2001) Regarded as the father of *information theory, he wrote the seminal 1948 paper *A Mathematical Theory of Communication* which introduced the idea of (Shannon) *entropy and showed it to be a constraint on how quickly *information can be transmitted. *See* SHANNON'S THEOREM.

Shannon's Theorem (Shannon's noiseless coding theorem) The theorem shows how *entropy is a *lower bound to how efficiently a source can be encoded. For a *memoryless source, producing source words with *random variable X, then $H(X) \leq l(C)$, where $H(X)$ is the entropy of X, C is any *binary, uniquely decipherable encoding of X, and $l(C)$ is the mean *length of the *codewords. *See also* HUFFMAN CODING.

shear A transformation in which points move *parallel to a fixed line or plane and move a distance proportional to their distance from the fixed line or plane. *Areas and *volumes are preserved under a shear in two and three dimensions respectively. When the fixed

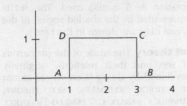

Original rectangle

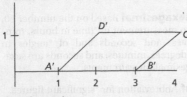

Sheared rectangle

line or plane is parallel to one of the sides, a shear transforms *rectangles into *parallelograms and *cuboids into *parallelepipeds.

shearing force (mechanics) The *internal force perpendicular to the axis of a thin rod or beam. *See* DEFORMATION.

sheet *See* HYPERBOLOID OF ONE SHEET, HYPERBOLOID OF TWO SHEETS.

SHM An abbreviation for SIMPLE HARMONIC MOTION.

shortest path algorithm (Dijkstra's method) An *algorithm to solve the *shortest path problem. In essence it is a procedure which labels *vertices with a minimum distance from the starting position, in order of ascending minimum distances. At each stage therefore you do not have to consider the multiple possible routes to get to any of the vertices already labelled. The procedure stops once the finish point has been labelled in this manner, even though there may be other points not yet labelled—for which the shortest path would be longer than the one of interest.

shortest path problem The problem in *graph theory to find the shortest *connected route between two points or vertices, i.e. the minimum sum of the edges on all possible connected routes between the two vertices.

SI *See* SI UNITS.

SIAM The Society for Industrial and Applied Mathematics, founded in 1951, is a professional association dedicated to the use of mathematics in industry. It also supports many student programmes and SIAM student chapters internationally.

side One of the lines joining two adjacent vertices in a *polygon. One of the faces in a *polyhedron.

Sierpinski triangle (Sierpinski gasket) A *fractal, named after the Polish mathematician Waclaw Sierpinski. It is formed as follows: given an *equilateral triangle, remove the central quarter of the triangle, thus forming three smaller equilateral triangles; repeat this for each of these three triangles, and then repeat indefinitely. Its *Hausdorff dimension equals $\log_2 3$.

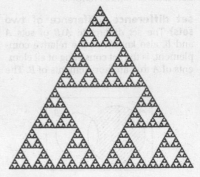

Sierpinski triangle

sieve of Eratosthenes The following method of finding all the *primes up to some given number N. List all the positive integers from 2 up to N. Leave the first number, 2, but delete all its multiples; leave the next remaining number, 3, but delete all its multiples; leave the next remaining number, 5, but delete all its multiples, and so on. The integers not deleted when the process ends are the primes.

(⊕) SEE WEB LINKS

- An interactive animation of the sieve of Eratosthenes used to obtain prime numbers.

sifting property Two similar properties of the *Kronecker delta and *Dirac delta function (which was why *Dirac chose the δ notation).

$$a_{ij} = \sum_{k=1}^{n} a_{ik}\delta_{kj},$$

$$f(a) = \int_{-\infty}^{\infty} f(x)\delta(x-a)\mathrm{d}x.$$

sigma Greek letter equivalent of s, written σ, commonly used to denote the *standard deviation of a *distribution or a set of *data. The capital $\sum$ is used to denote the sum in the *summation notation.

sigma algebra A sigma algebra or σ-algebra, X, of a set S is a family of *subsets which contains the *empty set, the set itself, the *complement of any set in X, and all *countable *unions of sets in X. The *measurable sets in a *measure space form a sigma algebra, as do the set of *events in a *probability space.

sigma function In *number theory, the function $\sigma(n)$ which is the sum of the factors of a positive integer n. For a *perfect number $\sigma(n) = 2n$. The sigma function is an *arithmetic function.

sign A symbol denoting an operation, such as $+$, $-$, $\times$ and $\div$ in arithmetic or $\circ$ to denote *composition of functions, etc.

The positive or negative nature of a quantity is called its sign.

signed area A term used to describe the *area represented by a definite *integral. If an *integrand $f(x)$ changes signs in an interval $a \le x \le b$, then the definite integral $\int_a^b f(x)\ \mathrm{d}x$ equals the area above the x-axis and below the graph, minus the area below the x-axis and above the graph. Instead, the total area between the graph and the x-axis equals $\int_a^b |f(x)|\ \mathrm{d}x$.

signed minor A synonym for COFACTOR.

signed rank test A synonym for WILCOXON SIGNED RANK TEST.

significance level See HYPOTHESIS TESTING.

significance test See HYPOTHESIS TESTING.

significant figures To count the number of significant figures in a given number, start with the first non-zero digit from the left and, moving to the right, count all the digits thereafter, counting final zeros if they are to the right of the decimal point. For example, 1.2048, 1.2040, 0.012048, 0.0012040 and 1204.0 all have 5 significant figures. In *rounding or *truncation of a number to n significant figures, the original is replaced by a number with n significant figures.

Note that final zeros to the left of the decimal point may or may not be significant: the number 1,204,000 has at least 4 significant figures, but without more information there is no way of knowing whether or not any more figures are significant. When 1,203,960 is rounded to 5 significant figures to give 1,204,000, an explanation that this has 5 significant figures is required. This could be made clear by writing it in *scientific notation: 1.2040×10^6.

S

To say that $a = 1.2048$ to 5 significant figures means that $1.20475 \leq a < 1.20485$.

sign test A *non-parametric method to test the *null hypothesis that a *sample is selected from a *population with *median m. If the null hypothesis is true, the sample of size n is expected to have an equal number of *observations above and below m, and the probability that r of them are greater than m has the *binomial distribution $B(n, 0.5)$. The null hypothesis is rejected if the value lies in the critical region determined by the chosen significance level.

The test may also be used to compare the medians of two populations from paired data.

signum function The real function, denoted by sgn, defined by

$$\operatorname{sgn} x = \begin{cases} -1, & \text{if } x < 0, \\ 0, & \text{if } x = 0, \\ 1, & \text{if } x > 0. \end{cases}$$

The name and notation come from the fact that the value of the function depends on the sign of x. This notation is also used for the *parity of *permutations, that is,

$$\operatorname{sgn} \sigma = \begin{cases} 1 & \text{if } \sigma \text{ is even,} \\ -1 & \text{if } \sigma \text{ is odd.} \end{cases}$$

similar (of figures) Two geometrical figures are similar if they are of the same shape but not necessarily of the same size. This includes the case when one is a mirror-image of the other, so the three triangles shown in the figure are all similar. For two similar triangles, there is a correspondence between their vertices such that corresponding angles are equal and the *ratios of corresponding sides are equal. In the figure, $\angle A = \angle P$, $\angle B = \angle Q$, $\angle C = \angle R$, and $QR/BC = RP/CA = PQ/AB$.

similar (of matrices) Two *square matrices $\mathbf{A}$ and $\mathbf{B}$ of the same order are

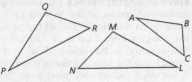

similar triangles

similar if there exists an *invertible matrix $\mathbf{P}$ such that $\mathbf{A} = \mathbf{P}^{-1}\mathbf{B}\mathbf{P}$. Similar matrices have the same *determinant, *trace, *characteristic, and *minimal polynomials. Two matrices which represent the same *linear map, with respect to different *bases, are similar. Similarity is an *equivalence relation.

simple curve A continuous plane curve that does not intersect itself.

simple extension A *field extension $K:F$ which is an extension of the *base field F by a single element a, that is, $K = F(a)$. Such an a is called a primitive element. A finite *degree extension, over a *characteristic zero base field, is simple.

simple fraction (common fraction, vulgar fraction) A *fraction in which the *numerator and *denominator are positive integers, as opposed to a *compound fraction. *Compare* IMPROPER FRACTION, MIXED NUMBER, PROPER FRACTION.

simple graph A *graph with no loops or multiple edges. *Compare* MULTIGRAPH.

simple group A non-trivial *group which has no *normal subgroups other than itself and the subgroup consisting of the *identity element. A finite *abelian group is simple if and only if it is *cyclic of prime *order. A_5 is the smallest non-abelian simple group. The classification theorem of finite simple groups was a landmark collaborative result of mathematics, occupying much of the 1960s,

1970s, and 1980s. *See* JORDAN-HÖLDER THEOREM, SPORADIC GROUP.

simple harmonic motion Suppose that a particle is moving in a straight line so that its *displacement x at time t is given by $x = A\sin(\omega t + \alpha)$, where $A > 0$, ω, and α are constants. Then the particle is exhibiting simple harmonic motion, with *amplitude A, *period $2\pi/\omega$, and *phase α. This equation gives the general solution of the differential equation $\ddot{x} + \omega^2 x = 0$.

An example of simple harmonic motion is the motion of a particle suspended from a fixed support by a spring (*see* HOOKE'S LAW). Also, the motion of a *simple pendulum performing oscillations of small amplitude is approximately simple harmonic motion.

simple interest Suppose that a sum of money P is invested, attracting interest at i per cent a year. When simple interest is given, the interest due each year is $(i/100)P$ and so, after n years, the amount becomes

$$P\left(1 + \frac{ni}{100}\right).$$

When points are plotted on graph paper to show how the amount increases, they lie on a straight line. Most banks and building societies in fact do not operate in this way but use the method of *compound interest.

simple pendulum In a *mathematical model, a simple pendulum is represented by a particle of mass m, suspended by a light string of constant length l from a fixed point. The particle, representing the bob of the pendulum, is free to move in a specified vertical plane through the point of suspension, acted on by gravity and tension in the string. Suppose the string makes an angle θ with the vertical at time t. The equation of motion is $\ddot{\theta} + (g/l)\sin\theta = 0$. When θ is small for all time, $\sin\theta \approx \theta$ and the

equation becomes $\ddot{\theta} + \omega^2\theta = 0$, where $\omega^2 = g/l$. It follows that the pendulum performs approximately *simple harmonic motion with period $2\pi\sqrt{l/g}$.

simple pole A *pole of order one.

simple root *See* ROOT.

simplest form *See* IRREDUCIBLE FRACTION.

simplex A generalization of a point, a line segment, a triangle, and a tetrahedron to higher dimensions. *See* K-SIMPLEX, SIMPLICIAL COMPLEX.

simplex method An algebraic method of solving the *standard form of a *linear programming problem which allows the solution of multivariable problems, which could not be tackled graphically. The process is as follows, using a two-variable, two-constraint problem to illustrate. To maximize $P = 2x + 3y$, subject to $x + 2y \leq 20$, $2x + y \leq 15$, $x \geq 0$, $y \geq 0$:

(i) introduce *slack variables, s and t, to rewrite each inequality as an equation, giving $x + 2y + s - 20 = 0$ and $2x + y + t - 15 = 0$;

(ii) introduce all other slack variables into each equation with a coefficient of zero, giving $x + 2y + s + 0t - 20 = 0$ and $2x + y + 0s + t - 15 = 0$;

(iii) rewrite the objective function in standard form, including all slack variables with a coefficient of zero, giving $P - 2x - 3y + 0s + 0t = 0$.

The simplex tableau is a table which shows the current values for each variable in each constraint, with the last row, known as the objective row, showing the values in the equation derived from the objective function. The simplex algorithm is the process by which that table is manipulated in order to find the optimal solution. In the solution, the algorithm will produce a series of tableaux.

The initial tableau for this problem will be:

basic variable	x	y	s	t	value
s		1	2	1	0	20
t		2	1	0	1	15
P		-2	-3	0	0	0

The optimality condition is that when the objective row shows zero in all columns representing basic variables and no negative entries, then it represents an optimal solution.

The standard procedure to achieve this condition is to repeat the following set of steps until all the basic variables, x and y, have been set to zero.

(i) Choose the variable with the largest negative entry in the objective row. The column it is in is called the pivotal column and it is known as the entering variable. In the example y is the entering variable, and the pivotal column is the third from the left.

(ii) For each row of the tableau, divide the value (in the last column) by the entering variable. In the example this gives $20/2 = 10$ and $15/1 = 15$. The smallest of these determines the pivotal row of the tableau at this stage and the leaving variable, which is s in the example. The pivot is the cell in both the pivotal column and the pivotal row, i.e. 2 in this case.

(iii) Divide all entries in the pivotal row by the pivot, so the row reads s, 0.5, 1, 0.5, 0, 10 in the example.

(iv) Add or subtract multiples of this row to (or from) each other to make the other entries in the pivotal column zero. So it has to be subtracted once from the t row, and added three times to the P row, leaving the tableau looking like this:

basic variable	x	y	s	t	value
s	0.5	1	0.5	0	10
t	1.5	0	-0.5	1	5
P	-0.5	0	1.5	0	30

The process would now be repeated, and x will now be the entering variable, and the values at step 2 are 20 for s, and $10/3$ for t, so t is the leaving variable, and at step 3 the pivotal row will read t, 1, 0, $-1/3$, $2/3$, $10/3$.

Repeating step (iv) gives:

basic variable	x	y	s	t	value
s	0	1	2/3	1/3	25/3
t	1	0	-1/3	2/3	10/3
P	0	0	4/3	1/3	95/3

For a many-variable problem this would have to be repeated for each variable in turn, but in this simple example the optimality condition is now satisfied and the maximum value of P is $95/3$ occurring when $s = 4/3$ and $t = 1/3$, corresponding to $x = 10/3$ and $y = 25/3$. In this simple problem, the solution could have been found much more quickly by the *vertex method, but the simplex method is an important tool because it can be applied to more complicated problems.

simplex tableau *See* SIMPLEX METHOD.

simplicial complex In *combinatorial topology, the decomposing of a space into simplices (plural of *simplex) this is a generalization of a *triangulation. A simplicial complex is a collection of simplices such that every face of a simplex is also in the collection, and if two simplices are not disjoint, then their intersection is a face of each simplex.

simplify Reduce an expression by algebraic manipulation. For example, $8x - 2x$ can be simplified to $6x$ and $3(2x + 5y) + 2(x - 3y)$ can be simplified to $8x + 9y$ by multiplying out the brackets and collecting *like terms.

simply connected A *path-connected *topological space is simply connected if any *closed curve can be shrunk continuously to (that is, is *homotopic to) a point in the space. This is equivalent to the *fundamental group being *trivial. Intuitively, this means that there are no holes in the space. A *disc in two dimensions and a *sphere in three dimensions are simply connected, while the *annulus, *circle, and *torus are not.

Simpson's paradox The *paradox in which considering the rates of occurrence in a two-way table of the characteristics in two groups separately can lead to the opposite conclusion from when the two groups are combined. For this to occur, it is necessary that there be a substantial difference in the proportions of one category within two groups. For example, if 80% of students applying for science are accepted and 40% of students applying for arts are accepted, irrespective of gender in both cases, then there is no discrimination. However, if 75% of boys apply for science courses and only 30% of girls apply for science, the overall success rate of applications for boys and girls would look discriminatory regarding gender. With the above proportions, and 1000 boys and 1000 girls altogether, the two-way table would be:

	Accept	Fail
Boys	700	300
Girls	520	420

That is, overall 70% of boys would be accepted compared with only 52% of girls, but the rates of acceptance from each group by type of course were identical.

Conversely, it may be the case that a drug trial seems to suggest, for a group as a whole, that the drug reduces mortality rates. However, when the *data are stratified by gender, say, it may prove to be the case that the drug increases mortality rates for both genders. An example of such data is presented below.

Columns 2–4 suggest the drug trial was a success, reducing the mortality rate, while columns 5–7 and 8–10 show it was separately detrimental to both men and women. This paradox is possible here, as men and women have very different mortality rates, and the treated and control groups are not evenly distributed by gender. So, the seeming improvement only reflects that women treated with the drug have a better mortality rate than men who are untreated.

Simpson's rule A rule to find approximations for the definite *integral

$$\int_a^b f(x)\,\mathrm{d}x.$$

Divide $[a, b]$ into n equal subintervals of length h by the *partition

$$a = x_0 < x_1 < x_2 < \ldots < x_{n-1} < x_n = b,$$

where $x_{i+1} - x_i = h = (b-a)/n$, where n is even. Denote $f(x_i)$ by f_i, and let P_i be the point (x_i, f_i). Take an arc of the *parabola through the points P_0, P_1, and P_2, an arc of a parabola through P_2, P_3, and P_4, similarly, and so on. This is why n is necessarily even. The resulting Simpson's rule gives

	WHOLE GROUP			MEN			WOMEN		
	Recover	Die	Mortality	Recover	Die	Mortality	Recover	Die	Mortality
Treated	20	20	50%	2	8	80%	18	12	40%
Control	16	24	60%	9	21	70%	7	3	30%

Tables illustrating Simpson's paradox

$$\frac{1}{3} h(f_0 + 4f_1 + 2f_2 + 4f_3 + 2f_4 + \ldots$$
$$+ 2f_{n-2} + 4f_{n-1} + f_n)$$

as an approximation to the integral. In general, Simpson's rule is much more accurate than the trapezium rule, and the *error is bounded by

$$\frac{(b-a)^5}{2880n^4} \max_{a \le x \le b} |f''''(x)|.$$

The rule is named after the English mathematician Thomas Simpson (1710–61), though the rule is due to *Newton, as Simpson himself acknowledged.

Simson line Given a triangle and a point, P, on its *circumcircle, the feet of the perpendiculars from P to each of the sides of the triangle (or their extensions) are *collinear and define the Simson line. The converse of this is true as well, that if the feet of the perpendiculars from a point P to each of the sides of the triangle (or their extensions) are collinear then P lies on the circumcircle. The line is named after the Scottish mathematician Robert Simson (1687–1768).

simulation An attempt to replicate a physical procedure mathematically, where the system being studied is too complicated for explicit analytic methods to be used. Because of the complexity, simulation is often carried out by computer, perhaps using *Monte Carlo methods. As simulation is often only an approximation to the physical procedure, the use of *sensitivity analysis is crucial.

simultaneity Two events are simultaneous if they occur at the same time. A paradoxical consequence of *relativity is that two observers cannot agree about the simultaneity of events. Indeed, the *Lorentz-Fitgerald contraction is a consequence of an *observer measuring between events which are not simultaneous in the *rest frame of the object being measured.

simultaneous linear differential equations Two simultaneous *linear differential equations in variables $x(t)$ and $y(t)$ take the form

$$\frac{dx}{dt} = Ax + By, \qquad \frac{dy}{dt} = Cx + Dy.$$

These may be solved by differentiating one or other of the equations and substituting an expression for dx/dt or dy/dt into the other equation to form a second order differential equation in x or y. Such systems of differential equations are of importance in the *linear theory of equilibria.

Alternatively, the equations can be rewritten as a single matrix equation

$$\frac{d}{dt} \begin{pmatrix} x \\ y \end{pmatrix} = \mathbf{M} \begin{pmatrix} x \\ y \end{pmatrix} \text{ with } \mathbf{M} = \begin{pmatrix} A & B \\ C & D \end{pmatrix}.$$

This then has the solution

$$\begin{pmatrix} x(t) \\ y(t) \end{pmatrix} = e^{\mathbf{M}t} \begin{pmatrix} x(0) \\ y(0) \end{pmatrix}.$$

(*See* EXPONENTIAL OF A MATRIX.)

simultaneous linear equations The solution of a set of m linear equations in n unknowns can be investigated by the method of *Gaussian elimination (or *Gauss-Jordan elimination) that transforms the *augmented matrix to *echelon form (or *reduced echelon form). The number of non-zero rows in the echelon form cannot be greater than the number of unknowns, and three cases can be distinguished:

(i) If the echelon form has a row with all its entries zero except for a non-zero entry in the last place, then the set of equations is *inconsistent.

(ii) If case (i) does not occur and, in the echelon form, the number of non-zero rows is equal to the number of unknowns, then the set of equations has a unique solution.

(iii) If case (i) does not occur and, in the echelon form, the number of non-zero rows is less than the number of unknowns, then the set of equations has more than one solution. In this case scalar *parameters can be introduced for the variable of each column that does not contain a leading 1. The other variables can then be determined by *backward substitution.

sine *See* TRIGONOMETRIC FUNCTION.

sine rule *See* HYPERBOLIC PLANE, SPHERICAL TRIANGLE, TRIANGLE.

singleton A *set containing just one *element.

Singmaster's conjecture There is an integer n such that all numbers in Pascal's triangle appear fewer than n times (excepting the 1s at the start and end of each row). It is known that there are infinitely many numbers that appear (at least) 6 times, but currently 3003 is the only number known to appear 8 times. The conjecture was made by David Singmaster in 1971.

singular A square matrix is singular if it has zero *determinant. A square matrix is singular if and only if it is not *invertible.

singularity (fluid dynamics) A singularity in a fluid is a point where the flow

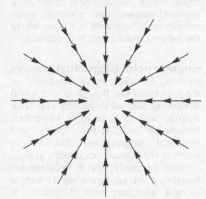

A point sink

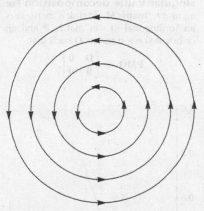

A point vortex

velocity is not defined, such as with a *source, *sink, or *vortex. *Streamlines may only meet at *stagnation points and *singularities.

singular point (singularity) (geometry) A point on a *curve where there is not a unique *well-defined *tangent. It may be an isolated point, or a point where the curve cuts itself such as a

A point source

*node, or a *cusp where there is a repeated tangent. More generally, a singularity on a *manifold is a point where the tangent space is not well defined.

singular point (singularity) (complex analysis) A *holomorphic function f on the punctured disc $0 < |z-a| < r$ is said to have an isolated singularity at a. Such singularities occur in three categories: *removable singularities, *poles, and *essential singularities, classified in terms of the *Laurent expansion at a.

More generally, for a holomorphic function f on an *open set U with a being a *boundary point of U, then a is singular if there is no *analytic continuation of f to a.

singular value decomposition For an $m \times n$ *matrix $\mathbf{M}$ of *rank r, there exists an *orthogonal $m \times m$ matrix $\mathbf{P}$ and an orthogonal $n \times n$ matrix $\mathbf{Q}$ such that

$$\mathbf{PMQ} = \begin{bmatrix} \mathbf{D} & \mathbf{0} \\ \mathbf{0} & \mathbf{0} \end{bmatrix},$$

where $\mathbf{D}$ is an $r \times r$ *diagonal matrix with decreasing positive entries. The *pseudoinverse $\mathbf{M}^+$ of $\mathbf{M}$ is given by

$$\mathbf{M}^+ = \mathbf{Q} \begin{bmatrix} \mathbf{D}^{-1} & \mathbf{0} \\ \mathbf{0} & \mathbf{0} \end{bmatrix} \mathbf{P}.$$

Compare SPECTRAL THEOREM.

sinh *See* HYPERBOLIC FUNCTION.

sink (fluid dynamics) *See* SOURCE.

sink (networks) The vertex in a *network towards which all flows are directed.

SIR epidemiology model In an SIR epidemiology model, there are three mutually exclusive subsets of a constant population as a disease spreads: the fractions of the population that are susceptible S, infected I, and recovered R. A first such model was due to Kermack and McKendrick in 1927. The equations governing the model are

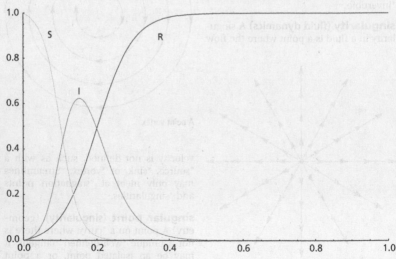

Graphs of S, I, and R in an epidemiology model

$$\frac{dS}{dt} = -\alpha SI,$$

$$\frac{dI}{dt} = \alpha SI - \beta I,$$

$$\frac{dR}{dt} = \beta I,$$

where α is a positive constant relating to how contagious the disease is and the level of contact between people and β is a positive constant relating to the rate of recovery. If cases grow, then the *maximum value of I is an *increasing function of $R_0 = \alpha/\beta$, which is the expected number of people an infected person infects.

Such an SIR model makes accurate predictions for a disease spreading through a homogeneous closed population, but clearly the model does not take into account different demographics within a population, movement of people, mutation of the disease, catching the disease more than once, and responses from governments and society. There are many other so-called 'compartmental models' addressing some of these aspects.

SI units The units used for measuring physical quantities in the internationally agreed system Système International d'Unités are known as SI units. There are seven base units and other supplementary units and prefixes (such as *kilo-) to create further units. For further details, *see* APPENDIX 4.

skew field A synonym for DIVISION RING.

skew lines Two *lines in 3-dimensional *space that do not *intersect and are not *parallel.

skewness The amount of asymmetry of a *distribution. One measure of skewness is the coefficient of skewness, which is defined to be equal to $\mu_3/(\mu_2)^{3/2}$, where μ_2 and μ_3 are the second and third *moments about the mean. This measure is zero if the distribution is

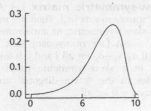

Skewed to the left

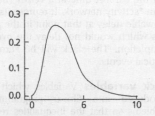

Skewed to the right

symmetrical about the mean. If the distribution has a long tail to the left, as in the top figure, it is said to be skewed to the left and to have negative skewness, because the coefficient of skewness is negative. If the distribution has a long tail to the right, as in the second figure, it is said to be skewed to the right and to have positive skewness, because the coefficient of skewness is positive. A skew distribution is one that is skewed to the left or skewed to the right.

skew-symmetric function A *function f on n variables $x_1, x_2, \ldots, x_n$ is skew-symmetric, or alternating, if f changes sign when two variables x_i and x_j are swapped. Equivalently,

$$f\left(x_{\sigma(1)}, x_{\sigma(2)}, \ldots, x_{\sigma(n)}\right) = \text{sgn}(\sigma) f(x_1, x_2, \ldots, x_n)$$

for all *permutations σ of n variables; here sgn(σ) denotes the *parity of σ, 1 if σ is *even and −1 if σ is odd. *See* ELEMENTARY SYMMETRIC POLYNOMIAL.

skew-symmetric matrix Let $\mathbf{A}$ be the *square matrix $[a_{ij}]$. Then the matrix $\mathbf{A}$ is skew-symmetric, or antisymmetric, if $\mathbf{A}^T = -\mathbf{A}$ (see TRANSPOSE); that is to say, if $a_{ij} = -a_{ji}$ for all i and j. It follows that in a skew-symmetric matrix the entries in the *main diagonal are all zero: $a_{ii} = 0$ for all i.

slack In *critical path analysis, slack is the difference between the latest time and the earliest time at a vertex (node) in an *activity network. It represents the allowable delay at that point in the process which would not delay the overall completion. The slack will be zero for *critical events.

slack variables Variables which are introduced into *linear programming problems so that the inequalities representing the constraints can be replaced by equations. Each equation will introduce one slack variable, for example if $2x + 3y \le 30$ then we introduce a slack variable s satisfying $2x + 3y + s - 30 = 0$. This allows the algebraic method of solution known as the *simplex method to be used.

slant asymptote See ASYMPTOTE.

slant height See CONE.

slash A synonym for SOLIDUS.

slide rule A mechanical device used for mathematical calculations such as multiplication and division. In the simplest form, one piece slides alongside another piece, each piece being marked with a *logarithmic scale. To multiply x and y, the reading 'x' on one scale is placed opposite the 'l' on the other scale, and the required product 'xy' then appears opposite the 'y'. Slide rules have now been superseded by calculators and computers.

(⊕) SEE WEB LINKS

- Descriptions of various slide rules and how they work.

sliding–toppling condition Where a *rigid body is at rest on a plane, whether it will slide or topple first can be determined by calculating what the minimum force is for it to topple, assuming it does not slide, and then calculating what the minimum force is for it to slide, assuming it does not topple first. The lower of these two minima will tell which happens first.

slope A synonym for GRADIENT.

small circle See GREAT CIRCLE.

Smith, Adrian Frederick Melhuish (1946–) British statistician who conducted the Inquiry into Post-14 Mathematics Education for the UK Secretary of State for Education and Skills in 2003–04. He has been President of the Royal Statistical Society and in 2008 was appointed as Director General for Science and Research in the Department for Innovation, Universities and Skills. He became the director of the *Alan Turing Institute in 2018.

Smith normal form An $m \times n$ *matrix, with *integer entries is *equivalent to a matrix of the form

$$\begin{bmatrix} d_1 & 0 & \ldots & 0 & \ldots & \ldots & 0 \\ 0 & \ddots & 0 & 0 & \ldots & \ldots & 0 \\ \vdots & 0 & d_r & 0 & \ldots & \ldots & 0 \\ 0 & \vdots & 0 & 0 & 0 & \ldots & 0 \\ \vdots & \vdots & \ldots & \ldots & \ldots & \ldots & 0 \\ 0 & 0 & \ldots & \ldots & \ldots & \ldots & 0 \end{bmatrix}$$

where r is the *rank of the matrix, the d_i are positive integers, and d_k divides d_{k+1} for $1 \le k < r$. The d_i are called invariant factors. This result applies equally when the entries are taken from any *principal ideal domain. The Smith normal form is key to the *Structure Theorem for *modules.

smooth The term 'smooth' is variously used to describe *functions. Commonly, the term implies a function has *derivatives of all orders or is C^∞. More weakly, the term may simply mean that the function is *differentiable everywhere.

smoothly hinged Implies that the *force at the hinge can be treated as a single force at an unknown angle, rather than requiring the consideration of a *couple as well.

smoothness condition *See* PARAMETERIZED SURFACE.

smooth surface (in mechanics) *See* CONTACT FORCE.

snowflake curve The *Koch curve or any curve constructed in a similar way.

software *See* COMPUTER.

SOHCAHTOA A mnemonic for the trigonometric ratios, short for 'sine = opposite ÷ hypotenuse, cosine = adjacent ÷ hypotenuse, tangent = oppositive ÷ adjacent'. *See* TRIGONOMETIC FUNCTIONS.

solid A 3-dimensional geometric figure such as a *cube or *cylinder.

solid angle The 3-dimensional analogue of 2-dimensional *angle. Just as an angle is bounded by two lines, a solid angle is bounded by the generators of a cone.

A solid angle is measured in steradians: this is defined to be the area of the intersection of the solid angle with a *sphere of unit radius. Thus the 'complete' solid angle at a point measures 4π steradians, this being the *surface area of

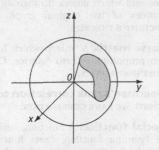

The shaded region subtends a solid angle at the origin

the unit sphere. The steradian is the *SI unit for measuring solid angle.

solid of revolution Suppose that a plane region is rotated through one revolution about a line in the plane that does not cut the region. The 3-dimensional region thus obtained is a solid of revolution. *See also* VOLUME OF A SOLID OF REVOLUTION.

solidus The symbol / denoting division. Also call 'slash' or 'forward slash'.

soluble group A synonym for SOLVABLE GROUP.

solution A solution of a set of *equations is an element that satisfies the equations. For a set of equations in n unknowns, a solution may be considered to be an n-tuple or a *column vector.
See also CONSISTENT, INEQUALITY.

solution of triangles A *triangle (up to *congruence) can be specified in various ways. The different information given is represented by the acronyms:

SSS	three sides' lengths are specified;
SAS	two sides and the angle they *subtend are specified;
SSA	two sides and an angle, different from the one they subtend, are specified;
ASA	a side and the two angles *adjacent to it are specified;
AAS	a side, an adjacent angle and the opposite angle are specified.

In each case, the triangle can be entirely determined by applications of the *sine and *cosine rules and knowing that the angles sum to 180 degrees.

solution set (solution space) The solution set of a set of equations is the set consisting of all the solutions. It may be referred to as the 'solution space' if it has further structure, for example that of an *affine space or *vector space. *See also* INEQUALITY.

solvable by radicals A *polynomial (with integer coefficients) is solvable by radicals if its roots can each be expressed using standard arithmetic operations and square roots, cube roots, etc. Polynomial equations of *degree up to 4 have long been known to be solvable by radicals. It was shown by *Abel and later *Galois that this is not generally the case for higher degrees; in fact, a polynomial is solvable by radicals if and only if its *Galois group is a *solvable group.

solvable group A group G is solvable (or soluble) if there exists a sequence of subgroups

$$\{e\} = G_0 \lhd G_1 \lhd \ldots \lhd G_{n-1} \lhd G_n = G$$

such that the *quotient groups G_{i+1}/G_i are *abelian groups. For a finite group, this is equivalent to the group's *composition factors being *cyclic of prime *order. It is significant to Galois theory that S_5 is not solvable. See GALOIS GROUP, SOLVABLE BY RADICALS.

sorting algorithms An *algorithm which puts the elements of a list in a certain order, usually alphabetical or numerical. Commonly used examples include the *bubble sort, bi-directional bubble sort, selection sort, shell sort, merge sort, insertion sort, heap sort, *quick sort, and combinations of these. How efficient these are depends on the size of the list to be sorted, the *algorithmic complexity of the procedure, and the available memory space in the *computer doing the sort.

source (fluid dynamics) A point source is a type of *singularity in *fluid mechanics. In a planar flow a point source at the origin may be modelled, using *polar coordinates, with flow velocity

$$\mathbf{u} = \frac{Q}{2\pi r}\mathbf{e}_r$$

where $Q > 0$ and $\mathbf{e}_r = (\cos\theta, \sin\theta)$ is the unit radial vector. Q is then the *flux across any *simple *closed curve around the origin. The *complex potential is

$$w = \frac{Q}{2\pi}\log z.$$

Note this is a *multifunction (a consequence of the flow not being *irrotational). When $Q < 0$, the above singularity is a sink. For diagrams of a source and sink, see SINGULARITY (FLUID DYNAMICS).

source (networks) The vertex in a *network away from which all flows are directed.

source (in transportation problems) See TRANSPORTATION PROBLEMS.

space A set of points with a structure which defines further characteristics of the space and the relationship between its points, particularly in 'spatial' ways regarding *distance, *dimension, etc. See AFFINE SPACE, EUCLIDEAN SPACE, INNER PRODUCT, MEASURE SPACE, METRIC SPACE, TOPOLOGICAL SPACE, VECTOR SPACE.

space-filling curve See PEANO CURVE.

space–time See MINKOWSKI SPACE.

span The span of a set S in a *vector space V is the smallest *subspace of V that contains S. Equivalently it is the set of *linear combinations of elements of S.

spanning set See BASIS.

spanning tree A *subgraph which is a *tree and which passes through all the *vertices of the original graph. See KIRCHOFF'S THEOREM.

sparse matrix A *matrix which has a high proportion of zero *entries. Compare DENSE MATRIX.

Spearman's rank correlation coefficient See RANK CORRELATION.

special function A function, such as the *gamma function, *zeta function, *Bessel functions, and the *error function, which cannot be written in terms of *elementary functions. This is not simply a case that no one has found a way to write these functions in terms of elementary functions; it can be mathematically proven to be impossible. A first result in this area was due to *Liouville.

special linear group *See* MATRIX GROUPS.

special relativity *See* RELATIVITY.

spectral theorem Given a *symmetric matrix **M** with real entries, there exists an *orthogonal matrix **P** such that $\mathbf{P}^{T}\mathbf{MP}$ is diagonal. Equivalently, **M** is *diagonalizable with respect to an *orthonormal basis. This result generalizes to a *Hermitian matrix **M** when **U***MU** is real and diagonal for some *unitary matrix **U** and to a *self-adjoint linear map of a finite-dimensional *inner product space which is likewise diagonalizable with respect to an orthonormal basis. The result is of importance as *Hessian, *covariance, and *inertia matrices are symmetric, and symmetric matrices can also be associated with *quadratic forms. Under further technical restrictions, versions of the spectral theorem apply for infinite-dimensional *Hilbert spaces.

spectrum The spectrum of a *square matrix **A** is the sets of its *eigenvalues λ or equally those λ for which $\mathbf{A}-\lambda\mathbf{I}$ is not *invertible. More generally, for a continuous (*see* CONTINUOUS FUNCTION) *linear map $T:V \to V$ on a *Banach space V, the spectrum, denoted $\sigma(T)$, is the set of values $\lambda \in \mathbb{C}$ such that is $T-\lambda I$ has no continuous inverse. This will include any eigenvalues but may also include other values.

speed In mathematics, it is useful to distinguish between *velocity and speed. First, when considering motion of a particle in a straight line, specify a positive direction so that it is a *directed line. Then the velocity of the particle is

positive if it is moving in the positive direction and negative if it is moving in the negative direction. The speed of the particle is the *absolute value of its velocity. In higher dimensions, when the velocity is a vector **v**, the speed is the magnitude |**v**| of the velocity.

speed of light In a vacuum, light travels at $299{,}792{,}458$ ms^{-1}. This is the same for other *electromagnetic waves.

sphere In 3-dimensional space, the sphere with centre C and radius r is the *locus of all points whose distance from C is equal to r. If C has Cartesian coordinates (a,b,c), this sphere has equation

$$(x-a)^2 + (y-b)^2 + (z-c)^2 = r^2.$$

The interior of a sphere of radius r has volume $\frac{4}{3}\pi r^3$, and the surface area equals $4\pi r^2$.

In *topology, the n-dimensional sphere is denoted as S^n and is often identified with the unit sphere in $\mathbb{R}^{n+1}$, that is,

$$S^n = \{(x_1, x_2, \ldots, x_{n+1})|$$
$$(x_1)^2 + (x_2)^2 + \ldots + (x_{n+1})^2 = 1\}.$$

The spheres described above are topologically S^2.

spherical angle An *angle formed by two *great circles meeting on the surface of a sphere, measured as the angle between their *tangents at the point of intersection. *See* SPHERICAL TRIANGLE.

spherical cap *See* ZONE.

spherically symmetric A real function on $\mathbb{R}^3$ is spherically symmetric if it is solely a function of $r = |\mathbf{r}|$, where **r** denotes the *position vector of a point from the origin. Equivalently, the function is constant on any *sphere centred at the origin.

spherical polar coordinates Suppose that three mutually perpendicular directed lines Ox, Oy, and Oz, intersecting at the point O, and forming a

S

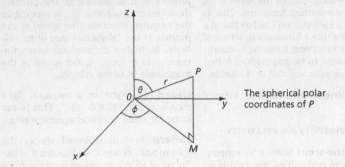

The spherical polar coordinates of P

right-handed system are taken as coordinate axes. For any point P, let M be the projection of P on the xy-plane. Let $r = |OP|$, let θ be the angle $\angle zOP$ in radians ($0 \leq \theta \leq \pi$) and let ϕ be the angle $\angle xOM$ in radians ($0 \leq \phi < 2\pi$). Then (r, θ, π) are the spherical polar coordinates of P. (The point O gives no value for θ or ϕ, but is simply said to correspond to $r = 0$.) The Cartesian coordinates (x,y,z) of P in terms of r, θ, and ϕ are given by

$$x = r\sin\theta\cos\phi, \qquad y = r\sin\theta\sin\phi,$$
$$z = r\cos\theta.$$

Compare CYLINDRICAL POLAR COORDINATES.

spherical triangle A triangle on a sphere, with three vertices and three sides that are arcs of *great circles. The angles of a spherical triangle do not add up to 180°. In fact, the sum of the angles can be anything between 180° and 540°. Consider, for example, a spherical triangle with one vertex at the North Pole and the other two vertices on the equator of the Earth. The area of a spherical triangle on a sphere of radius r equals $r^2(\alpha + \beta + \gamma - \pi)$, where the triangle's angles are α, β, γ, measured in *radians; this is Girard's Theorem.

Given a spherical triangle on a sphere of unit radius, with angles α, β, γ and sides of length a, b, c, the (spherical) sine rule states that

$$\frac{\sin\alpha}{\sin a} = \frac{\sin\beta}{\sin b} = \frac{\sin\gamma}{\sin c},$$

the cosine rule states that

$$\cos a = \cos b \cos c + \sin b \sin c \cos\alpha,$$

with similar versions for β and γ, and the dual cosine rule states that

$$\cos a = -\cos\beta\cos\gamma + \sin\beta\sin\gamma\cos a.$$

Note that when a, b, c are small enough that the approximations $\sin a \approx a$ and $\cos a \approx 1 - a^2/2$ apply, the sine rule and cosine rule approximate to the Euclidean versions.

spherical trigonometry (spherical geometry) The application of the methods of *trigonometry and *geometry to the study of such matters as the angles, sides, and areas of *spherical triangles and other figures on a sphere. *See also* ELLIPTIC PLANE.

spheroid A spheroid is a *surface of revolution formed by rotating an *ellipse about one of its axes. Equivalently an *ellipsoid is a spheroid if two of its axes are of equal length. The spheroid is oblate if the axis of rotation is the minor axis and prolate if the axis of rotation is the major axis. The *Earth is approximately an oblate spheroid.

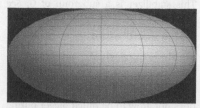

Oblate spheroid with vertical axis

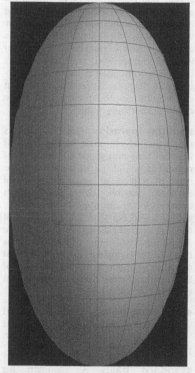

Prolate spheroid with vertical axis

spiral *See* ARCHIMEDEAN SPIRAL, EQUIANGULAR SPIRAL.

spline *Interpolating with *polynomials of high *degree, as is necessary with *Lagrangian interpolation for many *data points, can be computationally difficult, and the interpolating polynomial can fluctuate significantly between data points. Instead, it may be better to interpolate using different low-degree polynomials between the different data points. Given data points $(x_0,y_0), \ldots (x_n, y_n)$, where $a = x_0 < \ldots < x_n = b$, a cubic spline is a function $f(x)$ satisfying

- $f(x)$ is defined by some *cubic polynomial $p_i(x)$ on each interval $[x_i, x_{i+1}]$;
- $f(x_i) = y_i$ at each data point;
- $f(x)$ has a continuous (*see* CONTINUOUS FUNCTION) second *derivative.

Higher-degree splines can be similarly defined.

splitting field Given a polynomial $f(x)$ with coefficients from a *field F, its splitting field K is the smallest field in which $f(x)$ factorizes into linear factors. The splitting field is unique up to *isomorphism. For example, the splitting field of $x^2 + 1$ over $\mathbb{Q}$ is $\mathbb{Q}[i]$ as

$$x^2 + 1 = (x - i)(x + i).$$

sporadic group The classification theorem of *simple groups includes certain *denumerable families: the *cyclic groups of prime order, the *alternating groups (from A_5 onwards), and families of 'Lie type'. Beyond these families, there are 26 exceptional or "sporadic" simple groups; the search for a comprehensive list of such groups was a significant research effort for algebraists in the 1960s and 1970s.

spread A synonym for DISPERSION.

spring A device, usually made out of wire in the form of a *helix, that can be extended and compressed. It is *elastic and so resumes its natural length when the forces applied to extend or compress it are removed. How the *tension in the spring varies with the *extension may be complicated. In the simplest *mathematical model, it is assumed that *Hooke's law holds, so that the tension is proportional to the *extension.

spring constant A synonym for STIFFNESS.

spurious correlation Where two variables possess a *correlation but the link between them is not direct but through another variable, often time or size. For example, the recorded number of burglaries and the number of teachers in towns in the UK show a strong positive correlation, but this is because both of them have a positive correlation with the size of the town.

square The *regular *quadrilateral, i.e. with four equal sides and four right angles.

square (as a power) A number or expression raised to the *power of 2.

square-integrable See L^p.

square matrix A *matrix with the same number of *rows as *columns.

square number Any number which is the *square of an *integer.

square root A square root of a real number a is a number x such that $x^2 = a$. If a is negative, there is no such real number. If a is positive, there are two such numbers, one positive and one negative. For $a \geq 0$, the notation $\sqrt{a}$ is used to denote quite specifically the non-negative square root of a. For *complex numbers, square root is a *multifunction so that a *branch needs to be chosen.

squaring the circle See CONSTRUCTION WITH RULER AND COMPASS.

squeeze theorem See SANDWICH THEOREM.

SSA See SOLUTION OF TRIANGLES.

SSS See SOLUTION OF TRIANGLES.

stabilizer For a *group action of a group G on a set S, the stabilizer of $s \varepsilon S$ consists of those $g \varepsilon G$ such that $g \cdot s = s$.

This stabilizer is a subgroup of G. See also ORBIT-STABILIZER THEOREM.

stable equilibrium See EQUILIBRIUM.

stable numerical analysis In *numerical analysis, the stability of an *algorithm is the extent to which an approximation error in the calculation does not grow to be much bigger during the calculation. Given the great number of iterations that may be involved, this is a very real possibility. Since numerical analysis is concerned with finding approximate solutions to problems for which no analytical solution exists, it is an important consideration as to whether rounding or truncation errors in the algorithm are damped or amplified. Algorithms which can be shown not to magnify approximation errors are called numerically stable.

stadium paradox One of *Zeno's paradoxes which relate to the issue of whether time and space are made up of minute indivisible parts. It considers two rows of objects of equal size—of the smallest possible size—moving with equal velocity in opposite directions, with a speed equal to the smallest unit of length per smallest unit of time. The argument at the centre of the paradox is that relative to one another, they move at one unit of space in half of the unit of time, contradicting the proposition that there is a smallest unit of time.

stagnation point A point in a *fluid where the flow velocity is zero. Different *streamlines can only meet at stagnation points and *singularities.

standard basis A synonym for CANONICAL BASIS.

standard deviation The positive square root of the *variance, a commonly used measure of the *dispersion of *observations in a *sample. For a *normal distribution $N(\mu, \sigma^2)$, with *mean μ and *standard deviation σ, approximately

95% of the distribution lies in the interval between $\mu - 2\sigma$ and $\mu + 2\sigma$.

The standard deviation of an *estimator of a population parameter is the *standard error.

standard error The standard deviation of an *estimator of a population parameter. The standard error of the *sample mean, from a sample of size n, is $\sigma/\sqrt{n}$, where σ^2 is the *population *variance.

standard form (of a linear programming problem) Where the objective function $\sum a_i x_i$ is to be maximized, and all constraints other than the non-negativity conditions can be rewritten in the form $\sum b_j x_j \leq c_j$. If the objective function is to be minimized, then the problem can be converted to standard form by making the objective function $-\sum a_i x_i$ and maximizing this, with the proviso that the optimal solution must be multiplied by -1 at the end. *Linear programming problems in standard form can be solved by the *simplex method.

standard form (of a number) A synonym for SCIENTIFIC NOTATION.

standardize To transform a *random variable so that it has a *mean of zero and *variance of 1. If $E(X) = \mu$ and $\text{Var}(X) = \sigma^2$ the standardized variable Z will have $E(X) = 0$ and $\text{Var}(X) = 1$. The most common application of this is in the normal distribution where probabilities for $X \sim N(\mu, \sigma^2)$ are evaluated using the $N(0, 1)$ tables, applying the transformation

$$Z = \frac{X - \mu}{\sigma}.$$

standard normal distribution See NORMAL DISTRIBUTION.

stars and bars A counting technique in *combinatorics, so named because of the *notation used. For example: how many *triples (x,y,z) of non-negative integers are there such that $x + y + z = 100$?

Each such triple can be represented as $(\star \cdots \star | \star \cdots \star | \star \cdots \star)$ where there are 100 stars, separated by 2 bars into x, then y, then z stars. So there are $\binom{102}{2}$ ways of choosing where to place the 2 bars, and so this number of triples. If we insist x, y, z are positive, then the answer is $\binom{99}{2}$, as 3 of these choices have been made.

state (state space) See MARKOV CHAIN, STOCHASTIC PROCESS.

statement In *logic, the fundamental property of a statement is that it makes sense and is either true or false. For example, 'there is a real number x such that $x^2 = 2$' makes sense and is true; the statement 'if x and y are positive integers then $x - y$ is a positive integer' makes sense and is false. The *liar paradox is not a statement, though, as it cannot be meaningful true or false.

static friction See FRICTION.

statics A part of *mechanics concerned with the *forces acting on a *rigid body or system of *particles at rest with forces acting in *equilibrium.

stationary distribution A stationary distribution for a *Markov chain is a *probability vector $\mathbf{p}$ such that $\mathbf{pM} = \mathbf{p}$, where $\mathbf{M}$ is the transition matrix. An irreducible chain has a stationary distribution if and only if all the states are *recurrent and have finite mean recurrence time. In this case, the stationary distribution is unique and $p_i = 1/m_i$, where m_i is the mean recurrence time of the ith state.

stationary point (in one variable) A point on the graph $y = f(x)$ at which f is *differentiable and $f'(x) = 0$. The term is also used for the number c such that $f'(c) = 0$. The corresponding value $f(c)$ is a stationary value.

S

(i) If $f'(x) > 0$ to the left of c and $f'(x) < 0$ to the right of c, then c is a *local maximum;

(ii) if $f'(x) < 0$ to the left of c and $f'(x) > 0$ to the right of c, then c is a *local minimum;

(iii) $f'(x)$ has the same sign to the left and to the right of c, then c is a *point of inflexion.

stationary point (in two variables) (x, y) is a stationary point of a function $f(x,y)$ if the surface $z = f(x,y)$ has a horizontal *tangent plane above (x,y). Equivalently, if $\partial f/\partial x = 0$ and $\partial f/\partial y = 0$. Now let

$$r = \frac{\partial^2 f}{\partial x^2}, \quad s = \frac{\partial^2 f}{\partial x \partial y}, \quad t = \frac{\partial^2 f}{\partial y^2}.$$

If $rt > s^2$ and $r < 0$, the stationary point is a local maximum; if $rt > s^2$ and $r > 0$, the stationary point is a local minimum; if $rt < s^2$, the stationary point is a *saddle-point. If $rt = s^2$, then the stationary point is said to be degererenate. *See also* HESSIAN.

stationary value *See* STATIONARY POINT.

statistic A *function of a *sample; in other words, a quantity calculated from a set of observations. Often a statistic is an *estimator for a population parameter. For example, the sample *mean, sample *variance, and sample *median are each a statistic. The sum of the observations in a sample is also a statistic, but this is not an estimator.

statistical equilibrium A synonym for STATIONARY DISTRIBUTION.

statistical mechanics In *mechanics, when a system involves a large number of particles, statistical methods are used to analyse the *equations of motion rather than in terms of individual particles. *Maxwell and *Boltzmann were pioneers in this area, and the Maxwell-Boltzmann distribution describes the distribution of speeds of a large number of gas molecules. This applies for low-density or high temperature gases, but otherwise *quantum theoretical aspects become relevant and Bose-Einstein statistics are instead necessary.

statistical model A statistical description of an underlying system, intended to match a real situation as nearly as possible. The *model for a *population is fitted to a *sample by estimating the *parameters in the model. It is then possible to perform *hypothesis testing, construct *confidence intervals and draw *inferences about the population.

statistical tables Sets of tables showing values of *probabilities or often cumulative probabilities of common *distributions for certain values of their parameters. Alternatively, the tables may give the values of the *random variable for specific percentages of the *distribution, giving only the values commonly used as significance levels in *hypothesis tests. Such tables have largely been superseded by *computer software packages.

statistics Quantitative *data on any subject, usually collected with a purpose such as recording performance of bodies for comparison, or for public record and which can then be analysed by a variety of statistical methods. Statistics also refers to the study of collecting, analysing, and displaying such data, a discipline which has grown enormously since the start of the 20th century. Statistics might be seen as the reverse of *probability; in probability, a random scenario or process is understood, and conclusions are made about data that might arise from that process while, with statistics, data have been collected and conclusions are drawn about the underlying random processes (*see* INFERENCE).

steady (steady state) The term steady in general relates to a scenario that is unchanging with *time. In *fluid mechanics, a flow is steady if it is unvarying with time. A 'steady state' solution to a *PDE involving spatial and temporal variables is one which is independent of time; on occasion, this is naturally the solution that the system converges to as time progresses.

Steinitz exchange lemma Let V be a *vector space and $v_1, \ldots, v_m, w_1, \ldots, w_n \in V$. If the v_i are *linearly independent and the w_i span V, then, with some possible reordering of the w_i,

$$\{v_1, \ldots, v_m, w_{m+1}, \ldots, w_n\}$$

spans V. Thus, each v_i can be introduced one at a time in exchange for some w_j and consequentially $m \leq n$. This implies all *bases of V have the same *cardinality and that the *dimension of V is *well defined.

stem-and-leaf plot A method of displaying *grouped data, by listing the observations in each group, resulting in something like a *histogram. For the observations 45, 25, 67, 49, 12, 9, 45, 34, 37, 61, 23, grouped using class intervals 0–9, 10–19, 20–29, 30–39, 40–49, 50–59 and 60–69, a stem-and-leaf plot is shown on the left. If, as in this case, the class intervals are defined by the digit occurring in the 'tens' position, the diagram is more commonly written as shown on the right.

0–9	9			0	9		
10–19	12			1	2		
20–29	23	25		2	3	5	
30–39	34	37		3	4	7	
40–49	45	45	49	4	5	5	9
50–59				5			
60–69	61	67		6	1	7	

stem-and-leaf plot

step function A step function is a *linear combination of *indicator functions of bounded intervals. That is a function of the form

$$\varphi(x) = \sum_{k=1}^{n} c_k \mathbf{1}_{I_k}(x)$$

where the I_k are bounded intervals. This step function can then be assigned (in a *well-defined manner) the *integral

$$\int_{-\infty}^{\infty} \varphi(x) \, dx = \sum_{k=1}^{n} c_k l(I_k)$$

where $l(I)$ denotes the length of the interval I. Consequently step functions are a natural starting point for theorems of integration.

steradian See SOLID ANGLE.

stereographic projection Suppose that a *sphere, centre O, touches a plane at the point S, and let N be the opposite end of the diameter through S. If P is any point (except N) on the sphere, the line NP meets the plane in a corresponding point P'. Conversely, each point in the plane determines a point on the sphere, so there is a *one-to-one correspondence between the points of the sphere (except N) and the points of the plane. This means of mapping sphere to plane is called stereographic projection. Circles not through N on the sphere's surface are mapped to circles in the plane; circles through N are mapped to straight lines. The mapping is *conformal; that is, angles at which curves intersect are preserved by the projection.

Points near N are mapped to points with large *modulus, and so it is natural to associate N with infinity in the *extended complex plane. In this way we identify the extended complex plane with the *Riemann sphere.

(((●))) SEE WEB LINKS

• A full description of stereographic projection with two videos to illustrate how it looks.

Stereographic projection

Stevin, Simon (1548–1620) Flemish engineer and mathematician, the author of a popular pamphlet that explained *decimal representations simply and was influential in bringing about their everyday use, replacing sexagesimal fractions. He made discoveries in statics and hydrostatics and was also responsible for an experiment in 1586 in which two lead spheres, one ten times the weight of the other, took the same time to fall 30 feet. This probably preceded any similar experiment by *Galileo.

Stewart, Ian (1945–   ) English mathematician who has published many popular mathematics books, including *Does God Play Dice* and *17 Equations That Changed the World*.

stiffness A parameter that describes the ability of a spring to be extended or compressed. It occurs as a constant of proportionality in *Hooke's law. *See also* YOUNG'S MODULUS OF ELASTICITY.

Stirling number of the first kind The number $s(n,r)$ of ways of partitioning a set of n elements into r *cycles. For example, the set $\{1,2,3,4\}$ can be partitioned into two cycles in the following ways:

$[1,2,3][4]$, $[1,3,2][4]$, $[1,2,4][3]$, $[1,4,2][3]$, $[1,3,4][2]$, $[1,4,3][2]$, $[2,3,4][1]$, $[2,4,3][1]$, $[1,2][3,4]$, $[1,3][2,4]$, $[1,4][2,3]$.

So $s(4, 2) = 11$. Clearly $s(n,1) = (n-1)!$ and $s(n, n) = 1$. It can be shown that

$$s(n+1, r) = s(n, r-1) + ns(n, r)$$

Rather like the *binomial coefficients, the Stirling numbers occur as coefficients in certain identities. They are named after the Scottish mathematician James Stirling (1692–1770).

Stirling number of the second kind The number $S(n,r)$ of ways of *partitioning a set of n elements into r non-empty subsets. For example, the set $\{1,2,3,4\}$ can be partitioned into two non-empty subsets in the following ways:

$\{1,2,3\}\cup\{4\}$, $\{1,2,4\}\cup\{3\}$, $\{1,3,4\}\cup\{2\}$, $\{2,3,4\}\cup\{1\}$, $\{1,2\}\cup\{3,4\}$, $\{1,3\}\cup\{2,4\}$, $\{1,4\}\cup\{2,3\}$.

So $S(4, 2) = 7$. Clearly, $S(n,1) = 1$ and $S(n,n) = 1$. It can be shown that

$$S(n+1, r) = S(n, r-1) + rS(n, r)$$

Rather like the *binomial coefficients, the Stirling numbers occur as coefficients in certain identities. They are named after the Scottish mathematician James Stirling (1692–1770).

Stirling's formula An estimate for the *factorial function, named after the Scottish mathematician James Stirling (1692–1770), though it was known earlier to *De Moivre. It states that $n!$ and $\sqrt{2\pi n}(n/e)^n$ are *asymptotically equal. The estimate is always an underestimate, but for $n = 100$ is 99.9% accurate. *Bounds for $n!$ exist such as

$$\sqrt{2\pi n}\left(\frac{n}{e}\right)^n < n! < \sqrt{2\pi n}\left(\frac{n}{e}\right)^n \exp\left(\frac{1}{12n}\right).$$

stochastic differential equation (SDE) A *differential equation involving terms that are *stochastic (or random) processes, so that the solution is itself a stochastic process. Such differential equations might be used, for example, to *model financial markets or *Brownian motion. The stochastic analysis involved in their study can be very technical and complicated.

stochastic matrix A *square matrix is row stochastic if all the entries are non-negative and the entries in each *row add up to 1. It is column stochastic if all the entries are non-negative and the entries in each *column add up to 1. A square matrix is stochastic if it is either row stochastic or column stochastic, such as a *transition matrix. It is doubly stochastic if it is both row stochastic and column stochastic.

stochastic process **(stochastic model)** A family $\{X(t), t \varepsilon T\}$ of *random variables, where T is some *index set. Often the index set is the set $\mathbb{N}$ of *natural numbers, and the stochastic process is a sequence $X_1, X_2, X_3, \ldots$ of random variables. For example, X_n may be the outcome of the nth *trial of some experiment, or the nth in a set of *observations. The possible values taken by the random variables are often called states, and these form the state space. The state space is said to be discrete if it is *countable, and continuous if it is uncountable. *See* MARKOV CHAIN, POISSON PROCESS.

stochastic variable A synonym for RANDOM VARIABLE.

Stokes, Sir George Gabriel (1819–1903) Irish mathematician who was Lucasian Professor of Mathematics at Cambridge University. His prolific mathematical research output arose mainly from study of physical processes such as *hydrodynamics and hydrostatics, wind pressure systems, and spectrum analysis. As well as being one of Britain's foremost scientists Stokes was unusually active in public life, as a member of Parliament for Cambridge University

1881–92, secretary of the Royal Society 1854–84, and then its President 1885–90.

Stokes' Theorem

$$\iint_{\Sigma} \text{curl} \mathbf{F} \cdot \mathbf{n} \ dS = \int_{C} \mathbf{F} \cdot d\mathbf{r}$$

where $\mathbf{F}$ is a differentiable *vector field defined on a *surface Σ which has a boundary *curve C, that is, that the *flux integral of the *curl of $\mathbf{F}$ over the surface is equal to the *work integral of $\mathbf{F}$ around the boundary of the surface. The surface and bounding curve need to be compatibly oriented; this means that when travelling along the boundary in the direction of the tangent $d\mathbf{r}$ and standing in the direction of the unit normal $\mathbf{n}$ the surface Σ is on your left.

Stokes' Theorem (Stokes-Cartan theorem) (generalized form) If M is a smooth *orientable *manifold with boundary ∂M and ω is a *differential form on M, then

$$\int_{M} d\omega = \int_{\partial M} \omega.$$

When M is 1-,2-, or 3-dimensional, this gives the *Fundamental Theorem of Calculus, *Stokes' theorem, or the *divergence theorem respectively.

Stone space A *compact, *Hausdorff, *zero-dimensional *topological space. Examples include finite *discrete spaces and the *Cantor set. Such spaces are of interest due to Stone's representation theorem for *Boolean algebras. The clopen subsets (*see* CLOPEN SET) of any topological space form a Boolean algebra—together with union, intersection, complement—and to every Boolean algebra a Stone space can be associated.

Stone-Weierstrass Theorem Weierstrass' Approximation Theorem shows any *continuous real-valued function on a closed bounded *interval can be arbitrarily and *uniformly approximated by *polynomials; that is, there is a

S

sequence of polynomials which uniformly converges to the function.

The theorem was considerably generalized by Stone to the following: let X be a *compact *Hausdorff space and let $C(X)$ denote the *algebra of continuous real-valued functions on X with the uniform topology; then any subalgebra A of $C(X)$ is *dense in $C(X)$ if for any distinct points x, y in X there exists f in A such that $f(x) \neq f(y)$.

straight angle An angle equal to half a turn, measuring 180 *degrees or π *radians, and equal to two *right angles.

straight edge A tool used for drawing straight lines, but without any measurement of distance, in geometrical constructions. *See* CONSTRUCTION WITH RULER AND COMPASS.

straight line *See* LINE (in two dimensions), LINE (in three dimensions).

strain *Hooke's law for a spring states that $T = \lambda x/l$, where T is *tension, l natural length, x is *extension, and λ denotes *Young's modulus. The strain is then the dimensionless quantity x/l. In more advanced *models in solid mechanics strain is represented as a rank two *tensor at each point of a body.

strategy A combination of moves for a player in a *game. In some games, a strategy can be decided by the current state of the game—such as in noughts and crosses. In other games, like poker, a variation in approach over a number of games is required so the other players cannot identify your hand simply by your moves. In game theory this is modelled by a probability distribution on the available choices. *See* MIXED STRATEGY.

stratified sample Where a *population is not homogeneous, but contains groups of individuals (the strata) for which the characteristic of interest is likely to be noticeably different, it is likely that a stratified sample will provide a better *estimator of the characteristic than a simple random *sample, in the sense that the estimates will be closer to the true value more often. The procedure is to take a random sample from within each group, proportional to the size of the group. This removes one major source of variation in the simple random sample, namely how many of each group is taken.

stream function For an *inviscid, *incompressible, planar flow, with flow velocity $\mathbf{u}$, a stream function is a scalar function ψ such that $\mathrm{curl}(\psi\mathbf{k}) = \mathbf{u}$. The stream function is constant along *streamlines. *See* COMPLEX POTENTIAL, VELOCITY POTENTIAL.

streamline For a *fluid with flow velocity $\mathbf{u}$, a streamline is a *curve such that at every point of the curve a *tangent vector is parallel to the flow velocity. In a *steady flow, the streamlines are the paths followed by particles, but this is not generally true in an unsteady flow. *See* STREAM FUNCTION.

stress The *force per unit area across a surface of a body. Different types of stress are shear, tensile, and compressive, and all bodies undergo it. As with *elastic strings and springs most bodies revert to their original shapes once the stress is removed, but if the stress is too great they may remain in the deformed state or even break (*see* DEFORMATION). In more advanced *models in solid mechanics stress is represented as a rank two *tensor.

strict Adjective implying that it is more restrictive than the *relation itself. *Inequalities involving $<$ and $>$ are called strict, compared with 'weak' inequalities involving $\leq$ and $\geq$. *Proper subsets are said to be strictly *contained in the set.

strictly decreasing *See* DECREASING FUNCTION, DECREASING SEQUENCE.

strictly determined game A *matrix game is strictly determined if there is an entry in the matrix that is the smallest in its row and the largest in its column. If the game is strictly determined then, when the two players R and C use *conservative strategies, the pay-off is always the same and is the value of the game.

The game given by the matrix on the left below is strictly determined. The conservative strategy for R is to choose Row 3, and the conservative strategy for C is to choose Column 1. The value of the game is 5. The game on the right is not strictly determined:

$$\begin{bmatrix} 2 & 8 & 4 \\ 3 & 1 & 4 \\ 5 & 6 & 7 \end{bmatrix}, \quad \begin{bmatrix} 2 & 8 & 4 \\ 6 & 1 & 3 \\ 5 & 4 & 7 \end{bmatrix}.$$

strictly increasing See INCREASING FUNCTION, INCREASING SEQUENCE.

strictly monotonic See MONOTONIC FUNCTION, MONOTONIC SEQUENCE.

string See ELASTIC STRING, INEXTENSIBLE STRING.

string A synonym for WORD.

strong convergence A synonym for CONVERGENCE ALMOST SURELY.

stronger statement If a statement P implies a statement Q and Q does not imply P, then Q is the weaker statement and P is the stronger statement. For example, x is even is a weaker statement than x equals 2, but a stronger statement than x is an integer. *Compare* EQUIPOLLENT, A FORTIORI.

strong law of large numbers Given a *random variable X with finite *variance, the sample *mean of the first n observations, that is,

$$\frac{1}{n}\sum_{i=1}^{n}X_i$$

*converges almost surely to the mean $\mu = E(X)$. *Compare* WEAK LAW OF LARGE NUMBERS.

structuralism A *philosophy of mathematics, dating to the 1960s, which maintains that mathematical objects are entirely defined by their position in mathematical structures. As such, mathematical objects do not have any intrinsic properties. Structuralism contrasts with some philosophies that would place a primacy on foundational theory (e.g. *set theory, *category theory).

Structure Theorem (for modules) For a *finitely generated *module M over a *principal ideal domain (or *Euclidean domain) R, there exists an integer $r \geq 0$ and non-zero, non-*unit elements d_1, $d_2, \ldots, d_k$ in R such that d_i divides d_{i+1} for $1 \leq i < k$ and such that M is *isomorphic, as a module, to

$$R^r \oplus \frac{R}{d_1 R} \oplus \frac{R}{d_2 R} \oplus \cdots \oplus \frac{R}{d_k R}.$$

r is the torsion-free *rank of M and the d_i are the invariant factors. This decomposition is unique up to multiplying the invariant factors by units. When $R = \mathbb{Z}$, then the classification of finite *abelian groups follows. When $R = F[x]$, where F is a field, the *rational canonical form follows, and when $R = \mathbb{C}[x]$, a similar version of the structure theorem gives the *Jordan normal form. *See also* SMITH NORMAL FORM.

strut A rod in *compression.

Student See GOSSET, WILLIAM SEALY.

Student's t-distribution A synonym for T-DISTRIBUTION.

Sturm–Liouville equation A second-order *differential equation

$$\frac{d}{dx}\left[p(x)\frac{dy}{dx}\right] + [\lambda\omega(x) - q(x)]y = 0$$

where $\omega(x)$ is called the 'weighting function'. The values of λ for which there is a solution are the *eigenvalues and the corresponding solutions are the *eigenfunctions.

subadditive A *function $f:\mathbb{R}\rightarrow\mathbb{R}$ is subadditive if $f(x+y) \leq f(x)+f(y)$ for all x,y.

subdivision (of a graph) An edge joins two vertices of a graph. If another vertex is introduced onto that edge, it creates two edges, so in the figures below, the insertion of the vertex C has subdivided the edge AB into two edges, namely AC and CB. The extra vertex C is necessarily of degree 2. A subdivision of a graph is a graph modified by the addition of one or more vertices of degree 2 into an existing edge or edges.

ACB is a subdivision of AB

subdivision (of an interval) A synonym for PARTITION (of an interval).

subdivision (of a surface) A subdivision of a closed *surface X is a finite collection of *vertices and *edges in X such that (i) every edge begins and ends in a vertex, (ii) edges intersect in vertices, (iii) the *complement of the vertices and edges has *connected components which are *homeomorphic to an open disc; these are known as faces. The *Euler characteristic of X the surface then equals $V-E+F$, where V, E, F denotes the number of vertices, edges, and faces respectively. See TRIANGULATION.

subfield Let F be a *field with operations of addition and multiplication. If S is a subset of F that forms a field with the same operations, then S is a subfield of F. For example, the set $\mathbb{Q}$ of rational numbers forms a subfield of the field $\mathbb{R}$ of real numbers.

subgraph A subset of the vertices and edges of a *graph which themselves form a graph.

subgroup Let G be a *group with a given operation. If H is a subset of G that forms a group with the same operation, then H is a subgroup of G. For example, $\{1, i, -1, -i\}$ forms a subgroup of the group of all non-zero *complex numbers with multiplication. See LAGRANGE'S THEOREM, SYLOW'S THEOREMS.

subgroup generated by a set Let S be a subset of a *group G. The subgroup generated by S, written $\langle S \rangle$, is the smallest subgroup of G that contains S. See CYCLIC GROUP, FINITELY GENERATED.

subject (in logic) The category which a *categorical proposition is about. For example, in the statement 'some trees are evergreen', the subject is 'trees'.

submatrix A submatrix of a *matrix A is obtained from A by deleting from A some number of rows and some number of columns. For example, suppose that A is a 4×4 matrix and that $A = [a_{ij}]$. Deleting the first and third rows and the second column gives the submatrix

$$\begin{bmatrix} a_{21} & a_{23} & a_{24} \\ a_{41} & a_{43} & a_{44} \end{bmatrix}.$$

See DETERMINANTAL RANK.

subring Let R be a *ring with operations of addition and multiplication. If S is a subset of R that forms a ring with the same operations, then S is a subring of R. For example, the set $\mathbb{Z}$ of all integers forms a subring of the ring $\mathbb{R}$ of all real numbers, and the set of all even integers forms a subring of $\mathbb{Z}$.

subscript One or more indices written to the lower right of a symbol, as with the n in a_n to denote a term in a *sequence or as with the ij in a_{ij} to denote an entry in a *matrix. Compare SUPERSCRIPT.

subsequence A subsequence $\{x_{n_k}\}_{k \geq 1}$ of a sequence $\{x_n\}_{n \geq 1}$ is an sequence selected from $\{x_n\}$ in order. Specifically the map $k \mapsto n_k$ is a *strictly increasing map from $\mathbb{N}$ to $\mathbb{N}$. If $\{x_n\}$ *converges to L, then all its subsequences also converge to L. See BOLZANO-WEIERSTRASS THEOREM.

subset The set A is a subset of the set B if every element of A is an element of B. When this is so, A is *contained in B, written $A \subseteq B$, and B contains A, written $B \supseteq A$. The following properties hold:

(i) For all sets A, $\emptyset \subseteq A$ and $A \subseteq A$, where $\emptyset$ denotes the *empty set.

(ii) For all sets A and B, $A = B$ if and only if $A \subseteq B$ and $B \subseteq A$.

(iii) For all sets A, B and C, if $A \subseteq B$ and $B \subseteq C$, then $A \subseteq C$.

See also ALGEBRA OF SETS, POWER SET, PROPER SUBSET.

subspace *See* N-DIMENSIONAL SPACE.

subspace topology For a *topological space X and subset A, the sets $A \cap U$, where U is an *open set of X, form a topology on A called the subspace (or relative or induced) topology. When X is a *metric space, the subspace topology on A is that induced by the metric restricted to pairs of points of A.

substitution The replacement of a term in an expression or equation with another which is known to have the same value. This includes replacing variables with their numerical value to evaluate a formula. *See also* INTEGRATION.

substitution group A synonym for PERMUTATION GROUP.

subtend The *angle that a line segment or arc with end-points A and B subtends at a point P is the angle APB. For example, one of the *circle theorems may be stated as follows: the angle subtended by a diameter of a circle at any point on the circumference is a right angle.

subtraction The *binary operation which is the inverse operation to *addition which calculates the difference between two numbers or quantities. So $7 - 2 = 5$ as $7 = 5 + 2$; the first equation follows from the second by subtracting 2 or adding -2, the additive *inverse of 2.

sufficient condition A condition which is enough for a statement to be true. For example, for a number to be even, it is sufficient that the number be divisible by 10. Note that if A is a *necessary condition for B, B is a sufficient condition for A, and vice versa, so in the above example for a number to be divisible by 10 it is necessary that the number is even.

sufficient estimator An *estimator of a parameter θ that gives as much information about θ as is possible from the *sample. The sample *mean is a sufficient estimator of the population mean of a normal distribution.

sufficiently large (sufficiently small) A proposition $P(x)$, where x is a *real number or *natural number, is said to be true 'for sufficiently large x' if there exists M such that $P(x)$ is true for all $x > M$. It may also be said that $P(x)$ is eventually true. Likewise, $P(x)$ is said to be true 'for sufficiently small x' if there exists $\varepsilon > 0$ such that $P(x)$ is true whenever $|x| < \varepsilon$.

sufficient statistic (for a parameter) A *statistic that contains all of the information in a *sample which is relevant to a point estimate of the parameter. When a sufficient statistic exists the *maximum likelihood estimator will be a function of the sufficient statistic.

sum The outcome of an *addition, whether it be of *vectors, *matrices, *sequences, or real-valued *functions. *See also* SERIES.

summation notation The finite *series $a_1 + a_2 + \ldots + a_n$ can be written, using the capital Greek letter *sigma, as

$$\sum_{r=1}^{n} a_r.$$

Similarly, for example,

$$1^2 + 2^2 + \ldots + 10^2 = \sum_{r=1}^{10} r^2,$$

$$1 + x + x^2 + \ldots + x^{n-1} = \sum_{r=0}^{n-1} x^r.$$

(The letter 'r' used here is a *dummy variable and so could be replaced by any other letter.) In a similar way, the infinite series $a_1 + a_2 + \ldots$ can be written as

$$\sum_{r=1}^{\infty} a_r.$$

The infinite series may not have a sum to infinity or *limit but, if it does, the same abbreviation is also used to denote this limit

sum to infinity See GEOMETRIC SERIES, SERIES.

sup Abbreviation for *supremum.

superadditive A *function $f : \mathbb{R} \to \mathbb{R}$ is superadditive if $f(x+y) \geq f(x) + f(y)$ for all x, y.

superposition principle If y_1 and y_2 are solutions of a *homogeneous linear differential equation $Ly = 0$, then $c_1 y_1 + c_2 y_2$ is also a solution, by definition of L being linear. The same may also be true of an infinite sum

$$\sum_{k=0}^{\infty} c_k y_k$$

of solutions y_k, though issues of *convergence and *differentiability may need addressing carefully. Given a homogeneous linear *boundary value problem, a general solution may be sought as an infinite sum of *separable solutions with the coefficients c_k being determined using *Fourier analysis.

superscript One or more indices written to the upper right of a symbol, as with the n in $f^{(n)}$ to denote a higher *derivative or as with the x in 2^x to denote an *exponent. Compare SUBSCRIPT.

supersink (in a network) If a network has two or more *sinks then a supersink can be introduced connected to all the sinks, with the flow on each edge equal to the total inflow to the source to which it connects.

supersource (in a network) If a network has two or more *sources then a supersource can be introduced connected to all the sources, with the flow on each edge equal to the total outflow from the source to which it connects.

supplementary angles Two angles that add up to two right angles. Each angle is the supplement of the other.

supplementary unit See SI UNITS.

support For a *function $f : X \to \mathbb{R}$ on a *topological space (or *metric space) X, the support of f is the *closure of the subset $\{x \in X \mid f(x) \neq 0\}$.

supremum (suprema) See BOUND.

surd An *irrational number in the form of a root of some number (such as $\sqrt{5}$, $\sqrt[3]{2}$ or $4^{2/3}$) or a numerical expression involving such numbers.

surface A set of points (x, y, z) in *Cartesian space, represented as a graph as $z = f(x, y)$, or defined implicitly as $g(x, y, z) = 0$, or described as a *parameterized surface. See also ATLAS, CHART, MANIFOLD.

surface (of a solid) The boundary of a three-dimensional geometric shape. For example, the surface of a solid cylinder comprises two discs and a curved surface.

surface (in topology) A 2-dimensional *manifold. The most familiar examples are the *surfaces of solids, but examples like the *non-orientable *Klein bottle and the *real projective plane are also surfaces. See CLASSIFICATION THEOREM FOR SURFACES, MANIFOLD.

surface area See SURFACE INTEGRAL.

surface integral For a smooth *surface Σ, and continuous (*see* CONTINUOUS FUNCTION) scalar field f on Σ and a continuous vector field $\mathbf{F}$ on Σ, a surface integral may take either of the forms $\int_\Sigma f \, dS$ or $\int_\Sigma \mathbf{F} \cdot d\mathbf{S}$. If f is identically 1, then the integral $\int_\Sigma f \, dS$ defines the surface area of Σ. The integral $\int_\Sigma \mathbf{F} \cdot d\mathbf{S}$ defines the *flux of $\mathbf{F}$ through Σ. If $\mathbf{r}(u,v)$, where $(u,v) \, \varepsilon \, U \subseteq \mathbb{R}^2$, is a parametrization of Σ, then the surface integrals are defined by

$$\int_\Sigma f \, dS = \iint_U f\big(\mathbf{r}(u,v)\big)\left|\frac{\partial \mathbf{r}}{\partial u} \times \frac{\partial \mathbf{r}}{\partial v}\right| du \, dv,$$

$$\int_\Sigma \mathbf{F} \cdot d\mathbf{S} = \iint_U \mathbf{F}\big(\mathbf{r}(u,v)\big) \cdot \left(\frac{\partial \mathbf{r}}{\partial u} \times \frac{\partial \mathbf{r}}{\partial v}\right) du \, dv,$$

where $\times$ denotes the *vector product. *See* DIVERGENCE THEOREM, STOKES' THEOREM.

surface of revolution Suppose that an arc of a plane *curve is rotated through one *revolution about a line in the plane that does not cut the arc (except possibly at the arc's ends). The surface of the 3-dimensional figure obtained is a surface of revolution. *Cylinders, *cones, *spheres and *spheroids, *catenoids, *tori and *hyperboloids of one sheet and *hyperboloids of two sheets can all be generated as surfaces of revolution. *See also* AREA OF A SURFACE OF REVOLUTION.

surjection (surjective mapping) A synonym for ONTO MAPPING.

suvat A term describing *equations of motion with constant acceleration.

syllogism An argument consisting of three *categorical propositions where the truth of the first two requires the third (the conclusion) also to be true. For example, 'All men are mortal; Socrates is a man; therefore Socrates is mortal.'

Any syllogism must relate to three entities (which can be a group, an individual entity, or a concept—here 'men',

Socrates', and 'mortality'), each of which appears exactly twice.

Sylow's theorems Three theorems due to the Norwegian mathematician Peter Sylow which provide partial converses to *Lagrange's theorem. Let G be a group of order $p^a m$, where p is a *prime which does not divide m and a is a positive integer. Then:

1ST THEOREM: G has a subgroup of order p^a. Any such subgroup is called a Sylow p-subgroup.

2ND THEOREM: Any two Sylow p-subgroups are *conjugate in G.

3RD THEOREM: If there are n_p Sylow p-subgroups in G, then n_p divides m and p divides $n_p - 1$.

Sylow's theorems have more significance when G's order is divisible by several primes, but imply nothing useful about groups whose orders are a power of a prime.

Sylvester's equation Given *square matrices A, B, C of the same order, Sylvester's equation is the matrix equation $AX + XB = C$. The equation has a unique solution X unless the matrices A and $-B$ have an *eigenvalue in common.

Sym Sym(S) denotes the *symmetric group of a set S.

symbol A single letter or sign which is used to represent an *operation or *relation, a *function, a *number, or a quantity. *See* NOTATION.

symbolic logic The area of mathematics relating to deductive argument within symbolic abstractions. Its two main branches are *propositional logic and *predicate logic.

symmetrical about a line A plane figure is symmetrical about the line l if, whenever P is a point of the figure, so too is P', where P' is the mirror-image of P in the line l. The line l is called a line of symmetry; and the figure is said to have bilateral symmetry and is *invariant under *reflection in the line l. The graph of an *even function is symmetrical about the y-axis.

symmetrical about a point A plane figure is symmetrical about the point O if, whenever P is a point of the figure, so too is P', where O is the midpoint of $P'P$. The point O is called a centre of symmetry; and the figure is said to have *rotational symmetry of order 2 about O because the figure appears the same when rotated through half a revolution about O. The graph of an *odd function is symmetrical about the origin.

symmetric design A *block design in which the number of blocks and the numbers of points in each block are the same.

symmetric difference For sets A and B (subsets of some *universal set), the symmetric difference, denoted by $A + B$, is the set $(A \backslash B) \cup (B \backslash A)$. The notation $A \triangle B$ is also used. Its logical equivalent is *exclusive disjunction or 'xor'. The set is represented by the shaded regions of the *Venn diagram shown. The following properties hold, for all A, B, and C (subsets of some universal set E):

(i) $A + A = \varnothing$, $A + \varnothing = A$, $A + A' = E$, $A + E = A'$.

(ii) $A + B = (A \cup B) \backslash (A \cap B)$
$= (A \cup B) \cap (A' \cup B')$.

(iii) $A + B = B + A$, the commutative law.

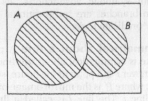

The symmetric difference $A \triangle B$

(iv) $(A + B) + C = A + (B + C)$, the associative law.

(v) $A \cap (B + C) = (A \cap B) + (A \cap C)$, the operation $\cap$ is distributive over the operation $+$.

symmetric function A *function f on n variables $x_1, x_2, \ldots, x_n$ is symmetric if

$$f\left(x_{\sigma(1)}, x_{\sigma(2)}, \ldots, x_{\sigma(n)}\right) = f(x_1, x_2, \ldots, x_n)$$

for all *permutations σ of n variables. See ELEMENTARY SYMMETRIC POLYNOMIAL.

symmetric group For any set X, a permutation of X is a *one-to-one onto mapping from X to X. The set of all of these, with *composition as the operation, forms a *group called the symmetric group of X, denoted Sym(X). If X has n elements, there are $n!$ permutations of X and the symmetric group is denoted S_n. See ALTERNATING GROUP, CAYLEY'S REPRESENTATION THEOREM.

symmetric matrix Let $\mathbf{A}$ be the square matrix $[a_{ij}]$. Then $\mathbf{A}$ is symmetric if $\mathbf{A}^T = \mathbf{A}$ (see TRANSPOSE); that is, if $a_{ij} = a_{ji}$ for all i and j.

symmetric relation A *binary relation $\sim$ on a set S is symmetric if, for all a and b in S, whenever $a \sim b$ then $b \sim a$. See EQUIVALENCE RELATION.

symmetry (of a geometrical figure) A figure may have reflective symmetry in a line (axis) or plane or may have *rotational symmetry. The term also refers to a transformation of the figure which maps the figure to itself; these symmetries form the *symmetry group under *composition. As an example, a *square is symmetrical about four lines and has rotational symmetry of order 4 about its centre, and so the square has 8 symmetries in all.

symmetry group (of a geometrical figure) The *symmetries of a figure form its symmetry *group under *composition. The symmetry group of a *regular polygon is a *dihedral group, of a *rectangle is V_4, of the *tetrahedron is A_4, of the *cube and *octahedron is S_4, and of the *dodecahedron and *icosahedron is A_5.

symplectic geometry An area of *geometry arising naturally from the study of *configuration spaces in *Hamiltonian mechanics. Symplectic *manifolds necessarily have even *dimensions.

symplectic group *See* MATRIX GROUPS.

synthetic geometry In contrast to *analytic geometry, which employs coordinates, theorems in geometry are derived from *axioms making no use of coordinates, as is the case in *Euclid's *Elements*.

systematic error An error which is not *random and is therefore liable to introduce *bias into an *estimator. For example, $\sum_{r=1}^{n} \dfrac{(x_i - \bar{x})^2}{n}$ has a tendency to underestimate the population *variance. In a production process, if one of ten prongs on a rotating drum is broken then every tenth item on the production line will be faulty in addition to any random faults which may occur.

systematic sampling Choose a number at random from 1 to n, and then take every nth value thereafter. This means that all units have an equal chance of being selected, but it does not constitute

a *random sample, as not all combinations are equally likely. There is a particular danger with this method, especially for 'production line sampling' as any faults with a particular machine in the system will affect items regularly, and the systematic method of choosing the items to be sampled means that you are very likely either to miss them all or hit them all (or some fraction of them).

Système International d'Unités *See* SI UNITS.

system of particles A collection of *particles whose motion is under investigation. In a given mathematical model, the particles may be able to move freely relative to each other, or they may be constrained by light rods, springs, or strings.

systems analysis The area of mathematics concerned with the analysis of large *complex systems, particularly where the systems involve interactions.

Szemerédi's theorem The theorem shows that a suitably dense subset of the *natural numbers contains arbitrarily long *arithmetic progressions. The set $A \subseteq \mathbb{N}$ needs to satisfy

$$\limsup_{n \to \infty} \frac{|A \cap \{1, 2, \ldots, n\}|}{n} > 0.$$

The theorem implies *Van der Waerden's theorem but does not imply the *Green-Tao theorem, as the primes do not have positive density. *See also* ERDŐS CONJECTURE.

S

T Symbol for *transpose of a matrix, written as a superscript, so $\mathbf{M}^{\mathrm{T}}$ is the transpose of $\mathbf{M}$.

T Symbol for 'true' in *logic and in *truth tables.

T Abbreviation for *tera-.

τ *See* TAU.

T_n *See* SEPARATION AXIOMS.

tableau *See* SIMPLEX METHOD.

tables A set of values of *functions, set out in a tabular form for a range of arguments such as log tables, trigonometric tables, *special functions, and statistical tables. With the advent of calculators and powerful computers, such tables are now rarely used.

tacnode *See* DOUBLE POINT.

tail A tail of a sequence $(x_n)_{n=1}^{\infty}$ is a *subsequence of the form $(x_n)_{n=N}^{\infty}$ for some N; that is, it is the remainder of the sequence past some given index. A sequence converges to a *limit l if every *neighbourhood of l contains a tail of the sequence.

tangent *See* TRIGONOMETRIC FUNCTION.

tangent (tangent line) (to a curve) Let P be a point on a *curve. Then the tangent line, or simply tangent, to the curve at P is the line through P that *touches the curve at P. *See also* GRADIENT (of a curve), TANGENT SPACE.

tangent bundle To each point P of a smooth *manifold M is associated a *tangent space $T_P(M)$. The union of these tangent spaces

$$TM = \bigcup_{P \in M} \{P\} \times T_P M$$

can itself be given the structure of a manifold. TM is known as the tangent bundle and has twice the *dimension of M. There is a natural (first coordinate) projection $\pi: TM \rightarrow M$ and $T_P(M) = \pi^{-1}(P)$ is called the fibre over P.

Tangent bundles are examples of vector bundles which more generally are manifolds comprising *vector spaces (the fibres) associated with points of the base manifold M. The topology of a manifold can be investigated by the use of vector bundles and other spaces naturally associated with the manifold.

tangent field A *differential equation gives information about the rate of change of the dependent variable but not the actual value of the variable. A graph showing the direction indicators of the tangents from which families of approximate solution curves can be drawn are called tangent fields, or direction fields. Curves along which the direction indicators have the same gradient are called isoclines, so in the figure, all the direction indicators at the same value of y are parallel and any horizontal line is an isocline.

The direction field of $\frac{dy}{dx} = -y$ is shown in the figure. The general solution to this differential equation is $y = Ae^{-x}$ and solution curves are shown for the cases where $A = -2$, -0.5 and 1. In the case

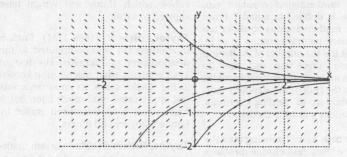

tangent field of $\frac{dy}{dx} = -y$

where an analytical or algebraic solution is available the tangent fields are not especially useful, but they are powerful tools when no exact solution can be obtained.

tangential Being a *tangent or in the direction of a *tangent vector.

tangent plane Let P be a point on a *smooth surface. The tangent plane at P is the plane through P that *touches the surface. Equally, it is the plane containing all the *tangent lines to smooth *curves in the surface that pass through P. If $\mathbf{r}(u,v)$ is a parameterization of the surface about $P = \mathbf{r}(u_0, v_0)$, then the tangent plane is the plane through P parallel to $\partial\mathbf{r}/\partial u(u_0, v_0)$ and $\partial\mathbf{r}/\partial v(u_0, v_0)$. *See* TANGENT SPACE.

tangent rule *See* TRIANGLE.

tangent space A tangent vector to a point P on a smooth *curve is any *vector parallel to the *tangent line. If $\mathbf{r}(s)$ is a parameterization of the curve by *arc length, then $\mathbf{t} = \mathbf{r}'(s)$ is a unit tangent vector (*see also* SERRET-FRENET FORMULAE). More generally, a tangent vector to a point P in a smooth *manifold X is a tangent vector to a smooth curve in the manifold passing through P.

The tangent space at P is the *vector space whose elements are the tangent vectors at P. So, if $\mathbf{r}(u,v)$ is a parameterization of a surface about $P = \mathbf{r}(u_0, v_0)$, then the tangent space is spanned by $\partial\mathbf{r}/\partial u(u_0, v_0)$ and $\partial\mathbf{r}/\partial v(u_0, v_0)$. Note that the tangent space is parallel to the *tangent plane but passes through the origin. The tangent space at P is denoted $T_P(X)$.

tangent vector *See* TANGENT SPACE.

tanh *See* HYPERBOLIC FUNCTION.

Taniyama-Shimura conjecture *See* MODULARITY THEOREM.

Tarski, Alfred (1901-83) Polish mathematician, especially influential in *logic, *proof theory and *model theory, though he published widely. In 1924 he published the famous *Banach-Tarski paradox.

Tartaglia, Niccolò (1499-1557) Italian mathematician remembered as the discoverer of a method of solving *cubic equations, though he may not have been the first to find a solution. His method was published by *Cardano, to whom he had communicated it in confidence. Born Niccolò Fontana, his adopted nickname means 'stammerer'.

tau The mathematical constant tau, denoted τ, is defined to equal 2 times *pi. Some mathematicians consider tau to be a more natural constant than pi, noting that it is the number of *radians in a complete *revolution and that 2π appears in many mathematical formulae such as in the *pdf of the standard *normal distribution, in *Stirling's formula, and in *Cauchy's residue theorem.

tautochrone As well as being the solution to the *brachistochrone problem, the *cycloid has another property. Suppose that a cycloid is positioned as in the diagram for the brachistochrone problem. If a particle starts from rest at any point of the cycloid and travels along the curve under the force of gravity, the time it takes to reach the lowest point is independent of its starting point. So the cycloid is also called the tautochrone (from the Greek for 'same time').

tautology A *compound statement that is true for all possible truth values of its components. For example, $p \Rightarrow (q \Rightarrow p)$ is a tautology, as can be seen by calculating its *truth table.

taxicab norm (Manhattan norm) The sum of the *absolute values of the components of a *vector. The name derives from the distance a taxi has to drive in a rectangular street grid to get from the origin to a particular point. The taxicab norm is $p = 1$ in the family of *p-norms.

taxicab number The number 1729. In 1919, *Hardy visited an ill *Ramanujan and commented that his taxicab number, 1729, was rather dull. Ramanujan contradicted Hardy, saying it was, in fact, the smallest number that could be written as the sum of two cubes in two different ways, namely $1729 = 9^3 + 10^3 = 1^3 + 12^3$. The term also relates to the smallest numbers which can be written in n different ways as the sum of two

cubes, which Hardy and Wright later showed to exist.

Taylor, Brook (1685–1731) English mathematician who contributed to the development of *calculus. His text of 1715 contains what has become known as the *Taylor series. Its importance was not appreciated until much later, but it had in fact been discovered earlier by James *Gregory and others.

Taylor, Richard (1962–) British mathematician and former student of Andrew *Wiles, who worked with him to complete the proof of *Fermat's Last Theorem.

Taylor series *See* TAYLOR'S THEOREM.

Taylor's Theorem Applied to a suitable function f, Taylor's Theorem gives a polynomial that is an approximation to $f(x)$.

Theorem Let f be a real function on an open interval I, such that the derived functions $f^{(r)}$ $(r = 1, \ldots, n)$ are *continuous functions and suppose that $a \varepsilon I$. Then, for all x in I,

$$f(x) = f(a) + \frac{f'(a)}{1!}(x-a) + \frac{f''(a)}{2!}(x-a)^2 + \ldots + \frac{f^{(n-1)}(a)}{(n-1)!}(x-a)^{n-1} + R_n,$$

where R_n denotes the remainder term R_n.

Two possible forms for R_n are

$$R_n = \frac{1}{(n-1)!} \int_a^x f^{(n)}(t)(x-t)^{n-1} dt \text{ and}$$

$$R_n = \frac{f^{(n)}(c)}{n!}(x-a)^n,$$

where c lies between a and x. By taking $x = a + h$, where $a + h\varepsilon I$, the formula

$$f(a+h) = f(a) + \frac{f'(a)}{1!}h + \frac{f''(a)}{2!}h^2 + \ldots + \frac{f^{(n-1)}(a)}{(n-1)!}h^{n-1} + R_n$$

is obtained. This enables $f(a + h)$ to be determined up to a certain degree of accuracy, the remainder R_n giving the *error. Suppose now that f is infinitely differentiable in I and that $R_n \to 0$ as $n \to \infty$; then an infinite series can be obtained whose sum is $f(x)$. In such a case, it is customary to write

$$f(x) = f(a) + \frac{f'(a)}{1!}(x-a) + \frac{f''(a)}{2!}(x-a)^2 + \ldots.$$

This is the Taylor series (or expansion) for f at (or about) a. The special case with $a = 0$ is the Maclaurin series for f. Note that the Taylor series of an infinitely differentiable function $f(x)$ can converge without converging to $f(x)$; it is important that the remainder term tends to 0. For example, the function

$$f(x) = \begin{cases} \exp(-1/x^2), & x \neq 0, \\ 0, & x = 0, \end{cases}$$

is infinitely differentiable at 0 with $f^{(n)}(0) = 0$ for all n. Thus, the Taylor series converges, but to the zero function, rather than to $f(x)$.

The Taylor series for a real function $f(x,y)$ of two variables, which has *partial derivatives of all orders, states that

$$f(a+h, b+k)$$
$$= f(a,b) + \left(f_x(a,b)h + f_y(a,b)k \right)$$
$$+ \frac{1}{2!} \left(f_{xx}(a,b)h^2 + 2f_{xy}(a,b)hk \right.$$
$$\left. + f_{yy}(a,b)k^2 \right) + \ldots.$$

In *complex analysis, a function $f(z)$ which is *holomorphic at a point a has a Taylor series

$$f(z) = \sum_{k=0}^{\infty} \frac{f^{(k)}(a)}{k!} (z-a)^k$$

which is convergent in a *neighbourhood of a. See also CAUCHY'S FORMULA FOR DERIVATIVES.

Tchebyshev Alternative spelling of CHE-BYSHEV.

t-distribution Introduced by William *Gosset in 1908, the t-distribution is the continuous probability *distribution of a *random variable formed from the ratio of a random variable with a standard *normal distribution and the square root of a random variable with a *chi-squared distribution divided by its degrees of freedom v. Alternatively, it is formed from the square root of a random variable with an *F-distribution in which the numerator has one degree of freedom. Its *probability density function is

$$f(x) = \frac{\Gamma\left(\frac{v+1}{2}\right)}{\sqrt{v\pi}\,\Gamma\left(\frac{v}{2}\right)} \left(1 + \frac{x^2}{v}\right)^{\frac{v+1}{2}},$$

which gives the *Cauchy distribution when $v = 1$. The graph of f is similar in shape to a standard normal distribution, but is less peaked and has fatter tails. The distribution is used to test for significant differences between a sample mean and the population mean and can be used to test for differences between two sample means. The table gives, for different degrees of freedom, the values corresponding to certain values of α, to be used in a *one-tailed or *two-tailed *t-test, as explained below.

Corresponding to the first column value α, the table gives the one-tailed value $t_{\alpha,v}$ such that $\Pr(t > t_{\alpha,v}) = \alpha$, for the t-distribution with v degrees of freedom. Corresponding to the second column value 2α, the table gives the two-tailed value $t_{\alpha,v}$ such that $\Pr(|t| > t_{\alpha,v}) = 2\alpha$. *Interpolation may be used for values of v not included.

α	2α	$v = 1$	2	3	4	6	8	10	15	20	30	60	∞
0.05	0.10	6.34	2.92	2.35	2.13	1.94	1.86	1.81	1.75	1.72	1.70	1.67	1.64
0.025	0.05	12.71	4.30	3.18	2.78	2.45	2.31	2.23	2.13	2.09	2.04	2.00	1.96
0.01	0.02	31.82	6.96	4.54	3.75	3.14	2.90	2.76	2.60	2.53	2.46	2.39	2.33
0.005	0.01	63.66	9.92	5.84	4.60	3.71	3.36	3.17	2.95	2.84	2.75	2.66	2.58

Values for one- and two-tailed t-test

telescoping series (method of differences) A series $\{a_r\}$ which can be expressed as a sum of differences of successive terms of another series $\{b_r\}$, i.e. that

$$a_r = b_r - b_{r-1} \text{ so } \sum_{r=a}^{n} a_r = b_n - b_0., \text{ as}$$

all terms between the first and last cancel out. For example, if $a_r = \frac{1}{r(r+1)} = \frac{1}{r} - \frac{1}{r+1}$, then $\sum_{r=1}^{n} a_r = \sum_{r=1}^{n} \left(\frac{1}{r} - \frac{1}{r+1} \right)$

$= 1 - \frac{1}{n+1} = \frac{n}{n+1}$.

temperature A physical property of matter that quantitatively measures hotness, usually on the *Celsius or *Kelvin scales. It is a scalar quantity whose *gradient determines heat *flux (*see* FOURIER'S LAW), and the *heat in a given body will be in proportion to its temperature. *See also* HEAT EQUATION, NEWTON'S LAW OF COOLING.

tend to To have as a limit. For example, the sequence $a_n = \frac{1}{n}$ tends to 0 as n tends to infinity.

tension An *internal force that exists at each point of a string or spring. If the string or spring were cut at any particular point, then equal and opposite forces applied to the two cut ends would be needed to maintain the illusion that it remains uncut. The magnitude of these applied forces is the magnitude of the tension at that point. When a particle is suspended from a fixed point by a light string, the tension in the string exerts a force vertically upwards on the particle, and it exerts an equal and opposite force on the point of suspension. *See* HOOKE'S LAW, POTENTIAL ENERGY.

tensor Tensors may be considered as generalizations of *vectors and *matrices. A type (p,q) tensor T on a real *vector space V is a *multilinear map

$$T : (V^*)^p \times V^q \to \mathbb{R}$$

where V^* denotes the *dual space. The order of the tensor is $p + q$, and the tensor may also be considered as a $(p+q)$-dimensional *array once bases are chosen. If $p = 0$, the tensor is called covariant, and if $q = 0$, the tensor is called contravariant. If $V = \mathbb{R}^n$, then the tensor is called Cartesian.

Let $e_1, \ldots, e_n$ be a *basis of V and $f_1, \ldots f_n$ be the dual basis. When $p = 0$, $q = 1$, T is a *functional, and when $p = 1$, $q = 0$, T is a *vector $v = (Tf_1, \ldots, Tf_n)$. When $p = q = 1$, then T represents a linear map $T : V \to V$ as follows: T is represented by the *matrix (a_{ij}), where

$$T\Big((f_i, e_j) \Big) = f_i(Te_j) = f_i \left(\sum_{k=1}^{n} a_{kj} e_k \right) = a_{ij}.$$

More properly, considered as a tensor, these entries should be written a_j^i to denote that one of these indices is covariant, one contravariant. *See also* EINSTEIN'S NOTATION.

Tensors appear widely throughout mathematics and physics; *grad is a $(0,1)$ tensor, an *inner product is a $(0,2)$ tensor, as is the *first fundamental form, the *vector product is a $(1,2)$ tensor, *stress and *strain are tensors, and *differential forms are tensors. The first fundamental form on a *surface instead might be considered as assigning an inner product, a tensor, to each tangent space; this is an example of a tensor field, important in physics and general *relativity particularly.

tensor product Given *vector spaces V and W (over the same *field F), with *bases $v_1, \ldots, v_m$ and $w_1, \ldots, w_n$ respectively, the tensor product $V \otimes W$ may be defined as the vector space *spanned by the formal symbols $v_i \otimes w_j$. This then defines a vector space of *dimension mn.

More naturally, $V \otimes W$ may be defined without reference to bases, as the *dual space of the space of bilinear maps $B: V \times W \to F$ via $(v \otimes w)(B) = B(v,w)$.

Further, Hom(V,W) is then isomorphic to $V^* \otimes W$ via $(f \otimes w)(v) = f(v)w$.

tera- Prefix used with *SI units to denote multiplication by 10^{12}. Abbreviated as T. For binary tera-, *see* KILO (BINARY).

term *See* SEQUENCE, SERIES.

terminal speed (terminal velocity) When an object falls to Earth from a great height, its speed, in certain *mathematical models, *tends to a value called its terminal speed. One possible mathematical model gives the equation $m\ddot{\mathbf{r}} = -mg\mathbf{j} - c\dot{\mathbf{r}}$, where m is the mass of the particle and $\mathbf{j}$ the unit vector in the direction vertically upwards. The second term on the right-hand side, in which c is a positive constant, arises from the *air resistance. The velocity corresponding to $\ddot{\mathbf{r}} = 0$ is equal to $(-mg/c)\mathbf{j}$, which is called the terminal velocity. The terminal speed is the magnitude mg/c of this velocity. If, instead, the equation $m\ddot{\mathbf{r}} = -mg\mathbf{j} - c|\dot{\mathbf{r}}|\dot{\mathbf{r}}$ is used, the terminal speed is $\sqrt{mg/c}$ and the terminal velocity is $-\sqrt{mg/c}\mathbf{j}$. As the particle falls, its velocity tends to the terminal velocity as t increases, irrespective of the initial conditions.

terminating decimal *See* DECIMAL REPRESENTATION.

ternary relation *See* RELATION.

ternary representation The representation of a number to *base 3.

tessellation In its most general form, a tessellation is a covering of the plane with non-overlapping shapes. Often the shapes are *polygons and the pattern is in some sense repetitive. A tessellation is regular if it consists of *congruent *regular polygons. There are just the three possibilities shown here: the polygon is either an equilateral triangle, a square, or a regular hexagon.

A tessellation is semi-regular if it consists of regular polygons, not all

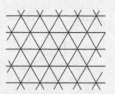

Tessellation with equilateral triangles

congruent. It can be shown that there are just eight of these, one of which has two forms that are mirror-images of each other. They use *triangles, *squares, *hexagons, *octagons, and *dodecagons. For example, one consisting of octagons and squares and another consisting of hexagons and triangles are shown here:

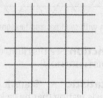

Tessellation with squares

Regular pentagons cannot tessellate the Euclidean plane (*see* EUCLIDEAN SPACE). There are currently 15 known ways to tessellate the plane with congruent pentagons, the latest being found in 2015. Tesselations of the *sphere and *hyperbolic plane are also studied, where regular pentagonal tessellations do exist.

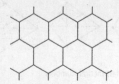

Tessellation with regular hexagons

See also PENROSE TILINGS.

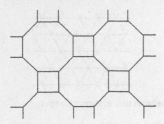

Tessellation with octagons and squares

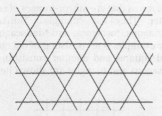

Tessellation with triangles and hexagons

test function *See* DISTRIBUTION (SCHWARTZ DISTRIBUTION).

test statistic A *statistic used in *hypothesis testing that has a known distribution if the null hypothesis is true.

tetra- Prefix denoting four (as does quad-), for example, in *tetrahedron, a *polyhedron with four faces.

tetrahedral number An integer of the form $\frac{1}{6}n(n+1)(n+2)$, where n is a positive integer. This number equals the sum of the first n *triangular numbers.

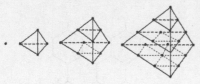

First four tetrahedral numbers

The first few tetrahedral numbers are 1, 4, 10, and 20, and the reason for the name can be seen from the figure.

tetrahedron (tetrahedra) A solid figure bounded by four triangular faces, with four vertices and six edges. A *regular tetrahedron has *equilateral triangles as its faces, and so all its edges have the same length.

TeX A markup language, pronounced 'tec' and created by Donald *Knuth in 1978 for the purpose of typesetting mathematics. Knuth made TeX freely available and it or variants of it, such as LaTeX, are widely used in mathematics and the physical sciences.

t-formulae A range of trigonometric identities that are useful as substitutions for integrations, often as half-angle formulae. If $t = \tan\frac{1}{2}A$, then

$$\sin A = \frac{2t}{1+t^2}, \qquad \cos A = \frac{1-t^2}{1+t^2},$$
$$\tan A = \frac{2t}{1-t^2}.$$

Thales of Miletus (about 585 BC) Greek philosopher frequently regarded as the first mathematician in the sense of being the one to whom particular discoveries have been attributed. Thales' theorem states that the *angle in a *semicircle is a *right angle. He also developed methods of measuring heights by means of shadows and calculating the distances of ships at sea.

theorem A mathematical statement established by means of a *proof. A theorem is usually a result of more significance than a *proposition. In *logic, the term refers to any statement that can be derived from agreed *axioms. For example, the *axiom of choice is *not* a theorem of the *ZF axioms.

Theorema Egregium Translating as 'remarkable theorem', it states that the *Gaussian curvature of a surface is an intrinsic property of the surface, that is, the curvature is expressible in terms of the *first fundamental form and so *invariant under *isometries.

theorem proving Automated theorem proving deals with generating *proofs of mathematical theorems using *computers. It is related to the simpler problem of proof verification, where computers certify that an existing proof is indeed valid.

there exists The meaning of the existential *quantifier whose symbol is ∃.

theta function A range of *special functions of a complex variable (*see* COMPLEX NUMBER). The *roots of all *polynomial equations can be expressed in terms of theta functions.

Thom, René Frédéric (1923–2002) French mathematician, awarded the *Fields Medal in 1958 for work in *topology but best known for his development of the mathematics in *catastrophe theory, where incremental changes can bring about sudden changes or the disappearance of *equilibria.

Thomson, William See KELVIN, LORD.

3blue1brown A mathematical YouTube channel begun by Grant Sanderson in 2015 that addresses a wide range of mathematical topics using animated videos. In the videos there are animated pi-shaped characters, a tutor who is brown, and three students who are blue, though the name originally comes from the heterochromia in Grant's right eye.

(⊕) SEE WEB LINKS
- The 3blue1brown YouTube channel.
- The official 3blue1brown website.

three-body problem The motion of two massive particles, being acted on by *gravity, was resolved by *Newton when he showed that a planet travels in an *ellipse around the sun. However the motion of three (or more) massive particles cannot be described using *elementary functions. Further, the motion of the three bodies is generally non-repeating and *chaotic. Nonetheless, between 1907 and 1912, the Finnish mathematician Karl Sundman provided a solution to the problem in terms of *infinite series.

three-door problem A synonym for MONTY HALL PROBLEM.

thrust When a body or system of masses expels mass, the body experiences a *force of equal *magnitude in the opposite *direction according to *Newton's third law of motion. This force is known as thrust.

tie A rod in tension, usually in a *light framework.

tiling A synonym for TESSELLATION.

time In the real world, the passage of time is a universal experience which clocks of various kinds have been designed to measure. In a *mathematical model, time is represented by a real variable, usually denoted by t, with $t = 0$ corresponding to some suitable starting point. An *observer associated with a *frame of reference has the capability to measure the duration of time intervals between events occurring in the problem being investigated.

Time has the dimension T, and the SI unit of measurement is the *second.

See SIMULTANEITY, TIME DILATION.

time dilation In special *relativity, if an observer O' is travelling at speed v relative to an observer O, then two events on the timeline of O' which are separated by time T (as measure by O') will be measured by O as being $T/\sqrt{1 - v^2/c^2}$ apart. This is greater than T. Here c denotes the speed of light, so the difference will be small unless v is comparable with c. Somewhat counterintuitively, the same time dilation applies when O' measures events on the timeline of O.

time series A sequence of *observations taken over a period of time, usually at equal intervals. Time series analysis is concerned with identifying the factors that influence the variation in a time series, perhaps with a view to predicting what will happen in the future. For many time series, two important components are seasonal variation, which is *periodic change usually in yearly cycles, and trend, which is the long-term change in the seasonally adjusted figures.

Tonelli's theorem A *measurable function $f(x,y)$ such that the *multiple integral $\int_R \int_R |f(x,y)|\, dy\, dx$ converges, is integrable on $\mathbb{R}^2$. *See* FUBINI'S THEOREM.

tonne The mass of 1000 *kilograms.

topological data analysis A modern approach to the study of large, high-dimensional *data sets, applying methods from *algebraic topology. The approach is *robust to noise. Its motivation is to rigorously study the 'shape' of the data; the main tool is called persistent homology.

topological group A topological group G is a *group, which is also a *topological space, such that multiplication $G \times G \to G$ given by $(g_1, g_2) \mapsto g_1 g_2$ and inversion $G \to G$ given by $g \mapsto g^{-1}$ are both *continuous functions. Examples include $\mathbb{R}$, $\mathbb{Q}$, the circle S^1, and the groups listed under *matrix groups. A topological group which is a (topological) *manifold is necessarily a *Lie group.

topologically distinguishable In a *topological space X, points x and y in X are topologically indistinguishable if they have exactly the same *neighbourhoods. Otherwise, they are distinguishable. Generally, 'distinguishable' is stronger than 'distinct' but not as strong as being *separated. *Compare* FUNCTIONALLY SEPARABLE, SEPARATION AXIOMS.

topological space A set X with an associated set T of subsets of X that includes the empty set and the whole set X, and which is closed under arbitrary *union and finite *intersection. The members of T are the *open sets of X, and T is known as the topology. A *metric space is a topological space, but not all topologies are *metrizable (i.e. arise from a metric), for example, the topology consisting of just the empty set and X. *See* CONTINUOUS FUNCTION, SEPARATION AXIOMS.

A sequence x_n in a topological space X has limit l if every *neighbourhood of l contains a *tail of the sequence x_n. In a general topological space, sequential convergence does not determine the topology. A set A may have a *limit point x which is not the limit of any sequence in A. *See* NET.

topological vector space A space which is both a *topological space and a *vector space in which addition and scalar multiplication are continuous. These spaces are the most general spaces used in *functional analysis. They include all *normed vector spaces, so, in particular, *Hilbert spaces and *Banach spaces are examples of topological vector spaces.

topology The area of mathematics concerned with the general properties of shapes and space, and in particular with the study of properties that are not changed by continuous distortions (*see* CONTINUOUS FUNCTION) such as stretching. Also the name given to the family of *open sets in a *topological space (or *metric space). *See* ALGEBRAIC TOPOLOGY, COMBINATORIAL TOPOLOGY, DIFFERENTIAL TOPOLOGY, GENERAL TOPOLOGY, GEOMETRIC TOPOLOGY.

toppling *See* SLIDING–TOPPLING CONDITION.

torque A synonym for MOMENT.

torsion *See* SERRET-FRENET FORMULAE.

torsion element In a *module M, over an *integral domain R, an element $m \varepsilon M$ for which $r \cdot m = 0$ for some non-zero $r \varepsilon R$. The torsion elements of M form the torsion submodule of M, and M is said to be torsion-free if 0 is the only torsion element. *See* STRUCTURE THEOREM.

torus (tori) Suppose that a circle of radius a is rotated through one revolution about a line, in the plane of the circle, a distance b from the centre of the circle, where $b > a$. The resulting surface or solid is called a torus, the shape of a 'doughnut'. The surface area of this torus equals $4\pi^2 ab$, and its volume equals $2\pi^2 a^2 b$. More generally, the term 'torus' applies to an *orientable closed *surface with g holes, called the *genus of the torus. *See* CLASSIFICATION THEOREM FOR SURFACES.

torus (higher dimensions) As a *surface, the torus is topologically $S^1 \times S^1$, where S^1 denotes the *circle. The n-dimensional torus is $(S^1)^n$, and such tori are important in the theory of *Lie groups. A *compact, *connected *abelian Lie group is a torus.

Also a compact, connected Lie group G will have a maximal torus T. Any two maximal tori are *conjugate subgroups, and every element of G lies in some maximal torus.

total differential Given a function $f(x,y,z)$, its total differential, by the chain rule, is

$$df = \frac{\partial f}{\partial x} dx + \frac{\partial f}{\partial y} dy + \frac{\partial f}{\partial z} dz.$$

Compare EXACT DIFFERENTIAL.

total expectation law *See* EXPECTED VALUE.

total force The (vector) sum of all the *forces acting on a *particle, a system of particles, or a *rigid body.

totally bounded (of a metric space) Relating to a *metric space which, for any $\varepsilon > 0$, can be presented as the *union of finite many *balls of radius ε. A *subspace of a totally bounded metric space is totally bounded. A metric space is *compact if and only if it is *complete and totally bounded.

totally disconnected (of a space) Relating to a *topological space in which the only *connected subsets are the empty set and the one-point sets. The *rational numbers and the *Cantor set are examples of totally disconnected spaces.

total order A *partial order $\leq$ on a set X is a total order if it is *connected; that is, for all $x, y \in X$ either $x \leq y$ or $y \leq x$. *Well-orders are total orders; the usual order on the reals is a total order without being a well-order; *divides on the positive integers is not a total order as neither of 6 and 9 divides the other.

total probability law Let $A_1, A_2, \ldots, A_n$ be *mutually exclusive events whose *union is the whole *sample space of an experiment. Then the total probability law is the following formula for the probability of an event B:

$$\Pr(B) = \Pr(B|A_1)\Pr(A_1) + \Pr(B|A_2)\Pr(A_2) + \ldots + \Pr(B|A_n)\Pr(A_n).$$

total relation A synonym for CONNECTED RELATION.

totient function (Euler's phi function) For a positive integer n, let $\phi(n)$ be the number of positive integers less than n that are *coprime to n. For example, $\phi(12) = 4$, since four numbers, 1, 5, 7, and 11, are coprime to 12. This function ϕ, defined on the set of positive integers, is the totient function. It can be shown that, if the prime decomposition of n is $n = p_1^{\alpha_1} p_2^{\alpha_2} \ldots p_r^{\alpha_r}$, then

$$\varphi(n) = p_1^{\alpha_1-1} p_2^{\alpha_2-1} \ldots p_r^{\alpha_r-1}(p_1 - 1)$$
$$(p_2 - 1)\ldots(p_r - 1),$$
$$= n\left(1 - \frac{1}{p_1}\right)\left(1 - \frac{1}{p_2}\right)\ldots\left(1 - \frac{1}{p_r}\right).$$

Euler proved the following extension of *Fermat's Little Theorem: if n is a positive integer and a is any integer such that $(a, n) = 1$, then $a^{\phi(n)} \equiv 1$ (mod n). The totient function is an *arithmetic function.

touch Two *curves that *intersect at a point touch at that point if they also have a common *tangent line. This means their order of *contact is at least 1. This generalizes to two smooth *manifolds of any dimension touching.

tower law Let L, K, F be *fields with $F \subseteq K \subseteq L$. If $\{k_i\}$ is a *basis for K over F, and $\{l_j\}$ is a basis for L over K, then $\{k_i l_j\}$ is a basis for L over F. Hence, the degree of L over F equals the product of the degree of L over K with the degree of K over F.

Tower of Brahma See TOWER OF HANOI.

Tower of Hanoi Imagine three poles with a number of discs of different sizes initially all on one of the poles in decreasing order from the bottom upwards. The problem is to transfer all the discs to one of the other poles,

moving the discs individually from pole to pole, so that one disc is never placed above another of smaller diameter. A version with 8 discs, known as the Tower of Hanoi, was invented by Edouard Lucas and sold as a toy in 1883, but the idea may be much older. The toy referred to a legend about the Tower of Brahma which has 64 discs being moved from pole to pole in the same way by a group of priests, the prediction being that the world will end when they have finished their task.

If t_n is the number of moves it takes to transfer n discs from one pole to another, then $t_n = 2^n - 1$, so the original Tower of Hanoi puzzle takes 255 moves, and the Tower of Brahma takes $2^{64} - 1$.

trace The trace of a square *matrix is the sum of the entries in the *main diagonal. It satisfies trace($\mathbf{AB}$) = trace($\mathbf{BA}$), where $\mathbf{A}$ is $m \times n$ and $\mathbf{B}$ is $n \times m$. For an $n \times n$ matrix $\mathbf{A}$, then $-$trace($\mathbf{A}$) is the coefficient of x^{n-1} in the *characteristic polynomial of $\mathbf{A}$.

tractrix (tractoid) If a massive particle is placed at $(0,1)$ and pulled on an inelastic unit-length string by someone walking from the origin along the x-axis, then the curve traced out by the particle is called the tractrix. The tractrix can be parameterized as

$$x = t - \tanh t, \quad y = \operatorname{sech} t, \quad (t \geq 0).$$

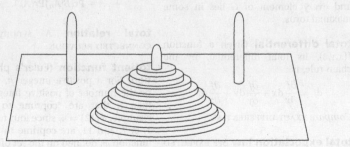

Initial position in Tower of Hanoi

The *surface of revolution formed by rotating the tractrix about the x-axis is called the tractoid. It has constant *Gaussian curvature -1.

trail A *walk in which no edge is repeated. *See also* PATH.

trajectory The path traced out by a *projectile. *See also* PHASE PLANE.

transcendental number A real or complex number that is not a *root of a *polynomial equation with integer *coefficients. In other words, a number is transcendental if it is not an *algebraic number. The first known transcendental number was *Liouville's constant, though *Cantor's work later showed 'most' complex numbers are transcendental, as the algebraic numbers are *countable. It can, though, be very difficult to show specific numbers are transcendental. In 1873, Hermite showed

that e is transcendental; and it was shown by Lindemann, in 1882, that π is transcendental.

transfinite number A *cardinal number relating to an *infinite set such as *aleph-null and the cardinality of the *real numbers.

transformation A term loosely synonymous with *function, particularly a function mapping a *set to itself. A linear transformation is a *linear map.

transformation group A *subgroup of an *automorphism group.

transformation matrix *See* MATRIX OF A LINEAR MAP.

transition map Given a *chart $\varphi_S : S \to \mathbb{R}^2$ of a subset S of a *surface X, we can define a function $f : S \to \mathbb{R}$ to be differentiable if $f \circ \varphi_S^{-1} : \varphi_S(S) \to \mathbb{R}$ is differentiable. Note that this is a map from

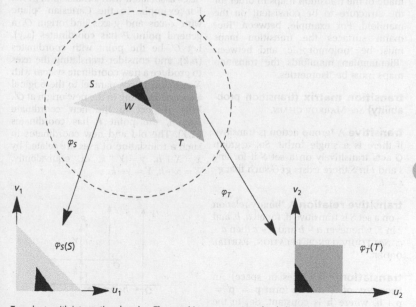

Two charts with intersecting domains. The transition maps connect the darkest regions

a subset of $\mathbb{R}^2$ to $\mathbb{R}$, and we defined differentiability in terms of the local coordinates. The issue arises that this definition might not be consistent between different sets of coordinates.

Given a second chart $\varphi_T : T \to \mathbb{R}^2$ such that $S \cap T = W \neq \varnothing$, then the transition maps for these two charts are

$$\varphi_T \varphi_S^{-1} : \varphi_S(W) \to \varphi_T(W) \quad \text{and}$$
$$\varphi_S \varphi_T^{-1} : \varphi_T(W) \to \varphi_S(W).$$

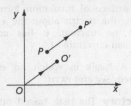

Effect of a translation on O and P

Provided the transition maps, which are maps between subsets of $\mathbb{R}^2$, are differentiable, the preceding definition of differentiability for f can be consistently used for any local coordinates. These ideas can be generalized to define differentiability of maps between any two *manifolds.

For a topological manifold the transition maps are automatically *continuous functions. But if a manifold has further structure, stronger assumptions must be made of the transition maps in order for the structure to be consistent on the manifold. For example, between *Riemann surfaces the transition maps must be *holomorphic, and between *Riemannian manifolds the transition maps must be *isometries.

transition matrix (transition probability) *See* MARKOV CHAIN.

transitive A *group action is transitive if there is a single *orbit. So, a group G acts transitively on a set S if for any s and t in S there exists $g \varepsilon G$ such that $g \cdot s = t$.

transitive relation A *binary relation $\sim$ on a set S is transitive if, for all a, b, and c in S, whenever $a \sim b$ and $b \sim c$ then $a \sim c$. *See* EQUIVALENCE RELATION, PARTIAL ORDER.

translation (of Cartesian space) An *isometry of $\mathbb{R}^n$ of the form $\mathbf{p} \mapsto \mathbf{p}' = \mathbf{p} + \mathbf{h}$, where $\mathbf{h}$ is constant. So, in the plane, point P with coordinates (x,y) is

mapped to the point P' with coordinates (x', y'), where $x' = x + h_1$, $y' = y + h_2$. Thus the origin O is mapped to the point O' with coordinates (h_1, h_2), and the point P is mapped to the point P', where the *directed line segment $\overrightarrow{PP'}$ has the same direction and length as $\overrightarrow{OO'}$. *See* EUCLIDEAN GROUP.

translation of axes (in Cartesian space) The change of *coordinates and *axes associated with a translation in $\mathbb{R}^n$. For example, in the Cartesian *plane with x-axis and y-axis and origin O, a general point P has coordinates (x,y). Let O' be the point with coordinates (h,k), and consider translating the axes to produce a new coordinate system with X-axis and Y-axis parallel to the original axes and meeting in the new origin at O'. With respect to the new coordinate system, the point P has coordinates (X,Y). The old and new coordinates in such a translation of axes are related by $x = X + h$, $y = Y + k$, or, equivalently, $X = x - h$, $Y = y - k$.

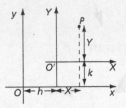

Coordinates a translation apart

This procedure can be useful for investigating the essential nature of curves and transformations. For example, the effect of the map $(x,y) \mapsto (2-y, 2-x)$ may not be immediately clear, but if we translate the origin to $(1,1)$ then the map becomes $(X,Y) \mapsto (-Y, -X)$, which we recognize as reflection in the line $X + Y = 0$ or equally the line $x + y = 2$.

transportation problem A problem in which units of a certain product are to be transported from a number of factories to a number of retail outlets in a way that *minimizes the total cost. For example, suppose that there are m factories and n retail outlets and that the transportation costs are specified by an $m \times n$ matrix $[c_{ij}]$, where c_{ij} is the cost, in suitable units, of transporting one unit of the product from the ith factory to the jth retail outlet. Suppose also that the maximum number of units that each factory can supply and the minimum number of units that each outlet requires are specified. By introducing suitable variables, the problem of minimizing the total cost can be formulated as a *linear programming problem.

transpose The transpose of an $m \times n$ *matrix is the $n \times m$ matrix obtained by interchanging the *rows and *columns. The transpose of $\mathbf{A}$ is denoted by $\mathbf{A}^T$, $\mathbf{A}^t$, or $\mathbf{A}'$. Thus, if $\mathbf{A} = [a_{ij}]$, then $\mathbf{A}^T = [a_{ij}']$, where $a_{ij}' = a_{ji}$; that is,

$$\mathbf{A} = \begin{bmatrix} a_{11} & a_{12} & \cdots & a_{1n} \\ a_{21} & a_{22} & \cdots & a_{2n} \\ \vdots & \vdots & \ddots & \vdots \\ a_{m1} & a_{m2} & \cdots & a_{mn} \end{bmatrix},$$

$$\mathbf{A}^T = \begin{bmatrix} a_{11} & a_{21} & \cdots & a_{m1} \\ a_{12} & a_{22} & \cdots & a_{m2} \\ \vdots & \vdots & \ddots & \vdots \\ a_{1n} & a_{2n} & \cdots & a_{mn} \end{bmatrix}.$$

The following properties hold, for matrices $\mathbf{A}$ and $\mathbf{B}$ of appropriate orders:

(i) $(\mathbf{A}^T)^T = \mathbf{A}$.
(ii) $(\mathbf{A} + \mathbf{B})^T = \mathbf{A}^T + \mathbf{B}^T$.
(iii) $(k\mathbf{A})^T = k\mathbf{A}^T$.
(iv) $(\mathbf{AB})^T = \mathbf{B}^T\mathbf{A}^T$.

transposition A *permutation of a set which swaps two elements but fixes all others or, equivalently, a *cycle of length 2. Any permutation of a *finite set can be written as a product of a number of transpositions; this expression is far from unique, though the *parity of the number is *well defined.

transversal A *line that intersects a given set of two or more lines in the plane. When a transversal intersects a given pair of lines, eight *angles are formed; the four angles between the pair of lines are interior angles and the four others are exterior angles.

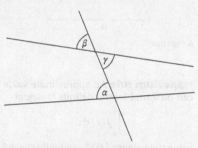

Corresponding and alternate angles

The figure shows a transversal intersecting two lines. The angles α and β are called corresponding angles and α and γ are called alternate angles. The given two lines are parallel if and only if corresponding angles are equal: they are also parallel if and only if alternate angles are equal.

The term 'transversal' also relates to a line intersecting to other lines in higher-dimensional spaces.

transverse axis *See* HYPERBOLA.

transverse component *See* RADIAL AND TRANSVERSE COMPONENTS.

transverse wave A *wave where the displacement occurs in a plane *perpendicular to the direction of the wave's propagation. *Electromagnetic waves, surface waves on a *fluid, and vibrations on a drum or stringed instrument are examples. *See* LONGITUDINAL WAVE.

trapezium (trapezia) A *quadrilateral with two *parallel sides. If the parallel sides have lengths a and b, and the distance between them is h, the area of the trapezium equals $\frac{1}{2}h(a + b)$.

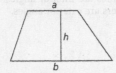

A trapezium

trapezium rule An approximate value can be found for the definite *integral

$$\int_a^b f(x)\,dx,$$

using the values of $f(x)$ at equally spaced values of x between a and b. Divide the *interval $[a, b]$ into n equal subintervals of length h by the *partition

$$a = x_0 < x_1 < x_2 < \dots < x_{n-1} < x_n = b,$$

where $x_{i+1} - x_i = h = (b-a)/n$. Denote $f(x_i)$ by f_i, and let P_i be the point (x_i, f_i). If the line segment P_iP_{i+1} is used as an approximation to the curve $y = f(x)$ between x_i and x_{i+1}, the area under that part of the curve is approximately the area of the *trapezium shown in the figure, which equals $\frac{1}{2}h(f_i + f_{i+1})$. By adding up the areas of all the trapezia, the trapezium rule gives

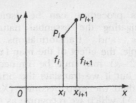

A single trapezium from the trapezium rule

$$\frac{1}{2}h(f_0 + 2f_1 + 2f_2 + \dots + 2f_{n-1} + f_n)$$

as an approximation to the value of the integral. The *error between this approximation and the definite integral is *bounded above by

$$\frac{(b-a)^3}{12n^2} \max_{a \le x \le b} |f''(x)|.$$

*Simpson's rule is significantly more accurate.

trapezoidal rule A synonym for TRAPEZIUM RULE.

travelling salesman problem (TSP) (in graph theory) This is a situation similar to the *minimum connector problem but the salesman wishes to return to the starting point (home) at the end, and so essentially the problem is to find a *closed *walk which visits every vertex and which minimizes the total *distance travelled. In practice there can be unusual situations in which the most efficient route is to go $A \rightarrow B \rightarrow A$ because to get to B from any other vertex, without going through A, is difficult. However, the problem is much easier to analyse if the assumption is made that every vertex is visited exactly once, and the problem reduces to finding the *Hamiltonian cycle of minimum length. The travelling salesman problem is an *NP problem with no known algorithm to solve it in *polynomial time, but upper and lower limits are straightforward to find.

traversable graph A *graph that can be drawn without removing pen from

Trees with five vertices or fewer

paper or going over the same edge twice. *Euler showed that a graph is traversable if and only if it has 0 or 2 *vertices with odd *degree. *See* EULERIAN GRAPH, EULERIAN TRAIL.

tree A *connected *graph with no *cycles. A disconnected graph with no cycles is called a forest. It can be shown that a connected simple graph with n vertices is a tree if and only if it has $n - 1$ edges. The figure shows all the trees with up to five vertices (*see* CAYLEY'S THEOREM).

Particularly in applications, one of the *vertices of a tree may be designated as the root, and the tree may be drawn with the vertices at different levels indicating their distance from the root. A rooted tree in which every vertex (except the root, of degree 2) has degree either 1 or 3, as shown below, is called a binary tree.

A rooted, binary tree

trefoil The simplest *knot which has 3 *crossings and which has left-hand and right-hand forms. For a figure, *see* CHIRALITY.

trend *See* TIME SERIES.

tri- Prefix denoting three, as in *triangle and *trisection.

trial (in statistics) A single *observation or experiment.

trial and improvement *See* FALSE POSITION.

triangle A *polygon with three *vertices. As demonstrated in the first figure, the *angles of a triangle ABC add up to 180°. By considering separate areas, illustrated in the second figure, the area of the triangle equals 'half base times height'.

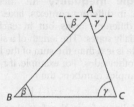

Proving the angle sum formula

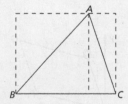

Determining the area of a triangle

If now A, B, and C denote the angles of the triangle, and a, b, and c the lengths of the sides opposite them, the following results hold:

(i) The area of the triangle equals $\frac{1}{2} bc \sin A$.

(ii) The sine rule:

$$\frac{a}{\sin A} = \frac{b}{\sin B} = \frac{c}{\sin C} = 2R,$$

where R is the radius of the *circumcircle.

(iii) The cosine rule:

$$a^2 = b^2 + c^2 - 2bc \cos A.$$

(iv) The tangent rule:

$$\tan \frac{B-C}{2} = \frac{b-c}{b+c} \cot \frac{A}{2}.$$

(v) Hero's formula: Let $s = \frac{1}{2}(a + b + c)$.
Then the area of the triangle equals

$$\sqrt{s(s-a)(s-b)(s-c)}.$$

See ACUTE, OBTUSE, OBLIQUE, EQUILATERAL, ISOSCELES TRIANGLE, RIGHT-ANGLED TRIANGLE, SCALENE TRIANGLE, SOLUTION OF TRIANGLES, SPHERICAL TRIANGLE.

triangle inequality An inequality which, in different contexts, takes somewhat different forms but in each case essentially states: 'The length of a side of a triangle is less than the sum of the lengths of the other sides.' For example, if z_1 and z_2 are complex numbers, then

$$|z_1 + z_2| \le |z_1| + |z_2|.$$

This follows from the fact that $|OQ| \le |OP_1| + |P_1Q|$, where P_1, P_2, and Q represent z_1, z_2, and $z_1 + z_2$ in the *complex plane.

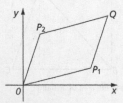

The parallelogram OP_1QP_2

Likewise, for *vectors **a** and **b**,

$$|\mathbf{a} + \mathbf{b}| \le |\mathbf{a}| + |\mathbf{b}|.$$

Finally, the inequality for *metric spaces

$$d(a, c) \le d(a, b) + d(b, c)$$

is also referred to as the triangle inequality. Here a,b,c are points in a metric space and d denotes the metric.

The triangle inequality generalizes naturally to finite and infinite series. For example, and the inequality

$$\left| \sum_{k=1}^{\infty} a_k \right| \le \sum_{k=1}^{\infty} |a_k|$$

for *absolutely convergent series. A similar inequality holds for *integrals

$$\left| \int_a^b f(x) \, dx \right| \le \int_a^b |f(x)| \, dx$$

provided that f and $|f|$ are integrable. *See also* REVERSE TRIANGLE INEQUALITY.

triangle of forces (in mechanics) If three *forces act at a point on a body in *equilibrium, their vector sum must be zero, and consequently the forces can be drawn in a closed triangle. This generalizes to a *polygon of forces for n forces acting in equilibrium.

triangularizable A *square matrix that is *similar to a *triangular matrix. This is the case if and only if the *minimal polynomial is a product of linear factors. So all complex matrices are triangularizable.

triangular matrix A *square matrix that is either lower triangular or upper triangular. It is lower triangular if all the entries above the main diagonal are zero, and upper triangular if all the entries below the main diagonal are zero. A triangular matrix is strictly triangular if the diagonal entries are also zero.

triangular number An integer of the form $\frac{1}{2}n(n+1)$, where n is a positive integer. The first few triangular numbers are 1, 3, 6, 10, and 15, and the reason for the name can be seen from the figure.

See also PENTAGONAL NUMBER, TETRAHEDRAL NUMBER.

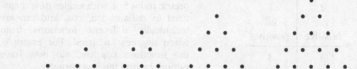

First five triangular numbers

triangulation (in geometry) The method of fixing a point by using directions from two known points to construct a triangle.

triangulation (in topology) A triangulation of a surface is a *subdivision in which all the *faces are *triangles (in the sense of having three *vertices and three curves as *edges). *See* SIMPLICIAL COMPLEX.

tridiagonal matrix A *square matrix with zero entries everywhere except on the *main diagonal and the neighbouring diagonal on either side.

trigonometric function Below, the trigonometric functions are defined, first where the *angle is measured in *degrees and secondly using *radians.

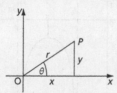

Adjacent x, opposite y, hypotenuse r.

Using degrees The basic trigonometric functions, cosine, sine, and tangent, are first introduced by using a right-angled triangle where $0 < \theta < 90$, but $\cos\theta°$, $\sin\theta°$, and $\tan\theta°$ can also be defined when θ is larger than 90 and when θ is negative. Let P be a point (not at O) with Cartesian coordinates (x,y). Suppose that OP makes an angle of $\theta°$ with the positive x-axis and that $|OP|=r$. Then the following are the definitions:

$$\sin\theta° = \frac{y}{r} = \frac{\text{opposite}}{\text{hypotenuse}},$$

$$\cos\theta° = \frac{x}{r} = \frac{\text{adjacent}}{\text{hypotenuse}},$$

$$\tan\theta° = \frac{y}{x} = \frac{\text{adjacent}}{\text{opposite}} \quad (x \neq 0).$$

It follows that $\tan\theta° = \sin\theta°/\cos\theta°$, and that $\cos^2\theta° + \sin^2\theta° = 1$. Some of the most frequently required values are given in the table.

The point P may be in any of the four quadrants. By considering the signs of x and y, the quadrants in which the different functions take positive values can be found and are shown in the figure.

θ	0	30	45	60	90
$\cos\theta°$	1	$\dfrac{\sqrt{3}}{2}$	$\dfrac{1}{\sqrt{2}}$	$\dfrac{1}{2}$	0
$\sin\theta°$	0	$\dfrac{1}{2}$	$\dfrac{1}{\sqrt{2}}$	$\dfrac{\sqrt{3}}{2}$	1
$\tan\theta°$	0	$\dfrac{1}{\sqrt{3}}$	1	$\sqrt{3}$	not defined

Common values

2	1
sin positive	all positive
3	4
tan positive	cos positive

Parity by quadrant

The following are useful for calculating values, when P is in quadrant 2, 3, or 4:

$$\cos(180 - \theta)^\circ = -\cos\theta^\circ,$$
$$\sin(180 - \theta)^\circ = \sin\theta^\circ,$$
$$\cos(180 + \theta)^\circ = -\cos\theta^\circ,$$
$$\sin(180 + \theta)^\circ = -\sin\theta^\circ,$$
$$\cos(-\theta)^\circ = \cos\theta^\circ,$$
$$\sin(-\theta) = -\sin\theta.$$

The functions cosine and sine are periodic, of period 360; that is to say, $\cos(360 + \theta)^\circ = \cos\theta^\circ$ and $\sin(360 + \theta)^\circ = \sin\theta^\circ$. The function tangent is periodic, of period 180; that is, $\tan(180 + \theta)^\circ = \tan\theta^\circ$.

Using radians In more advanced work, particularly involving *calculus, it is essential that angles are measured in radians. The functions sine, cosine, and tangent are defined by the same trigonometric ratios but, with angles now measured in radians, sin, cos, and tan are technically different functions from when degrees are used. For example, the functions cos and sin now have period 2π, and tan has period π.

For *identities connecting the trigonometric functions, *see* APPENDIX 14. These identities apply whether degrees or radians are used.

The other trigonometric functions, cotangent, secant, and cosecant, are defined as follows:

$$\cot x = \frac{\cos x}{\sin x}, \qquad \sec x = \frac{1}{\cos x},$$
$$\cosec x = \frac{1}{\sin x},$$

where, in each case, values of x that make the denominator zero must be excluded from the domain.

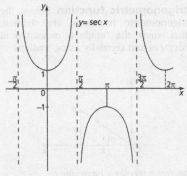

Graph of sec$x$

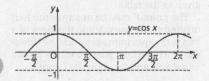

Graph of cos$x$

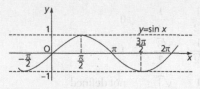

Graph of sin$x$

When radians are used, the derivatives of sin$x$ and cos$x$ are

$$\frac{d}{dx}(\sin x) = \cos x, \qquad \frac{d}{dx}(\cos x) = -\sin x.$$

The derivatives of the other trigonometric functions are found from these, by using the rules for differentiation, and are given in Appendix 7.

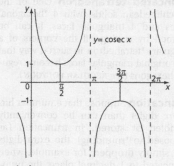

Graph of cosec$x$

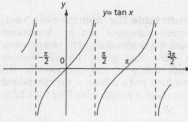

Graph of tan$x$

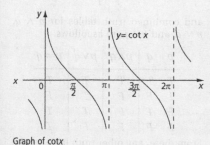

Graph of cot$x$

The *complex trigonometric functions can be defined in terms of the *trigonometric series expansions or in terms of the *complex exponential by

$$\cos z = \frac{\exp(iz) + \exp(-iz)}{2},$$

$$\sin z = \frac{\exp(iz) - \exp(-iz)}{2i}.$$

*Euler's formula $\exp(iz) = \cos z + i\sin z$ then holds for all complex z, though it does not follow that $\cos z$ is the real part of $\exp(iz)$ in general. Likewise, the identity $\cos^2 z + \sin^2 z = 1$ still holds for all complex z.

The trigonometric and *hyperbolic functions are then related by

$$\sin(iz) = i\sinh z, \qquad \cos(iz) = \cosh z$$

for all complex z.

trigonometric series expansions
For x measured in radians, the basic *trigonometric functions sine and cosine can be expressed by the following series expansions, derived as *Taylor series:

$$\sin x = x - \frac{x^3}{3!} + \frac{x^5}{5!} + \dots + (-1)^k \frac{x^{2k+1}}{(2k+1)!} + \dots,$$

$$\cos x = 1 - \frac{x^2}{2!} + \frac{x^4}{4!} + \dots + (-1)^k \frac{x^{2k}}{(2k)!} + \dots.$$

These series *converge for all real and complex values of x.

trigonometric tables *See* TABLES.

trigonometry The area of mathematics relating to the study of *trigonometric functions in relation to the measurements in triangles. *See* SPHERICAL TRIGONOMETRY.

trillion A million million (10^{12}). In Britain it used to mean a million cubed (10^{18}), but this is no longer common usage.

trim (trimmed mean) To discard extreme *observations in a *sample. The remaining values are regarded to be more typical of the behaviours, though it is important to realize that this is not always appropriate. The trimmed mean will be the arithmetic mean of the trimmed set.

trinomial Consisting of three terms, usually in the form $(a + b + c)$ or $ax^2 + bx + c$.

triple *See* N-TUPLE.

triple product (of vectors) *See* SCALAR TRIPLE PRODUCT, VECTOR TRIPLE PRODUCT.

triple root *See* ROOT.

trisect To divide into three equal parts.

trisection of an angle *See* CONSTRUCTION WITH RULER AND COMPASS.

trisector The trisectors of an *angle are the two *lines or *half-lines that divide the angle into three equal angles.

trivial The term is used variously in mathematics. It may refer to the easiness of a conclusion. It also refers to simple or straightforward examples: the trivial *group (or subgroup) consists only of the *identity element; the trivial set is the *empty set; the trivial *subspace consists only of the *zero vector; the trivial *topology on a set consists only of the set itself and the empty set; the trivial *factors of an integer n are ± 1, $\pm n$.

trivial solution The zero solution of a *homogeneous set of linear equations and, more generally, the zero solution for other homogeneous linear equations (or systems of equations), for example, $y = 0$ for the differential equation $dy/dx = y$. The term extends to relatively straightforward solutions for difficult equations; for example, the solutions -2, -4, -6, ... are known as the trivial solutions of the *zeta function.

truncated cube One of the *Archimedean solids, with 6 *octagonal faces and 8 triangular faces. It can be formed by cutting off the corners of a cube in such a way that the original square faces become regular octagons (*see* REGULAR POLYGON).

truncated tetrahedron One of the *Archimedean solids, with 4 *hexagonal faces and 4 triangular faces. It can be formed by cutting off the corners of a (regular) *tetrahedron in such a way that the original triangular faces become regular hexagons (*see* REGULAR POLYGON).

truncation Suppose that a number has more digits than can be conveniently handled or stored. In truncation (as opposed to *rounding), the extra digits are simply dropped; for example, when truncated to 1 decimal place, the numbers 1.875 and 1.845 both become 1.8. *See also* DECIMAL PLACES, SIGNIFICANT FIGURES.

truth table The *truth value of a *compound statement can be determined from the truth values of its components. A table that gives, for all possible truth values of the components, the resulting truth values of the compound statement is a truth table. The truth table for $\neg p$ is

p	$\neg p$
T	F
F	T

and combined truth tables for $p \wedge q$, $p \vee q$, and $p \Rightarrow q$ are as follows:

p	q	$p \wedge q$	$p \vee q$	$p \Rightarrow q$
T	T	T	T	T
T	F	F	T	F
F	T	F	T	T
F	F	F	F	T

From these, any other truth table can be completed. For example, the final column below, gives the truth table for the compound statement $(p \wedge q) \vee (\neg r)$, and is found by first completing columns for $p \wedge q$ and $\neg r$:

p	q	r	$p \wedge q$	$\neg r$	$(p \wedge q) \vee (\neg r)$
T	T	T	T	F	T
T	T	F	T	T	T
T	F	T	F	F	F
T	F	F	F	T	T
F	T	T	F	F	F
F	T	F	F	T	T
F	F	T	F	F	F
F	F	F	F	T	T

truth value A term whose meaning is apparent from the following usage: if a statement is true, its truth value is T (or TRUE); if the statement is false, its truth value is F (or FALSE).

t-test A test to determine whether or not a sample of size n with mean $\bar{x}$ comes from a *normal distribution with mean μ. The *statistic t given by

$$t = \frac{\sqrt{n}(\bar{x} - \mu)}{s}$$

has a t-distribution with $n-1$ degrees of freedom, where s^2 is the (unbiased) sample *variance. The t-test can also be used to test whether a sample mean differs significantly from a population mean and to test whether two samples have been drawn from the same population.

See T-DISTRIBUTION for a table for use in the t-test.

Tukey, John Wilder (1915–2000) American mathematician and statistician who published work in *topology and on the fast *Fourier transform, but it is for his work in *statistics that he will be best remembered. He worked initially with *time series and made contributions to the *analysis of variance and resolving difficulties about making *inferences about a set of parameter values from a single *sample. He is perhaps best known for his 1977 book *Exploratory Data Analysis* which changed the

basis on which *data are analysed and accelerated the move away from purely parametric statistics.

tuple *See* N-TUPLE.

Turing, Alan Mathison (1912–54) British mathematician and logician who conceived the notion of the *Turing machine and famously gave a negative answer to *Hilbert's *decision problem. During the Second World War, he was involved in cryptanalysis, the breaking of codes, and afterwards worked on the construction of some of the early digital computers and the development of their programming systems.

Turing machine A theoretical machine which operates according to extremely simple rules, invented by Turing with the aim of obtaining a mathematically precise definition of what is *computable. It has been generally agreed that the machine can calculate or compute anything for which there is an 'effective' *algorithm (*see* CHURCH'S THESIS).

turning point A point on the graph $y = f(x)$ at which $f'(x) = 0$ and $f'(x)$ changes sign. A turning point is either a *local maximum or a *local minimum. Some authors use 'turning point' as equivalent to *stationary point.

twin primes A pair of *prime numbers that differ by 2. For example, 29 and 31, 71 and 73, and 10 006 427 and 10 006 429, are twin primes. The conjecture that there are infinitely many such pairs remains an *open problem.

SEE WEB LINKS

• More information on twin primes, including a list of the twenty largest pairs identified.

two-person zero-sum game A *game with two players in which the total *payoff is zero, i.e. anything which one player gains is directly at the expense of the other player.

two-sample tests (in statistics) Any test which is to be applied to two *independent *samples, in contrast to paired-sample tests when the two samples are combined before applying a one-sample test to the resulting sample.

two-sided test *See* HYPOTHESIS TESTING.

two-tailed test *See* HYPOTHESIS TESTING.

Tychonoff space A *topological space that is both *completely regular and *Hausdorff. This *separation axiom lies between T_3 and T_4 and so is sometimes referred to as $T_{3.5}$. A space is Tychonoff if

and only if it can be *embedded in a *compact *Hausdorff space.

Tychonoff's theorem The product of two *compact *topological spaces is compact. More generally, this applies to arbitrary products of compact spaces when the product is given the Tychonoff topology.

Type I error A synonym for FALSE POSITIVE.

Type II error A synonym for FALSE NEGATIVE.

typical sequence (typical set) For an *iid sequence $(x_1, x_2, \ldots, x_n) \in \{0,1\}^n$ one would expect for a typical sequence that

$$P\big((X_1,X_2,\ldots,X_n)=(x_1,x_2,\ldots,x_n)\big) \approx 2^{-nH(X)}$$

where $H(X)$ denotes the Shannon *entropy of the distribution. For $\varepsilon > 0$, the typical set T_n^ε is the set of sequences $(x_1, x_2, \ldots, x_n)$ such that

$$H(X)-\varepsilon \le -\frac{1}{n}\log_2 p(x_1,x_2,\ldots,x_n) \le H(X)+\varepsilon.$$

It is then probable that a sequence will arise from the typical set despite, in most cases, T_n^ε being much smaller than 2^n in size.

UFD An abbreviation for UNIQUE FACTORIZA-TION DOMAIN.

UKMT (United Kingdom Mathematics Trust) Founded in 1996 to support the mathematical education of children in the UK. It runs the Junior, Intermediate, and Senior Mathematics Challenges and British Mathematical Olympiad, though these competitions predate the UKMT.

ultrametric A *metric d on a space X is an ultrametric if it further satisfies

$$d(x, z) \leq \max\{d(x, y), d(y, z)\},$$

which is stronger than the usual *triangle inequality.

unary operation A unary operation on a set S is a rule that associates with any element of S a resulting element of S. The following are examples of unary operations: the rule associating with any subset A of a set E its *complement A'; the rule associating with a *group element g its *inverse g^{-1}.

unbiased estimator See ESTIMATOR.

unbounded Not *bounded, with reference to *spaces, *sequences, and *functions.

unconditional statement A statement which is always true. For example, the inequality $x^2 + 2x + 3 > 0$ is unconditional, since it can be expressed as $(x + 1)^2 + 2 > 0$, which is positive for any real number x.

uncountable Not *countable, so the elements cannot be put into *one-to-one correspondence with a subset of the *natural numbers.

undecidable See DECIDABLE; see also HALTING PROBLEM.

underdetermined A *system of *equations for which there are more *variables than there are equations.

uniform A quantity in *mechanics may be called uniform when it is constant. For example, a particle may be moving with uniform *velocity or with uniform *acceleration. A uniform rod, lamina, or rigid body is one whose *density is uniform; that is, whose density is unvarying.

uniform convergence Describing a sequence of real-valued *functions on a given domain which *converge at the same rate for all points of the domain. More precisely, a sequence of functions f_n on a domain D converges uniformly to the function f if for every $\varepsilon > 0$ and for every x in D, there is an integer N for which $|f_n(x) - f(x)| < \varepsilon$ for all $n > N$. In particular, the sequence converges *pointwise on D. Uniform convergence generalizes naturally to functions taking values in a *metric space.

uniform distribution The uniform distribution on the interval $[a,b]$ is the continuous probability *distribution whose *probability density function f is given by $f(x) = 1/(b-a)$, where $a \leq x \leq b$. It has *mean $(a + b)/2$ and *variance $(b-a)^2/12$. There is also a discrete form: on the set 1, 2, ..., n, it is the probability distribution whose *probability mass function is given by $Pr(X = r) = 1/n$, for $1 \leq r \leq n$. For example, the random variable for the winning number in a lottery has a uniform distribution on the set of all the numbers entered in the lottery.

uniform gravitational force A *gravitational force, acting on a particular body, that is independent of the position of the body.

For example, the gravitational force acting on a particle of mass m near the Earth's surface, assumed to be a horizontal plane, can be taken to be $-mg\mathbf{k}$, where $\mathbf{k}$ is a unit vector directed vertically upwards and g is the magnitude of the acceleration due to gravity.

uniformization theorem Every *simply connected *Riemann surface is *conformally equivalent (or *biholomorphic) to precisely one of the *Riemann sphere, the *complex plane, or the open unit disc in $\mathbb{C}$. Consequently all Riemann surfaces that are *homeomorphic to the sphere are, in fact, biholomorphic; by contrast, there are uncountably many Riemann surfaces that are homeomorphic to the *torus but not biholomorphic. *See also* CLASSIFICATION THEOREM FOR SURFACES.

uniformly continuous A function $f : I \to \mathbb{R}$, defined on an *interval I, is uniformly continuous if, for every $\varepsilon > 0$, there exists a single $\delta > 0$ such that, for all $x, y \in I$ satisfying $|y - x| < \delta$, then $|f(y) - f(x)| < \varepsilon$. For continuity only, the value of δ typically depends on x as well as ε. The notion extends readily to *metric spaces, and continuous functions on a *compact space are uniformly continuous. *Absolutely continuous functions are uniformly continuous, but the converse does not hold (*see* CANTOR DISTRIBUTION).

uniform space *Uniform convergence, *uniform continuity, and *completeness can be generalized to *metric

spaces but not to general *topological spaces. Uniform spaces are a further generalization of metric spaces where these notions can be defined. A topological space is uniformizable if and only if it is *completely regular.

unimodal Describing a *distribution that has a unique *mode.

unimodular A *square matrix with *integer entries is unimodular if it has *determinant ± 1. This is equivalent to its *inverse having integer entries.

union The union of sets A and B is the set consisting of all objects that belong to A or B (or both), and it is denoted by $A \cup B$ (read as 'A union B'). Thus the term 'union' is used for both the resulting set and the operation, a *binary operation on the set of all subsets of a given set. The following properties hold:

(i) For all A, $A \cup A = A$ and $A \cup \varnothing = A$.
(ii) For all A and B, $A \cup B = B \cup A$; that is, $\cup$ is commutative.
(iii) For all A, B, and C, $(A \cup B) \cup C = A \cup (B \cup C)$; that is, $\cup$ is associative.

Given a family of sets A_i with *index set I, the union

$$\bigcup_{i \in I} A_i$$

consists of those elements x such that x is an element of A_i for at least one i in I.

Compare DISJOINT UNION. For the union of two events, *see* EVENT.

unique factorization domain (UFD) The *Fundamental Theorem of Arithmetic shows that an integer greater than 1 can be uniquely factorized into primes; a unique factorization domain is essentially a *ring in which the fundamental theorem holds. Specifically, an *integral domain R is a UFD if

(i) every non-zero element, which is not a *unit, can be written as a product of *irreducible elements;

(ii) if $p_1 p_2 \ldots p_m = q_1 q_2 \ldots q_n$, where the p_i and q_j are irreducible, then $m = n$ and (with possible reordering) for each i we have $p_i = u_i q_i$ for some unit u_i.

In a UFD, an element is *prime if and only if it is irreducible. If R is a UFD, then $R[x]$ is also a UFD. *Euclidean domains are UFDs.

Note in $R = \{a + b\sqrt{-3}\}$ that $4 = 2 \times 2 = (1 + \sqrt{-3})(1 - \sqrt{-3})$ are two essentially different factorizations of 4, and so R is not a UFD. By contrast, $\mathbb{R}[x]$, the ring of real polynomials in a variable x, is a UFD. Note that $2x^2 - 6x + 4 = (2x - 2)(x - 2) = (x - 1)(2x - 4)$, but these are essentially the same factorization as $(2x - 2) = 2(x - 1)$ and $(x - 2) = 2^{-1}(2x - 4)$, and 2 is a unit in $\mathbb{R}[x]$.

unit *See* SI UNITS.

unit (ring theory) An element u, in a *ring R with a 1, such that there exists v in R with $uv = 1 = vu$. The units of R form a *group under multiplication, denoted R^*. For R, the ring of $n \times n$ real *matrices, $R^* = GL(n, \mathbb{R})$, while for a *field F, then $F^* = F \setminus \{0\}$.

unitary matrix A *square complex matrix U such that $UU^* = I = U^*U$, where $U^* = \bar{U}^T$ denotes the *Hermitian conjugate of U. The $n \times n$ unitary matrices are the matrices which preserve the norm on $\mathbb{C}^n$. *Compare* ORTHOGONAL MATRIX.

unitary ratio *See* RATIO.

unit circle In the *plane, the unit *circle is the circle of radius 1 with its centre at the origin. In *Cartesian coordinates, it has equation $x^2 + y^2 = 1$. In the *complex plane, it represents those complex numbers z such that $|z| = 1$.

unit cube A cube of side 1 unit. The unit cube defined by the points $(0,0,0)$, $(1,0,0)$, $(0,1,0)$, $(0,0,1)$ can be used in 3-dimensional matrix transformations in an analogous manner to the *unit square.

unit matrix A synonym for IDENTITY MATRIX.

unit square A square of side 1 unit, specifically the unit square defined by $O(0,0)$, $I(1,0)$, $K(1,1)$, $J(0,1)$. It can be used to identify the *linear map from a matrix or find the matrix if the linear map is known because the images of points I and J form the columns of the matrix which performs that transformation. For example, in a rotation of $90°$ clockwise (about the origin), the image of I is $(0,-1)$ and the image of J is $(1,0)$ so the matrix is $\begin{pmatrix} 0 & 1 \\ -1 & 0 \end{pmatrix}$. If a matrix of a transformation is $\begin{pmatrix} 1 & 3 \\ 0 & 1 \end{pmatrix}$ then the image of I is just the first column, i.e. $(1,0)$, so it has not moved, and the image of J is $(3,1)$, so the transformation is a *shear parallel to the x-axis which moves $(0,1)$ to $(3,1)$.

unit vector A vector with *magnitude, or *length, equal to 1. For any non-zero vector $\mathbf{a}$, a unit vector in the direction of $\mathbf{a}$ is $\mathbf{a}/|\mathbf{a}|$.

unity A synonym for ONE.

univalent A *holomorphic function on an open subset (*see* OPEN SET) of $\mathbb{C}$ is said to be univalent if it is *injective. The *derivative of a univalent function is never zero.

universal covering space *See* COVERING SPACE.

universal gravitational constant A synonym for GRAVITATIONAL CONSTANT.

universal machine A synonym for TURING MACHINE.

universal quantifier *See* QUANTIFIER.

universal set In a particular piece of work, it may be convenient to fix the universal set E, a *set to which all the objects to be discussed belong. Then all the sets considered are *subsets of E.

unknot An unknotted *knot; that is, any knot in $\mathbb{R}^3$ which is *isotopic to the circle. *See* KNOT THEORY.

unknown A *variable or *function whose value is to be found.

unstable equilibrium *See* EQUILIBRIUM.

upper bound *See* BOUND.

upper limit *See* LIMIT OF INTEGRATION.

upper triangular matrix *See* TRIANGULAR MATRIX.

Urysohn's lemma *Disjoint *closed subsets of a *normal space are *functionally separable. The lemma shows that normal spaces are *Tychonoff spaces.

utility A measure of the total perceived value resulting from an outcome or course of action. This may be negative; for example, if an oil exploration company undertakes a survey showing that no oil can be extracted, the costs of undertaking the survey will be reflected in a negative utility for that outcome.

utility function A function which defines the *utility for the range of possible outcomes. If a probability distribution is known or can be estimated for those outcomes, then the *expected utility can be calculated.

V The Roman *numeral for 5.

V_4 Or sometimes just *V*, notation for the *Klein four-group. The letter 'v' is used from the German word 'vier' meaning 'four'.

valency A synonym for DEGREE (of a vertex of a graph).

validation (of a simulation model) The process of ensuring a *simulation is an adequate representation of the reality it is attempting to model.

Vallée-Poussin, Charles-Jean de la (1866–1962) Belgian mathematician who in 1896 proved the *prime number theorem independently of *Hadamard.

value *See* CONSTANT FUNCTION, FUNCTION.

value (of a matrix game) *See* FUNDAMENTAL THEOREM OF GAME THEORY.

Vandermonde's convolution formula The following relationship between *binomial coefficients:

$$\binom{m+n}{r} = \binom{m}{0}\binom{n}{r} + \binom{m}{1}\binom{n}{r-1} + \ldots + \binom{m}{r}\binom{n}{0}.$$

The formula may be proved by equating the coefficients of x^r on both sides of the identity $(1+x)^{m+n} = (1+x)^m(1+x)^n$.

Vandermonde's determinant For n real numbers $x_1, x_2, \ldots, x_n$, the *determinant

$$\begin{vmatrix} 1 & x_1 & x_1{}^2 & \ldots & x_1{}^{n-1} \\ 1 & x_2 & x_2{}^2 & \ldots & x_2{}^{n-1} \\ 1 & x_3 & x_3{}^2 & \ldots & x_3{}^{n-1} \\ \vdots & \vdots & \vdots & \ldots & \vdots \\ 1 & x_n & x_n{}^2 & \ldots & x_n{}^{n-1} \end{vmatrix} = \prod_{i>j}(x_i - x_j).$$

Note that the determinant is non-zero, and so the *matrix is *invertible, if and only if the x_i are *distinct.

Van der Waerden's Theorem A result in *Ramsey Theory, the theorem states that however the positive integers are each assigned with one of r colours, there are arbitrarily long *arithmetic progressions that are monochrome. The least number $W(r,k)$ by which point a monochrome progression of length k is guaranteed to occur is very difficult to compute in general. *See* SZEMERÉDI'S THEOREM.

vanish To become *zero or to tend to zero.

Var *See* VARIANCE.

variability A synonym for DISPERSION.

variable A variable, often denoted by a single letter which may represent any *element of a given *set. When variables x and y are related by a *function, so $y = f(x)$, then x is referred to as the independent variable and y as the dependent variable.

variance A measure of the *dispersion of a *random variable or of a *sample. For a random variable X, the population variance is the second moment about the population mean μ and is equal to $E((X-\mu)^2)$ (*see* EXPECTED VALUE). It is usually denoted by σ^2 or $\mathrm{Var}(X)$. For a sample, the sample variance, denoted by s^2, is the second moment of the data about the sample mean $\bar{x}$, but the denominator is usually taken as $n-1$ rather than n in order to make it an *unbiased estimator of the population variance. So

$$s^2 = \frac{\sum(x_i - \bar{x})^2}{n - 1}.$$

For computational purposes, notice that $\sum(x_i - \bar{x})^2 = \sum x_i^2 - n\bar{x}^2$.

variance, analysis of *See* ANALYSIS OF VARIANCE.

variation *See* BOUNDED VARIATION.

varies directly (varies inversely) *See* PROPORTION.

variety A main concept of *algebraic geometry, having a roughly equivalent role to that of a *manifold in *differential geometry. An affine variety in F^n, where F is a field, is a subset defined by *polynomial equations in $x_1, \ldots, x_n$. So *conics and *quadrics are examples of varieties. Projective varieties are similarly defined by *homogeneous polynomial equations in the coordinates. Commonly, the field will be assumed to be *algebraically closed (such as $\mathbb{C}$), as this results in a richer theory, such as *Bézout's theorem. It is also possible to define varieties abstractly, making no reference to an ambient space.

vector A mathematical object, possibly modelling a physical phenomenon, which has *magnitude and *direction. Examples from *applied mathematics include *force, *velocity, and gravitational field (*see* GRAVITY), but vectors are also used in *pure mathematics, particularly in *geometry.

A vector may be considered as the *position vector $\overrightarrow{OP}$ of a point P from an origin O or as a *translation vector, a movement of space. In this way the zero vector may considered as the position vector of the origin or as no translation. The distinction between a vector and a *coordinate vector may in some contexts be important, especially if more than one *coordinate system is being used; in this case the same vector may have different coordinates in the two systems.

With a given coordinate system, a vector can be identified with an ordered pair (x, y) or ordered triple (x, y, z), whether considered as a position vector or a translation vector. This definition can easily be generalized to other dimensions. *Addition and *scalar multiplication may be defined *componentwise. Other operations are important (*see* SCALAR PRODUCT, VECTOR PRODUCT).

vector analysis The area of mathematics concerned with the application of calculus to functions of vectors; for example, the extension of scalar differential equations of motion such as $v = \dfrac{dx}{dt}$ to the vector form $\mathbf{v} = \dfrac{d\mathbf{r}}{dt}$ where $\mathbf{r}(t)$ is the *position vector at time t.

vector bundle *See* TANGENT BUNDLE.

vector equation (of a line) *See* LINE (IN THREE DIMENSIONS).

vector equation (of a plane) Given a plane in 3-dimensional space, let $\mathbf{a}$ be the *position vector of a point A in the plane, and $\mathbf{n}$ a *normal vector to the plane. Then the plane consists of all points P whose position vector $\mathbf{p}$ satisfies $(\mathbf{p} - \mathbf{a}) \cdot \mathbf{n} = 0$. This is a vector equation of the plane. It may also be written $\mathbf{p} \cdot \mathbf{n} = \text{constant}$. By supposing that $\mathbf{p}$ has components x, y, z, that $\mathbf{a}$ has components x_1, y_1, z_1, and that $\mathbf{n}$ has components l, m, n, the first form of the equation becomes $l(x-x_1) + m(y-y_1) + n(z-z_1) = 0$, and the second form becomes the standard linear equation $lx + my + nz = $ constant.

vector field A vector field is a vector-valued function defined on $\mathbb{R}^3$ or some subset of $\mathbb{R}^3$, though the notion can easily be extended to other dimensions. A field will usually be assumed to have continuous derivatives (see CONTINUOUS FUNCTION). Vector fields are common in physical applied mathematics—examples include gravitational field (see GRAVITY), wind *velocity, heat *flux, *electric field, and *magnetic field—but may also be studied in their own right in *multivariable calculus. *Compare* SCALAR FIELD.

vector norm *See* NORM.

vector potential A vector potential for a *vector field F is a vector field A such that $F = $ curl$A$. If such a vector field exists, then div$F = 0$. Conversely, if div$F = 0$, then the question of whether F has a vector potential relates to the *topology of the *domain on which F is defined. *See* POTENTIAL.

vector product Let a and b be non-zero vectors, and let θ be the angle between them (θ in radians, with $0 \leq \theta < \pi$). The vector product $a \times b$ of a and b is defined as having magnitude $|a||b| \sin\theta$, and (if non-zero) its direction is perpendicular to a and b such that a, b and $a \times b$ form a *right-handed system. The notation $a \wedge b$ is also used for $a \times b$. If $a = a_1 i + a_2 j + a_3 k$, $b = b_1 i + b_2 j + b_3 k$, then $a \times b = (a_2 b_3 - a_3 b_2)i + (a_3 b_1 - a_1 b_3)j + (a_1 b_2 - a_2 b_1)k$. This can be written, using 3×3 determinant notation, as

$$a \times b = \begin{vmatrix} i & j & k \\ a_1 & a_2 & a_3 \\ b_1 & b_2 & b_3 \end{vmatrix}.$$

The following properties hold, for all vectors a, b, and c:

(i) $b \times a = -(a \times b)$.
(ii) $a \times (kb + lc) = k(a \times b) + l(a \times c)$.
(iii) $a \times b$ is *perpendicular to a and b.
(iv) If a and b are perpendicular *unit vectors, then $a \times b$ is a unit vector.

In fact, these four properties determine the vector product up to a choice of sign. Unlike the *scalar product, which can be defined in any dimension, a vector product with these properties only exists in 3 and 7 dimensions.

vector projection (of a vector on a vector) Given non-zero vectors a and b, let $\overrightarrow{OA}$ and $\overrightarrow{OB}$ be *directed line segments representing a and b, and let θ be the angle between them (θ in radians, with $0 \leq \theta \leq \pi$). Let C be the *projection of B on the line OA. The vector projection of b on a is the vector represented by $\overrightarrow{OC}$ The vector projection of b on a equals

$$\left(\frac{a.b}{a.a} \right) a.$$

vector space (linear space) A main mathematical structure in *linear algebra. A vector space V over a *field F consists of a set V with operations of addition +: $V \times V \to V$, and scalar multiplication $F \times V \to V$, such that V is an *abelian group under + and further

$$\alpha(\beta v) = (\alpha\beta)v, \quad (\alpha + \beta)v = \alpha v + \beta v,$$
$$\alpha(v + w) = \alpha v + \alpha w, \quad 1v = v,$$

where $\alpha, \beta \in F$ and $v, w \varepsilon V$. Elements of F are called scalars and elements of V are called vectors. Examples of real vectors spaces (where $F = \mathbb{R}$) include:

the space of $m \times n$ real *matrices;
the space of *polynomials with real *coefficients;
the solution space of *homogeneous *simultaneous linear equations;
the *kernel and *image of a *linear map;
the space Hom(V,W) of all linear maps between vector spaces V and W;
sequence spaces (see c and l^∞);

the space of *continuous real-valued functions on a *topological space.

A vector space V is finite-dimensional if it has a finite *basis, in which case all bases will contain the same number of elements, the space's *dimension. Once a basis is chosen, *coordinates may be uniquely assigned to vectors identifying V with $\mathbb{R}^{\dim V}$. *See* AFFINE SPACE, DUAL SPACE, INNER PRODUCT, MODULE, NORMED VECTOR SPACE, N-DIMENSIONAL SPACE, VECTOR SUBSPACE.

vector subspace A subset of a *vector space which is itself a vector space. Such a subset must contain the *zero vector and be *closed under *addition and *scalar multiplication.

vector sum The *binary operation that returns a vector of which the *length and *direction are those of the diagonal of the *parallelogram found by using the two vectors for the sides, i.e. to add vectors $\mathbf{p}$ and $\mathbf{q}$ shown below, if $\overrightarrow{OA} = \mathbf{p} = \overrightarrow{CB}$ and $\overrightarrow{OC} = \mathbf{q} = \overrightarrow{AB}$ then the vector $\mathbf{p} + \mathbf{q} = \overrightarrow{OB}$.

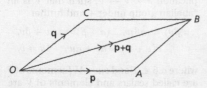

The vector sum $\mathbf{p} + \mathbf{q}$

vector triple product For vectors $\mathbf{a}$, $\mathbf{b}$, $\mathbf{c}$ in $\mathbb{R}^3$, the vector triple product is $\mathbf{a} \times (\mathbf{b} \times \mathbf{c})$. The use of brackets here is essential, since $(\mathbf{a} \times \mathbf{b}) \times \mathbf{c}$, another vector triple product, gives, in general, quite a different result. The vector $\mathbf{a} \times (\mathbf{b} \times \mathbf{c})$ is perpendicular to $\mathbf{b} \times \mathbf{c}$ and so lies in the plane determined by $\mathbf{b}$ and $\mathbf{c}$. In fact, $\mathbf{a} \times (\mathbf{b} \times \mathbf{c}) = (\mathbf{a}.\mathbf{c})\mathbf{b} - (\mathbf{a}.\mathbf{b})\mathbf{c}$.

velocity Consider a particle moving in a straight line, with a point O on the line taken as origin and one direction taken as positive. Let x be the *displacement of the particle at time t. The particle's velocity equals $\dot{x}$ or dx/dt, the *rate of change of x with respect to t. The velocity is positive when the particle is moving forwards (in a positive direction) and negative when moving backwards.

If we assign the *unit vector $\mathbf{i}$ in the line's positive direction then $\mathbf{r} = x\mathbf{i}$ and the velocity is seen to be a *vector equal to $\dot{x}\mathbf{i}$.

When the motion is in two or three dimensions, vectors are used explicitly. The velocity $\mathbf{v}$ of a particle P with position vector $\mathbf{r}$ is given by $\mathbf{v} = d\mathbf{r}/dt = \dot{\mathbf{r}}$. When Cartesian coordinates are used, $\mathbf{r} = x\mathbf{i} + y\mathbf{j} + z\mathbf{k}$, and then $\dot{\mathbf{r}} = \dot{x}\mathbf{i} + \dot{y}\mathbf{j} + \dot{z}\mathbf{k}$.

Velocity has the dimensions LT^{-1}, and the SI unit of measurement is the metre per second, abbreviated to 'ms^{-1}'.

velocity potential For an inviscid, *irrotational flow, with flow velocity $\mathbf{u}$ (in a *simply connected region) a velocity potential is a scalar function φ such that $\nabla\varphi = \mathbf{u}$. The velocity potential is unique up to a constant. *See* COMPLEX POTENTIAL; STREAM FUNCTION.

velocity ratio *See* MACHINE.

velocity–time graph A graph that shows *velocity plotted against time for a particle moving in a straight line. Let $x(t)$ and $v(t)$ be the displacement and velocity, respectively, of the particle at time t. The velocity–time graph is the graph $y = v(t)$, where the t-axis is horizontal and the y-axis is vertical with the positive direction upwards.

The gradient of the velocity–time graph at any point is equal to the *acceleration of the particle at that time. Also,

$$\int_{t_1}^{t_2} v(t)\, dt = x(t_2) - x(t_1),$$

is the *displacement of the particle during the time interval. Note that this is not necessarily the *distance travelled.

Venn, John (1834–1923) British logician who, in his work *Symbolic Logic* of 1881, introduced what are now called *Venn diagrams.

Venn diagram A method of displaying relations between subsets of some *universal set. The universal set E is represented by the interior of a rectangle, say, and subsets of E are represented by regions inside this, bounded by simple closed curves. For instance, two sets A and B can be represented by the interiors of overlapping circles, and then the sets $A \cup B$, $A \cap B$ and $A \setminus B = A \cap B'$, for example, are represented by the shaded regions shown in the figures.

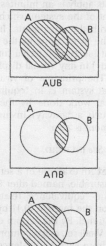

AUB

AnB

A\B

Given one set, A, the universal set is divided into two disjoint subsets A and A', which can be clearly seen in a simple Venn diagram. Given two sets A and B, the universal set E is divided into four disjoint subsets $A \cap B$, $A' \cap B$, and $A \cap B'$

and $A' \cap B'$. A Venn diagram drawn with two overlapping circles for A and B clearly shows the four corresponding regions. Given three sets A, B and C, the universal set E is divided into eight disjoint subsets $A \cap B \cap C$, $A' \cap B \cap C$, $A \cap B' \cap C$, $A \cap B \cap C'$, $A \cap B' \cap C'$, $A' \cap B \cap C'$, $A' \cap B \cap C$, and $A' \cap B' \cap C'$, and these can be illustrated in a Venn diagram as shown here.

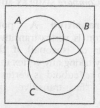

Venn diagram with three sets

Venn diagrams can be illustrative but should generally be avoided for careful proofs, because a diagram may only illustrate a special case. Four general sets, for example, should not be represented by four overlapping circles because they cannot be drawn in such a way as to make apparent the 16 disjoint subsets into which E should be divided. *See also* TRUTH TABLES.

verification (of a simulation model) The process of checking that the program output is consistent with the outcomes which should result from the random numbers generated in the *simulation.

vertex *See* CONE, ELLIPSE, HYPERBOLA, PARABOLA.

vertex (vertex-set) (of a graph) *See* GRAPH.

vertex method For a *linear programming problem where the decision variables are not required to take integer values, the optimal solutions will occur at one or more of the extreme points, i.e. vertices of the feasible region. The method of solution is to calculate the value of the objective function at each vertex, and the maximum or minimum (as required) of these values identifies any optimal solutions. Where more than one such vertex is found, all points on the boundary of the feasible region between them will also be an optimal solution. *Compare* SIMPLEX METHOD.

vertex-transitive A *polygon or *polyhedron is vertex-transitive (or vertex-regular or isogonal) if there is a *symmetry taking any vertex to any other vertex. So a *cuboid is vertex-transitive but not *face-transitive or *edge-transitive.

vertical angles (vertically opposite angles) A pair of non-adjacent angles formed by the intersection of two straight lines. *Compare* ADJACENT ANGLES.

Viète, François (1540–1603) French mathematician who moved a step towards modern algebraic *notation by using vowels to represent unknowns and consonants to represent known numbers. This practice, together with other improvements in notation, facilitated the handling of *equations and enabled him to make considerable advances in algebra. He also developed the subject of *trigonometry.

Viète's formulae Say that a cubic $z^3 + az^2 + bz + c$ has roots α, β, γ so that
$$z^3 + az^2 + bz + c = (z - α)(z - β)(z - γ),$$
then, by expanding the RHS and comparing coefficients,
$$α + β + γ = -a, \quad αβ + βγ + γα = b,$$
$$αβγ = -c.$$

These are *Viète's formulae and naturally generalize to higher degree *polynomials.

Viète's substitution Any *cubic equation can be put in the form $x^3 + ax + b = 0$ by means of a *translation. Viète's substitution $x = t - a/(3t)$ transforms this cubic equation in x into a *quadratic in t^3.

vinculum A horizontal line drawn over mathematical terms to group them together as would brackets. In modern notation, it survives to denote recurring decimal notation, for example, $\frac{1}{11} = 0.\overline{09}$, and vestigially in the notation for radicals such as $\sqrt{x + y}$.

Vinogradov's Theorem *See* GOLDBACH'S CONJECTURE.

virtual work principle (in mechanics) A virtual displacement in a mechanical system in which fixed *holonomic constraints apply is an infinitesimal displacement of the system which is compatible with the constraints. In such a system, the *work done by the forces of constraint (such as internal forces in a *rigid body) in any virtual displacement is zero. The principle of virtual work states that system is in *equilibrium if the virtual work done by the applied forces (such a *gravity) in any virtual displacement is zero.

viscous *See* INVISCID.

Vitali set An example of a set which is not *measurable, named after Giuseppe Vitali. The *equivalence relation ~ is defined on the interval [0,1] by $x \sim y$ if $x-y$ is *rational. By the *axiom of choice, a set, the Vitali set, exists comprising one element from each equivalence class. It can then be shown that if the Vitali set has measure zero, then [0,1] has measure zero, and if the Vitali set has positive measure, then [-1,2] has infinite measure. Both are contradictions, and so the Vitali set is non-measurable.

volume A measure of the 3-dimensional space enclosed by a solid.

volume of a solid of revolution Let $y = f(x)$ be the graph of a function f, a *continuous function on $[a,b]$, and such that $f(x) \geq 0$ for all x in $[a,b]$. The volume V of the solid of revolution obtained by rotating, through one revolution about the x-axis, the region bounded by the curve $y = f(x)$, the x-axis and the lines $x = a$ and $x = b$, is given by

$$V = \int_a^b \pi y^2 \mathrm{d}x = \int_a^b \pi \big(f(x)\big)^2 \mathrm{d}x.$$

Parametric form For the curve $x = x(t)$, $y = y(t)$ ($t \varepsilon [\alpha, \beta]$), the volume V is given by

$$V = \int_\alpha^\beta \pi y^2 \frac{\mathrm{d}x}{\mathrm{d}t} \mathrm{d}t = \int_\alpha^\beta \pi \big(y(t)\big)^2 x'(t) \mathrm{d}t.$$

Compare AREA OF A SURFACE OF REVOLUTION

Von Neumann, John (1903–57) Mathematician and polymath who made important contributions to a wide range of areas of pure and applied mathematics, physics, computer science, and economics. He was born in Budapest but lived in the United States from 1930, where he would become involved with the Manhattan Project. In applied mathematics, he was one of the founders of *optimization theory and the theory of *games. Within *pure mathematics, his work in *functional analysis is important, and through a *Hilbert space approach to *quantum theory he was able to unify the approaches of *Heisenberg and *Schrödinger. He was also involved crucially in the initial development of the modern electronic *computer and the important concept of the stored program.

vortex A region of fluid revolving around an axis in three dimensions or a point in the plane, such as arises in whirlpools and cyclones. A line vortex at the origin has flow velocity

$$\mathbf{u}(x,y) = \frac{\Gamma}{2\pi} \left(\frac{-y}{x^2 + y^2}, \frac{x}{x^2 + y^2} \right).$$

The flow has *circulation Γ but is, perhaps surprisingly, *irrotational, which highlights that *curl measures local rather than global rotation. For a diagram *see* SINGULARITY (FLUID DYNAMICS).

vorticity For a fluid with velocity $\mathbf{u}$, the vorticity ω equals curl$\mathbf{u}$. A flow with zero vorticity is called *irrotational. *See* CIRCULATION, CURL.

vulgar fraction A synonym for SIMPLE FRACTION.

W Symbol of *watt.

walk (in graph theory) The general name given to a sequence of adjacent edges in a *graph. A walk in which no edge is repeated is a *trail, and if no vertex is revisited in a trail, it is a *path. The special case where the path starts and finishes at the same vertex is an exception to this and is termed a *cycle.

Wallis, John (1616–1703) The leading English mathematician before *Newton. His most important contribution was in the results he obtained by using infinitesimals, developing *Cavalieri's method of indivisibles. He also wrote on mechanics. Newton built on Wallis's work in his development of calculus and his laws of motion. It was Wallis who first made use of negative and fractional indices. He was instrumental in the founding of the Royal Society in 1662.

Wallis' Product *See* PI.

wallpaper group A synonym for CRYSTALLOGRAPHIC GROUP.

Waring's problem Every positive integer can be written as the sum of at most 4 square numbers—this is *Lagrange's theorem—or as the sum of 9 or fewer cubes or as the sum of 19 or fewer fourth powers. Waring's problem (posed in 1770) asks whether, for any given power, there is such a maximum number of those powers needed to express all positive integers as a sum. This was proved true by *Hilbert in 1909.

warning limits The inner limits set on a *control chart in a production process. If the observed value falls between the warning and *action limits, then it is taken as a signal that the process may be off target, and another *sample is taken immediately. If it is also outside the warning limits, action will be taken, but if the second observation is within limits, the production is assumed to be on target. For the means of samples of size n in a process with *standard deviation σ with target *mean μ, the warning limits will be set at $\mu \pm 1.96 \frac{\sigma}{\sqrt{n}}$.

watt The *SI unit of *power, abbreviated to 'W'. One watt is equal to one *joule per *second.

wave A disturbance within a medium, such as sound or *electromagnetic waves. A sound wave transfers its energy from particles to adjacent particles and so needs a medium to propagate. Electromagnetic waves can also travel through a vacuum.

If the disturbance is given by the equation

$$y(x, t) = A \sin (kx - \omega t + \varepsilon),$$

then A is the wave's *amplitude, ε is the *phase, ω is the *angular frequency, $2\pi/\omega$ is the period, k is the wavenumber, $2\pi/k$ is the wavelength, and ω/k is the *speed of the wave's propagation. *See* WAVE EQUATION, LONGITUDINAL WAVE, TRANSVERSE WAVE.

wave equation The *hyperbolic *partial differential equation $\partial^2 y/\partial t^2 = c^2 \partial^2 y/\partial t^2$ satisfied by small transverse vibrations of a string, where c is the *speed of a wave's propagation, and first studied by *d'Alembert. The *general solution is $y(x,t) = f(x + ct) + g(x-ct)$, where f and g are arbitrary functions. The wave equation generalizes to three spatial dimensions as $\partial^2 f/\partial t^2 = c^2 \nabla^2 f$, as would be satisfied by the electric field of a light wave $E(\mathbf{x},t) = A \cos{(\omega t - \mathbf{u}.\mathbf{x}/c)}$, where ω denotes frequency and $\mathbf{u}$ is a unit vector parallel to the direction of the wave.

wave function A complex-valued function ψ which describes the state of a system in *quantum theory and which satisfies *Schrödinger's equation. The wave function is *normalized if it has unit *norm $\|\psi\|$, in which case $|\psi|^2$ is a *probability density function for the particle's position.

wavelength See WAVE.

wavenumber See WAVE.

weak convergence A synonym for CONVERGENCE IN DISTRIBUTION.

weaker statement See STRONGER STATEMENT.

weak inequality See INEQUALITY.

weak law of large numbers Given a *random variable X with finite *variance, the sample average of the first n observations; that is, $m_n = \frac{1}{n} \sum_{i=1}^{n} x_i$ *converges in probability to the mean $\mu = E(X)$. *Compare* STRONG LAW OF LARGE NUMBERS.

weakly hereditary property (of spaces) See HEREDITARY PROPERTY (OF SPACES).

Wedderburn's little theorem A theorem of *abstract algebra which states that every finite *division ring is a *field.

Weierstrass, Karl (Theodor Wilhelm) (1815–97) German mathematician who was a leading figure in introducing rigour to mathematical *analysis via *epsilon-delta notation. To show that intuition is not always reliable, he gave an example of a function that is *continuous at every point but nowhere *differentiable. He also proved the *isoperimetric inequality and disproved Dirichlet's principle. Weierstrass' approach to *complex analysis, via series, was in marked contrast to *Riemann's geometric approach, via *Riemann surfaces.

Weierstrass' Approximation theorem See STONE-WEIERSTRASS THEOREM.

Weierstrass' theorem A *continuous real-valued function on a *closed bounded interval is bounded and attains its bounds. This theorem can be generalized to a *compact domain.

weight The magnitude of the *force acting on a body due to *gravity. It is generally assumed that when a body of *mass m is near the Earth's surface it is acted upon by a *uniform gravitational force equal to $-mg\mathbf{k}$, where $\mathbf{k}$ is a *unit vector directed vertically upwards and g is the magnitude of the acceleration due to gravity. Thus, the weight of the body equals mg. Weight has *dimensions MLT^{-2}, the same as force, and the *SI unit of weight is the *newton.

weighted graph A *graph with a weight—a real number—associated to each edge of the graph, which might represent the cost, time, or distance of travel along an edge.

weighted mean See MEAN.

Weil conjectures Four influential *conjectures connecting *algebraic geometry and *number theory made by André Weil in 1949. They relate to the number of points on an algebraic *variety over a *finite field. The last of

w

the conjectures was proved by Deligne in 1974.

well-conditioned problem A problem which is not *ill-conditioned.

well defined A mathematical object is well defined if there is full information, with no ambiguity or contradiction, to meet the specifications of being such an object. For example, a well-defined *function needs a *domain, *codomain, and an assignment defined for each *argument with an *image in the codomain. A notion such as 'the smallest positive real number' is not well defined, as no such *real number exists. Well-definedness particularly applies for functions of *equivalence classes defined in terms of representatives; for example, squaring is well defined in *modular arithmetic as if $x \equiv y \pmod{n}$, then $x^2 \equiv y^2 \pmod{n}$, but *absolute value is not well defined.

well-formed formula (wff) In *logic, a well-formed formula is a finite sequence of symbols from a finite *alphabet that is part of a formal language and validly constructed from the rules of that language.

well ordered A well-ordered set is a set together with a *total order for which every non-empty subset has a *least element. The set of natural numbers is well ordered, a fact which is commonly referred to as the *well-ordering principle*. The statement that every set can be well ordered is equivalent to the *axiom of choice.

Weyl, Hermann (1885–1955) Highly influential German mathematician, who also published widely on theoretical physics and philosophy. He is perhaps most noted for results on *Lie groups and their *representation theory.

wff An abbreviation for WELL-FORMED FORMULA.

Whitehead, Alfred North (1861–1947) British mathematician and philosopher who produced *Principia Mathematica* with Bertrand *Russell in 1913.

Whitney embedding theorem Any smooth, m-dimensional *separable *manifold can be smoothly *embedded in $\mathbb{R}^{2m}$.

whole angle A synonym for FULL ANGLE.

whole number An *integer, though, in more informal language, a whole number may refer only to a positive integer or *natural number.

Wiener, Norbert (1894–1964) American mathematician and logician who founded cybernetics and who worked on guided missiles during the Second World War. *See* BROWNIAN MOTION.

Wilcoxon paired sample test If paired *observations from two *distributions are available and it is known, or can be assumed, that the difference in value between the two distributions is symmetrically distributed about the *median difference, then the *Wilcoxon signed rank test can be applied to the differences. In particular, looking for a shift in the population median is achieved by testing the *null hypothesis of zero difference.

Wilcoxon rank-sum test A non-parametric test (*see* NON-PARAMETRIC METHODS) for testing the *null hypothesis that two *independent samples of size n and m are from the same population. The observations are combined and ranked. If the null hypothesis is true, the sum U of the ranks of the observations in the sample of size n has *mean $n(n + m + 1)/2$ and has *variance $nm(n + m + 1)/12$. For even relatively small values of n and m, the value of U can be tested against a *normal distribution. This is also known as the Mann-Whitney U test.

Wilcoxon signed rank test A non-parametric test of the *null hypothesis that the *median is a specified value. It requires the assumption or knowledge that the *distribution being sampled is symmetric. It is based on ranking the observations by their distance above and below the median and comparing the total of the rankings above and below. It is a more powerful test than the *sign test which only counts the numbers of observations above and below the median and therefore discards quite a lot of information. The procedure is to rank the observations not equal to the hypothesized median by their distance from the median, and to sum the ranks above the median and below the median. There are tables of critical values for different values of n (the number of observations not equal to the hypothesized median) to conduct both one-tailed and two-tailed tests, but for even relatively small values of n, the distribution $N\left(\frac{1}{4}n(n+1), \frac{1}{24}n(n+1)(2n+1)\right)$ is a good approximation.

Wiles, Sir Andrew John (1953–) British mathematician famous for the proof of one of mathematics' longest standing problems—*Fermat's Last Theorem. He announced he had the proof in 1993; however, an error was found, and it was not until 1995 that, together with Richard *Taylor, he published a complete proof. He won the *Wolf Prize in 1996 and the *Abel Prize in 2016 'for his stunning proof of Fermat's Last Theorem by way of the modularity conjecture for semi-stable elliptic curves, opening a new era in *number theory'.

Wilson's theorem For a *prime number p, then $(p-1)!$ is *congruent to -1 *modulo p.

winding number Informally, the number of times a *closed planar *curve winds around a point anticlockwise. Thus, the winding number of a positively oriented circle is 1 around any point in its interior and 0 around any point in its exterior.

Given a positively oriented closed curve C in the *complex plane, the winding number of C about a point a not lying on C is given by

$$\frac{1}{2\pi i}\int_C \frac{dz}{z-a}.$$

within-subjects design A designed experiment where the same group of subjects (often people) are measured under different experimental conditions. *Compare* BETWEEN-GROUPS DESIGN.

Witten, Edward (1951–) American mathematician and theoretical physicist working on superstring theory. Winner of a *Fields Medal in 1990 for his work relating *knot theory and *quantum theory.

Wolf Prize Made by the Wolf Foundation in Israel which awards five or six yearly prizes across a number of scientific fields including mathematics. The prizes started in 1978.

word A word is a *finite string of symbols from an *alphabet. A binary word uses an alphabet {0,1}. When a *group has a *presentation, the alphabet consists of the *generators and their *inverses. Words can be combined using *concatenation.

word problem (group theory) Given a (finite) *presentation for a group in terms of *generators and *relations, the word problem is to find an *algorithm that determines whether two words in the generators represent the same group element. The general word problem is not *decidable but is decidable for certain types of group.

work The work done by a *force **F** during the time interval from $t = t_1$ to $t = t_2$ is equal to

$$\int_{t_1}^{t_2} \mathbf{F} . \mathbf{v} \, dt = \int_{\mathbf{r}_1}^{\mathbf{r}_2} \mathbf{F} \cdot d\mathbf{r},$$

where $\mathbf{v}$ is the *velocity of the point of application of $\mathbf{F}$. The second integral is an alternative form, where $\mathbf{r}$ denotes *displacement, so that $d\mathbf{r}/dt = \mathbf{v}$ and $\mathbf{r}_1$ and $\mathbf{r}_2$ are the initial and final displacements.

From the equation of motion $m\mathbf{a} = \mathbf{F}$ for a particle of mass m moving with acceleration $\mathbf{a}$, it follows that $m\mathbf{a} . \mathbf{v} = \mathbf{F} . \mathbf{v}$, and this then gives

$$(d/dt)\left(\frac{1}{2} m\mathbf{v}.\mathbf{v}\right) = \mathbf{F}.\mathbf{v}.$$

By integration, it follows that the change in *kinetic energy is equal to the work done by the force.

Suppose that a particle has displacement $x(t)\mathbf{i}$ is acted on by a constant force $F\mathbf{i}$ in the same direction. Then the work done during the time interval from $t = t_1$ to $t = t_2$ equals

$$\int_{t_1}^{t_2} F \; x'(t) \; dt = F(x(t_2) - x(t_1)).$$

This is usually interpreted as 'work = force × distance'.

The work done against a force $\mathbf{F}$ should be interpreted as the work done by an applied force equal and opposite to $\mathbf{F}$. When the force is *conservative, this is equal to the change in *potential energy. When a person lifts an object of mass m from ground level to a height z above the ground, work is done against the *uniform gravitational force and the work done equals the increase mgz in potential energy.

Work has the dimensions $ML^2 T^{-2}$, the same as energy, and the SI unit of work is the *joule.

work–energy principle The principle that the change in *kinetic energy of a particle during some time interval is equal to the *work done by the total force acting on the particle during the time interval.

wrt Abbreviation for 'with respect to'. In particular to identify the variable in an *integration or *differentiation process so $f'(x) = \frac{dy}{dx}$ is the differential of $f(x)$ wrt x, and $\int \pi(x^2 + k) \, dx$ is an integral wrt x.

X The Roman *numeral for 10.

x-axis One of the axes in a *Cartesian coordinate system.

xor Short for 'exclusive or'. *See* EXCLUSIVE DISJUNCTION.

y Abbreviation for *yocto-.

Y Abbreviation for *yotta-.

Yates' correction A continuity correction applied to a *chi-squared *contingency table test. Normally only applied for a 2 × 2 table, where the effect is greatest and by a happy coincidence the arithmetic required is the simplest. It involves diminishing the numerical size of the difference between observed and expected values by 0.5 before squaring, i.e. the chi-square statistic calculated is

$$\sum \frac{(|O - E| - 0.5)^2}{E}.$$

y-axis One of the axes in a *Cartesian coordinate system.

yocto- Prefix used with *SI units to denote multiplication by 10^{-24}. Abbreviated to y.

yotta- Prefix used with *SI units to denote multiplication by 10^{24}. Abbreviated to Y.

Young's inequality Given $x, y > 0$ and $a, b > 1$ such that $1/a + 1/b = 1$, then

$$xy \leq \frac{x^a}{a} + \frac{y^b}{b}.$$

When $a = b = 2$, this is equivalent to the *arithmetic-geometric mean inequality.

Young's modulus of elasticity The modulus of elasticity of elastic strings or springs made of the same material, but of different cross-sectional area, will vary, and Young's modulus of elasticity states the relationship as $E = \frac{\lambda}{A}$ giving *Hooke's law as $T = \frac{EA}{l}x$ where $E =$ Young's modulus of elasticity, $\lambda =$ modulus of elasticity, $A =$ cross-sectional area, $l =$ natural length, and $x =$ extension from natural length.

z Abbreviation for *zepto-.

Z Abbreviation for *zetta-.

ℤ The set of *integers. The symbol is also used to denote the set of integers as a *group or as a *ring. The use of the letter 'z' relates to the German word 'Zahlen' for 'numbers'.

$\mathbb{Z}_n$ The integers *modulo n, where $n \geq 2$; also this set considered as a *ring. *See* MODULO N ARITHMETIC, RESIDUE CLASS (modulo n).

Zariski topology An important, but not *Hausdorff, *topology in *algebraic geometry. In the Zariski topology on $\mathbb{C}^n$ a *closed subset is one of the form

$$V(I) = \{\mathbf{x} | f(\mathbf{x}) = 0 \text{ for all } f \in I\},$$

where $I \subseteq \mathbb{C}[x_1, x_2, \ldots, x_n]$. The Zariski topology on complex *projective space can be similarly defined using *homogeneous polynomials. The Zariski topology on any affine or projective *variety is then the *subspace topology induced by the ambient space.

z-axis One of the axes in a *Cartesian coordinate system.

Zeckendorf's theorem Any integer $n \geq 1$ can be uniquely written as a sum of non-consecutive *Fibonacci numbers $1, 2, 3, 5, 8, 13, 21, 34, \ldots$ For example, $300 = 233 + 55 + 8 + 3 + 1 = F_{13} + F_{10} + F_6 + F_4 + F_2$. *See* FIBONACCI ARITHMETIC.

Zeeman, Sir Erik Christopher (1925–2016) British mathematician who founded the mathematics department at the University of Warwick in 1964 and whose work in *topology and *catastrophe theory found wide applicability in areas as diverse as physics, the social sciences, and economics. The *LMS medal for the promotion of mathematics is named in his honour.

Zeno of Elea (5th century BC) Greek philosopher whose paradoxes, known through the writings of Aristotle, may have significantly influenced the Greeks' perception of magnitude and number. The four paradoxes of motion are concerned with whether or not space and time are fundamentally made up of minute indivisible parts. If not, the first two paradoxes, the Dichotomy and the Paradox of Achilles and the Tortoise, appear to lead to absurdities. If so, the third and fourth, the Paradox of the Arrow and the Paradox of the Stadium, apparently give contradictions. *See* ACHILLES PARADOX, ARROW PARADOX.

zepto- Prefix used with *SI units to denote multiplication by 10^{-21}.

zetta- Prefix used with *SI units to denote multiplication by 10^{21}.

Zermelo, Ernst (Friedrich Ferdinand) (1871–1953) German mathematician considered the founder of *axiomatic set theory. In 1908, he formulated a set of *axioms for *set theory, which attempted to overcome problems such as that posed by *Russell's paradox. These, with modifications by Fraenkel around 1922, have been the foundation on which much subsequent work in the subject has been built.

Zermelo-Fraenkel axioms (ZF axioms) An axiomatic system, in fact the canonical axioms, for describing what constitutes a *set. The *axioms state:

1. Two sets are equal if they have the same *elements.
2. There is a set with no members, the *empty set.
3. Given two sets, there is a set containing just those sets as elements.
4. Given two sets, there is a set whose elements are those of one or other set.
5. Given a set, there is a set whose elements are the subsets of the set.
6. The image of a set under a (definable) *function is a set.
7. There is no infinite descending sequence for set membership.
8. There is an *infinite set.

(Here the language of some of these axioms is necessarily somewhat informal.) The above are the ZF axioms. They circumvent issues such as *Russell's paradox by not permitting Russell's paradoxical 'set' to be a set. Both the *axiom of choice and the *continuum hypothesis are independent of the ZF axioms.

zero The *real number 0, which is the additive identity, i.e. $x + 0 = 0 + x = x$ for any real number x.

zero (of a function) See ROOT.

zero-dimensional (of a space) Describing a *topological space which has a *basis of sets that are *clopen. Every *discrete space is zero-dimensional; by contrast, the set of *rational numbers is zero-dimensional but does not have *isolated points. Being zero-dimensional is a *hereditary property of a space. See DIMENSION (TOPOLOGY).

zero-divisor A non-zero element x in a *ring such that there exist non-zero y, z with $yx = 0 = xz$. There are no zero-divisors within $\mathbb{Z}$, $\mathbb{R}$, or $\mathbb{C}$; generally, a *commutative ring with an *identity which has no zero-divisors is called an *integral domain. For example, if $\mathbf{A} = \begin{pmatrix} 1 & 0 \\ 0 & 0 \end{pmatrix}$ and $\mathbf{B} = \begin{pmatrix} 0 & 0 \\ 0 & 1 \end{pmatrix}$ then

$\mathbf{AB} = \mathbf{BA} = \mathbf{0}$, and so $\mathbf{A}$ and $\mathbf{B}$ are zero-divisors within the ring of 2×2 matrices. *Factorization cannot be applied when zero-divisors exist; for example, the equation $(x-1)(x+1) = 0$ over $\mathbb{Z}_8$ does not imply that $x = 1$ or $x = -1$, as $x = 3$ and $x = 5$ are also solutions because $2 \times 4 = 4 \times 6 = 0$ in $\mathbb{Z}_8$.

zero element An additive *identity element.

zero function The zero function is the *real function f such that $f(x) = 0$ for all x in $\mathbb{R}$. More generally, the term might refer to any *constant function on a *set mapping all its elements to some *zero element.

zero matrix The $m \times n$ zero *matrix $\mathbf{O}$ with all zero entries. A zero *column vector or *row vector is denoted by $\mathbf{0}$.

zero-sum game See MATRIX GAME.

zero vector See VECTOR.

zeta function The function $\zeta(s) = \sum_{n=1}^{\infty} \frac{1}{n^s}$ defined on the complex variables s. The sum is convergent, and gives an *analytic function, when $\mathrm{Re}\,s > 1$, and in particular $\zeta(2) = \frac{\pi^2}{6}$; $\zeta(4) = \frac{\pi^4}{90}$; $\zeta(6) = \frac{\pi^6}{945}$. This definition can then be *analytically extended to any $s \neq 1$. The zero function has 'trivial' zeros at $s = -2, -4, -6, \ldots$ and all other zeros are conjectured to be on the 'critical line' $\mathrm{Re}\,s = \frac{1}{2}$ (see RIEMANN HYPOTHESIS).

There are deep connections between the zeta function and the exact distribution of the primes, as Riemann found an expression for this distribution in terms of the zeros of the zeta function.

ZF Short for the *Zermelo-Fraenkel axioms or to denote an assumption of these axioms.

ZFC Short for the *Zermelo-Fraenkel axioms and the *axiom of choice, or to denote an assumption of these axioms.

zone A zone of a sphere is the part between two parallel planes. If the sphere has radius r and the distance between the planes is h, the area of the curved surface of the zone equals $2\pi rh$.

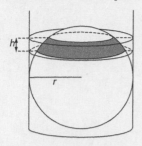

A circumscribing cylinder of a sphere is a right-circular cylinder with the same radius as the sphere and with its axis passing through the centre of the sphere. Take the circumscribing cylinder with its axis perpendicular to a zone's bounding planes, as shown. It is an interesting fact that the area of the curved surface of the zone is equal to the area of the part of the circumscribing cylinder between the same two planes.

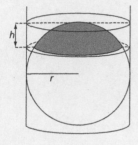

Zorn's lemma An equivalent statement to the *axiom of choice which is often more conveniently applicable in some areas of mathematics. The lemma states that: a *partially ordered set S, where every *totally ordered subset has an *upper bound, has a *maximal element. Zorn's lemma can be used to show that any *vector space has a *basis.

Appendix 1

Areas and volumes

(For unexplained notation, see under the relevant reference.)

*Rectangle, length a, width b:
Area $= ab$.

*Parallelogram:
Area $= bh = ab \sin \theta$.

*Triangle:
Area $= \frac{1}{2}$base $\times$ height $= \frac{1}{2}bc \sin A$.

*Trapezium:
Area $= \frac{1}{2}h(a + b)$.

*Circle, radius r:
Area $= \pi r^2$.
Length of circumference $= 2\pi r$.

*Right-circular *cylinder, radius r, height h:
Volume $= \pi r^2 h$,
Curved surface area $= 2\pi rh$.

*Right-circular *cone:
Volume $= \frac{1}{3}\pi r^2 h$,
Curved surface area $= \pi rl$.

*Frustum of a *cone:
Volume $= \frac{1}{3}\pi h(a^2 + ab + b^2)$,
Curved surface area $= \pi(a + b)l$.

*Sphere, radius r:
Volume $= \frac{4}{3}\pi r^3$,
Surface area $= 4\pi r^2$.

Surface area of a *cylinder:
Curved surface area $= 2\pi rh$.

Centres of mass

Triangular *lamina:
2/3 of the way along any median from its vertex.

Solid *hemisphere, radius r:
$3r/8$ from the centre, along the radius which is perpendicular to the base.

Hemispherical shell, radius r:
$\frac{1}{2} r$ from the centre, along the radius which is perpendicular to the base.

*Cone or *pyramid:
$\frac{3}{4}$ of the way from the vertex to the centroid of the base.

Conical shell:
2/3 of the way from the vertex to the centre of the base.

Circular *sector, radius r and angle 2θ:
$\frac{2r \, \sin\theta}{3\theta}$ from the centre of the circle.

Circular *arc, radius r and angle 2θ:
$\frac{r \, \sin\theta}{\theta}$ from the centre of the circle.

Moments of inertia

For uniform bodies of mass M.

Thin *rod, length $2l$, about perpendicular axis through centre:
$\frac{1}{3}Ml^2$.

Rectangular *lamina, about axis in plane bisecting edges of length $2l$:
$\frac{1}{3}Ml^2$.

Thin rod, length $2l$, about perpendicular axis through end:
$\frac{4}{3}Ml^2$.

Rectangular lamina, about edge perpendicular to edges of length $2l$:
$\frac{4}{3}Ml^2$.

Rectangular lamina, sides $2l$ and $2b$, about perpendicular axis through centre:
$\frac{1}{3}M(l^2 + b^2)$.

Hoop or cylindrical shell, radius r, about axis:
Mr^2.

Hoop, radius r, about a diameter:
$\frac{1}{2}Mr^2$.

***Disc or solid *cylinder, radius r, about axis:**
$\frac{1}{2}Mr^2$.

Disc, radius r, about a diameter:
$\frac{1}{4}Mr^2$.

Solid *sphere, radius r, about a diameter:
$\frac{2}{5}Mr^2$.

Spherical shell, radius r, about a diameter:
$\frac{2}{3}Mr^2$.

Elliptical lamina, axes $2a$ and $2b$, about axis of length $2a$:
$\frac{1}{4}Mb^2$.

Elliptical lamina, axes $2a$ and $2b$, about perpendicular axis through centre:
$\frac{1}{4}M(a^2 + b^2)$.

Solid *ellipsoid, axes $2a$, $2b$, and $2c$, about axis of length $2c$:
$\frac{1}{5}M(a^2 + b^2)$.

Ellipsoidal shell, axes $2a$, $2b$, and $2c$, about axis of length $2c$:
$\frac{1}{3}M(a^2 + b^2)$.

***Right-circular *cone, base radius r, about axis of cone:**
$\frac{3}{10}Mr^2$.

SI prefixes, units, and constants

SI prefixes

10	deka-	da	10^{-1}	deci-	d
10^2	hecto-	h	10^{-2}	centi-	c
10^3	kilo-	k	10^{-3}	milli-	m
10^6	mega-	M	10^{-6}	micro-	μ
10^9	giga-	G	10^{-9}	nano-	n
10^{12}	tera-	T	10^{-12}	pico-	p
10^{15}	peta-	P	10^{-15}	femto-	f
10^{18}	exa-	E	10^{-18}	atto-	a
10^{21}	zetta-	Z	10^{-21}	zepto-	z
10^{24}	yotta-	Y	10^{-24}	yocto-	y

SI base units

Quantity	Unit	Symbol
*time	second	s
*length	metre	m
*mass	kilogram	kg
electric current	ampere	A
*temperature	kelvin	K
amount of substance	mole	Mol
luminous intensity	candela	cd

Named SI supplementary units

Quantity	Unit	Symbol	Derivation
*angle	*radian	rad	dimensionless
*solid angle	*steradian	sr	dimensionless
*frequency	hertz	Hz	s^{-1}
*force	*newton	N	$kg\ m\ s^{-2}$
*pressure	pascal	Pa	$kg\ m^{-1}\ s^{-2}$
*energy	*joule	J	$kg\ m^2\ s^{-2}$
*power	*watt	W	$kg\ m^2\ s^{-3}$
electrical potential difference	volt	V	$kg\ m^2\ s^{-3}\ A^{-1}$
electrical resistance	ohm	Ω	$kg\ m^2\ s^{-3}\ A^{-2}$
electrical charge	coulomb	C	$A\ s$

electrical conductance	siemens	S	$kg^{-1} m^{-2} s^3 A^2$
capacitance	farad	F	$kg^{-1} m^{-2} s^4 A^2$
magnetic flux	weber	Wb	$kg m^2 s^{-2} A^{-1}$
magnetic flux density	tesla	T	$kg s^{-2} A^{-1}$
inductance	henry	H	$kg m^2 s^{-2} A^{-2}$
luminous flux	lumen	lm	cd
illuminance	lux	lx	$cd m^{-2}$

Mathematical constants

Name	Symbol	Value (to 10 dp)	irrational	transcendental
*pi, *Archimedes' constant	π	3.1415926536	Y	Y
*e, *Euler's number	e	2.7182818285	Y	Y
*Euler's constant	γ	0.5772156649	?	?
*golden ratio	φ	1.6180339887	Y	N
*Catalan's constant	G	0.9159655942	?	?
*Apéry's constant	$\zeta(3)$	1.2020569032	Y	?
*Liouville's constant		0.1100010000	Y	Y

Physical constants

Name	Symbol	Value (up to 10 sf)	Units
speed of light	c	299,792,458	ms^{-1}
universal gravitational constant	G	6.6743×10^{-11}	$kg^{-1} m^3 s^{-2}$
Planck's constant	h	$6.62607015 \times 10^{-34}$	$kg m^2 s^{-1}$
reduced Planck constant	$\hbar$	$1.054571817 \times 10^{-34}$	$kg m^2 s^{-1}$
elementary charge	e	$1.602176634 \times 10^{-19}$	C
Avogadro's constant	N_A	$6.02214076 \times 10^{23}$	mol^{-1}
absolute zero		-273.15	°C
permittivity of space	ε_0	$8.854187813 \times 10^{-12}$	$kg^{-1} m^{-3} s^4 A^2$
permeability of space	μ_0	$1.256637062 \times 10^{-6}$	$kg m s^{-2} A^{-2}$
Boltzmann constant	k	1.380649×10^{-23}	JK^{-1}

Geometry: equations of lines and planes

Equation of a line

If a line contains the point $\mathbf{p} = (x_1, y_1, z_1)$ and is parallel to the vector $\mathbf{a} = (a, b, c)$, then the line has equation

$$\frac{x - x_1}{a} = \frac{y - y_1}{b} = \frac{z - z_1}{c}.$$

This can be written parametrically as $\mathbf{r} = \mathbf{p} + t\mathbf{a}$, where $t \in \mathbb{R}$.

Alternatively, if $\mathbf{a}, \mathbf{b} \in \mathbb{R}^3$, where $\mathbf{a} \neq \mathbf{0}$ and $\mathbf{a} \cdot \mathbf{b} = 0$, then $\mathbf{r} \times \mathbf{a} = \mathbf{b}$ defines a line parallel to $\mathbf{a}$ through the point $\mathbf{a} \times \mathbf{b}/|\mathbf{a}|^2$.

Distance from a point to a line

The distance s from $ax + by + c = 0$ to $P(x_1, y_1)$ is $s = \dfrac{|ax_1 + by_1 + c|}{\sqrt{a^2 + b^2}}$. The distance of a point $\mathbf{c}$ from the line $\mathbf{r} \times \mathbf{a} = \mathbf{b}$ equals $|\mathbf{c} \times \mathbf{a} - \mathbf{b}|/|\mathbf{a}|$.

Equation of a plane

If a plane contains the point (x_1, y_1, z_1) and the vector $\begin{pmatrix} a \\ b \\ c \end{pmatrix}$ is normal to the plane, then the plane is given by $ax + by + cz = d$ (where $d = ax_1 + by_1 + cz_1$).

Distance from a point to a plane

The distance s from $ax + by + cz = d$ to $P(x_1, y_1, z_1)$ is $s = \dfrac{|ax_1 + by_1 + cz_1 - d|}{\sqrt{a^2 + b^2 + c^2}}$.

Distance between two lines

The distance between two lines with equations $\mathbf{r} \times \mathbf{a}_1 = \mathbf{b}_1$ and $\mathbf{r} \times \mathbf{a}_2 = \mathbf{b}_2$ equals

$$\frac{|\mathbf{a}_1 \cdot \mathbf{b}_2 + \mathbf{a}_2 \cdot \mathbf{b}_1|}{|\mathbf{a}_1 \times \mathbf{a}_2|}.$$

Angle between two planes

The angle between the planes $a_1 x + b_1 y + c_1 z = d_1$ and $a_2 x + b_2 y + c_2 z = d_2$ is $\arccos\left(\dfrac{a_1 a_2 + b_1 b_2 + c_1 c_2}{\sqrt{a_1^2 + b_1^2 + c_1^2} \sqrt{a_2^2 + b_2^2 + c_2^2}} \right)$; the planes are parallel if and only if $\dfrac{a_1}{a_2} = \dfrac{b_1}{b_2} = \dfrac{c_1}{c_2}$, and the planes are perpendicular if and only if $a_1 a_2 + b_1 b_2 + c_1 c_2 = 0$.

The circle

The equation of a circle with centre (a, b) and radius r is $(x - a)^2 + (y - b)^2 = r^2$.

In parametric form, $x = a + r\cos t$, $y = b + r\sin t$.

				*Rectangular hyperbola
	*Ellipse	***Parabola**	***Hyperbola**	

*Conic sections

	*Ellipse	***Parabola**	***Hyperbola**	***Rectangular hyperbola**
Cartesian form	$\frac{x^2}{a^2} + \frac{y^2}{b^2} = 1$	$y^2 = 4ax$	$\frac{x^2}{a^2} - \frac{y^2}{b^2} = 1$	$xy = c^2$
Parametric form	$(a\cos\theta, b\sin\theta)$	$(at^2, 2at)$	$(a\sec\theta, b\tan\theta)$ $(\pm a\cosh\theta, b\sinh\theta)$	$\left(ct, \frac{c}{t}\right)$
*Eccentricity	$e < 1,\ b^2 = a^2(1-e^2)$	$e = 1$	$e > 1,\ b^2 = a^2(e^2-1)$	$e = \sqrt{2}$
*Foci	$(\pm ae, 0)$	$(a, 0)$	$(\pm ae, 0)$	$(\pm\sqrt{2c}, \pm\sqrt{2c})$
*Directrices	$x = \pm\frac{a}{e}$	$x = -a$	$x = \pm\frac{a}{e}$	$x + y = \pm\sqrt{2c}$
*Asymptotes	None	None	$\frac{x}{a} = \pm\frac{y}{b}$	$x = 0, y = 0$

Appendix 6

Basic algebra

*Arithmetic series (first term a, common difference d, last term l with n terms)
$u_n = a + (n-1)d$, $S_n = \frac{1}{2}n(a+l) = \frac{1}{2}n\{2a + (n-1)d\}$.

*Geometric series (first term a, common ratio r, with n terms)

$$u_n = ar^{n-1}, \qquad S_n = \frac{a(1-r^n)}{1-r} \text{ if } r \neq 1, \qquad S_\infty = \frac{a}{1-r} \text{ if } |r| < 1.$$

$S_n = \frac{a(1-r^n)}{1-r}$, $S_\infty = \frac{a}{1-r}$ if $|r| < 1$.

*Logarithms and exponentials
$\log_a x = \frac{\log_b x}{\log_b a}$, $e^{x\ln a} = a^x$.

*Numerical integration
   *Trapezium rule:

$$\int_a^b y\,dx \approx \frac{1}{2}h\{(y_0 + y_n) + 2(y_1 + y_2 + \dots + y_{n-1})\}, \quad \text{where } h = \frac{b-a}{n}.$$

   *Simpson's rule, for even n:

$$\int_a^b y\,dx \approx \frac{1}{3}h\{(y_0 + y_n) + 4 \times (y_2 + \dots + y_{n-2}) + 2 \times (y_1 + \dots + y_{n-1})\},$$

$$\text{where } h = \frac{b-a}{n}.$$

Summations

$$\sum_{r=1}^{n} r = \frac{1}{2}n(n+1),$$

$$\sum_{r=1}^{n} r^2 = \frac{1}{6}n(n+1)(2n+1),$$

$$\sum_{r=1}^{n} r^3 = \frac{1}{4}n^2(n+1)^2.$$

*Newton–Raphson iterative method of solution
To solve $f(x) = 0$: $x_{n+1} = x_n - \frac{f(x_n)}{f'(x_n)}$.

*Complex numbers

$$e^{i\theta} = \cos\theta + i\sin\theta,$$

$$[r(\cos\theta + i\sin\theta)]^n = r^n(\cos n\theta + i\sin n\theta).$$

The roots of $z^n = 1$ are given by $z = e^{\frac{i2\pi k}{n}}$, for $k = 0, 1, 2, \dots, n-1$.

Derivatives

$f(x)$	$f'(x)$
k (constant)	0
x	1
x^k	kx^{k-1}
$\sin x$	$\cos x$
$\cos x$	$-\sin x$
$\tan x$	$\sec^2 x$
$\sec x$	$\sec x \tan x$
$\operatorname{cosec} x$	$-\operatorname{cosec} x \cot x$
$\cot x$	$-\operatorname{cosec}^2 x$
e^{kx}	ke^{kx}
$\ln x$	$1/x$
$a^x\ (a > 0)$	$a^x \ln a$
$\sin^{-1} x$	$\frac{1}{\sqrt{1-x^2}}$
$\cos^{-1} x$	$-\frac{1}{\sqrt{1-x^2}}$
$\tan^{-1} x$	$\frac{1}{1+x^2}$
$\sinh x$	$\cosh x$
$\cosh x$	$\sinh x$
$\tanh x$	$\operatorname{sech}^2 x$
$\coth x$	$-\operatorname{cosech}^2 x$
$\sinh^{-1} x$	$\frac{1}{\sqrt{x^2+1}}$
$\cosh^{-1} x$	$\frac{1}{\sqrt{x^2-1}}$
$\tanh^{-1} x$	$\frac{1}{\sqrt{1-x^2}}$

Integrals

Notes:

(i) The table gives, for each function f, an *antiderivative ϕ. The function $\phi(x) + c$, where c is an arbitrary constant, is also an antiderivative of f.

(ii) In certain cases, for example when $f(x)=1/x$, $\tan x$, $\cot x$, $\sec x$, $\operatorname{cosec} x$, and $\sqrt{x^2 - a^2}$, the function f is not continuous for all x. In these cases, a definite integral $\int_a^b f(x)$ can be evaluated as $\phi(b)-\phi(a)$ only if a and b both belong to an interval in which f is continuous.

$f(x)$	$\phi(x) = \int f(x)\,\mathrm{d}x$
$x^k \ (k \neq -1)$	$\frac{x^{k+1}}{k+1}$
$1/x$	$\ln\lvert x\rvert$
$\sin x$	$-\cos x$
$\cos x$	$\sin x$
$\tan x$	$-\ln\lvert\cos x\rvert = \ln\lvert\sec x\rvert$
$\sec x$	$\ln\lvert\sec x + \tan x\rvert$
$\operatorname{cosec} x$	$\ln\lvert\operatorname{cosec} x - \cot x\rvert = \ln\lvert \tan\frac{1}{2}x \rvert$
$\cot x$	$\ln\lvert\sin x\rvert$
$\sin^2 x$	$\frac{1}{2}\left(x - \frac{1}{2}\sin 2x\right)$
$\cos^2 x$	$\frac{1}{2}\left(x + \frac{1}{2}\sin 2x\right)$
$\sinh x$	$\cosh x$
$\cosh x$	$\sinh x$
$e^{kx} \ (k \neq 0)$	e^{kx}/k
$e^{ax}\sin bx$	$\frac{e^{ax}}{a^2+b^2}(a\,\sin bx - b\,\cos bx)$
$e^{ax}\cos bx$	$\frac{e^{ax}}{a^2+b^2}(a\,\cos bx + b\,\sin bx)$
$a^x \ (a > 0, \ a \neq 1)$	$a^x/\ln a$
$\ln x$	$x\ln x - x$
$\frac{1}{\sqrt{a^2-x^2}} \ (a > 0)$	$\sin^{-1}\frac{x}{a}$
$\frac{1}{a^2+x^2} \ (a > 0)$	$\frac{1}{a}\tan^{-1}\frac{x}{a}$
$\frac{1}{x^2-a^2} \ (a > 0)$	$\frac{1}{2a}\ln\lvert\frac{x-a}{x+a}\rvert$
$\frac{1}{\sqrt{x^2+a^2}} \ (a > 0)$	$\sinh^{-1}\frac{x}{a}$ or $\ln(x + \sqrt{x^2 + a^2})$

$\frac{1}{\sqrt{x^2-a^2}}$ $(a>0)$	$\cosh^{-1}\frac{x}{a}$ or $\ln(x+\sqrt{x^2-a^2})$
$\sqrt{x^2+a^2}$	$\frac{1}{2}x\sqrt{x^2+a^2}+\frac{1}{2}a^2\ln(x+\sqrt{x^2+a^2})$
$\sqrt{x^2-a^2}$	$\frac{1}{2}x\sqrt{x^2-a^2}-\frac{1}{2}a^2\ln\lvert x+\sqrt{x^2-a^2}\rvert$
$\sqrt{a^2-x^2}$ $(a>0)$	$\frac{1}{2}x\sqrt{a^2-x^2}+\frac{1}{2}a^2\sin^{-1}\frac{x}{a}$

Reduction Formulae

For $I_n = \int x^n e^x\ \mathrm{d}x$, where $n \geq 1$, then

$$I_n = x^n e^x - nI_{n-1}.$$

For $I_n = \int (\ln x)^n\ \mathrm{d}x$, where $n \geq 1$, then

$$I_n = x(\ln x)^n - nI_{n-1}.$$

For $I_n = \int \sin^n x\ \mathrm{d}x$, where $n \geq 2$, then

$$I_n = -\frac{\sin^n x \cos x}{n} + \frac{n-1}{n}I_{n-2}.$$

For $I_n = \int \cos^n x\ \mathrm{d}x$, where $n \geq 2$, then

$$I_n = \frac{\cos^n x \sin x}{n} + \frac{n-1}{n}I_{n-2}.$$

For $I_n = \int \tan^n x\ \mathrm{d}x$, where $n \geq 2$, then

$$I_n = \frac{\tan^{n-1} x}{n-1} - I_{n-2}.$$

For $I_{n,k} = \int x^k(x^2 + Ax + B)^n\ \mathrm{d}x$, where $n \geq 0$, $k \geq 2$, then

$$I_{n,k} = \frac{x^{k-1}(x^2+Ax+B)^{n+1}}{(k+2n+1)} - \frac{B(k-1)}{(k+2n+1)}I_{n,k-2} - \frac{A(k+n)}{(k+2n+1)}I_{n,k-1}.$$

For $I_n(a,b) = \int \frac{ax+b}{(x^2+Ax+B)^n}\ \mathrm{d}x$, where $n \geq 2$, then

$$I_n(a,b) = \frac{bA - 2aB + (2b - aA)x}{(n-1)(4B - A^2)(x^2+Ax+B)^n} + \frac{(2n-3)(2b-aA)}{(n-1)(4B-A^2)}I_{n-1}(0,1).$$

Appendix 9

Common ordinary differential equations and solutions

Type	Form	Solution
1st order, separable variables (*see* SEPARABLE FIRST-ORDER DIFFERENTIAL EQUATION)	Can be written as $f(y)\frac{dy}{dx} = g(x)$	$\int f(y)dy = \int g(x)dx$
1st order, *integrating factor	$\frac{dy}{dx} + P(x)y = Q(x)$	$ye^{\int P(x)dx} = \int \left(Q(x)e^{\int P(x)\,dx}\right)dx$
Homogeneous 1st order (*see* HOMOGENEOUS FIRST-ORDER DIFFERENTIAL EQUATION)	$dy/dx = f(x,y)$ in which the function f, of two variables, has the property that $f(kx,ky) = f(x,y)$ for all k.	Let $y = vx$ so $dy/dx = x\,dv/dx + v$. The differential equation for v as a function of x that is obtained is always *separable*.
*Simple harmonic motion	$\frac{d^2x}{dt^2} = -k^2x$	$x = a\cos(kt + \varepsilon)$.
2nd order with constant coefficients	$a\frac{d^2y}{dx^2} + b\frac{dy}{dx} + cy = 0$ where $am^2 + bm + cm = 0$ has 2 real roots, m_1 and m_2.	$y = Ae^{m_1x} + Be^{m_2x}$
	$a\frac{d^2y}{dx^2} + b\frac{dy}{dx} + cy = 0$ where $am^2 + bm + cm = 0$ has 2 complex roots, m_1 and $m_2 = \alpha \pm \beta i$.	$y = Ae^{m_1x} + Be^{m_2x}$ $= e^{\alpha x}\left(C\cos(\beta x) + D\sin(\beta x)\right)$.
	$a\frac{d^2y}{dx^2} + b\frac{dy}{dx} + cy = 0$ where $am^2 + bm + cm = 0$ has 1 repeated root m.	$y = (Ax + B)e^{mx}$

Appendix 10

Laplace Transforms

Listed below are functions $f(x)$ together with their Laplace transforms

$$\bar{f}(p) = \int_0^\infty f(x)e^{-px} \, dx.$$

$f(x)$	$\bar{f}(p)$	$f(x)$	$\bar{f}(p)$
1	$1/p$	df/dx	$p\bar{f}(p) - f(0)$
x^n	$n!/p^{n+1}$	d^2f/dx^2	$p^2\bar{f}(p) - pf'(0) - f(0)$
x^a	$\Gamma(a+1)/p^{a+1}$	$e^{ax}f(x)$	$\bar{f}(p-a)$
e^{ax}	$1/(p-a)$	xdf/dx	$-d\bar{f}/dp$
$\sin ax$	$a/(p^2 + a^2)$	$\int_0^x f(t)dt$	$\dfrac{\bar{f}(p)}{p}$
$\cos ax$	$p/(p^2 + a^2)$	$(f * g)(x)$	$\bar{f}(p)\bar{g}(p)$
$\delta(x)$	1	$f(x-a)H(x-a)$	$e^{-ap}\bar{f}(p)$

In the above, δ denotes the *Dirac delta function, Γ the *gamma function, H denotes the *Heaviside function, and $*$ denotes *convolution.

Series

$$e^x = 1 + \frac{x}{1!} + \frac{x^2}{2!} + \ldots + + \frac{x^n}{n!} + \ldots \quad \text{(for all } x)$$

$$\sin x = x - \frac{x^3}{3!} + \frac{x^5}{5!} - \frac{x^7}{7!} + \ldots + (-1)^n \frac{x^{2n+1}}{(2n+1)!} + \ldots \quad \text{(for all } x)$$

$$\cos x = 1 - \frac{x^2}{2!} + \frac{x^4}{4!} - \frac{x^6}{6!} + \ldots + (-1)^n \frac{x^{2n}}{(2n)!} + \ldots \quad \text{(for all } x)$$

$$\tan x = x + \frac{1}{3}x^3 + \frac{2}{5}x^5 + \ldots + (-1)^{n-1} \frac{2^{2n}(2^{2n}-1)B_{2n}}{(2n)!} x^{2n-1} + \ldots \quad \text{(for } |x| < \pi/2)$$

where B_n denotes the nth *Bernoulli number

$$\sec x = 1 + \frac{1}{2}x^2 + \frac{5}{24}x^4 + \ldots + (-1)^n \frac{E_{2n}}{(2n)!} x^{2n} + \ldots \quad \text{(for } |x| < \pi/2)$$

where E_n denotes the nth *Euler number

$$\sinh x = x + \frac{x^3}{3!} + \frac{x^5}{5!} + \frac{x^7}{7!} + \ldots + \frac{x^{2n+1}}{(2n+1)!} + \ldots \quad \text{(for all } x)$$

$$\cosh x = 1 + \frac{x^2}{2!} + \frac{x^4}{4!} + \frac{x^6}{6!} + \ldots + \frac{x^{2n}}{(2n)!} + \ldots \quad \text{(for all } x)$$

$$\tanh x = x - \frac{1}{3}x^3 + \frac{2}{5}x^5 - \ldots + \frac{2^{2n}(2^{2n}-1)B_{2n}}{(2n)!} x^{2n-1} + \ldots \quad \text{(for } |x| < \pi/2)$$

$$\text{sech } x = 1 - \frac{1}{2}x^2 + \frac{5}{24}x^4 - \ldots + \frac{E_{2n}}{(2n)!} x^{2n} + \ldots \quad \text{(for } |x| < \pi/2)$$

$$\ln(1+x) = x - \frac{x^2}{2} + \frac{x^3}{3} - \frac{x^4}{4} + \ldots + (-1)^{n-1} \frac{x^n}{n} + \ldots \quad (-1 < x \leq 1)$$

$$\tan^{-1}x = x - \frac{x^3}{3} + \frac{x^5}{5} - \frac{x^7}{7} + \ldots + (-1)^n \frac{x^{2n+1}}{2n+1} + \ldots \quad (-1 \leq x \leq 1)$$

$$\sin^{-1}x = x + \frac{1}{2}\frac{x^3}{3} + \frac{1\times 3}{2\times 4}\frac{x^5}{5} + \frac{1\times 3\times 5}{2\times 4\times 6}\frac{x^7}{7} + ... + \binom{2n}{n}\frac{x^{2n+1}}{2^{2n}(2n+1)} + ... \quad (-1 \le x \le 1)$$

$$(1+x)^\alpha = 1 + \frac{\alpha}{1!}x + \frac{\alpha(\alpha-1)}{2!}x^2 + ... + \frac{\alpha(\alpha-1)...(\alpha-n+1)}{n!}x^n + ... \quad (-1 < x < 1)$$

(This is a binomial expansion; when α is a non-negative integer, the expansion is a finite series and is then valid for all x.)

For an infinitely differentiable real function $f(x)$ Taylor's Theorem states

$$f(a+h) = f(a) + f'(a)h + \frac{f''(a)}{2!}h^2 + ... + \frac{f^{(n)}(a)}{n!}h^n + \frac{f^{(n+1)}(c)}{(n+1)!}h^{n+1},$$

where a, h are real, n is a positive integer, and c is some number in the range $a < c < c + h$.

If the remainder term (the last term above) converges to 0, the infinite Taylor series converges and

$$f(a+h) = f(a) + f'(a)h + \frac{f''(a)}{2!}h^2 + ... + \frac{f^{(n)}(a)}{n!}h^n + ...$$

The Taylor series for two variables states

$$\begin{aligned}
f(a+h, b+k) = f(a,b) &+ [f_x(a,b)h + f_y(a,b)k] \\
&+ \frac{1}{2!}[f_{xx}(a,b)h^2 + 2f_{xy}(a,b)hk + f_{yyy}(a,b)k^2] \\
&+ \frac{1}{3!}[f_{xxx}(a,b)h^3 + 3f_{xxy}(a,b)h^2k + 3f_{xyy}(a,b)hk^2 + f_{yy}(a,b)k^2] + ...
\end{aligned}$$

provided that the square-bracketed contributions converge to 0.

Appendix 12

Convergence tests for series

*Absolute convergence	If the series $\sum_{i=1}^{\infty} |a_i|$ converges then $\sum_{i=1}^{\infty} a_i$ also converges.
*Alternating series test	If $\sum a_n$ is a series of non-negative real terms, with $a_{i+1} \leq a_i$ and also $a_i \to 0$ as $i \to \infty$, then $\sum_{i=1}^{\infty} (-1)^i a_i$ converges.
*Abel's test	If $\sum a_n$ is a convergent infinite sequence, and $\{b_n\}$ is monotonically decreasing, i.e. $b_{n+1} \leq b_n$ for all n, then $\sum a_n b_n$ is also convergent.
*Comparison test	If $0 \leq a_i \leq b_i$ for all $i >$ some positive integer N, then: if $\sum_{i=1}^{\infty} b_i$ converges then $\sum_{i=1}^{\infty} a_i$ also converges and if $\sum_{i=1}^{\infty} a_i$ diverges then $\sum_{i=1}^{\infty} b_i$ also diverges.
*Dirichlet's test	If $\sum a_n$ is a series which has bounded partial sums, i.e. $\left| \sum_{n=1}^{m} a_n \right| < K$ for all values of m, and $\{b_n\}$ is decreasing and converges to zero, i.e. $b_n < b_{n-1}$ and $b_n \to 0$ then $\sum_{n=1}^{\infty} a_n b_n$ converges.
*Geometric series	If $\sum a_n$ is a sequence of real numbers, with $a_{i+1} = r \times a_i$ then $\sum_{i=1}^{\infty} a_i$ converges (to $\frac{a_1}{1-r}$) if $|r| < 1$ and diverges otherwise.
*Integral test	For $f(x)$, a non-negative, continuous, decreasing function defined for $x \geq N$, then $\sum_{n=N}^{\infty} f(n)$ converges if and only if $\int_N^{\infty} f(x) \, dx$ exists.
*Ratio test	For an infinite series $\sum a_n$, if the ratio $\left| \frac{a_{n+1}}{a_n} \right| \to k$ as $n \to \infty$ then when k is < 1 the series converges absolutely, and for $k > 1$ the series diverges. If $k = 1$, the ratio test is not sufficient to determine whether it converges or not.
*Root test (or Cauchy's test)	For a *series $\sum a_n$ of positive terms, set $L = \limsup a_n^{1/n}$. If $L < 1$, then the series *converges; if $L > 1$, then the series *diverges; if $L = 1$, then the series may or may not converge.

Common inequalities

*Triangle inequality	$\|\|x\| - \|y\|\| \le \|x + y\| \le \|x\| + \|y\|$ $\|x + x_2 + \ldots + x_n\| \le \|x_1\| + \|x_2\| + \ldots + \|x_n\|$
*Means	arithmetic mean $\ge$ geometric mean $\ge$ harmonic mean $$\frac{x_1 + x_2 + \ldots + x_n}{n} \ge \sqrt[n]{x_1 x_2 \ldots x_n} \ge \frac{n}{\frac{1}{x_1} + \frac{1}{x_2} + \ldots + \frac{1}{x_n}}$$ where $x_1, x_2, \ldots, x_n$ are positive real numbers
Abel's	$\left\| \sum_{i=1}^{n} a_i b_i \right\| \le A b_1$ where $\{b_i\}$ is a sequence of non-negative numbers with $b_{i+1} \ge b_i$ and $\{a_i\}$ a sequence of real or complex numbers, and A is the maximum modulus of the first n partial sums of $\{a_i\}$.
Bernoulli's	$(1 + x)^n \ge 1 + nx$ for integers $n > 1$ and any real number $x > -1$.
*Cauchy–Schwarz inequality for integrals	If $f(x)$, $g(x)$ are real functions then $\left\{ \int [f(x)g(x)] \mathrm{d}x \right\}^2$ $\le \left\{ \int [f(x)]^2 \mathrm{d}x \right\} \left\{ \int [g(x)]^2 \mathrm{d}x \right\}$ if all these integrals exist.
*Cauchy–Schwarz inequality for sums	If a_i and b_i are real numbers, $i = 1, 2, \ldots, n$ then $\sum_{i=1}^{n} a_i b_i \le \sqrt{(\sum_{i=1}^{n} a_i^2)(\sum_{i=1}^{n} b_i^2)}$.
*Chebyshev's inequalities	$\Pr\{\|X - \mu_x\| > k\sigma\} \le \frac{1}{k^2}$ If X is a random variable and $g(X)$ is always ≥ 0 then $\Pr\{g(X) \ge k\} \le \frac{E\{g(X)\}}{k}$.
Hölder's for integrals	$\int_a^b \|f(x).g(x)\| \mathrm{d}x \le \left\| \int_a^b \|f(x)\|^p \mathrm{d}x \right\|^{\frac{1}{p}} \times \left\| \int_a^b \|g(x)\|^q \mathrm{d}x \right\|^{\frac{1}{q}}$ for $\frac{1}{p} + \frac{1}{q} = 1$
Hölder's for sums	$\sum_{i=1}^{n} \|a_i b_i\| \le (\sum_{i=1}^{n} \|a_i\|^p)^{\frac{1}{p}} \times (\sum_{i=1}^{n} \|b_i\|^q)^{\frac{1}{q}}$ for $\frac{1}{p} + \frac{1}{q} = 1$
*Isoperimetric inequality	If p is the perimeter of a closed curve in a plane and the area enclosed by the curve is A, then $p^2 \le 4\pi A$.

Trigonometric formulae

$$\tan A = \frac{\sin A}{\cos A}, \quad \cot A = \frac{\cos A}{\sin A} = \frac{1}{\tan A},$$

$$\sec A = \frac{1}{\cos A}, \quad \operatorname{cosec} A = \frac{1}{\sin A},$$

$$\cos^2 A + \sin^2 A = 1, \quad \sec^2 A = 1 + \tan^2 A, \quad \operatorname{cosec}^2 A = 1 + \cot^2 A.$$

*Addition formulae

$$\begin{aligned}
\sin(A + B) &= \sin A \, \cos B + \cos A \, \sin B, \\
\sin(A - B) &= \sin A \, \cos B - \cos A \, \sin B, \\
\cos(A + B) &= \cos A \, \cos B - \sin A \, \sin B, \\
\cos(A - B) &= \cos A \, \cos B + \sin A \, \sin B, \\
\tan(A + B) &= \frac{\tan A + \tan B}{1 - \tan A \, \tan B}, \\
\tan(A - B) &= \frac{\tan A - \tan B}{1 + \tan A \, \tan B}.
\end{aligned}$$

*Double-angle formulae

$$\sin 2A = 2\sin A \, \cos A,$$

$$\cos 2A = \cos^2 A - \sin^2 A,$$

$$\cos 2A = 1 - 2\sin^2 A, \quad \sin^2 A = \frac{1}{2}(1 - \cos 2A),$$

$$\cos 2A = 2\cos^2 A - 1, \quad \cos^2 A = \frac{1}{2}(1 + \cos 2A),$$

$$\tan 2A = \frac{2\tan A}{1 - \tan^2 A}.$$

Tangent-of-half-angle formulae

Let $t = \tan \frac{1}{2}A$; then

$$\sin A = \frac{2t}{1 + t^2}, \quad \cos A = \frac{1 - t^2}{1 + t^2}, \quad \tan A = \frac{2t}{1 - t^2}.$$

Product formulae

$$\begin{aligned}
\sin A \, \cos B &= \frac{1}{2}\Big(\sin(A + B) + \sin(A - B)\Big), \\
\cos A \, \sin B &= \frac{1}{2}\Big(\sin(A + B) - \sin(A - B)\Big), \\
\cos A \, \cos B &= \frac{1}{2}\Big(\cos(A + B) + \cos(A - B)\Big), \\
\sin A \, \sin B &= \frac{1}{2}\Big(\cos(A - B) - \cos(A + B)\Big).
\end{aligned}$$

Factor formulae

$$\sin C + \sin D = 2\sin \frac{1}{2}(C + D)\cos \frac{1}{2}(C - D),$$
$$\sin C - \sin D = 2\cos \frac{1}{2}(C + D)\sin \frac{1}{2}(C - D),$$
$$\cos C + \cos D = 2\cos \frac{1}{2}(C + D)\cos \frac{1}{2}(C - D),$$
$$\cos C - \cos D = -2\sin \frac{1}{2}(C + D)\sin \frac{1}{2}(C - D).$$

Trigonometric values for some special angles

$\sin 0 = 0$, $\cos 0 = 1$, $\tan 0 = 0$;

$\sin 30° = \frac{1}{2}$, $\cos 30° = \frac{\sqrt{3}}{2}$, $\tan 30° = \frac{1}{\sqrt{3}}$;

$\sin 45° = \frac{1}{\sqrt{2}}$, $\cos 45° = \frac{1}{\sqrt{2}}$, $\tan 45° = 1$;

$\sin 60° = \frac{\sqrt{3}}{2}$, $\cos 60° = \frac{1}{2}$, $\tan 60° = \sqrt{3}$;

$\sin 90° = 1$, $\cos 90° = 0$; $\tan 90°$ is undefined.

Symmetry

$\sin(-\theta) = -\sin\theta$, $\cos(-\theta) = \cos\theta$, $\tan(-\theta) = -\tan\theta$;

$\sin(90° - \theta) = \cos\theta$, $\cos(90° - \theta) = \sin\theta$, $\tan(90° - \theta) = \cot\theta$;

$\sin(180° - \theta) = \sin\theta$, $\cos(180° - \theta) = -\cos\theta$, $\tan(180° - \theta) = -\tan\theta$.

Shifts and periodicity

A shift by 360° for sin and cos and a shift by 180° for tan, or by multiples of these values, leave the function value unchanged because these are the periods of those functions. In addition, the following hold:

$\sin(90° + \theta) = \cos\theta$, $\cos(90° + \theta) = -\sin\theta$, $\tan(90° + \theta) = -\cot\theta$;

$\sin(180° + \theta) = -\sin\theta$, $\cos(180° + \theta) = -\cos\theta$.

Appendix 15

Probability distributions

Discrete probability distributions

*Distribution	*Parameters	*Probability mass function	*Mean	*Variance	*Moment generating function
*Binomial	n, p	$P(X = r) = \binom{n}{r} p^r (1-p)^{n-r}, r = 0, 1, 2, \ldots, n$	$\mu = np$	$\sigma^2 = np(1-p)$	$M(t) = (1 + p(e^t - 1))^n$
*Geometric	p	$P(X = r) = p(1-p)^{r-1}, r = 1, 2, \ldots$	$\mu = \frac{1}{p}$	$\sigma^2 = \frac{(1-p)}{p^2}$	$M(t) = \dfrac{pe^t}{1-(1-p)e^t}, t < -\ln(1-p)$
*Negative binomial	k, p	$P(X = r) = \binom{r-1}{k-1} p^k (1-p)^{r-k},$ $r = k, k+1, k+2, \ldots$	$\mu = \frac{k(1-p)}{p}$	$\sigma^2 = \frac{k(1-p)}{p^2}$	$M(t) = \dfrac{(pe^t)^k}{[1-(1-p)e^t]^k}, t < -\ln(1-p)$
*Hypergeometric	n, N, M	$P(X = r) = \dfrac{\binom{M}{r}\binom{N-M}{n-r}}{\binom{N}{n}},$ for r with $\max(0, n-N+M) \le r \le \min(n, M)$	$\mu = n \cdot \frac{M}{N}$	$\sigma^2 = n \cdot \frac{M}{N}\left(1 - \frac{M}{N}\right)\left(\frac{N-n}{N-1}\right)$	$M(t) = \dfrac{\binom{N-M}{n}}{\binom{N}{n}} F(-n, -M; N-M-N+1, e^t)$
*Poisson	λ	$P(X = r) = e^{-\lambda}\frac{\lambda^r}{r!}, r = 0, 1, 2, \ldots$	$\mu = \lambda$	$\sigma^2 = \lambda$	$M(t) = \exp\{\lambda(e^t - 1)\}$

Continuous probability distributions

*Distribution	*Parameters	*Probability density function	*Mean	*Variance	*Moment generating function
*Uniform on $[a, b]$	a, b	$f(x) = \frac{1}{b-a}, a \le x \le b$	$\mu = \frac{(b+a)}{2}$	$\sigma^2 = \frac{(b-a)^2}{12}$	$M(t) = \frac{e^{tb} - e^{ta}}{(b-a)t}, t \ne 0$
*Normal	μ, σ	$f(x) = \frac{1}{\sigma\sqrt{(2\pi)}} e^{\frac{-(x-\mu)^2}{2\sigma^2}}, -\infty \le x \le \infty$	μ	σ^2	$M(t) = \exp\left(\mu t + \frac{1}{2}\sigma^2 t^2\right)$
*Exponential	λ	$f(x) = \lambda e^{-\lambda x}, x \ge 0$	$\mu = \frac{1}{\lambda}$	$\sigma^2 = \frac{1}{\lambda^2}$	$M(t) = \frac{\lambda}{\lambda - t}, t < \lambda$
*Gamma	λ, r	$f(x) = \frac{\lambda^r}{\Gamma(r)} x^{r-1} e^{-\lambda x}, x > 0$	$\mu = \frac{r}{\lambda}$	$\sigma^2 = \frac{r}{\lambda^2}$	$M(t) = \left(\frac{\lambda}{\lambda - t}\right)^r, t < \lambda$
*Cauchy		$\frac{1}{\pi(1+x^2)}$	undefined	undefined	undefined
*t-distribution	ν	$\frac{\Gamma\left(\frac{\nu+1}{2}\right)}{\sqrt{\nu\pi}\Gamma\left(\frac{\nu}{2}\right)} \left(1 + \frac{x^2}{\nu}\right)^{-(\nu+1)/2}$	0 for $\nu > 1$	$\frac{\nu}{\nu-2}$ for $\nu > 2$	undefined
*Chi–squared	n	$\frac{1}{2^{\frac{n}{2}}\Gamma\left(\frac{n}{2}\right)} x^{\frac{n}{2}-1} e^{-\frac{x}{2}}, x > 0$	$\mu = n$	$\sigma^2 = 2n$	$M(t) = \left(\frac{\frac{1}{2}}{\frac{1}{2}-t}\right)^{\frac{n}{2}} = \frac{1}{(1-2t)^{\frac{n}{2}}}, t < \frac{1}{2}$

Vector algebra and operators

Algebra: Given vectors $\mathbf{a}$, $\mathbf{b}$, $\mathbf{c}$ in $\mathbb{R}^3$, which may vary with t,

$(\mathbf{a} \cdot \mathbf{b})^2 + |\mathbf{a} \times \mathbf{b}|^2 = |\mathbf{a}|^2|\mathbf{b}|^2$.

$|(\mathbf{a} \cdot \mathbf{b}| \leq |\mathbf{a}||\mathbf{b}|$. (*Cauchy-Schwarz inequality)

$[\mathbf{a},\mathbf{b},\mathbf{c}] = [\mathbf{b},\mathbf{c},\mathbf{a}] = [\mathbf{c},\mathbf{a},\mathbf{b}] = -[\mathbf{b},\mathbf{a},\mathbf{c}] = -[\mathbf{a},\mathbf{c},\mathbf{b}] = -[\mathbf{c},\mathbf{b},\mathbf{a}]$. (*scalar triple product)

$\mathbf{a} \times (\mathbf{b} \times \mathbf{c}) = (\mathbf{a} \cdot \mathbf{c})\mathbf{b} - (\mathbf{a} \cdot \mathbf{b})\mathbf{c}$. (*vector triple product)

$(\mathbf{a} \times \mathbf{b}) \cdot (\mathbf{c} \times \mathbf{d}) = (\mathbf{a} \cdot \mathbf{c})(\mathbf{b} \cdot \mathbf{d}) - (\mathbf{a} \cdot \mathbf{d})(\mathbf{b} \cdot \mathbf{c})$. (*scalar quadruple product)

$$\frac{d}{dt}(\mathbf{a} \cdot \mathbf{b}) = \mathbf{a} \cdot \frac{d\mathbf{b}}{dt} + \frac{d\mathbf{a}}{dt} \cdot \mathbf{b}. \qquad \frac{d}{dt}(\mathbf{a} \times \mathbf{b}) = \mathbf{a} \times \frac{d\mathbf{b}}{dt} + \frac{d\mathbf{a}}{dt} \times \mathbf{b}.$$

Differential operator identities: Given vector fields $\mathbf{F}$, $\mathbf{G}$ and scalar fields f, g on $\mathbb{R}$, then

$\mathrm{curl}(\mathrm{grad}f) = \mathbf{0}$.

$\mathrm{div}(\mathrm{curl}\mathbf{F}) = 0$.

$\mathrm{grad}(fg) = f\,\mathrm{grad}g + g\,\mathrm{grad}f$.

$\mathrm{div}(f\mathbf{F}) = f\,\mathrm{div}\mathbf{F} + (\mathrm{grad}f) \cdot \mathbf{F}$.

$\mathrm{curl}(f\mathbf{F}) = f\,\mathrm{curl}\mathbf{F} + \mathrm{grad}f \times \mathbf{F}$.

$\nabla^2(fg) = f\,\nabla^2 g + 2\nabla f \cdot \nabla g + g\,\nabla^2 f$.

$\mathrm{curl}(\mathrm{curl}\mathbf{F}) = \mathrm{grad}(\mathrm{div}\mathbf{F}) - \nabla^2\mathbf{F}, \mathrm{div}(\mathbf{F} \times \mathbf{G}) = \mathbf{G} \cdot \mathrm{curl}\mathbf{F} - \mathbf{F} \cdot \mathrm{curl}\mathbf{G}$.

$\mathrm{curl}(\mathbf{F} \times \mathbf{G}) = \mathbf{F}\,\mathrm{div}\mathbf{G} - \mathbf{G}\,\mathrm{div}\mathbf{F} + (\mathbf{G} \cdot \nabla)\mathbf{F} - (\mathbf{F} \cdot \nabla)\mathbf{G}$.

$\mathrm{grad}(\mathbf{F} \cdot \mathbf{G}) = (\mathbf{F} \cdot \nabla)\mathbf{G} + (\mathbf{G} \cdot \nabla)\mathbf{F} + \mathbf{F} \times \mathrm{curl}\mathbf{G} + \mathbf{G} \times \mathrm{curl}\mathbf{F}$.

Curvilinear coordinates:

With $d\mathbf{r} = h_1 du_1 \mathbf{e}_1 + h_2 du_2 \mathbf{e}_2 + h_3 du_3 \mathbf{e}_3$ and $\mathbf{F} = f_1 \mathbf{e}_1 + f_2 \mathbf{e}_2 + f_3 \mathbf{e}_3$, we have

$$\frac{\partial(x,y,z)}{\partial(u_1,u_2,u_3)} = h_1 h_2 h_3.$$

$$\mathrm{grad}f = \frac{1}{h_1}\frac{\partial f}{\partial u_1}\mathbf{e}_1 + \frac{1}{h_2}\frac{\partial f}{\partial u_2}\mathbf{e}_2 + \frac{1}{h_3}\frac{\partial f}{\partial u_3}\mathbf{e}_3.$$

$$\mathrm{div}\ \mathbf{F} = \frac{1}{h_1 h_2 h_3}\left(\frac{\partial}{\partial u_1}(h_2 h_3 f_1) + \frac{\partial}{\partial u_2}(h_1 h_3 f_2) + \frac{\partial}{\partial u_3}(h_1 h_2 f_3)\right).$$

$$\mathrm{curl}\ \mathbf{F} = \begin{vmatrix} h_1\mathbf{e}_1 & h_2\mathbf{e}_2 & h_3\mathbf{e}_3 \\ \partial/\partial u_1 & \partial/\partial u_2 & \partial/\partial u_3 \\ h_1 f_1 & h_2 f_2 & h_3 f_3 \end{vmatrix}.$$

$$\nabla^2 = \frac{1}{h_1 h_2 h_3}\left(\frac{\partial}{\partial u_1}\left(\frac{h_2 h_3}{h_1}\frac{\partial}{\partial u_1}\right) + \frac{\partial}{\partial u_2}\left(\frac{h_1 h_3}{h_2}\frac{\partial}{\partial u_2}\right) + \frac{\partial}{\partial u_3}\left(\frac{h_1 h_2}{h_3}\frac{\partial}{\partial u_3}\right)\right).$$

Groups of order 15 or less

Order	Groups up to isomorphism	Order	Groups up to isomorphism
2	C_2	9	C_9, $C_3 \times C_3$
3	C_3	10	C_{10}, D_{10}
4	C_4, $C_2 \times C_2$	11	C_{11}
5	C_5	12	C_{12}, $C_2 \times C_6$, D_{12}, A_4, $C_3 \rtimes C_4$
6	C_6, D_6	13	C_{13}
7	C_7	14	C_{14}, D_{14}
8	C_8, $C_2 \times C_4$, $C_2 \times C_2 \times C_2$, D_8, Q_8	15	C_{15}

Relevant group theory:

- A group of prime order is cyclic.
- If a group has order $2p$, where p is an odd prime, the group is C_{2p} or D_{2p}.
- A group of order p^2, where p is prime, is C_{p^2} or $C_p \times C_p$.
- A group of order pq, where $p > q$ are primes and $q \nmid p-1$ is cyclic.

Prime numbers less than 1000

2	3	5	7	11	13	17	19	23	29
31	37	41	43	47	53	59	61	67	71
73	79	83	89	97	101	103	107	109	113
127	131	137	139	149	151	157	163	167	173
179	181	191	193	197	199	211	223	227	229
233	239	241	251	257	263	269	271	277	281
283	293	307	311	313	317	331	337	347	349
353	359	367	373	379	383	389	397	401	409
419	421	431	433	439	443	449	457	461	463
467	479	487	491	499	503	509	521	523	541
547	557	563	569	571	577	587	593	599	601
607	613	617	619	631	641	643	647	653	659
661	673	677	683	691	701	709	719	727	733
739	743	751	757	761	769	773	787	797	809
811	821	823	827	829	839	853	857	859	863
877	881	883	887	907	911	919	929	937	941
947	953	967	971	983	991	997			

Appendix 19

Symbols and abbreviations

Real and Complex Numbers

Symbol	Reference
$+$	*addition
$-$	*subtraction
$\times$, $\cdot$	*multiplication
$\div$, $/$	*division
$\pm$	*plus or minus
$\mp$	*minus or plus
$\sqrt{}$	*square root, *radical sign
$\sqrt{}$	*square root
$\sqrt[n]{}$	*nth root
$\lvert x \rvert$	*absolute value
$[x]$, $\lfloor x \rfloor$	*integer part, *floor
$\lceil x \rceil$	*ceiling
$\{x\}$	*fractional part
$\lvert x \rvert$	*absolute value
i	$\sqrt{-1}$
Re z	*real part
Im z	*imaginary part
$\lvert z \rvert$	*modulus
arg$z$	*argument
$\bar{z}$, z^*	*conjugate
$\mid$	*divides
(a,b), hcf(a,b), gcd(a,b)	*greatest common divisor
$[a,b]$, lcm(a,b)	*lowest common multiple

Equality and Inequality

Symbol	Reference
$=$	*equality
$:=$, $\stackrel{\text{def}}{=}$	defining equality
$\equiv$	*identity
$\equiv$	*congruence
mod	*modulo n arithmetic
$\neq$	not equal to (*see* INEQUALITY)
$<$	strictly less than (*see* INEQUALITY)
$>$	strictly greater than (*see* INEQUALITY)
$\leq$	less than or equal to (*see* INEQUALITY)
$\geq$	greater than or equal to (*see* INEQUALITY)
$\ll$	much less than
$\gg$	much greater than
$\approx$	approximately equal to
$\sim$	*asymptotically equal to

Important sets, spaces, and functions

Symbol	Reference
d	*divisor function
σ	*sigma function
μ	*Möbius function
φ	*totient function
$(a\|p)$	*Legendre symbol
$\mathbb{N}$	the set of *natural numbers
$\mathbb{Z}$	the set of *integers
$\mathbb{Q}$	the set of *rational numbers
$\mathbb{R}$	the set of *real numbers
$\mathbb{C}$	the set of *complex numbers
$\mathbb{Z}[i]$	the *Gaussian integers
$\mathbb{C}_\infty$	the *extended complex plane
$\mathbb{H}$	the set of *quaternions
$\mathbb{R}^n$	n-dimensional *Cartesian space
E^n	n-dimensional *Euclidean space
$\mathbb{Z}_n$	*modulo n arithmetic
$\mathbb{F}_q$	the *finite field with q elements
$\mathbb{Q}_p$	*p-adic numbers
$\mathbb{R}[x]$	real *polynomials in a variable x
$\mathbb{R}(x)$	real *rational functions in a variable x
$\mathbb{R}[x_1,\ldots,x_n]$	real *polynomials in many variables $x_1,\ldots,x_n$
$\mathbb{R}[[x]]$	formal real *power series in a variable x
$[a,b]$	*closed interval
(a,b)	*open interval
$[a,b)$, $(a,b]$	*half-open *interval

Set Theory and Logic

Symbol	Reference
¬	*negation, not
∧	'and' (*see* CONJUNCTION), 'min' (*see also* LATTICE)
∨	'or' (*see* DISJUNCTION), 'max' (*see also* LATTICE)
⇒	'implies' (*see* IMPLICATION)
⇔, iff	'if and only if' (*see* IMPLICATION)
∃	'there exists' (*see* QUANTIFIER)
∃!	'there uniquely exists'
∀	'for all', (*see* QUANTIFIER)
∈, ∉	*belongs to, does not belong to
⊆, ⊂	is a *subset of
⊂	is a *proper subset of (inconsistently used)
∪	*union
⊔, ⊎, ⊎	*disjoint union
∩	*intersection
$A + B$, $A \triangle B$	*symmetric difference
A', A^c, $\bar{A}$	*complement
$A \backslash B$, $A - B$	*set difference, *relative complement
∅	*empty set
$A \times B$	*Cartesian product
A^n	the Cartesian product $A \times A \times \ldots \times A$ (n times)
B^A	the set of functions from A to B
$|A|$, $n(A)$, $\#A$	*cardinality
$\wp(A)$	*power set
∼	*equivalence relation
$[a]$, $\bar{a}$	*equivalence class
$X/\sim$	set of *equivalence classes
$f: x \mapsto y$	*function
$f: S \to T$	*function
↪	*injection
↠	*surjection
$f \circ g$	*composition
f^{-1}	*inverse function
$f(A)$	*image
$f^{-1}(A)$	*pre-image
$\aleph_0$	*aleph-null, the *cardinal number of ℕ
$\aleph_n$	the nth infinite *cardinal number
ω	the *ordinal number of ℕ
c	the *cardinal number of ℝ

Vectors and Matrices

Symbol	Reference
$|\mathbf{v}|$	*magnitude of a vector
$\overrightarrow{AB}$	*directed line segment
$|AB|$	*length of line segment
$\mathbf{a} \cdot \mathbf{b}$	*scalar (or dot) product
$\mathbf{a} \times \mathbf{b}$, $\mathbf{a} \wedge \mathbf{b}$	*vector (or cross) product
$\mathbf{a} \cdot (\mathbf{b} \times \mathbf{c})$, $[\mathbf{a}, \mathbf{b}, \mathbf{c}]$	*scalar triple product
$\mathbf{a} \times (\mathbf{b} \times \mathbf{c})$	*vector triple product
$\mathbf{A}^T$, $\mathbf{A}^t$, $\mathbf{A}'$	*transpose
$\mathbf{A}^*$	*Hermitian conjugate
$\mathbf{A}^{-1}$	*inverse matrix
$\mathbf{I}$	*identity matrix
$|\mathbf{A}|$, $\det \mathbf{A}$	*determinant
δ_{ij}	*Kronecker delta
$J(\lambda, n)$	*Jordan block
$\chi_{\mathbf{A}}(x)$	*characteristic polynomial
$m_{\mathbf{A}}(x)$	*minimal polynomial
S^1	the *circle, as unit modulus complex numbers under multiplication
$GL(n, \mathbb{R})$, $SL(n, \mathbb{R})$	*invertible real $n \times n$ matrices (with det 1)
$O(n)$, $SO(n)$	*orthogonal $n \times n$ matrices (with det 1)
$U(n)$, $SU(n)$	*unitary $n \times n$ matrices (with det 1)
$AGL(n, \mathbb{R})$	*affine transformations
$PGL(n, \mathbb{R})$	*projective transformations

Algebra

Symbol	Reference
$*$, $\circ$	*binary operation
$(G, *)$	*group
$|G|$	*order of group
$o(g)$	*order of group element
S_n, $\mathrm{Sym}(S)$	*symmetric group
A_n	*alternating group
C_n	*cyclic group
D_{2n}	*dihedral group
V_4	*Klein four group
Q_8	*quaternion group
$(R, +, \times)$	*ring
R^*	group of *units of a ring under multiplication
PID	*principal ideal domain
UFD	*unique factorization domain
$\cong$	*isomorphic
$\leq$	is a *subgroup of, is a *subspace of
$\triangleleft$	is a *normal subgroup of, is an *ideal of

G/H	set of *cosets
$\lvert G : H \rvert$	*index of H in G
G/N, R/I	*quotient group, *quotient ring
$\langle S \rangle$	*span of S, *subgroup generated by S
$\langle\langle S \rangle\rangle$	*normal subgroup generated by S
dim	*dimension
ker	*kernel
Im	*image
Aut	*automorphism group
End	*endomorphism ring
Hom	space of linear maps/group of homomorphisms—*see* HOM
$(\_,\_)$	*inner product
$\times$	*direct product
$\rtimes$	*semi-direct product
$\oplus$	*direct sum
$\otimes$	*tensor product
$\wedge$	*exterior product
V^*	*dual space (algebraic)
V'	*dual space (analytic)
T^*	dual map, *adjoint
$K{:}F$	*field extension
$\lvert K{:}F \rvert$	degree of *field extension

Calculus

Symbol	Reference
f', $\frac{df}{dx}$	*derivative, *derived function
df	*differential, *differential form
$f', f'', \ldots f^{(n)}$, $\frac{df}{dx}$, $\frac{d^2f}{dx^2}$, $\ldots$, $\frac{d^nf}{dx^n}$	*higher derivatives
f_x, f_y, f_1, f_2, $\frac{\partial f}{\partial x}$, $\frac{\partial f}{\partial y}$	*partial derivatives
f_{xx}, f_{xy}, f_{11}, f_{12}, $\frac{\partial^2 f}{\partial x^2}$, $\frac{\partial^2 f}{\partial x \partial y}$	*higher partial derivatives, *mixed derivatives
$\dot{x}, \ddot{x}, \ldots$	*rate of change
$\propto$	*proportion
$\int$	*integral, *antiderivative
$\int_C f(s)\, ds$, $\int_C \mathbf{f}(\mathbf{r})\cdot d\mathbf{r}$	*line integral
$\int_C f(z)\, dz$,	*path integral
$\oint$	*line integral around a closed curve
$\iint, \iiint$	*multiple integral
grad, ∇	*del, *gradient, nabla
curl, $\nabla \times$	*curl
div, $\nabla \cdot$	*divergence
∇^2, Δ	*Laplacian
$\frac{\partial f}{\partial n}$	*directional derivative along outward-pointing normal

Analysis

Symbol	Reference
max	*maximum
min	*minimum
sup, lub	*supremum (least upper bound)
inf, glb	*infimum (greatest lower bound)
$\{x_n\}$, (x_n)	*sequence
lim	*limit
Σ	*summation notation, *series
Π	*product notation
$\rightarrow$	converges to, tends to (*see* LIMIT)
$\nearrow$, $\searrow$	*limit from the left and right
limsup	*limit superior
liminf	*limit inferior
$O(n)$	*big O notation
$o(n)$	*little o notation
$\hat{f}(s)$, $\mathcal{F}$	*Fourier transform
$\bar{f}(p)$, $\mathcal{L}$	*Laplace transform
$\delta(x)$	*Dirac delta function
$\|\cdot\|$	*norm
$\|\cdot\|_p$	*p-norm
c, c_0, l^p, l^∞	*sequence space
$C(X)$	*continuous real functions on X
$C^*(X)$	*continuous bounded real functions on X
$C^1(X)$	continuously *differentiable real functions on X
$C^\infty(X)$	infinitely *differentiable real functions on X
$C^\omega(X)$	*analytic real functions on X

Geometry and Topology

Symbol	Reference
i,j,k	*canonical basis for $\mathbb{R}^3$
$\parallel$	*parallel
$\perp$	*perpendicular
$\llcorner$	*right angle
$\angle$	*angle
$[x_1 : x_2 : \ldots : x_n]$	*homogeneous coordinates
$\mathbb{P}^n$	n-dimensional *projective space
T_pX	*tangent space
D/dt	*covariant derivative
κ	*curvature
τ	*torsion
t, n, b	*Serret-Frenet triad
K	*Gaussian curvature
H	*mean curvature

∂A	*boundary
$\bar{A}$	*closure
A°	*interior
$\cong$	*homeomorphic
$\simeq$	*homotopic
$\pi_1(X)$	*fundamental group
$X \# Y$	*connected sum
$\chi(X)$	*Euler characteristic
S^n	the n-dimensional *sphere
$(S^1)^n$	the n-dimensional *torus

Combinatorics, Probability, and Statistics

Symbol	Reference
$n!$	*factorial
$n!!$	double factorial (*see* FACTORIAL)
$\binom{n}{r}$, ^nC_r	*binomial coefficient, *combination
$\binom{n}{n_1, n_2, \ldots, n_k}$	*multinomial coefficient
^nP_r	*permutation
K_n	*complete graph
$K_{m,n}$	complete *bipartite graph
$E(X)$	*expected value, expectation
$E(X^k)$	*moment
μ	population *mean
$\bar{x}$	sample *mean
σ	*standard deviation
$\text{Var}(X)$	*variance
$\text{Cov}(X, Y)$	*covariance
$\text{Pr}(A)$	*probability
$\text{Pr}(A|B)$	*conditional probability
p_X	*probability mass function (pmf)
f_X	*probability density function (pdf)
F_X	*cumulative distribution function (cdf)
$G_X(s)$	*probability generating function (pgf)
$M_X(s)$	*moment generating function (mgf)
iid	*independent and identically *distributed

Physical Applied Mathematics

Common Symbol	Reference
m	*mass
m_0	*rest mass
I_P	*moment of inertia
$\mathbf{I}_P$	*inertia matrix
E	*energy
T	*kinetic energy
V	*potential energy
L	*Lagrangian
H	*Hamiltonian
ψ	*wave function
$\langle\_|$	*bra
$|\_\rangle$	*ket
ω	*angular speed
f, ν	*frequency
ω	*vorticity
k	*stiffness, spring constant
λ	*Young's modulus of elasticity
D/Dt	*convective derivative

Greek letters

Name	Lower case	Capital
Alpha	α	A
Beta	β	B
Gamma	γ	Γ
Delta	δ	Δ
Epsilon	ε	E
Zeta	ζ	Z
Eta	η	H
Theta	θ	Θ
Iota	ι	I
Kappa	κ	K
Lambda	λ	Λ
Mu	μ	M
Nu	ν	N
Xi	ξ	Ξ
Omicron	o	O
Pi	π	Π
Rho	ρ	P
Sigma	σ	Σ
Tau	τ	T
Upsilon	υ	Υ
Phi	φ	Φ
Chi	χ	X
Psi	ψ	Ψ
Omega	ω	Ω

Roman numerals

If smaller numbers follow larger numbers, then the numbers are added, but if a smaller number comes before a larger number it is subtracted, so CM is 900 and MC is 1100.

The basic values are I = 1, V = 5, X = 10, L = 50, C = 100, D = 500, and M = 1000.

	Roman		Roman
1	I	50	L
2	II	51	LI
3	III	55	LV
4	IV	60	LX
5	V	70	LXX
6	VI	80	LXXX
7	VII	90	XC
8	VIII	95	XCV
9	IX	99	XCIX
10	X	100	C
11	XI	101	CI
12	XII	105	CV
13	XIII	110	CX
14	XIV	150	CL
15	XV	200	CC
16	XVI	300	CCC
17	XVII	400	CD
18	XVIII	500	D
19	XIX	600	DC
20	XX	700	DCC
21	XXI	800	DCCC
22	XXII	900	CM
23	XXIII	1000	M
24	XXIV	1100	MC
25	XXV	1500	MD
30	XXX	1900	MCM
35	XXXV	2000	MM
40	XL	2020	MMXX
45	XLV		
49	XLIX		

Fields Medal winners

Year	Location	Winners
2018	Rio de Janeiro, Brazil	Caucher Birkar, Alessio Figalli, Peter Scholze, Akshay Venkatesh
2014	Seoul, South Korea	Artur Avila, Manjul Bhargava, Martin Hairer, Maryam *Mirzakhani
2010	Hyderabad, India	Elon Lindenstrauss, Ngô Bao Châu, Stanislav Smirnov, Cedric Villani
2006	Madrid, Spain	Andrei Okounkov, Grigori Perelman (who declined the award), Terence Tao, Wendelin Werner
2002	Beijing, China	Laurent Lafforgue, Vladimir Voevodsky
1998	Berlin, Germany	Richard Ewen Borcherds, William Timothy *Gowers, Maxim Kontsevich, Curtis T. McMullen. A silver plate was awarded to Andrew *Wiles as a special tribute.
1994	Zürich, Switzerland	Efim Isakovich Zelmanov, Jacques-Louis Lions, Jean Bourgain, Jean-Christophe Yoccoz
1990	Kyoto, Japan	Vladimir Drinfeld, Vaughan Frederick Randal Jones, Shigefumi Mori, Edward *Witten
1986	Berkeley, California, USA	Simon *Donaldson, Gerd Faltings, Michael *Freedman
1982	Warsaw, Poland	Alain Connes, William Thurston, Shing-Tung Yau
1978	Vancouver, British Columbia, Canada	Pierre Deligne, Charles Fefferman, Grigory Margulis, Daniel *Quillen
1974	Helsinki, Finland	Enrico Bombieri, David Mumford
1970	Nice, France	Alan Baker, Heisuke Hironaka, Sergei Petrovich Novikov, John Griggs Thompson
1966	Moscow, Russia	Michael Francis *Atiyah, Paul Joseph Cohen, Alexander *Grothendieck, Stephen Smale
1962	Stockholm, Sweden	Lars Hörmander, John *Milnor
1958	Edinburgh, Scotland, United Kingdom	Klaus Roth, René *Thom
1954	Amsterdam, Netherlands	Kunihiko Kodaira, Jean-Pierre *Serre
1950	Cambridge, Massachusetts, USA	Laurent Schwartz, Atle Selberg
1936	Oslo, Norway	Lars Ahlfors, Jesse Douglas

Millennium Prize problems

1. P versus NP

The class of problems for which an algorithm can find a solution quickly (in polynomial time) is termed P. The class of problems for which an algorithm can verify a solution quickly is termed NP. The question is whether all problems in NP are also in P. Most mathematicians and computer scientists expect that they are not, but the problem remains currently open.

2. The Hodge conjecture

The Hodge conjecture is a major unsolved problem in algebraic geometry. It arises out of attempts to approximate the shape of a given object by putting together simple geometric building blocks of increasing dimension.

The Hodge conjecture is that for projective algebraic varieties, Hodge cycles are rational linear combinations of algebraic cycles.

3. The *Riemann hypothesis

The distribution of prime numbers does not follow any regular pattern, but Riemann observed that it is closely related to the zeros of the (Riemann) zeta function

$$\zeta(s) = \sum_{n=1}^{\infty} \frac{1}{n^s} = 1 + \frac{1}{2^s} + \frac{1}{3^s} + \frac{1}{4^s} + \ldots.$$

The Riemann hypothesis is the conjecture that the real part of any non-trivial solution to $\zeta(s) = 0$ is $\frac{1}{2}$, i.e. it is in the form $s = \frac{1}{2} + iy$ for some real value y.

This conjecture was first stated in 1859 and was included in Hilbert's list of 23 unsolved problems published in 1900. While the statement has been confirmed for the first one and a half billion cases, it has resisted a general proof for over 150 years.

4. Yang–Mills existence and mass gap

Yang-Mills theory is now the foundation of most elementary-particle theory in quantum theory, generalizing the work of Maxwell in electromagnetism. Its predictions in relation to the strong interactions of elementary particles have been tested and verified experimentally, and confirmed by computer simulations, but the mathematics remains unexplained. Classical theory says that electromagnetism travels in waves at the speed of light with zero mass, while the Yang-Mills theory depends on the quantum particles having positive masses (hence the 'mass gap').

The problem is to establish rigorously the Yang-Mills quantum theory and that the mass of the least massive particle of the force field is strictly positive, that is, that each type of particle has a lower mass bound.

5. Navier–Stokes existence and smoothness

The Navier-Stokes equations describe the motion of fluids. Since 'fluids' is a collective term for any substance which is not solid, it embraces liquids and gases, meaning that everything from the behaviour of the contents of a glass of tap water to waves in the oceans to the turbulence created by the motion of a Formula 1 car taking a corner at 170 mph is governed by these equations. As with the Riemann hypothesis, the statement of these equations dates back to the 19th century, being first published by Claude Louis Navier in 1821 and 1822, and developed by Sir George Gabriel Stokes over the following years. Again, although these equations have been the subject of a huge amount of effort by the best mathematicians and physicists over an extended period, the complete solutions have yet to be found.

It has not yet been proven that solutions to the Navier–Stokes equations always exist in three dimensions (the existence), nor that if they do exist, then they do not contain any singularities (the smoothness). These two aspects comprise the Navier–Stokes existence and smoothness problem.

6. The *Birch and Swinnerton-Dyer conjecture

Much of mathematics is engaged with trying to produce generalizations of what can be done for particular cases; so, the quadratic formula generates solutions for all quadratic equations, and we know there is nothing left to be resolved in the case of equations of that type. Still valuable in terms of mathematics is the knowledge of when it is not possible to produce a generalization—so the negative proof in the early 1970s of Hilbert's tenth problem represented a very significant piece of mathematics. It was proved that there is no algorithm to determine whether a given polynomial Diophantine equation with integer coefficients has a solution in integers. However, it remains possible that there are some special cases for which it is possible to say something.

The Birch and Swinnerton-Dyer conjecture is that when the solutions are the points of an abelian variety, the size of the group of rational points is related to the behaviour of an associated zeta function $\zeta(s)$ near $s = 1$. In particular, it asserts that if $\zeta(1) = 0$ then there are an infinite number of rational points or solutions, and that if $\zeta(1) \neq 0$ then there are only a finite number of such points.

7. The Poincaré conjecture

In topology, a sphere with a 2-dimensional surface must be simply connected. It is also true that every 2-dimensional surface which is both compact and simply connected is topologically equivalent to a sphere. The Poincaré conjecture is that this is also true for spheres as 3-dimensional surfaces. These shapes are difficult, perhaps impossible, to visualize—the universe, for instance, is sometimes thought of as a 3-sphere. However, even without a physical representation of a 3-sphere, we can define properties of it which allow us to do useful mathematics on a 3-sphere. It is known that the equivalent statement is true for dimensions higher than three. Solving the Poincaré conjecture for three dimensions is central to the problem of classifying 3-manifolds. The conjecture has been proved—the Clay Mathematics Institute announced on 18 March 2010 that it was awarding the Millennium Prize for the resolution of the Poincaré conjecture to Grigory Perelman. Perelman declined the prize of one million dollars, without giving the institute any reason. However, on other occasions he has cited his belief that his contribution was no greater than that of Richard Hamilton, who had first suggested using the Ricci flow as a way to approach the problem.

Appendix 24

Historical timeline

c.2000 BC	Sumerians have *sexagesimal (base 60) positional numeral system
c.1300 BC	Decimal number system (see DECIMAL REPRESENTATION) recorded in China
c.500 BC	Pythagorean school (see PYTHAGORAS)
c.300 BC	*Euclid writes his *Elements*
287 BC	Birth of *Archimedes
276 BC	Birth of *Eratosthenes, who determined the Earth's circumference
c.262 BC	Birth of *Apollonius, author of *Conics*
262 BC	*Archimedes shows *pi lies between $3\frac{1}{7}$ and $3\frac{10}{71}$
c. AD 250	*Diophantus writes his *Arithmetica*
c. AD 300	*Chinese remainder theorem developed
c. AD 830	*al-Khwārizmī writes *Al-Jabr*, which provides the word 'algebra'
c.1150	*Bhāskara solves *Pell's equation
1202	*Fibonacci publishes his *Liber Abaci*
c.1525	Dürer writes on perspective and geometrical constructions
1535	*Tartaglia solves the *cubic equation
1540	Ferrari solves the *quartic equation
1545	*Cardano's *Ars Magna*
1572	*Bombelli publishes his *Algebra*, developing *complex numbers
1591	*Viète introduces letters for unknown variables
1609	*Kepler publishes first two of three laws (third law 1619)
1614	*Napier introduces *logarithms
1618	*Napier makes first reference to the number e
1624	*Briggs publishes his tables of *logarithms
1629	*Fermat develops methods to find *tangents (published 1636)
1632	*Galileo formulates *Galilean relativity
1636	*Fermat proves *Fermat's Little Theorem
1637	*Descartes, and independently *Fermat, introduce *Cartesian coordinates
1638	*Fermat claims, in a margin, to have a proof of *Fermat's Last Theorem
1654	*Fermat and *Pascal lay foundations of *probability
1656	*Wallis' product
1664	*Newton begins his work on *calculus
1668	*Gregory first proves *Fundamental Theorem of Calculus
1671	*Gregory discovers *Taylor series
1673	*Leibniz develops his theory of *calculus, publishes 1684
1678	*Hooke's Law
1687	*Newton publishes his *Principia*
1691	*Rolle publishes his theorem
1696	*Brachistochrone problem posed by Jean *Bernoulli
1715	Taylor develops *Taylor series

1722	*De Moivre states his theorem
1727	*Euler introduces e for the base of *natural logarithms
1734	Bishop George *Berkeley publishes *The Analyst,* critical of the foundations of *calculus
1735	*Euler solves the *Basel problem
1736	*Euler shows Königsberg Bridge problem unsolvable
1737	*Euler shows *e is *irrational
1738	Daniel *Bernoulli publishes *Hydrodynamica*
1742	*Goldbach makes his conjecture in a letter to *Euler
1744	*Euler works on *calculus of variations
1747	*D'Alembert derives and solves the *wave equation
1748	*Euler proves the identity $e^{i\theta} = \cos\theta + i\sin\theta$
1751	*Euler formula $V-E+F = 2$ for *polyhedra
1761	*Lambert shows *pi is *irrational
1762	*Lagrange states the *divergence theorem
1763	*Bayes proves his theorem
1763	*Euler generalizes *Fermat's little theorem and introduces the *totient function
1771	Lagrange's first version of *Lagrange's theorem
1777	*Euler popularizes use of i for $\sqrt{-1}$
1788	*Lagrange publishes his equations in *mechanics
1796	*Legendre conjectures the *prime number theorem
1799	*Gauss proves the *Fundamental Theorem of Algebra
1799	Ruffini gives an incomplete proof that quintics are not *solvable by radicals
1801	*Gauss publishes his *Disquisitiones arithmeticae*
1801	*Gauss relocates Ceres by means of the *least squares theorem
1807	*Fourier series introduced
1817	*Bolzano defines *continuity and proves the *intermediate value theorem
1821	*Cauchy publishes his *Cours d'analyse*
1822	Feuerbach's theorem of the *nine-point circle
1822	Navier and (1842) Stokes state the *Navier-Stokes equations
1824	*Abel proves the *quintic is not *solvable by radicals
1825	*Cauchy presents the *Cauchy integral theorem in *complex analysis
1826	*Crelle's Journal* first published
1828	*Green's Theorem proved
1828	*Gauss proves the *Theorema Egregium*
1829	*Lobachevsky's and *Bolyai's (1832) work on *hyperbolic geometry
1832	*Galois shot in duel
1833	*Hamiltonian mechanics
1835	*Dirichlet proves his theorem about *primes in *arithmetic progressions
1837	Wantzel proves doubling the cube and trisecting angles impossible
1838	Verhulst introduces logistic equation (*see* LOGISTIC MAP) as population model
1843	*Hamilton introduces the *quaternions
1844	*Liouville gives first example of a *transcendental number
1845	*Bertrand states his postulate
1846	*Liouville popularizes *Galois' work
1846	*Chebyshev proves *weak law of large numbers
1848	Bonnet proves special case of *Gauss-Bonnet theorem
1850	*Kelvin writes to Stokes with details of *Stokes' Theorem
1852	*Chebyshev proves Bertrand's postulate
1854	*Cayley gives *axioms for an abstract *group

1854	*Riemann presents work on the *Riemann integral
1854	*Riemann's theorem on *conditionally convergent series
1857	*Riemann's *Theory of Abelian Functions* popularizes *Riemann surfaces
1858	*Möbius introduces the *Möbius band
1858	*Cayley states the *Cayley-Hamilton Theorem
1859	*Riemann poses the *Riemann hypothesis
1861	*Weierstrass's first lectures on *epsilon-delta analysis
1869	*Lie begins his work on *Lie groups
1870	*Weierstrass proves the *isoperimetric inequality
1871	*Dedekind defines *rings, *fields, and *modules
1872	*Sylow publishes his three theorems
1873	*Maxwell publishes his equations
1873	*Hermite proves e is *transcendental
1874	*Cantor shows there are different *infinite *cardinalities
1877	*Boltzmann defines entropy
1877	*Frobenius shows there are 3 *associative *division algebras
1878	*Cantor states the *continuum hypothesis
1881	*Poincaré proves the *Hairy Ball Theorem
1882	*Lindemann proves π is *transcendental
1884	*Frege's *Grundlagen der Arithmetik*
1885	*Weierstrass proves *Weierstrass' Approximation Theorem
1885	Tait publishes table of *knots up to 10 *crossings
1887	*Jordan proves *Jordan Curve Theorem
1888	*Hilbert's basis theorem proved
1889	Fitzgerald and Lorentz (1892) propose the *Lorentz-Fitzgerald contraction
1889	*Jordan-Hölder Theorem proved
1891	Fedorov shows there are 17 *wallpaper groups
1895	*Poincaré releases his seminal *Analysis Situs*
1896	*Frobenius founds *representation theory
1896	*Prime number theorem proven by *Hadamard and de la Vallée-Poussin
1897	First *International Congress of Mathematicians in Zurich
1897	Hensel introduces *p-adic numbers
1898	Hurwitz shows there are four normed *division algebras
1899	*Hilbert axiomatizes *Euclidean geometry
1900	*Hilbert announces his 23 problems
1901	*Lyapunov proves the *Central Limit Theorem
1902	*Lebesgue introduces the *Lebesgue integral
1903	*Russell's paradox stated
1904	*Zermelo introduces the *axiom of choice
1904	Lorentz introduces *Lorentz transformations
1904	*Poincaré Conjecture made
1905	*Einstein publishes his theory of *special relativity
1906	*Metric spaces defined by Fréchet
1906	*Joukowski models lift on an aerofoil
1906	*Markov publishes first paper on *Markov chains
1908	Student's (*Gosset's) t-distribution
1908	*Zermelo lists his axioms of *set theory
1910	*Russell and Whitehead publish the first volume of *Principia Mathematica*
1910	*Brouwer proves his fixed point theorem
1910	Lotka and later Volterra (1926) publish their *predator-prey model

491 **Appendix 24**

1912	*Brouwer generalizes *Jordan Curve Theorem to higher dimensions
1914	*Topological spaces defined by *Hausdorff
1915	*Einstein publishes his theory of general *relativity
1915	*Noether's theorem
1916	*Bieberbach conjecture made
1918	*Hausdorff dimension (or *fractal dimension) defined
1922	Fraenkel suggests further axioms to create *Zermelo-Fraenkel axioms
1924	*Banach-Tarski paradox published
1925	Morse publishes first paper on *Morse theory
1925	*Heisenberg formulates his approach to *quantum theory
1925	*Fisher publishes influential *Statistical Methods for Research Workers*
1926	*Poincaré-Hopf theorem proved
1926	*Schrödinger formulates his approach to *quantum theory
1927	Kermack and McKendrick introduce the *SIR epidemiology model
1927	*Van der Waerden's Theorem proved
1927	*Noether publishes her *isomorphism theorems
1928	*Alexander introduces his polynomial *knot invariant
1928	*Von Neumann's *On the Theory of Games and Strategy* published, including *Minimax Theorem
1930	*Kuratowski's Theorem proved
1930	*Ramsey proves his theorem
1930	*Dirac publishes *The Principles of Quantum Mechanics*
1931	*Gödel's Incompleteness Theorem
1932	*Von Neumann's *Mathematical Foundations of Quantum Mechanics* published
1933	*Kolmogorov states the axioms of *probability
1933	*Neyman-Pearson lemma proved
1935	*Zorn introduces his lemma (Kuratowski had earlier in 1922)
1935	*Church provides negative answer to the *decision problem
1936	First *Fields Medals awarded
1937	*Neyman introduces the *confidence interval
1937	*Vinogradov makes progress on *Goldbach's conjecture
1937	*Turing writes *On Computable Numbers* (unaware of *Church's work)
1938	*Gödel proves the *continuum hypothesis and the *axiom of choice to be consistent with *Zermelo-Fraenkel axioms
1945	Eilenberg and Mac Lane define *categories
1945	Cartan generalizes *Stokes' Theorem
1947	Dantzig publishes his *simplex method
1948	*Turing introduces *LU decomposition in *Rounding-off Errors in Matrix Processes*
1948	Schwartz publishes on *distributions
1948	*Shannon's seminal paper *A Mathematical Theory of Communication* published
1949	Weil states his influential *Weil conjectures
1950	Hodge presents his conjecture at the 1950 *ICM
1950	*Hamming distance introduced in *Error Detecting and Error Correcting Codes*
1950	John Nash introduces *Nash equilibria
1951	*Arrow's Impossibility Theorem
1951	*Huffman describes his optimal codes
1955	*Roth's theorem proved
1955	*Word problem for *groups proven undecidable by Novikov

1955	*Taniyama-Shimura conjecture made
1963	*Lorenz attractor first described
1963	Atiyah-Singer Index Theorem proved
1963	Cohen proves *axiom of choice independent of *Zermelo-Fraenkel axioms and *continuum hypothesis independent of *ZFC axioms
1965	Birch-Swinnerton-Dyer conjecture made
1966	Robinson introduces *hyperreals
1967	Langlands conjectures made
1968	*Thom begins his work on *catastrophe theory
1973	*Chen makes progress on *Goldbach conjecture
1974	First *Penrose tilings discovered
1974	Deligne proves Weil conjectures
1975	*Mandelbrot coins the term *fractal
1975	*Szemerédi's theorem proved
1976	*Four Colour Theorem proven by Appel and Haken using computers
1976	May finds *chaos in discrete logistic equation (*see* LOGISTIC MAP)
1976	*Lakatos's *Proofs and Refutations* published posthumously
1978	*Mandelbrot set first defined
1978	Donald *Knuth introduces *TeX
1978	*RSA cryptography developed
1983	Faltings proves the *Mordell conjecture
1983	*Donaldson shows exotic differentiable structures exist in 4D
1984	Jones introduces his polynomial *knot invariant
1985	*Bieberbach conjecture proved by de Branges
1985	*abc conjecture stated
1995	*Wiles proves *Fermat's Last Theorem
1998	Hales proves the *Kepler conjecture. Finally accepted 2017
2000	*Millennium Prize problems announced
2002	*Catalan's conjecture proved by Mihăilescu
2003	Perelman proves the *Poincaré Conjecture
2004	*Green-Tao theorem proved
2013	Zhang makes advances on the *twin prime conjecture
2014	Miriam *Mirzakhani wins Fields Medal

Oxford Quick Reference

A Dictionary of Chemistry

Over 5,000 entries covering all aspects of chemistry, including physical chemistry and biochemistry.

'It should be in every classroom and library ... the reader is drawn inevitably from one entry to the next merely to satisfy curiosity.'

School Science Review

A Dictionary of Physics

Ranging from crystal defects to the solar system, 4,000 clear and concise entries cover all commonly encountered terms and concepts of physics.

A Dictionary of Biology

The perfect guide for those studying biology — with over 5,800 entries on key terms from biology, biochemistry, medicine, and palaeontology.

'lives up to its expectations; the entries are concise, but explanatory'

Biologist

'ideally suited to students of biology, at either secondary or university level, or as a general reference source for anyone with an interest in the life sciences'

Journal of Anatomy

Oxford Quick Reference

A Dictionary of Psychology
Andrew M. Colman

Over 9,500 authoritative entries make up the most wide-ranging dictionary of psychology available.

'impressive ... certainly to be recommended'
Times Higher Education Supplement

'probably the best single-volume dictionary of its kind.'
Library Journal

A Dictionary of Economics
John Black, Nigar Hashimzade, and Gareth Myles

Fully up-to-date and jargon-free coverage of economics. Over 3,500 terms on all aspects of economic theory and practice.

'strongly recommended as a handy work of reference.'
Times Higher Education Supplement

A Dictionary of Law

An ideal source of legal terminology for systems based on English law. Over 4,800 clear and concise entries.

'The entries are clearly drafted and succinctly written ... Precision for the professional is combined with a layman's enlightenment.'
Times Literary Supplement

A Dictionary of Education
Susan Wallace

In over 1,000 clear and concise entries, this authoritative dictionary covers all aspects of education, including organizations, qualifications, key figures, major legislation, theory, and curriculum and assessment terminology.

Oxford Quick Reference

A Dictionary of Sociology
John Scott

The most wide-ranging and authoritative dictionary of its kind.

'Readers and especially beginning readers of sociology can scarcely do better ... there is no better single volume compilation for an up-to-date, readable, and authoritative source of definitions, summaries and references in contemporary Sociology.'

A. H. Halsey, Emeritus Professor, Nuffield College,
University of Oxford

The Concise Oxford Dictionary of Politics and International Relations
Garrett Brown, Iain McLean, and Alistair McMillan

The bestselling A–Z of politics with over 1,700 detailed entries.

'A first class work of reference ... probably the most complete as well as the best work of its type available ... Every politics student should have one'

Political Studies Association

A Dictionary of Environment and Conservation
Chris Park and Michael Allaby

An essential guide to all aspects of the environment and conservation containing over 9,000 entries.

'from *aa* to *zygote*, choices are sound and definitions are unspun'
New Scientist

Oxford Quick Reference

The Oxford Dictionary of Art & Artists
Ian Chilvers

Based on the highly praised *Oxford Dictionary of Art*, over 2,500 up-to-date entries on painting, sculpture, and the graphic arts.

'the best and most inclusive single volume available, immensely useful and very well written'

Marina Vaizey, *Sunday Times*

The Concise Oxford Dictionary of Art Terms
Michael Clarke

Written by the Director of the National Gallery of Scotland, over 1,800 entries cover periods, styles, materials, techniques, and foreign terms.

The Oxford Dictionary of Architecture
James Stevens Curl and Susan Wilson

Over 6,000 entries and 250 illustrations cover all periods of Western architectural history.

'splendid ... you can't have a more concise, entertaining, and informative guide to the words of architecture.'

Architectural Review

'... definitions are not only elegantly concise, they often sparkle with sententious wit. Give me this pleasingly well written dictionary any day.'

Christopher Catling, *SALON: Society of Antiquaries of London Online Newsletter*

Oxford Quick Reference

The Kings and Queens of Britain
John Cannon and Anne Hargreaves

A detailed, fully-illustrated history ranging from mythical and pre-conquest rulers to the present House of Windsor, featuring regional maps and genealogies.

A Dictionary of World History

Over 4,000 entries on everything from prehistory to recent changes in world affairs. An excellent overview of world history.

A Dictionary of British History
Edited by John Cannon

An invaluable source of information covering the history of Britain over the past two millennia. Over 3,000 entries written by more than 100 specialist contributors.

Review of the parent volume
'the range is impressive ... truly (almost) all of human life is here'

Kenneth Morgan, *Observer*

The Oxford Companion to Irish History
Edited by S. J. Connolly

A wide-ranging and authoritative guide to all aspects of Ireland's past from prehistoric times to the present day.

'packed with small nuggets of knowledge' *Daily Telegraph*

The Oxford Companion to Scottish History
Edited by Michael Lynch

The definitive guide to twenty centuries of life in Scotland.
'exemplary and wonderfully readable'

Financial Times

OXFORD